The Urban Book Series

The Urban Book Series is a resource for urban studies and geography research worldwide. It provides a unique and innovative resource for the latest developments in the field, nurturing a comprehensive and encompassing publication venue for urban studies, urban geography, planning and regional development.

The series publishes peer-reviewed volumes related to urbanization, sustainability, urban environments, sustainable urbanism, governance, globalization, urban and sustainable development, spatial and area studies, urban management, transport systems, urban infrastructure, urban dynamics, green cities and urban landscapes. It also invites research which documents urbanization processes and urban dynamics on a national, regional and local level, welcoming case studies, as well as comparative and applied research.

The series will appeal to urbanists, geographers, planners, engineers, architects, policy makers, and to all of those interested in a wide-ranging overview of contemporary urban studies and innovations in the field. It accepts monographs, edited volumes and textbooks.

Indexed by Scopus.

Milena Vukmirović · Aleksandra Đukić ·
Eva Vaništa Lazarević ·
Aleksandra Krstić-Furundžić
Editors

Digital Dreams, Urban Realities

Crafting Sustainable, Inclusive, and Beautiful Cities

Springer

Editors
Milena Vukmirović
Department of Landscape Architecture and Horticulture
University of Belgrade—Faculty of Forestry
Belgrade, Serbia

Eva Vaništa Lazarević
Department of Urbanism
University of Belgrade—Faculty of Architecture
Belgrade, Serbia

Aleksandra Đukić
Department of Urbanism
University of Belgrade—Faculty of Architecture
Belgrade, Serbia

Aleksandra Krstić-Furundžić
Urban Laboratory Belgrade
Belgrade, Serbia

ISSN 2365-757X ISSN 2365-7588 (electronic)
The Urban Book Series
ISBN 978-3-031-76352-6 ISBN 978-3-031-76353-3 (eBook)
https://doi.org/10.1007/978-3-031-76353-3

This Springer imprint is published by the registered company Springer Nature Switzerland AG
The registered company address is: Gewerbestrasse 11, 6330 Cham, Switzerland

Preface

In an era of rapid urbanisation and digital transformation, pursuing sustainable, inclusive, and beautiful cities have never been more critical. *Digital Dreams, Urban Realities: Crafting Sustainable, Inclusive, and Beautiful Cities* is a comprehensive exploration of how the principles of the New European Bauhaus can guide us towards a harmonious integration of technology, nature, and culture. This book invites you to explore the forefront of urban innovation through a curated collection of papers from the 8th International Academic Conference on Places and Technologies. Held on the 19th and 20th of October 2023, in Belgrade, Serbia, this conference captured the essence of technology-driven urban evolution. Delving into various scales of urban development, the book presents research findings and case studies that illuminate the path towards ecological transition and resilience. The book is divided into four parts: macro scale, mezzo scale, micro scale, and education. Each of these parts addresses specific topics.

At the macro scale, the book explores the crucial roles that cities, small towns, villages, and regions play in driving ecological transition. It presents innovative approaches and case studies that emphasise environmental sustainability. The focus is on how different communities can contribute to creating sustainable, resilient, and beautiful urban environments through the involvement of local communities and the application of ecological principles in urban planning.

In pursuing sustainable, inclusive, and beautiful cities, it is essential to recognise the critical roles that cities, small towns, villages, and regions play in driving the ecological transition. This chapter delves into the transformative potential of various scales of human settlements, highlighting innovative approaches and case studies that illustrate effective strategies for achieving environmental sustainability. By exploring these detailed examples, we gain valuable insights into how different communities can fuel the ecological transition and foster sustainable, resilient, and vibrant urban environments.

A key focus is integrating natural and cultural heritage in urban development. The Action Plan for Sustainable Sombor in Serbia is a prime example of how engaging local communities and leveraging their knowledge can result in effective heritage-based policies. This approach not only enhances urban identity but also improves the quality of life for residents.

Revitalising regions like Southern and Eastern Serbia is essential for fostering environmental sustainability in rural areas. These areas, rich in natural resources and biodiversity, offer significant opportunities for sustainable development. By addressing social, economic, and spatial challenges, rural revitalisation can contribute to the broader ecological transition and enhance the quality of life for rural populations.

Furthermore, the rise of remote work and digital technologies presents new opportunities for rural development. The case study of a Digital Nomad Centre in Pécsdevecser, Hungary, illustrates how digital nomads can be integrated into rural communities, transforming them into vibrant centres of life and economic activity. This innovative approach revitalises rural areas and aligns with the New European Bauhaus principles by promoting sustainability, inclusivity, and beauty.

As urban environments face mounting challenges from climate change, pollution, and biodiversity loss, designing cities harmonising with nature becomes more pressing. This chapter delves into innovative approaches and methodologies for integrating ecological principles into urban planning and development. Exploring these diverse yet interconnected strategies underscores how designing with nature and embracing a more-than-human design culture can lead to sustainable, resilient, and livable urban environments.

In Granada, the formation of heat islands, poor air quality, and biodiversity loss have long been aggravated by high motorisation and insufficient attention from local and regional administrations. To address these issues, a comprehensive study explored how national and international cities have adopted ecosystem mechanisms to combat climate change. The research resulted in an inventory of urban elements that enhance green infrastructure, making cities more resilient to heat and biodiversity loss. This study systematically categorises urban spaces based on foundational traits, urban fabric, green infrastructure, and liveability, providing a roadmap for decarbonisation and improving urban health and well-being.

Contemporary urban landscapes also face increasing pressure on natural and semi-natural elements, threatening vital ecosystem services. The Geodesign framework emerges as a strategic planning tool to mitigate these pressures by preserving and enhancing green infrastructure. By examining the case of Belgrade, this research highlights the application of Geodesign to evaluate and improve ecosystem services within urban areas. This approach integrates geostatistical models and strategic planning to bolster urban resilience and ensure the provision of essential ecosystem services.

Waterfronts play a crucial role in urban planning, offering unique opportunities for dynamic development. The revitalisation of the Danube-Tisa-Danube (DTD) Canal banks in Serbia illustrates how a participatory design approach rooted in local identity can transform urban spaces. By surveying residents to understand their daily needs and mapping potential locations for active public spaces, this study sets guidelines for promoting vibrant, interactive riverfront areas. This approach addresses the residents' needs and fosters a distinctive urban identity, enhancing the overall quality of urban life.

Local and regional communities emerge as pivotal innovators in the quest for sustainable, inclusive, and climate-neutral cities. This chapter explores the transformative potential of grassroots initiatives and collaborative projects that drive ecological and social change. Through diverse case studies, it underscores the critical role of local and regional communities in innovating and implementing sustainable, inclusive, and resilient solutions. These grassroots efforts are integral to the broader ecological transition, aligning with the vision of the New European Bauhaus to create beautiful, sustainable, and inclusive living environments.

One significant challenge highlighted in recent workshops is the perceived high threshold of capacity and competence required for inclusive co-creation projects. Addressing this challenge involves improving project designs to facilitate radical collaboration among diverse stakeholders. This part draws on examples from several European projects, including + City x Change, CrAFt, NEB-STAR, and Re-Value, coordinated by the Norwegian University of Science and Technology (NTNU). These projects emphasise multi-actor cooperation and embody the principles of the New European Bauhaus, focusing on creating beautiful, sustainable, and inclusive communities.

Studying urban facilities in Granada's northern and southern districts provides insights into how spatial location influences social cohesion. Urban facilities act as collective nodes, fostering social interactions and linking community spheres. However, increasing social and spatial fragmentation necessitates strategically using public spaces and urban facilities to strengthen social capital. This research proposes methodologies to evaluate and enhance the social function of urban facilities, offering patterns and criteria for their optimal location and design to promote economic and social activation in vulnerable neighbourhoods.

Responsible tourism is another innovative approach to sustainable development, as illustrated by the Eibenthal Initiative in the Czech village of Eibenthal, Romania. This small settlement, deeply affected by the closure of a nearby coal mine, has undergone a revival through community-led projects aimed at attracting tourists and promoting local heritage. By adhering to the principles of the New European Bauhaus, the initiative has successfully fostered a resilient economy and enhanced community well-being through sustainable tourism practices.

The book addresses land-use planning strategies that promote sustainable urban renewal and transformation through inclusive and participatory approaches at the mezzo scale. It examines methods for enhancing energy efficiency, promoting green urban renewal, and preserving cultural heritage through sustainable practices to achieve carbon neutrality.

Integrating a holistic land-use planning perspective is essential for sustainable urban development. This chapter explores innovative strategies and methodologies that promote sustainable renewal and transformation of metropolitan areas through inclusive and participatory approaches. Examining various case studies provides valuable insights into how a holistic perspective and inclusive land use planning can drive sustainable urban renewal. The emphasis on co-creation and participatory design aligns with the principles of the New European Bauhaus, fostering resilient and vibrant urban communities.

The development of cultural tourism involving local communities plays a crucial role in fostering sustainable urban renewal. The TExTOUR Cultural Tourism Lab in Trebinje, Bosnia and Herzegovina is a pioneering example of how co-creation methods within Living Labs can support cultural heritage management and broader community-led urban development. The project addresses the challenges and benefits of urban renewal through cultural tourism by engaging local stakeholders in the co-creation process. This part examines the methodology and findings of the TExTOUR project, providing guidelines for enhancing local co-creation processes and promoting sustainable cultural tourism strategies.

Urban transformation is inherently challenging, particularly when motivating citizens to invest in sustainable renovation. The New European Bauhaus (NEB) movement offers a people-oriented framework emphasising sustainability, aesthetics, and inclusion. The case study of Nyhavna in Trondheim demonstrates the application of NEB values in urban transformation. This research focuses on developing an assessment framework for NEB values and testing it through the work of NTNU students. The findings highlight the importance of social components, co-creation methods, and inclusive design in creating sustainable, inclusive, and beautiful urban districts.

Carbon neutrality in buildings, spaces, and infrastructures is critical to sustainable urban development. This chapter explores various strategies and methodologies to enhance energy efficiency, promote green urban renewal, and preserve heritage through sustainable practices. Integrating these diverse approaches underscores the importance of carbon neutrality in buildings, spaces, and infrastructures. The emphasis on energy efficiency, green urban renewal, and sustainable restoration aligns with the principles of the New European Bauhaus, fostering resilient and sustainable urban environments.

Evaluating buildings' energy performance under changing climatic conditions is essential for developing sustainable architectural solutions. Assessing prefabricated and conventional single-family buildings in Thessaloniki, Greece, provides insights into how different construction methods and insulation levels perform in present and future climates. Using dynamic energy performance simulations with the Energy Plus tool, this study compares the energy demands of prefabricated buildings and conventional reinforced concrete and brick masonry structures. The findings highlight the impact of climate change on energy requirements and offer guidelines for improving building resilience and efficiency.

Green urban renewal strategies play a pivotal role in sustainable urbanisation policies. This part examines the comprehensive framework developed within the ESPON SUPER project, which assesses various forms of urbanisation for their sustainability. The project produced a handbook for planners and decision-makers, emphasising the principles of the New European Bauhaus. The research highlights the potential for sustainable urban development in Croatia and beyond by aligning green urban renewal strategies with these principles. The relationship between green urban renewal and the New European Bauhaus is explored, offering practical applications and enhancing the strategy's implementation potential.

Preserving the authentic appearance of heritage buildings while ensuring their sustainability is another crucial aspect of carbon-neutral urban development. The systematic approach to restoring the Žiča Monastery's Dining Room exemplifies how historic buildings can be sustainably renovated. This study focuses on improving energy efficiency and conservation while maintaining the structure's historical integrity. The research offers recommendations for the adaptive reuse of heritage buildings, balancing contemporary comfort standards with environmental preservation.

At the micro-scale, the book highlights strategies for transforming urban landscapes into areas that foster physical, mental, and social well-being. It underscores the importance of public spaces in promoting healthy lifestyles. It also explores approaches for extending the lifespan of existing housing stock, implementing energy-saving renovations, and integrating modern technologies to create affordable and sustainable social housing. The potential of digital and technological solutions in urban development is also discussed.

The design and utilisation of public spaces play a crucial role in promoting healthy lifestyles in urban environments. This chapter explores innovative strategies for transforming urban landscapes into areas that foster physical, mental, and social well-being. Examining these approaches underscores the importance of public spaces in encouraging healthy lifestyles. Integrating physical, mental, and social well-being into urban design aligns with sustainable and inclusive city planning principles, fostering environments where communities can thrive.

Urban spaces traditionally dominated by cars are undergoing a visionary shift towards more community-oriented designs. In Belgrade, parklets—small public spaces converted from parking spots—are emerging as a transformative solution. Initially temporary installations, parklets have proven their potential to enhance urban areas by making them more pedestrian-friendly and socially vibrant. This part discusses the evolution of parklets in Belgrade and their impact on urban spaces. It highlights the adverse effects of a car-centric urban design on the environment, public health, and social cohesion and presents a vision for a more sustainable and liveable city. By converting parking spaces into miniature urban areas, improving sidewalks, and introducing street life zones, the authors propose a multi-faceted approach to creating vibrant and inclusive communities. The potential benefits of this transformation include reduced air pollution, improved public health, and increased social interaction. The chapter emphasises the importance of political will, public participation, and innovative urban planning strategies in successfully implementing this visionary shift.

Beyond the physical redesign of spaces, the relationship between the built environment and health-oriented behaviour is complex and multifaceted. This part extends the discussion beyond physical health to include mental health, exploring how urban design can influence overall well-being. The authors develop and validate a hypothetical model that promotes healthy lifestyles within urban settings by conducting a comprehensive review of research from the past decade. This model integrates psychological and social aspects of behaviour, highlighting how interactions with the built environment affect health-seeking behaviours. The chapter concludes with recommendations for qualitative transformations and design interventions supporting physical and mental health, creating more holistic, health-promoting urban environments.

Creating affordable and sustainable social housing is a critical challenge for modern urban development. This chapter explores innovative strategies to extend the lifespan of existing housing stock, implement energy-saving renovations, and integrate modern technologies into urban design. Integrating these diverse approaches provides a comprehensive overview of strategies to create affordable, sustainable, and technologically advanced social housing. The focus on energy efficiency, renewable materials, and modern communication technologies aligns with contemporary urban development goals, fostering resilient and liveable communities.

Part I examines the potential for extending the lifespan of post-war multi-family residential buildings through comprehensive renovation. In Serbia, buildings constructed between 1946 and 1970 constitute a significant portion of the housing stock, yet they suffer from energy, spatial, and functional deficiencies. This study analyses multi-criteria renovation methods, including volumetric additions, to improve energy efficiency, spatial organisation, and residential comfort. The research presents two variants of comprehensive renovation for a selected building, aiming to achieve the highest possible energy class and extend the building's lifespan by improving the thermal envelope, increasing usable space, and enhancing vertical communication with elevators.

Part II focuses on sustainable modular renovations using renewable materials. This approach is particularly relevant in Bosnia-Herzegovina and Serbia, where many buildings from the Prosperous Years of Construction (1961–1980) are prime candidates for energy-saving upgrades. The research highlights the potential of timber modular systems for deep renovation, comparing the energy conditions of buildings before and after applying these systems. The study underscores the importance of using renewable materials to align with EU sustainability goals and the Renovation Wave Strategy, demonstrating significant energy savings and improved building performance.

The adaptability of urban interface design to 5G millimetre-wave mobile communication addresses the gap between urban reality and digital aspirations. The deployment of 5G technology faces challenges due to the high frequency and poor penetration of millimetre waves, which require line-of-sight transmission. This research explores how urban design can integrate the requirements of 5G coverage, proposing methodologies that incorporate millimetre wave channels as a new factor in urban planning. The study aims to balance the technological demands with the existing urban fabric, ensuring seamless integration of 5G infrastructure.

Digital and technological solutions are indispensable in the quest for sustainable urban development. This chapter explores the intersection of digital innovation, sustainable practices, and urban planning, highlighting their transformative potential. Examining these diverse yet interconnected approaches underscores the critical role of digital and technological solutions in fostering sustainable urban development.

Integrating serious games, BIM technology, and sustainable aesthetics aligns with contemporary urban planning goals, driving innovation and promoting resilient and liveable urban environments.

Part I delves into co-production within science and technology studies (STS) and its application to serious games in smart cities. Co-production emphasises the active involvement of diverse stakeholders in the design and implementation of technologies, challenging traditional notions of expertise. Through a case study from the Norwegian University of Science and Technology, this part examines how serious games have been designed to engage children in urban planning. It underscores the importance of recognising the complex interactions between scientific knowledge, societal practices, and technological innovation. The benefits and limitations of serious games are analysed, revealing their potential to enhance public participation and collaborative decision-making in urban development.

Building Information Modelling (BIM) technology represents a significant trend in digitalising the Architecture, Engineering, Construction, and Operations (AECO) sector. The application of BIM in Serbia is explored through a series of company presentations highlighting the advanced use of this technology in the market. This research showcases the profiles of BIM-intensive companies, their role in the AECO sector, and the range of BIM applications. The findings emphasise the importance of BIM in improving efficiency, accuracy, and collaboration within the construction industry, contributing to the overall digital transformation of the sector.

The aesthetic implications of sustainable architecture are critically examined in the final part. This study addresses the fragmentation often seen in sustainability, where ecological, technological, and aesthetic aspects are treated in isolation. The research advocates for a holistic approach that unites environmental responsibility, technological innovation, and humanistic values by integrating various philosophical traditions, from Romanticism to contemporary ecological discourses. This synthesis aims to create a deeper understanding of sustainability, promoting further research and dialogue on integrating technology and nature harmoniously.

The book's final part focuses on education, exploring innovative approaches in architectural and urban design education that integrate the principles of the New European Bauhaus. These curricula prepare students to create sustainable, inclusive, and aesthetically pleasing environments. The book also investigates how living labs and collaborative innovation ecosystems foster sustainable urban design and community development through active participation, innovative teaching methodologies, and integrating digital tools with personal engagement.

The New European Bauhaus initiative aims to bridge the gap between science, technology, art, and culture by fostering sustainable, inclusive, and beautiful living environments. This chapter examines innovative educational approaches integrating these principles into architectural and urban design education. Exploring these approaches provides valuable insights into how architectural education can embody the principles of the New European Bauhaus. Integrating theoretical knowledge, practical problem-solving, and sustainability-focused methodologies, these curricula prepare students to contribute effectively to creating sustainable, inclusive, and aesthetically pleasing built environments.

Architectural education must equip students with the knowledge and skills required to operate effectively in interdisciplinary teams dedicated to sustainability. The University of Belgrade's Undergraduate Studies of Architecture offers a course in structural design that combines theoretical knowledge with practical, contextual problem-solving. This approach, particularly in the final year, emphasises teamwork in studio settings to address real-world challenges. The survey results from 230 final-year students indicate that early team contextual learning enhances competence development. However, continuous improvement of educational methodologies is necessary to meet the evolving demands of architectural practice.

The introduction of the New European Bauhaus (NEB) project underscores the importance of the 2030 Agenda for Sustainable Development, especially Goal 11. As part of the European Green Deal, NEB advocates for ecological, digital, and social transitions in architecture and urbanism. At the University of Belgrade's Faculty of Architecture, programs are aligned with Sustainable Development Goals (SDGs), particularly within the Master's in Urbanism, Sustainable City. This part highlights the project-oriented approach of the "Sustainable City 1—Transformations" course and the Master's diploma project, demonstrating how the NEB agenda updates educational content towards ecological transition while fostering critical thinking, creativity, and innovation among students.

Incorporating the LEED (Leadership in Energy and Environmental Design) rating system into academic curricula presents unique challenges and opportunities. Initially introduced as a single course at the University of Belgrade—Faculty of Architecture, LEED has evolved to address various learning objectives across multiple study programs. This part presents methodologies that utilise LEED-based tools and criteria to empower students to assess the environmental impact of their design decisions from the early stages. The approach encourages integrating various design strategies to improve environmental features and effectively communicate these green features. The success of these methodologies is evidenced by the first prize win at The New Green Voice international design competition in 2022.

Living labs and collaborative innovation ecosystems are vital for advancing urban design and community development. This chapter explores how active participation, innovative teaching methodologies, and integrating digital tools with personal engagement can foster sustainable and inclusive urban environments. Integrating these diverse approaches underscores the significance of living labs and collaborative innovation ecosystems in urban design and community development. The focus on active participation, innovative teaching, and the synergy between digital and personal engagement aligns with contemporary goals of fostering resilient, inclusive, and vibrant urban environments.

The DANUrB+ Student Workcamp in Golubac, Serbia, exemplifies the role of community involvement in urban design. Conducted as part of the INTERREG Project in June 2022, this unique work camp aimed to raise awareness about open public spaces and urban design among students and local communities. The camp provided hands-on training in urban participation, emphasising mutual understanding, democratic awareness, and negotiation skills. By detailing the planning and implementation of the work camp, this part highlights the importance of on-site engagement in urban design education and community-led development.

In the context of modern technological advancements and the challenges posed by the COVID-19 pandemic, innovative teaching methodologies are essential for inspiring students in higher education. The elective course "Architecture in Context" at the University of Belgrade's Faculty of Architecture is a case study for creative education. This course combines innovative technologies, student involvement in scientific research and conferences, and collaboration between academia and practice. The study demonstrates how these techniques can enhance student motivation and produce higher-quality results, suggesting a path forward for the future of architectural education aimed at creating beautiful, sustainable cities.

The DANUrB Interreg DTP project illustrates the complementary roles of digital tools and analogue actions in creating a network of communities connected by the cultural heritage of the Danube and spanning seven countries; the project aimed to unite diverse cultures through a transnational brand. While digital tools like Facebook effectively disseminated information and generated interest, the project found that personal, local engagement was crucial for active stakeholder involvement and long-term co-creation. This part emphasises the necessity of combining digital platforms with on-site presence to foster collaboration, community engagement, and sustainable tourism development.

Digital Dreams, Urban Realities: Crafting Sustainable, Inclusive, and Beautiful Cities provides a comprehensive overview of the strategies and practices driving the ecological transition across different scales and contexts. By examining diverse case studies and research findings, this book offers valuable insights into how the New European Bauhaus principles can be applied to create sustainable, resilient, and inclusive communities.

By following a comprehensive approach, the book offers a holistic view of ecological transition within urban environments, making the book highly significant for various reasons. It provides a thorough overview of innovative strategies and methods for sustainable development, which can be invaluable for urban planners, architects, policymakers, and researchers. By presenting the latest innovative approaches, the book serves as a practical guide for implementing sustainable practices across different contexts. The book highlights practical applications through concrete case studies and examples, offering hands-on guidance for real-world implementation. This feature ensures that the theoretical aspects are grounded in pragmatic reality, making the strategies more accessible and applicable.

Moreover, the book's interdisciplinary approach connects science, technology, art, and culture, promoting a comprehensive method for addressing urban issues. This integration fosters a broader understanding and collaboration across various fields, essential for holistic urban planning and development. The book also emphasises the educational value by presenting innovative curricula and educational approaches that prepare future professionals for the challenges of sustainable urban development. These educational insights are crucial for equipping students with the necessary skills and knowledge to contribute effectively to sustainable practices.

Active community participation is another crucial aspect, as the book underscores the importance of involving local communities in the planning and development. This participatory approach not only enhances democratic governance but also ensures that development initiatives are more inclusive and reflective of community needs. Furthermore, the book focuses on creating a sustainable future by outlining strategies for developing sustainable, inclusive, and aesthetically pleasing urban environments. This future-oriented perspective is vital for addressing the long-term challenges our cities and communities face.

Additionally, the book offers practical recommendations and tools for planners and decision-makers aiming to implement sustainable practices within their communities. These actionable insights are essential for translating theory into practice and achieving tangible results. The global perspective provided through international examples and collaborations broadens the scope of learning, enabling readers to gain insights from diverse contexts and experiences. This international outlook fosters a more comprehensive understanding of sustainable urban development. It is also a valuable resource for policymakers, urban planners, architects, and researchers dedicated to crafting future cities.

Belgrade, Serbia

Milena Vukmirović
Aleksandra Đukić
Eva Vaništa Lazarević
Aleksandra Krstić-Furundžić

Introduction

In an era of rapid urbanization and digital transformation, pursuing sustainable, inclusive, and aesthetically pleasing cities is more critical than ever. *Digital Dreams, Urban Realities: Crafting Sustainable, Inclusive, and Beautiful Cities* explore how the principles of the New European Bauhaus can guide the harmonious integration of technology, nature, and culture within urban environments.

The New European Bauhaus (NEB) initiative, launched by the European Commission in September 2020, represents a transformative movement that seeks to merge worlds of art, culture, science, and technology to design and create more sustainable, inclusive, and beautiful living environments (Eckert 2022). Drawing inspiration from the historical Bauhaus movement, which revolutionized architecture and design in the early 20th century, the NEB aims to catalyze a new paradigm that aligns with the European Green Deal's goals of achieving climate neutrality by 2050 (European Commission 2024). Different authors have focused on the New European Bauhaus as part of the EU's efforts through European Green Deal policies and strategies, indicating its role in the broader context of EU climate policy and sustainability efforts.

As discussed by Bason (Bason et al. 2020), the New European Bauhaus initiative is integral to the Union's broader efforts to address climate and environmental challenges through the European Green Deal. It aims to foster a new cultural movement that combines sustainability with style, serving as an engine room for transformation in architecture and design (Ness 2021). The New Bauhaus encourages meeting societal needs with reduced consumption and carbon footprints, emphasizing regeneration over new construction, and inspiring creativity within the constraints of existing buildings' ethical and cultural dimensions (Ness 2021).

The New European Bauhaus initiative also redefines human-environment relations in the Anthropocene (Schwägerl and Crutzen 2018) by fostering a multidisciplinary approach to urban regeneration, sustainability, and aesthetics. It emphasizes integrating social, environmental, and cultural sustainability, aiming to absorb the impacts of industrial revolutions and current global challenges, such as the climate crisis and COVID-19 pandemic (Rosado-García et al. 2021). This initiative seeks to expand the concept of smart cities into smart territories, incorporating societal participation and inclusivity, and is supported by a diverse ecosystem of actors, including governments, academia, and industry. NEB's approach aligns with the broader discourse on the Anthropocene, which recognizes the profound human impact on the Earth system and the need for new theoretical constructs to address these challenges (Paul 2023). The initiative's focus on aesthetics is also significant as it aligns with the argument that appreciating beauty can create valuable and sustainable products, fostering a positive relationship between humans and their environment (Nowakowski 2024).

At its core, the New European Bauhaus is a creative and interdisciplinary initiative that promotes the co-creation of spaces, products, and experiences that reflect the values of sustainability, inclusiveness, and aesthetics. By fostering collaboration across various sectors, including architecture, urban planning, design, education, and social sciences, NEB encourages innovative approaches to address the challenges of climate change, resource efficiency, and social equity.

One of the NEB's fundamental principles is the idea of "beautiful, sustainable, and together." This triad emphasizes that the initiative is not merely about environmental sustainability but also about enhancing the quality of life and ensuring that the benefits of green transformation are accessible to all members of society. The NEB envisions a future in which people live in harmony with nature, public spaces are designed to foster community engagement, and sustainable practices are integrated seamlessly into everyday life.

The initiative also places strong emphasis on participatory processes, inviting citizens, artists, designers, researchers, and policymakers to contribute to dialogue and co-create solutions. Through various pilot projects, funding opportunities, and educational programs (Sadowski 2021), the NEB seeks to bring its vision to life, with a particular focus on bridging the gap between science and culture and between tradition and innovation.

The New European Bauhaus is more than just a policy initiative. This movement aspires to inspire change across Europe and beyond, creating a future in which sustainability, beauty, and inclusiveness are harmoniously intertwined in all aspects of life.

The visionary concept of the New European Bauhaus (NEB) served as the foundational theme for the 2023 International Academic Conference on Places and Technologies held in Belgrade. This conference brought together leading academics, practitioners, and policymakers to explore the theme "Keeping up with Technologies to Imagine and Build Together Sustainable, Inclusive, and Beautiful Cities." The event was a significant gathering that underscored the importance of integrating new technologies with NEB's core values of sustainability, inclusiveness, and aesthetics in urban development.

This conference highlights the critical need for interdisciplinary approaches to shape the future of cities. It provides a platform for discussions on how emerging technologies can be leveraged to create urban spaces that not only address the environmental and social challenges of the 21st century, but also enhance the quality of life through thoughtful and inclusive design. The discussions were deeply rooted in

NEB's principles, with a strong emphasis on collaboration across different sectors and citizens' active involvement in the co-creation process.

Eminent authors and researchers who presented their work at the conference were subsequently invited to contribute to this book titled *Digital Dreams, Urban Realities: Crafting Sustainable, Inclusive, and Beautiful Cities*, which aims to extend the dialogue initiated at the conference, delving deeper into the ways in which digital technologies can be harmonized with the NEB's vision to craft cities that are not only technologically advanced but also sustainable and inclusive.

The book comprises a series of articles and research papers that delve into diverse aspects of urban development in the context of digital transformation. Each contribution demonstrates a shared dedication to the principles of NEB, offering valuable insights into how digital innovations can be leveraged to create urban environments that are aesthetically pleasing, just and resilient.

The subsequent monograph represents significant milestones in the ongoing endeavor to bring NEB's vision to fruition globally. By encouraging a collaborative and interdisciplinary approach to urban development, these initiatives aim to inspire novel ways to envision cities' futures, ensuring that they remain dynamic, inclusive, and sustainable for future generations.

The structure of this book is deeply inspired by the seminal work "S, M, L, XL" (Koolhaas and Mau 1997), which explores architectural and urban design projects through the lens of scale ranging from small (S) to extra-large (XL). Similarly, this book is organized into sections that correspond to different scales of urban planning and design: macro, mezzo, and micro. These sections provide a comprehensive examination of the ways in which urban environments are conceptualized, designed, and implemented at various scales, from large metropolitan areas to individual public spaces. Additionally, a dedicated section on education explores the critical role that contemporary education in urban planning and design, and architecture plays in crafting sustainable, inclusive, and beautiful cities. This section emphasizes the importance of equipping future professionals with the skills, knowledge, and creative thinking necessary to address the complex challenges of the 21st century, aligning with the principles of the New European Bauhaus. Accordingly, the book comprises four separate, yet interconnected sections: macro, mezzo, microscale, and education. Each section provides an in-depth analysis of various aspects of urban development, highlighting how different scales and approaches contribute to the ultimate goal of sustainable urban futures.

The macroscale section of the book investigates the essential roles of cities, small towns, villages, and regions as engines of ecological transition. It also examines the innovative strategies employed by communities across various geographical contexts to foster sustainable development. Considerable emphasis is placed on integrating natural and cultural heritage into urban planning, which is crucial for achieving environmental sustainability and enhancing residents' quality of life.

For example, the Action Plan for Sustainable Sombor in Serbia illustrates how engaging local communities and leveraging their knowledge can create compelling heritage-based policies. These policies strengthen urban identity and align development closely with the ecological and cultural characteristics of a region.

Similarly, revitalization efforts in Southern and Eastern Serbia highlight the importance of addressing social, economic, and spatial challenges in rural areas, demonstrating how rural revitalization can contribute significantly to broader ecological transitions.

Furthermore, the rise of remote work and digital technologies has created new opportunities for rural development. The Digital Nomad Centre case study in Pécsdevecser, Hungary, illustrates how integrating digital nomads into rural communities can transform these areas into vibrant life and economic activity centers. This approach revitalizes rural areas and aligns with the principles of the New European Bauhaus by promoting sustainability, inclusivity, and beauty.

The macro-scale section of the book demonstrates that ecological transition is not confined to urban centers; instead, it is a multifaceted process that involves the active participation of cities, towns, villages, and regions. By integrating natural and cultural heritage into urban planning and leveraging digital innovations, communities can drive sustainable development that is both resilient and culturally enriched. This section highlights the critical role of local and regional actors in shaping the future of sustainable urban environments.

The mezzo-scale section of the book delves into essential land use planning strategies for sustainable urban renewal and transformation. This section emphasizes the importance of inclusive and participatory approaches in enhancing energy efficiency, promoting green urban renewal, and preserving cultural heritage. The pursuit of carbon neutrality within urban environments has emerged as a central theme with practical strategies and methodologies discussed to guide this endeavor.

For example, the development of the TExTOUR Cultural Tourism Lab in Trebinje, Bosnia and Herzegovina illustrates how co-creation methods within living laboratories can support cultural heritage management and broader community-led urban development. This collaborative approach addresses the challenges of urban renewal and ensures that urban growth is sustainable and inclusive.

This section also explores the methodologies for evaluating the energy performance of buildings under changing climatic conditions. By comparing the energy demands of different construction methods, this study provides actionable guidelines for improving building resilience and efficiency, which are critical for aligning urban development with the principles of the New European Bauhaus.

The mezzo-scale section emphasizes that sustainable urban renewal requires a balance between cultural preservation and environmental innovation. Inclusive land-use planning, combined with advanced methodologies for energy efficiency, provides a pathway toward achieving carbon neutrality in urban environments. The case studies presented here illustrate that through co-creation and participatory design, urban renewal can be both culturally enriching and environmentally sustainable, paving the way for resilient and vibrant cities.

The microscale segment of the book focuses on transforming urban spaces to promote physical, mental, and social wellbeing. This part of the book examines the crucial role of public spaces in fostering healthy lifestyles and explores innovative methods to enhance these areas.

For instance, Belgrade's initiative, which involves transforming parking spaces into parklets, demonstrates that small public areas can significantly improve pedestrian friendliness and social vibrancy in urban areas. This initiative supports broader sustainability goals by reducing car dependency and encouraging active transportation, thereby enhancing quality of urban life.

This section also delves into the intricate relationship between the built environment and health-oriented behavior, exploring how urban design can impact mental health and overall well-being. Practical recommendations for designing interventions that promote holistic health emphasize the importance of creating urban environments that support and enhance community wellbeing.

Furthermore, this section discusses strategies for extending the lifespan of the existing housing stock through energy-saving renovations and the integration of modern technologies. It provides a comprehensive overview of how affordable, sustainable, and technologically advanced social housing can be developed.

The microscale section demonstrates that urban space transformation is more than just a physical redesign; it is about enhancing community well-being. This section highlights how urban environments can be made livable, sustainable, and inclusive by focusing on public spaces, energy-efficient renovations, and integration of modern technologies. The strategies outlined in this section serve as a blueprint for cities that wish to foster holistic health and social well-being through thoughtful urban design.

The section of the book devoted to education underscores the pivotal role that architectural and urban design education play in shaping the next generation of leaders in sustainable urban development. This part of the book showcases innovative curricula and educational methodologies that embody the principles of the New European Bauhaus, thereby preparing students to create sustainable, inclusive, and aesthetically pleasing environments.

A central theme of this section is the incorporation of living laboratories and collaborative innovation ecosystems into education. These approaches foster active participation and personal engagement, enabling students to apply theoretical knowledge to real-world situations.

For instance, the University of Belgrade's approach to architectural education, which emphasizes teamwork and practical problem-solving, equips students with the necessary skills to tackle the complex challenges of sustainable urban development. Furthermore, the book examines how the New European Bauhaus initiative is being integrated into educational programs, particularly in light of the European Green Deal and the 2030 Agenda for Sustainable Development. By aligning educational content with these global initiatives, the book ensures that future professionals are well-prepared to lead ecological transitions in urban environments.

The education section highlights that the future of sustainable urban development hinges on the quality of education that tomorrow professionals receive today. Integrating the principles of the New European Bauhaus into architectural and urban design education underscores the importance of preparing students to think critically, work collaboratively, and design innovatively. The methodologies and curricula presented here serve as a model for educational institutions striving to equip their students with tools to lead ecological transitions in urban environments.

Digital Dreams, Urban Realities: Crafting Sustainable, Inclusive, and Beautiful Cities provides an extensive examination of the various scales of urban development, ranging from macro to micro, and their role in achieving the ultimate goal of ecological transition and urban resilience. This book presents a comprehensive perspective of the challenges and opportunities confronting urban planners, architects, policymakers, and researchers striving to create sustainable, inclusive, and aesthetically pleasing urban landscapes through case studies, practical strategies, and educational methodologies.

The book's interdisciplinary approach connects science, technology, art, and culture and advocates a comprehensive method for tackling urban difficulties. By blending theoretical knowledge with practical applications, Digital Dreams, Urban Realities serve as both a practical handbook and a visionary call for action to craft future cities. This volume's diverse contributions make it an indispensable resource for anyone devoted to creating urban environments that are not only sustainable and resilient but also inclusive and beautiful.

References

Bason C, Conway R, Hill D, Mazzucato M (2020) A new Bauhaus for a green deal

Eckert J (2022) Extending the new European Bauhaus—an educational initiative that is much needed to transform our society, vol 16. Springer, Cham. 10.1007/978-3-030-79879-6_9

European Commission (2024) The new European Bauhaus. European Urban Initiative. Accessed: Aug 26. Available: https://www.urban-initiative.eu/new-european-bauhaus-topic-first-call (Online)

Koolhaas R, Mau B (1997) S, M, L, XL, 2nd edn. Monacelli Press, USA

Ness D (2021) The shift from new build to regeneration. Can the new Bauhaus transform architecture and design to meet global challenges? AGATHON—Int J Arch Art Des 09:22–31

Nowakowski P (2024) Beauty and utility in architecture, interior design and in the new European Bauhaus concepts. Buildings 14(4):870. 10.3390/BUILDINGS14040870

Paul E (2023) Age of the Anthropocene. In: Australia in the Anthropocene. Palgrave Macmillan, Singapore, pp 9–60. 10.1007/978-981-19-8178-4_2

Rosado-García MJ, Kubus R, Argüelles-Bustillo R, García-García MJ (2021) A new European Bauhaus for a culture of transversality and sustainability. Sustainability 13(21):11844. 10.3390/SU132111844

Sadowski K (2021) Implementation of the new European Bauhaus principles as a context for teaching sustainable architecture. Sustainability 13(19):10715. 10.3390/SU131910715

Schwägerl C, Crutzen P (2018) The Anthropocene the human era and how it shapes our planet. Synergetic Press

Contents

Part I
Macro Scale: Cities, Small Towns, Villages, and Regions as Engines for the Ecological Transition

Macro Scale: Crafting the Big Picture

The macro-scale represents the broadest and most impactful level of analysis and intervention in urban planning and architecture. This scale encompasses vast geographical areas, from the metropolitan region to transnational urban systems. At this level, overarching systems, strategies, and structures that govern the development and functioning of urban environments are conceived, debated, and ultimately realized. Urban and regional planning at this scale is crucial for guiding growth, ensuring connectivity, and balancing the needs of diverse populations across sprawled urban landscapes (Hall and Tewdwr-Jones 2020). The importance of macro-scale urban systems is further emphasized in the context of global cities, where regional planning intersects with global economic networks (Sassen 2001).

The macro-scale is the foundation upon which sustainable, inclusive, and beautiful future cities must be built. Here, we confront the grand challenges and opportunities of urbanization, shaping the long-term development of cities and regions in resilient, equitable, and environmentally friendly ways. Strategic macro-scale planning is essential for managing these complex dynamics (Campbell and Fainstein 2003). Additionally, Friedmann highlights the critical role of public policy and governance in this process, where the macro scale becomes a key arena for translating knowledge into action for sustainable urban futures (Friedmann 1987). At this scale, urban and regional planning goes beyond individual projects or neighborhoods; it involves crafting coherent frameworks that guide growth, ensure connectivity, and balance the needs of diverse populations across sprawling urban landscapes.

The mentioned macro scale book section delves into the critical components of macro-scale spatial design, exploring how land-use planning, infrastructure development, environmental management, and economic strategies collectively form the backbone of sustainable urban growth. The United Nations Global Report on Human Settlements (Habitat 2009) highlights the importance of these elements in achieving sustainable urbanization, particularly in rapidly growing regions. Moreover, Calthorpe's work on urbanism in the age of climate change underscores the urgency of integrating sustainable strategies at the macro scale to address environmental challenges (Calthorpe 2011). We examine the role of policymaking and governance in shaping the macro scale, where strategic decisions made by governments and regional authorities set the trajectory for urban development (Habitat 2009).

Central to the New European Bauhaus vision is the ambition to craft sustainable, resilient, profoundly inclusive, and beautiful cities. This initiative seeks to bridge the worlds of science, technology, and arts, creating urban environments where ecological and social challenges are met with innovative, human-centered solutions (European Commission 2021). At the macro scale, this means rethinking how all cities and regions are designed to embody the principles of the New European Bauhaus, where functionality and aesthetics are seamlessly integrated and where every citizen has access to the benefits of a well-planned urban environment.

The New European Bauhaus calls for a holistic approach to urban design that considers the interconnectedness of ecological sustainability, social inclusion, and cultural richness (European Commission 2021). Forman's work on urban ecology supports this approach, emphasizing the need for urban systems that integrate natural processes at a large scale (Forman 2014). Similarly, the concept of "splintering urbanism" underscores the difficulties in sustaining cohesive urban infrastructure across extensive areas, stressing the necessity of integrated planning to prevent the fragmentation of urban systems (Graham and Marvin 2001). At the macro scale, this translates into planning strategies prioritizing green infrastructure, equitable access to resources, and preserving and celebrating cultural heritage. The design of large-scale urban projects, such as sustainable districts or regional green belts, must align with these values, ensuring that urban growth enhances the natural environment and quality of life for all residents.

Large-scale urban projects, from revitalizing entire districts to creating eco-cities, highlight the transformative potential of macro-scale interventions. These projects, often spanning decades, have the power to reshape urban environments, bringing about social, economic, and environmental benefits that reverberate across regions and nations. Harvey's investigation into the urban revolution highlights the importance of such initiatives in altering urban landscapes and promoting social justice (Harvey 2012). The socio-spatial dynamics at the macro scale, as discussed by Soja, are crucial for understanding the impacts of large-scale urban planning (Soja 2000). Within the New European Bauhaus framework, such projects are not merely about expansion or modernization but about crafting spaces that inspire, connect, and sustain communities across generations.

The macro-scale section of the book delves into how cities, small towns, villages, and regions can act as engines for ecological transition—a concept central to the New European Bauhaus vision (European Commission 2021). By engaging local communities in Sombor, Serbia, and integrating their knowledge of natural and cultural heritage, we see how macro-scale planning can be both inclusive and sustainable. The revitalization of rural areas in Southern and Eastern Serbia underscores the challenges and opportunities of regional frameworks in promoting environmental sustainability, showcasing the need for comprehensive planning at this scale.

Moreover, the case of establishing a digital nomad center in rural Hungary highlights how large-scale strategic interventions can breathe new life into declining regions, aligning economic development with the goals of sustainability and inclusivity. The micro-scale section of the book also explores the "Design with Nature" approach, advocating for a more-than-human design culture that prioritizes ecosystem services, as demonstrated in the mapping of Granada City and the revitalization of the Danube-Tisa-Danube Canal.

As we explore the macro scale, we are reminded that this level of planning is not just about managing the complexities of urban growth but also about envisioning and crafting a sustainable, inclusive, and beautiful future. Mike Davis's critical perspective on global urbanization challenges, particularly in informal settlements, highlights the importance of ensuring that macro-scale planning is inclusive and equitable (Davis 2006). At this scale, the dreams of resilient cities and thriving communities are translated into tangible realities, setting the stage for more detailed design work that will follow at the mezzo- and micro-scales. The following chapters provide insights and case studies that illustrate how macro-scale thinking is essential for shaping the cities of tomorrow, ensuring that they are well-equipped to meet the challenges of the 21st century and beyond.

References

Calthorpe P (2011) Urbanism in the age of climate change. Island Press/Center for Resource Economics, London

Campbell S, Fainstein S (eds) (2003) Readings in planning theory, 2nd edn. Wiley Blackwell, USA

Davis M (2006) Planet of slums. Verso, London, New York

European Commission (2021) "New European Bauhaus, Beautiful, sustainable, together", Brussels. Accessed: Aug 28, 2024. Available: https://new-european-bauhaus.europa.eu/system/files/2021-09/COM(2021)_573_EN_ACT.pdf (Online)

Forman RT (2014) Urban ecology: science of cities. Cambridge University Press, Cambridge

Friedmann J (1987) Planning in the public domain: from knowledge to action. Princeton University Press, USA

Graham S, Marvin S (2001) Splintering urbanism: networked infrastructures, technological mobilities.... Routledge, UK

Hall P, Tewdwr-Jones M (2020) Urban and regional planning. Routledge, UK

Harvey D (2012) "The creation of the urban commons", rebel cities: from the right to the city to the urban revolution, pp 67–88

Sassen S (2001) The global city: New York, London, Tokyo. Princeton University Press, USA. Accessed: Aug 28, 2024. Available: http://www.jstor.org/stable/j.ctt2jc93q (Online)

Soja EW (2000) Postmetropolis: critical studies of cities and regions. Wiley-Blackwell, USA

UN Habitat (2009) Planning sustainable cities

Chapter 1
Engaging Local Communities' Knowledge and Ideas for Natural and Cultural Heritage Integration: The Case of Action Plan for Sustainable Sombor, Serbia

Danijela Milovanović Rodić, Aleksandra Djukić, Branislav Antonić, and Ana Šabanović

Abstract The protection and sustainable use of natural and cultural heritage for enhancing urban identity and local community's quality of life, as well as positioning Serbian cities within the global eco-cultural tourism network, remain underappreciated in national and local development agendas. Following growing interest in tourism that integrates tangible and intangible heritage with natural assets, there's a need to reassess this approach, especially for small and medium towns facing decline. Across Europe, there's a heightened acknowledgement of heritage protection's role in sustainable urban development, leading to numerous national and pan-European initiatives over recent decades. These efforts initiated the creation of new methodologies and different types of pilot projects for their validation. Their common denominator is that the capacity of all involved in conceiving and executing heritage-based policies should be enhanced. This paper outlines the methodology, development process, and content of an Action Plan focused on the integration and sustainable use of the natural and cultural heritage of Sombor. Developed collaboratively with local stakeholders, the primary objective was to ensure that it is deeply rooted in contextually relevant knowledge and authentic ideas, thereby facilitating its implementation in a more efficient and effective manner. Additionally, it aimed to foster connections among local initiatives, enhancing their capacities for further collaborations.

Keywords Collaborative planning · Urban development · Action plan · Integrated approach · Heritage · Local knowledge

D. M. Rodić (✉) · A. Djukić · B. Antonić · A. Šabanović
University of Belgrade—Faculty of Architecture, Beograd, Serbia
e-mail: danijela@arh.bg.ac.rs

M. Vukmirović et al. (eds.), *Digital Dreams, Urban Realities*, The Urban Book Series,
https://doi.org/10.1007/978-3-031-76353-3_1

1.1 Introduction

The Action Plan Integrating Natural and Cultural Heritage for Sustainable Sombor 2022–2027: network of eco-cultural routes is developed within the international project DANUrB + (2020–22) and its strategic framework expressed in the form of the Development Strategy of the Danube Region based on natural and cultural heritage developed within previous DANUrB (2017–19). Both DANUrB projects and their products, as this Action plan is, are built upon the concept of eco-cultural tourism and the idea of shaping new authentic routes that promote local identity by protecting and integrating natural and cultural heritage along the Danube.

The plan, alongside its ten integrated projects, is informed both by territorial analysis and stakeholder initiatives. While solutions presented in the plan may not be comprehensive and exhaustive, they serve as well as a methodological roadmap for further initiative development. The pilot action implemented in accordance with the Action plan's values and goals exemplifies successful heritage revaluation aimed to inspire future actions and contribute to local identity reinforcement through stakeholder engagement and knowledge transfer.

Due to the applied approach, it can be considered that the most significant result in this process is in the domain of social capital. All involved, regardless of the sector, whether they are from—the locals, public representatives and planners—gathered new ideas and knowledge and built a common understanding of the mutual benefits and possibilities opened up by the integration of their different goals and interests. Shaping the network of interconnected people and projects can serve as invaluable support for future planning endeavours, providing a substantial resource upon which further planning activities can rely (Milovanović Rodić et al. 2023).

1.2 Theoretical Framework

1.2.1 Collaborative and Integrated Approach for Sustainable Development

Since the 1990s, the collaborative approach to urban planning has been intensely debated, critiqued, reevaluated, practised, and promoted. Its widespread adoption and acceptance by a large number of planning theorists and practitioners are driven by the increasing complexity of developmental issues and the growing number of conflicts among interest groups in decision-making and implementation processes concerning development (Milovanović Rodić and Maruna 2022).

Arguments in favour of the collaborative planning model and motives for its application stem from philosophical, practical, and politico-ideological spheres. Arnstein's discussion (1969) on types and forms of decision-making is considered a precursor and catalyst for the development of the concept, whose theoretical elaboration occurred over the subsequent decades (Friedmann 1973; Habermas 1984;

Giddens 1986; Foster 1988; Sager 1994; Innes 1995; Innes and Booher 2003; Healey 1997; Healey 2006). It recognises the importance of engaging diverse stakeholders, including professionals from public administration, the private sector, and civil society, to collectively plan and manage the common affairs of cities. By fostering collaboration and partnerships, this approach enables the integration of various knowledge, ideas, influences, and resources within the processes of tracking the sustainable future of cities and their communities. It is based on the view that through developed governance mechanisms, urban planning can address the complexities and challenges of contemporary society, promoting sustainable development that meets the needs of both present and future generations.

The concept of sustainable development entails adopting an integrated approach to planning (UN 1992). It emerged in response to the increasingly adverse effects of conflicts between the demands and goals of economic development and the need for environmental preservation. An integrated approach implies that the sectoral approach to planning should evolve into a comprehensive system for managing developmental processes within space. "The idea of sustainability encompasses not only awareness of the environment and sustainable economy but, more importantly, social integration and new paths in urban governance, i.e., the participatory role of every citizen" (Pušić 2001). The essence of the concept of social sustainability is reflected in the realisation of a city where people feel a sense of belonging and willingness, rather than compulsion, to stay in that place, to improve and develop it. "There is no sustainable city without the creative activities of citizens. The precondition is an active citizen. To be active, the citizen must be interested. To be interested, they must be informed. For the citizen to be informed, the system must be democratic" (Pušić 2001).

The integrated approach is founded on the principles of good governance, which emphasise inclusivity, transparency, accountability, and participatory decision-making (UN-HABITAT 2002; UNESCAP 2009; UN 2015). It encompasses multiple dimensions, including economic, social, environmental, and cultural aspects, and addresses themes such as urban mobility, social inclusion, resilience, employment, and more. Furthermore, it integrates different policies, strategies, and plans at various spatial levels, ranging from the neighbourhood to the regional and national levels. The integration extends to various administrative levels of government and involves diverse actors and institutions from the public, commercial, NGO, and civil sectors. Additionally, an integrated approach incorporates different forms of funding, such as local, regional, and national budgets, public–private partnerships, and international sources of finance (Milovanović Rodić et al. 2017; Milovanović Rodić et al. 2018).

1.2.2 Engaging Local Knowledge and Ideas

Understanding planning as a value-laden activity, which is not solely based on knowledge generated through expert-driven "scientific approaches" but also on knowledge generated in interactive processes within a society during which "ways of thinking, ways of valuing, and ways of acting are actively constructed by all participants in the process" (Healey 1997, p. 29), dates back to the 1970s. Today, it has been articulated in the form of argumentative, communicative, or interpretative approaches to planning, which are based on the following principles:

1. All forms of knowledge and thinking are socially constructed, i.e., contextually sensitive,
2. Expert knowledge and theories need to be integrated with local "lay" knowledge and "practical reasoning",
3. Communication and knowledge development can take various forms—from rational, systematic analysis to expressionistic expression of attitudes, thoughts, and feelings;
4. Power relations are manifested not only in the distribution of material resources but also in the value system, understanding of reality, and perception of one's role in development management processes (Sager 1994; Innes 1995; Innes and Booher 2003; Healey 1997; Milovanović-Rodić 2013).

The basic idea is that all meanings, thoughts, and actions are socially conditioned. Different interests in space lead to different strategies and relationships towards space and place quality. Therefore, it is important to develop planning processes in a way that allows:

1. Involvement of all stakeholders and the collection of various knowledge and ideas about development,
2. Insight into the collected data by all stakeholders to enable a broader and different perspective than individual ones; this can facilitate changes in the perception and attitudes of participants in the process;
3. The generation of a common understanding of reality and a desirable future (Milovanović Rodić et al. 2023).

Engaging local communities knowledge and ideas in natural and cultural heritage responsible urban planning allows invaluable insights into valuable natural resources, cultural landmarks, and heritage sites that may otherwise go unnoticed or undervalued. Integrating this knowledge into urban planning processes, planners can foster ownership, promote sustainability, and enhance innovation and the authenticity of development projects.

Effective and meaningful community engagement requires the development of techniques that can support dialogue and knowledge and ideas exchange. In the following section, the methodology for developing an Action plan for sustainable urban development based on the previously stated premises will be presented.

1.3 The Case of the Action Plan for Sombor

Sombor Region is located in the northwestern corner of Serbia, on the border with Hungary and Croatia (the Danube). The region consists of the city of Sombor and 15 nearby villages. The whole region has been experiencing economic and demographic decline for 30 years. However, Sombor is the best-preserved historic medium-sized city in Serbia, and the region is rich in cultural (tangible and intangible) and natural heritage. Additionally, there are many active craft and hospitality enterprises, ethnic and religious events, local associations, and small cultural and leisure manifestations in the city and villages. Many of them have been well-established at the regional and national levels but without significant cooperation across long national borders in the region. Furthermore, most of these entities have not been properly interconnected within the Sombor Region to make an integrated approach to potential external partners.

1.3.1 Designing Planning Process for Engaging Local Knowledge and Ideas

The main aim of the Action Plan Integrating Natural and Cultural Heritage for Sustainable Sombor 2022–2027: Network of Eco-Cultural Routes was to protect and improve the cultural (tangible and intangible) heritage and natural environment in accordance with local culture and available resources, knowledge, and skills by engaging and supporting people in integrating their efforts.

The specific objectives of the plan were (Milovanović Rodić et al. 2023):

(a) Identification, valorisation, protection, enhancement and promotion of natural and cultural heritage as a resource for the development of sustainable tourism and improvement of the quality of life of local communities;
(b) Raising the capacity of local stakeholders to formulate and launch development projects based on natural and cultural heritage,
(c) Connecting stakeholders and their projects in order to improve existing and enforce the development of new eco-cultural products, services and destinations,
(d) Improving the capacity of local stakeholders for cooperation and access to funds at the local and international levels.

The Action plan was developed within four interrelated phases (Table 1.1) in continuous cooperation with representatives of the local self-government and with the active participation of about 15 stakeholders from public, private and NGO sectors, as well as from the spheres of culture, tourism, heritage preservation, ecology and public administration. Cooperation with stakeholders in the process of developing the plan took place in the phase of data collection and understanding of the territory and in the phase of generating project ideas. In each phase, participation from the stakeholders was made in two ways: at workshops in the form of a round table held

Table 1.1 Form of stakeholder participation within the process of action plan development (Milovanović Rodić et al. 2023)

The planning process	Form of stakeholder participation	The outcomes
Phase of understanding the territory	Questionnaire 1	Questionnaire results: swot components, key values, components of vision and goals
	Workshop 1 (Fig. 1.1)	Verification, addition and integration of the contextual analysis of the results obtained by the questionnaire Discussion on the problems and potentials for the development of Sombor
Project formulation phase	Questionnaire 2	Project ideas
	Workshop 2 (Fig. 1.1)	Discussion of project ideas, time frame, activities, sources of funding Discussion of synergistic effects of projects
Phase of designing the action plan	Conversations with individual stakeholders	Preparation of the Draft Plan based on the contextual analysis, the results of the 1st and 2nd workshops)
Action plan finalization stage	Plan available to the public and Plan sent to all workshop participants	Action plan

on the premises of the city administration and by filling out an online questionnaire. Each workshop was preceded by a questionnaire that aimed to inform and interest potential participants and encourage them to join the following workshop. In addition, the content of the questionnaires was discussed, supplemented, and corrected at the workshops that followed them.

The general public was also enabled to submit information about the territory, their perception of problems and potential for development, as well as project ideas by filling out a digital online or analogue questionnaire posted on social networks and placed on the local government's official pages.

1.3.2 Designing the Questionnaire for Engaging Local Knowledge and Ideas

The initial questionnaire was designed to (a) inform the citizens of Sombor about the goals, methodology, and expected outcomes of the Action Plan, (b) motivate them to participate in its development, (c) enable the creation of a current and contextualised

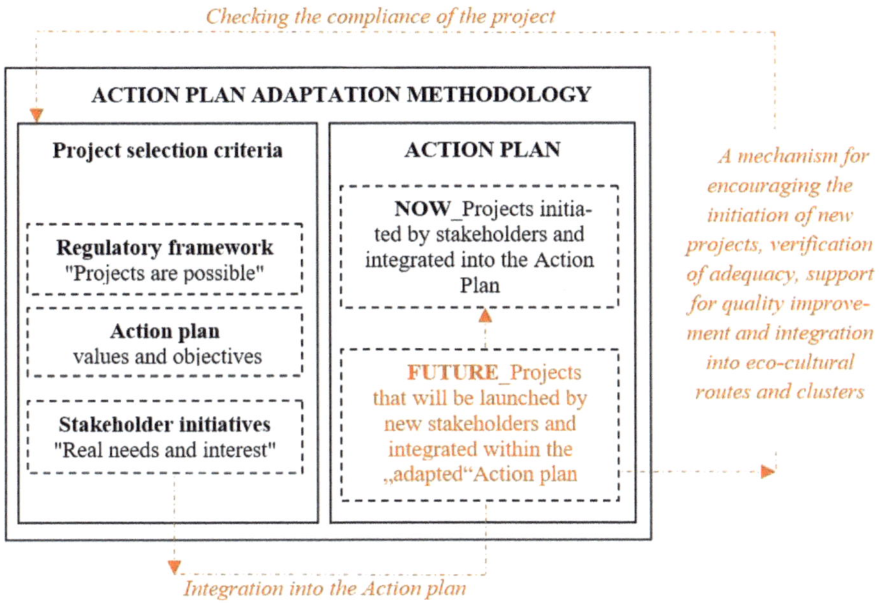

Fig. 1.1 Methodology for the action adaptation

shared knowledge base, and (d) enable a two-way exchange of values, ideas, concepts, and attitudes about the desirable and feasible development of the territory.

The questionnaire consisted of 22 questions grouped into four sections:

1. Information about the respondent
2. Desired future
3. Understanding the territory
4. Further involvement

Its purpose, content and structure stem from the understanding that collecting information about the respondent's demographics, understanding their preferences and desired future for the city, and identifying their priorities play a crucial role in ensuring that the Action Plan for nature and heritage protection reflects the needs and aspirations of the community it serves.

The inclusion of questions regarding personal information may provide insights into potential disparities or specific needs within the community—certain development initiatives may need to be tailored differently based on those different demographics' groups. For example, age demographics can inform housing needs, employment opportunities, and access to services. Education level can impact employment opportunities, income levels, and overall community engagement. Working Length of Stay in the City can provide insights into the level of attachment individuals have to the community—long-term residents may have different perspectives and priorities compared to newcomers. The location of the workplace can help identify commuting patterns or transportation needs. Understanding whether respondents are employed full-time, part-time, self-employed, or unemployed can provide insights into the economic dynamics of the territory. It can also help identify areas where employment opportunities may need to be expanded or diversified. Different sectors contribute differently to the economy and may have different development needs. The collection of such personal information has to be conducted ethically and with due consideration for privacy and confidentiality concerns.

The second group of questions aimed to identify the values that, according to the respondents' opinions, should be the cornerstones of development. The question "What kind of city do you want to live in?" helps identify the respondent's values and aspirations for the city's future. This allows planners to gauge the importance placed on aspects such as healthy lifestyles, environmental preservation, cultural identity, economic prosperity, and communal infrastructure development. Respondents ranked the offered values and priorities, with the option to add any missing values according to their opinions (Table 1.2).

The third group of questions aimed to identify priorities regarding nature and heritage protection by ranking the offered statements and adding missing ones. Respondents could prioritise the aspects they deem most important for the city's development by 2030. This helps in understanding the community's collective vision, guiding the formulation of specific goals and strategies within the Action Plan.

Finally, the respondents were invited to get involved in the further process of developing the plan and leave their contacts.

It was possible to fill out the questionnaire online. The digital questionnaire was sent by e-mail to 54 stakeholders' addresses (stakeholders analyses result). Besides that, the invitation was sent publicly on social networks and the official website of the City of Sombor. 25 stakeholders responded and called to join the workshop, and received responses were used for its preparation.

The first workshop was held in the form of a round table after a brief presentation of both the aim and role of the Action plans and the data and opinions obtained through the questionnaire. Seventeen stakeholders participated in the workshop and discussed the current problems and challenges in Sombor Region.

The second questionnaire was intended (a) to continue the process of collaboration with interested stakeholders, (b) to enable them to formulate and present their project ideas—explain the reasons, potential participants, expected effects, required resources and timeframe, and (c) create a base of project ideas that will enable their exchange, provoke discussion and inspire their integration, even foster new collaboration and partnership in the future (Table 1.3).

The questionnaire had questions grouped into three sections:

1. Information about the respondent
2. Project idea proposal
3. Further involvement in the development of project ideas.

The questionnaire was filled out by five stakeholders who presented their project ideas in detail.

Table 1.2 The 1st questionnaire: desired future, understanding of the territory, and call for further involvement

I Information about the respondent	
Understanding preferences-respondent's correlation	Gender, Age, Level of education, Length of stay (living and working) in the city, Location of the living and workplace, Labor status, Working sector (public/private/civil), Working domain
II Desired future	
Values identification What kind of city do you want to live in?	1. Healthy citizens and good conditions for healthy lifestyles 2. Preserved nature and high quality of the environment 3. Good governance and developed communal infrastructure 4. Local identity and preserved cultural heritage 5. A thriving city with developed economic activities
Priorities identification How do you see the city of Sombor in 2030? Rank offered statements and add missing one	1.1 Diverse and accessible cultural amenities 1.2 Diversity of educational facilities 1.3 Sports and recreational facilities 2.1 High quality tap water, ground and surface water 2.2 Developed network of green spaces and rich biodiversity 2.3 Protected and unpolluted agricultural land 2.4 Clean air and developed usage of renewable energy sources 3.1 Democracy and transparency in decision making processes 3.2 Efficient and integrated transportation system 3.3 Development of small and medium-sized local enterprises 3.4 Resilience, safety and security 3.5 Youth employment support programs 4.1 Preserved, protected, and revitalized tangible heritage 4.2 Preserved intangible cultural heritage 4.3 Multiculturalism 5.1 Development based on agriculture 5.2 Development based on tourism 5.3 Development based on craftsmanship and trade 5.4 Development based on industry 5.5 Development based on foreign investments 5.6 Development based on strengthening local enterprises
III Understanding the territory	
Key problems	1. What are the key problems in your city? 2. What are the key problems in your field of activity?

(continued)

Table 1.2 (continued)

Attitudes about the local administration's transparency and effectiveness Motivation and possibilities of personal involvement in the process of formulating solutions	3. Are you aware of the measures taken by the local administration to solve some of these problems? If the answer is "No", please state the reason: 3.1 I don't know the procedures 3.2 I don't understand the proposal (plan, project, decision, etc.) 3.3 I cannot gain insight into the proposals 3.3.1 Public hearings are organized during my working hours I don't have or can't use the software 3.3.3 I do not have access to the Internet 3.3.4 Something else 4. Do you think that personally engagement and involvement in solving some of those problems can be purposeful? If the answer is "No", please state the reason: 4.1 I don't have time 4.2 I don't believe that my engagement can really change something 4.3 I don't know what the procedure is 4.4 I don't know what my rights are—what I can demand and get
Key potentials and resources for development	5. What are the most significant development potentials of the city? 6. What are the most significant elements of the natural and cultural heritage that need to be protected and preserved? 7. What are the potentials for the development of your activity?
IV Further involvement	
	1. Are you interested in getting involved in the further development of this Action Plan? If the answer is „Yes, please leave contacts

Table 1.3 The 2nd questionnaire: Stakeholders' project ideas

I Information about the respondent
II Project ideas
1. Briefly describe your idea—explain the reasons, potential participants, expected effects and timeframe 2. State how the implementation of the idea will affect the improvement of your life and work 3. State what your community (personal or professional community), and perhaps other citizens of Sombor, will gain from the realization of your idea 4. What resources are needed for the realization of this idea? 5. List and specify everything else you consider important for understanding and implementing your idea
III Further involvement

The second workshop was held in the form of a roundtable. The 12 stakeholders participated in the workshop. The goal of the workshop was to encourage the participants to present existing project ideas (those that were submitted by questionnaire were presented) and conceive new ones.

1.3.3 Sensitive and Receptive to Local Initiatives and Local Specificities Action Plan Able to Adapt

Based on the results of the workshop, ten integrated, compatible, mutually synergistic projects that embrace the concept of sustainable development were formulated:

1. **Green Network** of bicycle and pedestrian paths, public transport and water routes, roads, bus stations, and accommodation
2. **New Places to Stay**—developing new and improving existing accommodation facilities.
3. **Arboterum_a Seed for the Green Future of a City**—a place for a gene pool of Indigenous species that will: (a) strengthen the eco-awareness of the local community, especially children, (b) enable their physical and mental development, (c) be a new space for education and recreation, (d) be a new tourist attraction.
4. **On the Land of Mythical Bodrog—**rearranging the river Island "Langcet" in Bački Monoštor will: (a) enable a meaningful stay in nature and provide an active connection with nature, especially for young people (b) enable their physical and mental development, (c) improve the inhabitants' quality of life, (d) be a new tourist attraction.
5. **BeeCool—**a beekeeping camp in Bački Monoštor to (a) enable education and the transfer of knowledge about beekeeping to all ages, (b) enable a stay in nature that contributes to physical and mental development, (c) improve the quality of local inhabitants' life as a new space for education and recreation, (d) encourages cooperation among local beekeepers.
6. **Common Shopwindow_Together to the Bayer**—a new organisational and spatial model for increasing the visibility of small local producers in order to place local products on the market and improve the stability and volume of business.
7. **Green School—**a pilot model for the development of curricular and extracurricular activities in cooperation with local associations and companies within the eco-cultural network in order to improve the health of school children, improve the quality of teaching and improve the awareness and knowledge of all involved about the importance of natural and cultural heritage for the development of communities.
8. **Stapar Rose Centre—**an adaptation of the Stapar village cooperative house into a village cultural and social centre in order to improve the quality of life of

the local community, increase the visibility and accessibility of local products and the tourist attractiveness of the village.

9. **Old New School**—abandoned school reconstruction in the village of Bački Monoštor into a teaching, research, social and cultural village centre in order to improve the quality of life of the locals, increase tourist attractiveness and attract pupils, students and researchers.
10. **Eco-Cultural Network's Good Governance**—establishing an agency that (a) will support the realisation and coordination of the projects and (b) will encourage the articulation of new projects in accordance with the Action Plan goals.

Figure 1.1 illustrates how the Action plan could be further developed by adding new ones due to mechanisms for future project generation and development. The proposed model should be flexible enough to allow different modalities of stakeholder initiatives to be involved and integrated within the Action plan after its formalisation.

This planning methodology can be an instrument that supports the overall development of the community.

The model is open to innovation, sensitive, and receptive to existing local initiatives and local specificities, but it is also stimulating for the creation of new initiatives that have a high degree of self-organisation and adaptability and a developed system of mutual learning. Such planning adapts to changes in development contexts and enables different practices in the environments in which it is practised; that is, it has mechanisms that will ensure its own evolution.

1.4 Conclusion

Cities serve as catalysts for social and cultural advancement, stimulating their own growth as well as that of regions and states. In order to enhance their position within the global network of cities, urban centres are increasingly focusing on entrepreneurial activities. By actively participating in initiatives, cities gain support in project development, facilitating access to local, regional, national, and international funding opportunities. They also benefit from labelling, promotion, and inclusion in EU projects aimed at advancing the development of Danube cities.

Through workshops and questionnaires, numerous ideas were generated and subsequently integrated into a comprehensive Action Plan consisting of ten projects. Crucially, the establishment of a planning culture is emphasised, emphasising effective communication, empowerment of stakeholders, integration, and networking. This approach aims to build a cohesive system where different elements harmonise and collaborate seamlessly. By embracing these principles, cities can effectively drive their own progress and foster sustainable development.

Acknowledgements Supported by the Science Fund of the Republic of Serbia within project number 7744648, Empowering Participation through Culture and Architecture: Activating Public Resources FOR and WITH a Community—EPICA (2022–2024).

References

Foster J (1988) Planning in the face of power. University of California Press

Friedmann J (1973) The transactive style of planning. Contemporary movements in planning theory, pp 12–33

Giddens A (1986) The constitution of society: outline of the theory of structuration. Polity Press

Habermas J (1984) The theory of communicative action. Beacon Press, Boston

Healey P (1997) Collaborative planning shaping places in fragmented societies. Palgrave, London

Healey P (2006) Transforming governance: challenges of institutional adaptation and a new politics of space. Eur Plan Stud 14:299–319

Innes J (1995) Planning theory's emerging paradigm: communicative action and interactive practice. J Plan Educ Res 14(3):183–189

Innes JE, Booher DE (2003) Collaborative policymaking: governance through dialogue. In: Deliberative policy analysis: understanding governance in the network society, pp 33–59

Milovanović-Rodić D (2013) Redefining the model of citizen participation in urban planning in Serbia according to the communicative-collaborative paradigm. In: Doctoral dissertation, Faculty of Architecture, University of Belgrade, Belgrade, Serbia

Milovanović Rodić D, Maruna M (2022) Kolaborativni pristup za utvrđivanje i odbranu javnog interesa u neoliberalnom, politizovanom i polarizovanom urbanom pejzažu. U potrazi za javnim interesom: dometi urbanizma. Serbia, Univerzitet u Beogradu - Arhitektonski fakultet, Belgrade, pp 64–83

Milovanović Rodić D, Čolić R, Maruna M (2017) Multilevel governance instruments for achieving balanced urban-rural development. In: Keeping up with technologies in the context of urban and rural synergy: book of conference proceedings, Sarajevo

Milovanović Rodić D, Čolić R, Maruna M (2018) The role of university in a policy making process: introducing integrated urban projects for effective urban governance in Serbia. In: Integrated urban planning, TU Delft

Milovanović Rodić D, Šabanović A, Antonić B, Đukić A (2023) Collaborative approach for integrating natural and cultural heritage for sustainable urban development: the case of action plan for Sombor within DANURB+. In: 8th International academic conference on places and technologies, Belgrade, Serbia

Pušić L (2001) Održivi grad: ka jednoj sociologiji okruženja, Beograd

T Sager (1994) Communicative planning theory

UN (1992) Agenda 21, United Nations

UN (2015) Transforming our world: the 2030 agenda for sustainable development, UN

UNESCAP (2009) What is good governance?

UN-HABITAT (2002) The global campaign for good urban governance. Concept Paper

Chapter 2
Rural Areas in Regional Frameworks and Environmental Sustainability: Opportunities and Challenges for Their Revitalisation as a Tool for Ecological Transition in the Region of Southern and Eastern Serbia

Milica Igić, Milena Dinić Branković, Jelena Đekić, Milica Ljubenović, and Mihailo Mitković

Abstract Rural areas occupy a large part of Serbia's population; they are home to a significant number of people, and they have a key role in the food production chain. As their comparative advantage, natural resources, biodiversity and unpolluted environment can be defined. Decades of negligence and marginalisation contributed to the deterioration of rural areas in a social, economic and spatial sense. Based on the analyses of the specific characteristics of rural areas within the Region of Southern and Eastern Serbia, this chapter examines opportunities and challenges for their revitalisation in the context of environmental sustainability. Starting from the assumption that the revitalisation of rural areas is a precondition for their development and that rural development can contribute to even regional and overall sustainable development, the impact of the revitalisation of rural areas on ecological transition will also be discussed within this chapter. This chapter aims to investigate how the process of revitalisation of rural areas within the Region of Southern and Eastern Serbia can foster ecological transition and improve environmental sustainability. By revitalising rural areas, it is possible to improve their sustainable development, improve quality of life, and preserve biodiversity, contributing to achieving a better place for life for the rural population.

Keyword Rural areas · Ecological transition · Revitalisation of rural areas · Environmental sustainability

---

M. Igić (✉) · M. Dinić Branković · J. Đekić · M. Ljubenović · M. Mitković
Faculty of Civil Engineering and Architecture, University of Niš, Niš, Serbia
e-mail: milica.igic@gaf.ni.ac.rs

M. Vukmirović et al. (eds.), *Digital Dreams, Urban Realities*, The Urban Book Series,
https://doi.org/10.1007/978-3-031-76353-3_2

2.1 Introduction

Rural areas are characterised by a dynamic historical development, where influences of different cultures and religions over the centuries had a great impact on their creation and shaping. In the case of underdeveloped countries, rural areas are characterised by a more religious and less open society—a traditional and static society that is underdeveloped in the technological and cultural sense (Blanche et al. 2021). During the last decades, rural areas in Serbia, especially those in the Region of Southern and Eastern Serbia, faced an expressed trend of demographic emptying and economic stagnation, which negatively affected all the other structures of these areas (Babović et al. 2016). Although negative demographic growth is considered the biggest development problem in rural areas, there are other problems within socio-demographic, economic, physical and functional, as well as environmental structure, that negatively affect the quality of life and rural development in these areas. Rural areas are facing one of the biggest contemporary challenges—climate change, and their negative impact represents a great threat to all structures of rural areas (Zelenakova et al. 2015). According to many authors (Ringler and Meinzen-Dick 2019; Steiner and Shenggen 2019), along with the ongoing trends in the EU, the process of revitalisation of rural areas is recognised as one of the effective tools for the improvement of their structure.

This chapter discusses rural areas within the region of Southern and Eastern Serbia and investigates opportunities and challenges for their revitalisation in the context of improvement of environmental sustainability. The starting point of this research is that rural areas represent an important resource for fostering balanced regional development. Rural areas in Serbia are facing a decline, so it is important to first revitalise them to improve their existing structure and contribute to their development. Based on that, this research aims to explore the impact of the revitalisation of rural areas, on a regional level, on ecological transition and overall sustainable development. Since the rural areas have a share of more than 90% of the total territory, they are home to almost half of the population, and taking into account their current state, their improvement must be considered a priority. Taking all of the above into consideration, the revitalisation of rural areas can be observed as an instrument for fostering a process of ecological transition and improving sustainable development.

2.2 Methodology

The methodological framework in this chapter is based on the analysis of the specific characteristics of rural areas and their current state in the Region of Southern and Eastern Serbia, as well as the investigation of the opportunities and challenges for their revitalisation in the context of environmental sustainability. Within the first part of the chapter, the background is set up through analyses of the European and Serbian context regarding rural areas. Following, analyses of specific characteristics of all structures of rural areas in the Region of Southern and Eastern Serbia and their impact on environmental structure are presented. The second part of the chapter discusses opportunities and challenges for the revitalisation of these areas and the impact of this process on the region's ecological transition.

The Region of Southern and Eastern Serbia represents the most underdeveloped region, whose GDP is the lowest compared to other regions; it is characterised by a very heterogeneous structure, rich biodiversity and natural resources. Its territory occupies almost 1/3 of the total country territory but still has the lowest population density; rural areas are declining, so the quality of life is at a very low level. Rural areas in this Region are facing expressed negative impact of climate change, which affects all the structures of rural areas. Based on that, for this research, rural areas from the Region of Southern and Eastern Serbia are selected for analysis.

2.3 Background: Rural Areas and Rural Policies in the EU

Observed globally, rural areas occupy a significant part of the territory of each country, and their definition was the subject of a large number of research. However, even today, there is no universal definition of rural areas. Rural areas are often determined by their negative definition as non-urban areas (Vasilevska 2006). The classification that is in use today, at the global level, for determining rural areas is according to the Organisation for Economic Co-operation and Development (OECD) criterion (1994), where settlements with a population density of less than 150 inhabitants/km^2 are defined as rural. Rural areas have an important role in the overall development of each country; in that regard, one of the priorities of development policies is certainly rural development, which can be determined as a process of improving the quality of life and economic structure of rural areas (Mosley 2003). Besides improving economic structure, rural development has an impact on improving the environmental, social and spatial structure of rural areas. Revitalisation of rural areas can be defined as a process of their transformation into places for the rural population to live and work by considering all their specific characteristics (Steiner and Shenggen 2019). The process of revitalisation can also be observed as a long-term process that improves rural development. It is based on the use of new and diverse economies and adequate answers to the challenges modern society is facing. According to contemporary planning theory and practice, the revitalisation of rural areas contributes to the regional development and decentralisation processes and also improves the activating of natural resources (Milić 2011; Leuba 2017). These processes are mutually connected because the revitalisation of rural areas is one of the key factors for improving overall rural development.

According to the EU Commission data (2021), rural areas occupy about 83% of the European territory, and they are home to approximately 30% of the total population. They represent areas that are characterised by an underdeveloped economy, a low degree of life quality, and a lower GDP than the European average. It is common for rural areas that areas in peripheral zones are less developed than those near urban areas. The reason is frequent migrations from peripheral to urban zones on the one hand and the spreading of the urban territory on the other. Underdeveloped rural areas are also facing infrastructural problems, and the availability of basic public services is not developed (Dijkstra and Poelman 2014).

Since Serbia is a candidate for EU membership, development, strategic, and planning documents regarding rural areas should follow the adopted EU documents. Regarding the model in rural policies, according to (Hodge and Midmore 2010), we can differentiate three: (a) sectoral, (b) multisectoral, (c) territorial and (d) local approach. The approach that is based on the specific characteristics of an area is the territorial-spatial approach, and it is the most widely accepted and in use because it takes into account the specific characteristics of the rural area itself. One of the oldest and most important EU policies, on which the current Rural Development Policy is based, is the Common Agricultural Policy (CAP). The Common Agricultural Policy 2023–27 is currently in effect (European Commission 2022). The main goal of the current Common agricultural policy is securing the future of agriculture production and forestry, as well as achieving the objectives of the European Green Deal.

Current CAP policy is based on ten key objectives: (1) ensure a fair income for farmers; (2) increase competitiveness; (3) improve the position of farmers in the food chain; (4) climate change action; (5) environmental care; (6) preserve landscapes and biodiversity; (7) support generational renewal; (8) vibrant rural areas; (9) protect food and health quality; (10) fostering knowledge and innovation. Within CAP policy, priorities are defined to improve agriculture production and increase its contribution to the goals of the European Green Deal. Determined priorities within current CAP policy regard the increase in funding of measures to support climate and biodiversity, improve environmentally–friendly farming and make it mandatory for all the countries to increase their green ambitions compared to the previous programming period.

OECD presented the Rural Agenda for Climate Action (2021), which defined focus areas for climate action to achieve environmentally sustainable rural economies with reduced emissions while overcoming the challenges faced by these areas. Within the agenda, it is determined that rural areas have an important role in achieving global zero because emissions per capita were three times higher in remote rural areas than in urban areas in OECD countries in 2018. This is caused by the specific processes and use of traditional technologies in agriculture production.

2.4 Rural Areas in Serbia: Legislative and Strategic Framework

Based on the OECD criterion (1994), approximately 94.1% of the territory of Serbia is characterised as a rural area, within which approximately 47.8% of the total population lives, and the average population density is about 47 inh/km^2. Within the rural territory of the Republic of Serbia, there are 4158 settlements, which is about 72% of the total number of settlements in Serbia (Draft Spatial Plan draft 2021). The current division of settlements is into urban and other, and it is determined by the Statistical Office of the Republic of Serbia. This division was adopted based on the administrative decision of the local self-government units to declare a certain settlement a city and has been in use since the 1981 census (Mitrović 2015). Considering the current division, all urban settlements are urban, and all others are automatically defined as rural settlements.

The territory of the Republic of Serbia consists of five statistical regions. The division into regions is made according to the administrative territory because there are no clearly defined regional entities. The regions are classified as NUTS 2 territorial level, and this division was created in 2010 to statistically divide the territory following the territorial division in the EU. Based on the data from the Census from 2022 (Statistical Office of the Republic of Serbia 2023), the Region of Southern and Eastern Serbia has the lowest population density in other settlements; it is the most underdeveloped region in which the lowest number of people live in rural areas in relation to the territory it occupies. The share of other rural settlements is the highest in Southern and Eastern Serbia. Rural areas have been considered within the Draft Spatial Plan of the Republic of Serbia for the period 2021–2035 (2021), which is currently in the public insight process, and its adoption is expected in the following period. Within the Plan, rural areas in the Republic of Serbia are determined as areas with great potential for future development and as areas that play a significant role in overall regional development. One of the main problems that rural areas in Serbia are facing is certainly the demographic decline, so the number of villages with smaller numbers of inhabitants is increasing. According to the Draft Plan (2021), over 20% of rural settlements have less than 100 inhabitants, and 42% have up to 500 inhabitants—which in total represents more than half of rural settlements. Table 2.1 shows basic statistical data regarding rural areas for each of the regions.

Within the Draft Plan, rural areas are classified into six regions:

- *The Danube rural region*—is characterised by high population density and the potential for the development of non-agricultural activities
- A region with high agricultural potential is characterised by fertile soil, a favourable climate, and a qualification structure for agricultural development.
- *Regions of rigid agrarian structures are* characterised by large surfaces of agricultural land, a large share of mountainous areas, and an unfavourable age and qualification structure of the population for the development of agriculture.
- *Region of rural distortions*—it is characterised by negative demographic growth, expressed heterogeneity and uneven development between border and peripheral zones and areas around urban centres.
- *Mountain rural region*—it is characterised by rich biodiversity, natural resources and cultural and historical heritage, with great potential for the development of tourism and non-agricultural activities. On the other hand, it is also characterised by pronounced depopulation, an underdeveloped economy, poor urban–rural connections and undeveloped infrastructure.

Table 2.1 Main statistical data of the regions in Serbia (Authors according to data from Statistical Office of the Republic of Serbia 2023)

Region	Total area/ km^2/	Settlements urban	Settlements other	Population urban	Population other	Density rural/inh/ km^2
Belgrade region	3234	16	141	1,383,875	297,530	92
Vojvodina region	21,614	52	415	1,075,836	664,394	31
Region of Šumadija and Western Serbia	26,493	53	2059	911,119	908,199	34
Region of Southern and Eastern Serbia	26,248	46	1927	749,952	656,098	25
Region of Kosovo and Metohija*	10,910	26	1423	*514,755	*1,069,685	*98

*The last available census data for the Region of Kosovo and Metohija is from 1981 (Statistical Office of the Republic of Serbia 2013a, b)

- *A multi-problem rural area*—it is characterized by an altitude of more than 500 m, rich biodiversity and a large area under forests.

As the economy in rural areas in Serbia is mainly based on agriculture production, the rural development policy in Serbia is part of the agricultural policy. With the development and diversification of non-agricultural economies in rural areas, there is a different approach, and rural development policies in Europe no longer rely only on the development of agriculture. The legislative framework for rural areas in Serbia represents the *Law on Agriculture and Rural Development from 2016* (2016). Within the Law, it is defined that the agricultural policy and the rural development policy of the Republic of Serbia are implemented through the Strategy of agriculture and rural development of the Republic of Serbia, the National Program for Agriculture and the National Program for Rural Development, as well as through the IPARD programs. The law defines the following main development goals:

- Growth in the production and stability of producers' income
- Growth of competitiveness through adaptation to market requirements and technical–technological improvement
- Sustainable management of resources and protection of the environment
- Improvement of the quality of life in rural areas and reduction of poverty

- Effective management of public policies and improvement of the institutional framework for the development of agriculture and rural areas.

Agriculture and Rural Development Strategy of the Republic of Serbia for the period 2014–2024 (2014) deals with rural areas and the economy in the context of the possibilities for agricultural production development and exploitation of resources. Within the Strategy, it is also recognised that climate change and the consequences it causes have a great impact on both rural and agricultural development. Draft of the National Rural Development Program for the period 2022–2024 (2022), which is currently in the public insight process, represents the document whose goal is to provide implementation of the goals set within the Strategy. The goal of the Program is to improve current agricultural production and increase its competitiveness, to improve the management of the rural environment, risks and the process of adaptation to climate change, and to improve the spatial and social structure of rural areas.

Rural areas in different regions face similar problems, but still, due to some of the specificities of each district, it is important to analyse them on the local level. In Serbia, currently, there is a centralised planning system which observes rural areas only on the national level, and as a consequence, regional development is not even. Problems that rural areas are facing are very complex; they are the result of decades of underestimation, devastation and stagnation. Economic underdevelopment causes migration to urban areas, which leads to pronounced demographic emptying. Living standards are at a very low level, the poverty of the rural population is increasing, and the profitability of agriculture production has decreased during the last decade. Social exclusion of the rural population in decision-making and planning processes increases their inactivity, which is a huge threat when planning future development. According to Igić et al. (2023), the settlement network is insufficiently developed and dysfunctional, which disables their efficient functional connections. Due to the traditional ways of agricultural production and land cultivation, there is also an increase in harmful CO_2 emissions, which negatively affects GHGs and the environment itself. Common for all rural areas in Serbia is the negative impact of climate change, which has been pronounced in recent years.

2.5 Setting the Context: Rural Areas Within the Region of Southern and Eastern Serbia

The region of Southern and Eastern Serbia is one of the five regions in the Republic of Serbia and occupies an area of close to 30% of the total area. It consists of 9 districts, which can be classified as NUTS 3 areas. On its territory, there are nine cities, which also represent administrative centres of the districts and 53 municipalities. Of the total number of settlements, 46 are defined as urban and 1927 as "other" settlements (Statistical Office of the Republic of Serbia 2010). To comprehensively analyse rural areas and take into account all their specific characteristics, an analysis of each of the structures of rural areas will be presented below.

Socio-demographic structure: According to the Population Census from 2022 data (2023), 656,098 inhabitants live in the rural territory of this Region, which is 12.2% less than the number of inhabitants compared to the previous Census from 2011. The demographic structure of the settlements in this region is characterised by pronounced negative demographic growth and a decrease in population in almost all districts. Within this Region, the share of the elderly population in rural areas is also expressed because migrations of the younger—working population are becoming more frequent. As for the gender structure, the number of men and women in the population structure is almost the same in rural settlements in all districts (Statistical Office of the Republic of Serbia 2023). These are settlements that are not located in specific locations, and so the availability of jobs for the male and female population is relatively even.

Economic structure: since large areas of fertile land and a moderate continental climate are one of the comparative advantages of rural areas in this Region, in the case of most other—rural settlements, the main, and often the only, economic activity is agricultural production. The hilly and mountainous terrain is also suitable for animal husbandry, which has been developed in some mountain villages. According to data from the Agricultural Census in 2012 (2013) (data from the 2023 census is not yet officially available), about 58% of the agricultural land is in use, the area of forest land occupies 17%, unused land 17%, while 7% is the share of the "other" land. In the Region of Southern and Eastern Serbia, there are 187,744 registered agriculture households.

According to the Regulation (2018), the territory of 918 settlements within 40 municipalities is located at an altitude of 500 m or higher, and these settlements are marked as settlements with difficult conditions for working in agriculture. Besides them, 678 settlements within 18 municipalities were also determined as settlements with difficult conditions for working in agriculture, because their territory is located within the borders of the national park or is located in a devastated area (where the number of employees is less than 100 per 1000 inhabitants).

Aside from agriculture, tourism is also recognised as a profitable activity. Based on the census data, 112 agriculture households were registered in the Region of Southern and Eastern Serbia, which, in addition to agricultural production, are also engaged in tourism. There is no accurate data regarding tourist offers within agriculture households in rural settlements because not all of them are officially registered. According to Igić et al. (2019), within the Region of Eastern and Southern Serbia, at the beginning of 2019, there were tourist offers in 150 rural settlements, which involved over 300 households.

Physical and functional structure: problems faced by rural settlements in this region are uneven development, non-existing or insufficiently developed hierarchy and network of settlements, as well as underdeveloped urban–rural connections, both at the level of municipalities, districts and the Region itself. In a large number of rural settlements, there is a low level of availability of public services, such as basic health care, education, service and commercial activities, which means that the quality of life is at a very low level. One of the problems is also public transport, which has very few departures to certain settlements—or no departures at all, so the mobility of residents who do not have their vehicles is endangered. Considering that one of the main economic activities is agriculture, the huge problem is the low availability of supporting facilities such as veterinary stations, agricultural pharmacies and agriculture processing facilities. As a consequence, some of the modern tools used in agricultural production, land cultivation or livestock feeding are not available to everyone. As for the communal infrastructure of rural settlements in this region, it is characterised by a lack of a generally underdeveloped network—the network of public water supply, sewage system, and waste management, which affects the health and quality of life of residents.

Environmental structure: due to activities mainly caused by residents or by the negative impact of climate change, there is a significant increase in pollution of the environment and natural landscapes. This Region is characterised by rich natural resources, but their exploitation and management are inadequate and inefficient. The potential of natural resources, in general, plays a major role in the development of rural settlements and the improvement of their quality of life. Rich biodiversity contributes to the reduction of pollution, but due to the constant uncontrolled expansion of construction land, untouched natural landscapes are endangered. The natural environment provides an opportunity for the development of certain economic activities that can contribute to the diversification of the economy in rural areas. Watercourses represent one of the greatest potentials of the environmental structure in this Region on the one hand, but on the other, they are threatened in rural areas—rivers and small streams (Dinić Branković et al. 2020). Due to frequent periods of drought, water courses are endangered, and this adversely affects a large number of rural activities—in the first place, agriculture and animal husbandry.

Given the fact that this research explores the impacts of ecological transition, the focus will be primarily on the environmental structure of rural areas and the impacts that it has on the other structures and vice-versa. Based on that, Fig. 2.1 shows the impact of socio-demographic, economic, physical, and functional structures on environmental structures and the other way around.

Rural population represents the main and the most important resource of rural areas. Their activities often have a negative impact on environmental structure, especially because of the inadequate and inefficient management and exploitation of natural resources. This is the consequence of the low level of awareness of the rural population regarding the importance of biodiversity, natural resources and the environment. On the other hand, a polluted environment and the consequences of climate change have a negative impact on residents' health and quality of life in rural areas. This also has a negative impact on the rural economy because it represents a direct threat to agriculture production and husbandry, and it limits their performance and development. Since the dominant rural economy is agriculture, which is mainly done traditionally, CO_2 emissions have a huge negative impact on the environment. Uncontrolled and inefficient exploitation of water resources for agricultural production is a threat to the environment and represents a huge problem during drought periods. Different rural non-agricultural activities that are developed or have the potential for development represent possible threats to environmental pollution.

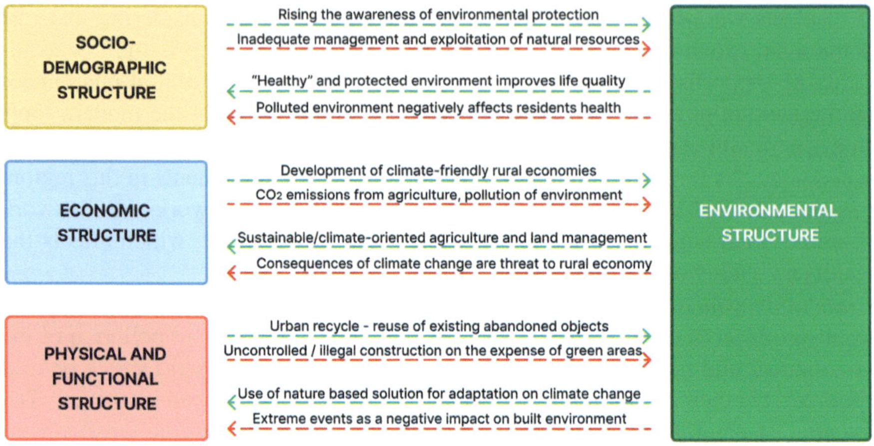

Fig. 2.1 Influences of different structures of rural areas in the Region of Southern and Eastern Serbia on the environmental structure and vice-versa (Authors)

In the case of physical and functional structures, there is a large stock of abandoned public buildings in the settlements, which are often neglected because of unsolved ownership problems. The great problem is uncontrolled and illegal construction, which leads to the city's territory spreading on account of the green areas and arable land. How the settlements mainly emerged spontaneously; in some public areas, there is a problem with overheating and also a problem with the management of rain runoff. Regarding the built structures, consequences of climate change, such as extreme weather conditions, have a negative impact on living standards and quality of life.

2.6 Discussion: Revitalization of Rural Areas and Ecological Transition

Considering the mutual impacts of different structures of rural areas on the environmental structure, we can conclude that problems from all structures must be taken into account, not only when planning rural development but also when planning the revitalisation of each of the structures of the rural areas. According to Rotondo et al. (2022), the ecological transition can be observed as a process that follows criteria for environmental sustainability and is based on the implementation of technological innovations as some of the main objectives of the ecological transition can be recognised: implementation of the measures for the adaptation and mitigation of climate change and preserving the natural environment—efficient management and exploitation of forests and agricultural land. Since the environmental structure represents one of the main focuses of ecological transition, the revitalisation of all structures of rural areas is observed in the context of the environment and their causal-consequential influences.

Based on the analysis of the structures of rural areas in the Region of Southern and Eastern Serbia, it is obvious that these areas are in poor condition; they are marginalised and stagnated in the social, cultural, spatial and technological sense. Pronounced depopulation, devastated economy on the one hand, and rich natural resources and unpolluted natural landscape on the other are currently the main characteristics of these areas. The territory of the Region of Southern and Eastern Serbia is very diverse and has heterogeneous structures, and each of the districts has its own specifics. Because of that, it is important to plan this process for smaller territorial units rather than on a national level, as it is currently because it does not offer efficient and comprehensive solutions.

An approach that will be used to explore opportunities and challenges for the revitalisation of rural areas in this Region is territory-based and provides the possibility to comprehensively consider all the characteristics of these areas. By taking into account problems and potentials faced by all the structures, the influence of all-natural and created factors is also included. As the main goal of the revitalisation process, quality of life can be determined. Improvement of life quality in rural areas must be imperative to create better places for a decent life for the rural population. The importance of rural areas in the food production chain must also be considered, and rural economy improvement must be one of the main focuses to "keep" rural areas alive. When planning the revitalisation of rural areas, it is important to preserve a unique rural ambience—built, cultural or natural- because territorial integrity plays an important role in empowering the local community. To plan an efficient revitalisation process, climate change and its consequences must be observed as one of the greatest threats, and measures to combat them must also be one of the priority focuses. For this process to be sustainable, it is necessary to activate the local population as the greatest resource in these areas because they can be seen as the main carriers of technological innovations.

In Serbia, there is a policy for rural development, strategy, and national programs, but they are all adopted at the national level without considering the particular potentials and problems of rural areas in different parts of the country. In the Region of Southern and Eastern Serbia, four territorial levels can be observed (Fig. 2.2): (1) regional, (2) district, (3) municipal, and (4) local–level of the settlement. For successful revitalisation, actions must be taken at each of the territorial levels of the region because each is dependent on the other (Fig. 2.2).

Within this Region, there are nine districts, and some of them, based on geographical reference and functional connections represent specific spatial units: (1) Braničevo and Danube district along the Danube River; (2) Bor and Zaječar district within Timok area; (3) Niš, Pirot and Toplica districts in the central part of the Region and (4) Jablanica and Pčinja district in the southern part of the Region. Within the determined spatial units, development agencies have been established that could significantly contribute to the revitalisation of rural areas. Strategic planning and coordination with development agencies should be carried out on the regional level. On the level of districts, planning and development documents should be adopted

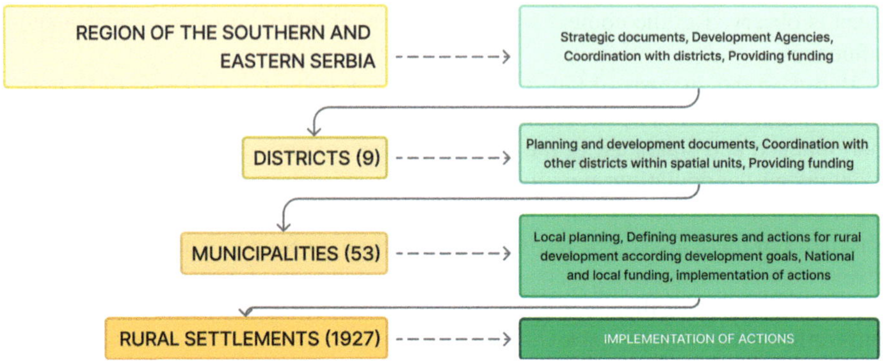

Fig. 2.2 Proposed actions at different territorial levels within the revitalisation process in the Region of Southern and Eastern Serbia (Authors)

for all the municipalities on their territory because a large number of municipalities are underdeveloped and have no adequate capacities for planning rural development. Planning for measures and activities at the local level could be done at the municipality level for all the rural settlements on its territory. Implementation of the proposed precise action must be carried out mainly at the level of settlements or municipalities where necessary.

The rural population makes up about 47% of the total population in the Region of Southern and Eastern Serbia. To reduce the trend of demographic emptying, it is necessary to improve the quality of life. This is possible only if the standard of living is improved because the most common are migrations caused by the lack of basic living conditions. The main opportunities and challenges for the revitalisation of this structure are presented in Table 2.2. By revitalising the socio-demographic structure, it is possible to contribute to its more favourable structure—age, gender, and qualification, and that way, stop rural areas from declining within this Region. The activation of rural communities could also be increased by revitalisation, which is a basis for planning future development and digitalisation. By raising awareness, it is possible to contribute to the creation of more resilient communities and improve healthy lifestyles, which is a synonym for rural image in modern society, and also reduce environmental pollution within the territory of this Region.

Table 2.2 Main opportunities and challenges for the revitalisation of socio-demographic structure in the Region of Southern and Eastern Serbia (Authors)

Socio-demographic structure	
Main opportunities	Main challenges
• Heterogenous population structure • Tradition in agriculture production	• Depopulation • Unfavourable age and gender structure
• Healthy lifestyles • Traditional way of life	• Low awareness regarding the environment • Inactive and digitally illiterate community

Almost all settlements have a long tradition of agriculture, and in addition to that, there are over 180,000 registered agriculture households within the territory of this region. Since this is a dominant rural economy, its improvement according to contemporary trends and global challenges is obligatory. The main opportunities and challenges for the revitalisation of this structure are presented in Table 2.3. By revitalising the economic structure, it is possible to improve the current status of agricultural production and introduce the use of modern technologies and measures to increase the yields and profitability of agricultural production. By introducing new, non-agricultural sustainable activities, the diversification of the rural economy and the economic dependence on only agriculture will be improved. New ways of agricultural production, such as organic production, can improve its competitiveness and position in the market. One of the tools for increasing competitiveness is also the development of entrepreneurship. This could create new job opportunities within agricultural households. Education in the domain of the use of contemporary measures in agriculture can contribute to creating more climate-friendly production and increase the resilience of the economy within this Region.

The built core of the rural areas and surrounding arable and forest land represent an inseparable whole, so in the revitalisation process, it is important to achieve and maintain territorial integrity. This is also significant for the overall development and improving homogeneity. The main opportunities and challenges for the revitalisation of this structure are presented in Table 2.4. By revitalising the physical and functional structure, it is possible to keep and protect the traditional rural character and environment. Abandoned buildings in the settlements within this Region could be "recycled–re-used" to prevent new construction when not necessary. Also, by improving the planning system and monitoring its implementation, illegal expansion of built areas at the expense of green areas and agricultural land could be reduced. Applying efficient adaptation to the consequences of climate change by implementing some of the contemporary urban and architectural concepts makes it possible to increase the resilience of built areas. By improving communal infrastructure, pollution of the environment could be reduced, and quality of life would be increased.

Due to uncontrolled human activities and expressed consequences of climate change, natural resources and the environment are endangered, which, as a consequence, have reduced the quality of life and endangered the health and lives of the rural population and built environment. The main opportunities and challenges for the

Table 2.3 Main opportunities and challenges for the revitalisation of economic structure in the Region of Southern and Eastern Serbia (Authors)

Economic structure	
Main opportunities	Main challenges
• Huge areas of agricultural land • Large number of agricultural households	• Traditional ways of agricultural production • Underdeveloped agricultural infrastructure
• Existence of non-agricultural economies • Dynamic and diverse terrain topography • Rich hydrographic network	• Low awareness regarding the environment • Production for self needs • Inadequate management of agricultural land

Table 2.4 Main opportunities and challenges for the revitalisation of physical and functional structure in the Region of Southern and Eastern Serbia (Authors)

Physical and functional structure	
Main opportunities	Main challenges
• Large stock of abandoned objects • Unique rural ambience • Territorial integrity of constructed and agricultural land	• Illegal construction • Underdeveloped infrastructure • Negative impact of climate change on built areas and public open spaces
• Rich network of public open spaces	• Lack of planning documents

revitalisation of this structure are presented in Table 2.5. By revitalising the environmental structure, it is possible to prevent damage to the natural landscape and protect the resources. By activating and efficient exploitation of renewable energy sources, sustainability could be improved, which would significantly affect the quality of life in rural areas. Special emphasis must be placed on combating climate change and its consequences since it negatively affects all the structures of rural areas. One of the instruments is building capacities for early warning and adaptation, where the rural population can be recognised as the main carrier of activities. Bringing nature within built structures by implementing nature-based solutions represents an effective tool towards ecological transition and sustainable development (Igić et al. 2020).

There are possibilities for funding the revitalisation process within the national funds determined through the draft National Program for Rural Development. Within the Draft National Program, four main goals are defined, and for each of them, a set of measures for achieving them are defined: (1) Improvement of competitiveness of agricultural and food sector, market chain and risk management; (2) Improvement of management of natural resources, preservation of biodiversity and mitigation to the consequences of climate change; (3) Improvement of rural economy and social structure of rural areas; and (4) Creation and transfer of knowledge and innovations. It can be seen that all goals are defined towards improving each of the structures of rural areas. This is very important because it enables the revitalisation process, it can contribute to an efficient implementation of revitalisation projects, and it is available to all interested residents or rural organisations. Besides this fund, there is a

Table 2.5 Main opportunities and challenges for the revitalisation of environmental structure in the Region of Southern and Eastern Serbia (Authors)

Environmental structure	
Main opportunities	Main challenges
• Biodiversity • Renewable energy sources • Rich natural resources, favourable climate • Unpolluted environment • Water resources	• Illegal construction • Underdeveloped infrastructure • Devastation of forest land • Lack of adaptation measures for climate change

possibility for funding through IPARD programs, and it is determined that measures from the National Program and IPARD do not overlap.

Based on the analyses conducted and knowing the key focus points for ecological transition, we could say that the revitalisation process of rural areas represents an important factor in fostering ecological transition and sustainability. For planning future development, rural areas must first undergo the process of revitalisation to improve their current poor structure by overcoming all the problems and developing all the potential. Each of the structures of rural areas must be considered both separately and in interrelation with others because only in that way can a comprehensive revitalisation process be planned.

2.7 Conclusions

Rural areas in the Region of Southern and Eastern Serbia represent significant resources for future overall development. Based on the conducted analyses, their territory and economy are devastated, and many uncontrolled activities within them pose a threat to the natural environment. All the structures of rural areas face a large number of problems that affect one another. Mutual, causally consequential connections between problems and potentials within different structures have an important role in planning their future development because they are interconnected, and one triggers the others. Because of that, it is concluded that the problems that rural areas are facing are multi-layered and very complex and that they must be observed on smaller territorial units to provide a comprehensive approach. When planning future development, it is important to overcome existing problems, which is possible through the implementation of the revitalisation process.

By analysing the characteristics of rural areas, it can be concluded that they have a very heterogeneous structure, which on one side represents their comparative advantage but on the other represents a threat to future development. To contribute to the ecological transition, it is necessary to activate and better exploit the rural population and natural resources since they are the greatest rural resource. Through the revitalisation of all structures of the rural areas, it is possible to adequately respond to all global challenges of today, improve the process of digitisation of rural areas and contribute to ecological transition.

Taking everything into account, it is clear that rural areas have an important role in sustainable development on the regional level. Revitalisation of these areas has a positive impact on the quality of life, rural economy, built area, environment, and natural resources. According to the findings of this research, it can be concluded that the revitalisation process of rural areas can be determined as a tool for improving ecological transition and contributing to sustainable rural development.

References

Babović S, Lović Obradović S, Prigunova I (2016) Depopulation of villages in southeastern Serbia as hindrance to economic development. J Geogr Inst “Jovan Cvijic”, SASA 66(1):61–74

Belanche D, Casalo L, Angeles´Rubio M (2021) Local place identity: a comparison between residents of rural and urban communities. J Rural Stud 82:242–252

Dijkstra L, Poelman H (2014) A harmonised definition of cities and rural areas: the new degree of urbanisation. https://ec.europa.eu/regional_policy/sources/work/2014_01_new_urban.pdf. Accessed 28 Jan 2024

Dinić Branković M, Igić M, Mitković P et al (2020) Exploring the potentials of small urban streams in creating blue-green infrastructure in the City of Niš, Serbia. In: Proceedings of the XIV International conference STEPGRAD, University of Banja Luka - Faculty of Architecture, Civil Engineering and Geodesy, Banja Luka, 11–12 June 2020

European Commission (2021) A long-term vision for the EU's rural areas—towards stronger, connected, resilient and prosperous rural areas by 2040. https://data.consilium.europa.eu/doc/document/ST-10404-2021-ADD-2/en/pdf Accessed 1 Feb 2024

European Commission (2022) Common agricultural policy for 2023–2027. 28 CAP strategic plans at a glance. https://agriculture.ec.europa.eu/common-agricultural-policy/cap-overview/cap-2023-27_en. Accessed 03 Feb 2024

Hodge I, Midmore P (2010) Models of rural development and approaches to analysis evaluation and decision-making. Économie Rurale 307:23–38

Igić M, Mitković M, Mitković P et al (2019) Rural tourism development as a catalyst of urban-rural synergy in the region of Southern and Eastern Serbia. In: Aleksandra Krstić-Furundžić and Aleksandra Đukić (ed) Handbook of research on urban-rural synergy development through housing, landscape, and tourism. IGI Global, Hershey, pp 172–192

Igić M, Mitković P, Dinić Branković M, Đekić J et al (2020) Improvement of life quality using nature-based solutions - case study settlements in south-eastern Serbia. In: Proceedings of the 7th international academic conference on places and technologies, University of Belgrade—Faculty of Architecture, Belgrade, 29–30 Oct 2020

Igić M, Dinić Branković M, Lj V, Živković J (2023) Development problems and potentials of rural settlements—a case study of rural settlements on the territory of the City Municipality Pantelej, Niš. Facta Univ Ser Archit Civil Eng 21(3):397–414

Leuba P (2017) A better future for Europe's rural areas. https://rm.coe.int/a-better-future-for-europe-s-rural-areas-governance-committee-rapporte/168074b728. Accessed 22 Jan 2024

Milić B (2011) Ruralni razvoj—praktikum za lokalne aktere (Rural development - a practicum for local actors). SKGO, Beograd

Mitrović M (2015) Sela u Srbiji, promena strukture i problemi održivog razvoja (Villages in Serbia, structural change and problems of sustainable development). Statistical Office of the Republic of Serbia, Belgrade

Moseley MJ (2003) Rural development: principles and practice. SAGE Publications Ltd., London

Nacrt Prostornog plana Republike Srbije za period 2021–2035 (Draft Spatial Plan of the Republic of Serbia for the period 2021–2035) (2021) https://www.mgsi.gov.rs/sites/default/files/PPRS%20Nacrt.pdf. Accessed 25 Jan 2024

Nacrt Nacionalnog programa ruralnog razvoja za period 2022–2024 (Draft National Rural Development Program for the period 2022–2024) (2022) http://www.minpolj.gov.rs/nacionalni-program-ruralnog-razvoja-za-period-2022-2024-godine/?script=lat. Accessed 27 Jan 2024

OECD (1994) Tourism strategies and rural development. https://one.oecd.org/document/OCDE/GD(94)49/en/pdf. Accessed 22 Jan 2024

OECD (2021) The rural agenda for climate action. https://www.oecd.org/regional/rural-development/Rural-Agenda-for-Climate-Action.pdf Accessed 31 Jan 2024

Pravilnik o određivanju područja sa otežanim uslovima rada u poljoprivredi (Rulebook on determining areas with difficult working conditions in agriculture) (2018) Official Gazette of the Republic of Serbia 102/18, Belgrade

Ringler C, Meinzen-Dick Ruth S (2019) Environment: revitalizing, restoring, and improving rural areas. https://ebrary.ifpri.org/utils/getfile/collection/p15738coll2/id/133141/filename/133349.pdf. Accessed 26 Jan 2024

Rotondo F, Perchinunno P, L'Abbate S, Mongelli L (2022) Ecological transition and sustainable development: integrated statistical indicators to support public policies. Sci Rep 12(18513). https://doi.org/10.1038/s41598-022-23085-0

Statistical Office of the Republic of Serbia (2010) Administrative territorial division and NUTS levels. https://www.stat.gov.rs/sr-latn/oblasti/registar-prostornih-jedinica-i-gis/administrativno-teritorijalna-podela-i-nstj-nivoi-1-2-3/. Accessed 25 Jan 2024

Statistical Office of the Republic of Serbia (2013) 2011 Census of population, households and dwellings in the Republic of Serbia: comparative overview of the number of population in 1948, 1953, 1961, 1971, 1981, 1991, 2002, 2011, Book 20: Population, Belgrade

Statistical Office of the Republic of Serbia (2013) Census of agriculture 2012. https://publikacije.stat.gov.rs/G2015/Pdf/G201514010.pdf. Accessed 25 Jan 2024

Statistical Office of the Republic of Serbia (2023) 2022 census of population, households and dwellings: age and sex, data by settlements, in Book 2: Population, Belgrade

Steiner A, Shenggen F (2019) Rural revitalization: tapping into new opportunities. In 2019 global food policy report, https://ebrary.ifpri.org/utils/getfile/collection/p15738coll2/id/133135/filename/133352.pdf. Accessed 25 Jan 2024

Strategija poljoprivrede i ruralnog razvoja Republike Srbije za period 2014–2024, godine (Agriculture and rural development strategy of the Republic of Serbia for the period 2014–2024) (2014) Official Gazette of the Republic of Serbia 85/14, Belgrade

Vasilevska L (2006) Ruralni razvoj u regionalnim okvirima (Rural development in regional frameworks) Zadužbina Andrejević, Belgrade

Zakon o poljoprivredi i ruralnom razvoju (Law on Agriculture and Rural Development) (2016) Official Gazette of the Republic of Serbia number 41/2009, 10/2013 and 101/2016, Belgrade

Zelenakova M, Purcz P, Hlavata H, Blištan P (2015) Climate change in urban versus rural areas. Proc Eng 119:1171–1180

Chapter 3
A Revitalization Approach: Establishing a Digital Nomad Centre in Rural Hungary

Noémi Kókai and Donát Rétfalvi

Abstract With the rise of remote work and the increase in the accessibility of digital technologies, a new generation of employees was born, whom we call digital nomads, who can work from anywhere in the world with adequate internet access. While previously digital nomads preferred urban environments, recently, especially as a result of the pandemic, interest in rural areas is also growing. The research examines the possibilities of involving digital nomads in the field of village development. This research aims to establish a co-working house in a Hungarian rural area, setting an example for future domestic initiatives. Through the case study, it outlines the impact of co-working spaces on rural communities if we treat them not only as isolated workplaces but as new centres of life. The research includes a detailed examination of the village Pécsdevecser and its region, taking into account its unique characteristics and potential opportunities and investigating adaptation approaches, with a primary emphasis on creating a fitting rural co-working model. Based on these findings, a comprehensive program plan and development strategy are formulated, outlining how digital nomads can be integrated into the village's life and contribute to its development.

Keywords Digital nomads · Rural co-working · Community engagement

3.1 Introduction

Digital nomads are a growing population of mobile workers who are not tied to a physical location and can work from anywhere with an internet connection (Hannonen 2020). Whilst digital nomads are often associated with urban environments and urban co-working spaces, there is a growing trend for digital nomads to seek out rural areas for a different work-life balance. There is a growing number of co-living and co-working spaces in the countryside that cater to digital nomads (Nagel 2020;

N. Kókai (✉) · D. Rétfalvi
University of Pecs, Pecs, Hungary
e-mail: kokai.noemi@pte.hu

M. Vukmirović et al. (eds.), *Digital Dreams, Urban Realities*, The Urban Book Series,
https://doi.org/10.1007/978-3-031-76353-3_3

Foertsch 2022). Rural work is becoming more prominent as a result of the digital transformation and then the Covid-19 pandemic (Foertsch 2023).

3.1.1 Benefits and Challenges of Inviting Digital Nomads into Rural Areas

Opening rural villages to foreign digital nomads has many benefits for both the local community and the nomads, including economic development, innovation and social cohesion. Many rural areas are facing depopulation and economic decline. The arrival of digital nomads can bring fresh impetus and energy to these areas, helping to rehabilitate them and create a more sustainable future. These spaces provide a collaborative environment for nomads to work together and share ideas while also connecting with the local community through events and activities. By attracting new economic activity, nomads can help create jobs and support entrepreneurship; they often have expertise in areas such as technology and entrepreneurial skills that may be lacking in the local community (Füzi 2015; Jiwasiddi et al. 2022). Another positive impact is a reduction in commuter traffic and associated transport emissions and avoidable health stress, and people do not have to commute daily thanks to co-working spaces (Sutton-Parker 2021; Ohnmacht et al. 2020). This can stimulate the local economy by people staying local as consumers of local goods and services, and the nomads also buying local goods and using local services (Wang and Cecez-Kecmanovic 2019). The possibility that they can now work in the co-working office can also encourage those who have moved away to return to their hometown (Kremkau et al. 2021).

The emergence of digital nomads in rural areas is accompanied by a number of challenges. One of these challenges is the strain on local infrastructure, as rural areas do not always have the infrastructure to support a growing population. The mass influx of people can put pressure on local infrastructure, including housing, transport and public services, challenging local communities and policymakers. Villages need to adapt to accommodate digital nomads while preserving their unique identity and sense of community. In terms of economic impact, digital nomads can contribute to local economies, but this impact may be limited if they are only in a place for a short period of time. Digital nomads can also contribute to the growth of the gig economy, as they often rely on short-term contracts and freelance work (Yuen Thompson 2018; Woodcock and Graham 2019; Sutherland and Jarrahi 2017). However, rural digital nomads have been observed to stay in one place longer than urban digital nomads. A new trend and movement known as slowmads has also emerged. They make conscious choices about their lifestyle and travel patterns, taking into account environmental impacts. They often prefer public transport, such as trains or buses, rather than flying and aim to stay in one place for as long as possible (Smercina 2019; Almeida et al. 2023).

3.1.2 Rural Co-Working Scene in Europe

Rural co-working spaces are increasingly recognised worldwide as an integral part of the community (Görmar 2021; P-IRIS2021). It can be assumed that remote work will become widely available worldwide over the years, leading to the emergence of many innovative collaborations in the future.

The rural co-working landscape is evolving differently across Europe. Certain countries like Germany, Switzerland, Italy, and Norway are more advanced in their development; this is attributed to national programs and networks, compared to countries such as the Czech Republic or Hungary, where the co-working scene is still emerging (Kremkau et al. 2021). Even beyond the European Union, certain Balkan countries such as Bosnia Herzegovina and Serbia showcase some excellent exemplary initiatives. For instance, Mokrin House in Banat, Serbia, stands out as a prime example of the transformative impact a rural co-working space can have on both the village and its surroundings. Since its establishment in 2016, Mokrin has evolved into a genuine Global Village, illustrating the potential of rural co-working.

In the leading EU countries in the field, the government and the municipalities have recognised the importance of rural co-working. More and more municipalities understand that co-working spaces will, in the future, form an important part of the provision of services of general interest and will be a weighty argument in the competition for new residents and commercial settlements (Bosworth et al. 2023; Krasilnikova 2024). In the rest of the mentioned countries and regions where the practice of rural co-working does not yet exist, private initiatives usually play an important role—they network, provide information and give impetus. A wide range of concepts and usage options are emerging—from pop-up offers to the conversion of vacant buildings (Kremkau et al. 2021; Bertelsmann 2020).

3.1.3 New Places of Work Have Emerged

When talking about the rural co-working scene, the question of economic viability cannot be left out. Rural co-working applies alternative business models, which are different from the urban co-working ones, to ensure the long-term survival of a co-working space. A study by Bertelsmann Stiftung and CoWorkLand conducted in 2020 identified seven different business models in rural areas—the classic Co-working space, the Commuter Port, the Workation space, the Retreat space, and the New Village Center, by first interviewing more than 200 individual operators, users, and experts to assess the rural co-working situation in Germany, and then expanding their findings with the state of the international scene in Europe (Kremkau et al. 2021; Bertelsmann 2020) Table 3.1.

The different co-working business models have many different effects and distinguishing features. The classification is intended to facilitate an understanding of the mode of action and requirements of the individual typologies:

Table 3.1 New places of work

Typology	Space[1]	Dissemination[2]	Resilience[3]	Community[4]	Catchment[5]	Region[6]
1. Coworking classic	●○○○	●●●●○	●●○○○	●●●●○	●●●○○	●●○○○
2. Commuter port	●●○○○	●○○○○	●●○○○	●●○○○	●●●○○	●○○○○
3. Bottom Hub	●○○○○	●●●○○	●●○○○	●●●○○	●●○○○	●●●○○
4. Retreat	●●●○○	●○○○○	●●●○○	●●○○○	●●●●○	●●○○○
5. Workation	●●●○○	●○○○○	●●○○○	●○○○○	●●●○○	●○○○○
6. New village centre	●●●○○	●○○○○	●●●●○	●●●○○	●●○○○	●●●●○
7. Housing and work projects	●●●○○	●○○○○	●●●●○	●●●○○	●●○○○	●●●●○

[1]*Space* The space required as well as the required variability of working spaces
[2]*Dissemination* Examples available for the business model
[3]*Resilience* Indicates the resilience of the business model and target groups
[4]*Community* Focus on the community aspect
[5]*Catchment area* Commuting distances
[6]*Region* Effects on the region
Source Categorization by CoWorkLand and BertelsmannStiftung, 2021. Edit: The Authors

3.1.4 Co-Working Situation in Hungary

Hungary has a comparatively low level of urbanisation and can be described as a mainly rural country. A quarter of Hungary's total population lives in the metropolitan region of Budapest (<1.7 million). Hungary has only eight cities with populations exceeding 100,000, while most cities have fewer than 50,000 residents (ca. 5 million residents). Approximately one-third of the population still lives in rural regions in villages and smaller communities (>2.8 million).

Obtaining statistical data on co-working spaces is quite challenging. In 2021, 87 co-working offices were identified in Hungary (Horeczki 2022). As of 2024, there are still over 80 co-working spaces in Hungary, predominantly concentrated in Budapest. Approximately two-thirds of these are situated in the capital, while rural areas accommodate between 20 and 30, often located in larger rural cities. Precise figures are difficult to ascertain due to the continual fluctuation in the number of community offices in recent years, with ongoing changes still occurring Table 3.2.

Compared to the number of people living in rural areas, the number of rural co-working spaces is significantly low. Co-working spaces have appeared in two villages in recent years, both on the northern shore of Lake Balaton: PortusHome, established in Vászoly in 2021, and Parókia Coworking, opened in Káptalantóti in 2022.

Table 3.2 Distribution of co-working offices in Hungary

Typology	Population	Location	Quantity
Capital	<1.7 million	Budapest	60–70
County seat	150.001–250.000	Debrecen	10–15
		Szeged	
		Miskolc	
		Pécs	
		Győr	
City	100.001–150.000		5–10
Large town	50.001–100.000		0–5
Town	10.001–50.000		0–5
Small town	5001–10.000		1
Village	<5.000		2
			Total 80–100

Source Internet search. Edit: The Authors

3.1.5 *Identifying the Gaps*

Based on the available information, gathered from a limited number of studies and online sources—indicative of the significant underdevelopment in this field- the following observations can be made:

The co-working scene in Hungary is underdeveloped, particularly in rural regions. The majority of co-working spaces are concentrated in Budapest and larger urban centres; there are almost no rural co-working places. There is a lack of national or regional co-working associations and networking initiatives among co-working spaces. All founders of rural co-working spaces in the country, regardless of their operational model, are private initiatives and founders; there are no known co-working spaces founded by economic development agencies or municipalities. Most co-working places are classic co-working spaces or bottom hubs; a few retreats and workstations have also been established in recent years. Commuter ports are not yet known in Hungary, nor are alternative living and working projects. No places like New Village Centres exist in Hungary.

Unlike the leading countries in the field of rural co-working, like Germany or Switzerland, the inherent role of co-working spaces in rural development has not yet been recognised. The role of co-working spaces in rural development has not been explored. There have been many village rehabilitation initiatives (Horeczki and Póla 2023; Horeczki and Egyed 2021), but none investigated the co-working spaces as catalysators of revitalization in the Hungarian context. No real examples to be followed, that are decisive from the point of view of village development.

3.2 Methodology

The purpose of this research is to establish a co-working house in a Hungarian rural area as an example which could be followed in the future and could be a leading approach for future domestic developments of the transformative power that co-working spaces can have in rural areas if they are thought of not just as isolated workplaces but as new centres of life. The research focuses on a small Hungarian village situated in Baranya County as the prospective location for the co-working space.

Every village has different features and thus needs a unique approach based on its strengths and weaknesses so that development opportunities can be drawn on and leveraged effectively. The research begins with a detailed examination of the region and the specific village under consideration, taking into account its unique characteristics and potential opportunities, investigating adaptation approaches, with a primary emphasis on creating a fitting rural co-working model. As part of the methodology's analysis process, the 15-min settlement concept (Papadopoulos et al. 2023; Logan et al. 2022) is applied to the territory to ensure accessibility and convenience for residents.

Based on these findings, a comprehensive program plan and development strategy are formulated, outlining how digital nomads can be integrated into the village's life and contribute to its development.

3.3 Case Study: Pécsdevecser, Hungary

Pécsdevecser is a tiny Swabian village in Baranya County, just over 20 km from Pécs in the direction of Villány. The name of Pécsdevecser was first mentioned in the certificates in 1332. By the end of the Turkish era, the village was depopulated; it was populated again only from 1733, with German settlers, and later Hungarians also settled in the village. According to census data, in 1910, 298 out of 313 inhabitants (95.2%), but even in 1941, 232 out of 245 inhabitants (94.69%) were Germans. The German influence can still be felt. Based on the 2022 census data, the Swabian identity is still significant, as 42.2% of the population declared themselves to be of German nationality. Today, barely 100 people live in the settlement.

In terms of tourism, the region offers a multitude of tourist opportunities stemming from its proximity to various attractions and popular destinations, such as Pécs, the Villány wine region, Siklós and Harkány can be reached within 20–30 min, Fig. 3.1 where not only domestic but also foreign target audiences come regularly for tourism and leisure purpose. According to 2016 data from KSH, the Siklós microregion, which includes Pécsdevecser, emerges as the number one leading tourism destination in the County, ***also benefiting*** from its close geographical proximity to Croatia. Additionally, Pécs, serving as the cultural hub of Baranya, offers visitors a rich supply of historical and cultural experiences, but nature enthusiasts can also discover the

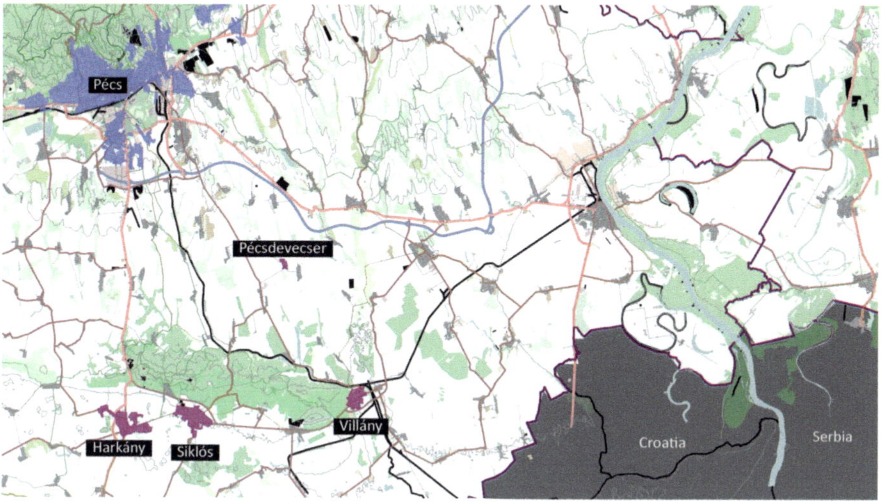

Fig. 3.1 Region overview map. *Source* The Authors

natural territories of the Mecsek Mountains. With such a diverse range of attractions and easy access to neighbouring countries, Pécsdevecser stands as an ideal destination for travellers seeking cultural immersion and outdoor exploration.

3.3.1 Characteristics of a Vital Village

The 15-Minute City concept seeks to create self-sustaining neighbourhoods where all the essential functions of daily life, including life, work, commerce, healthcare, education and entertainment, can be reached within a comfortable 15-min walk or bike ride. This approach aims to decentralise urban functions and services, promoting accessibility and reducing reliance on long commutes or extensive travel to meet basic needs (Papadopoulos et al. 2023; Logan et al. 2022).

There is a concept similar to the 15-min city but tailored for rural villages, known as the "15-min village" or "15-min rural community." The 15-min village concept aims to enhance the liveability, resilience, and self-sufficiency of rural communities by fostering local connections, preserving heritage, and ensuring access to essential services and amenities within a manageable distance. While the term "15-min village" may not be widely recognised in scientific literature, the underlying principles of creating compact, walkable, and self-sufficient communities in rural areas have been discussed in various contexts.

Our analysis includes the following three pillars: an overview of population changes over the years, transport connections within the region and with the main cities, available range of services. The analysis takes into account the characteristics of the 15-min village concept and explores the problems of the village and its region accordingly.

3.3.2 Analysis

3.3.2.1 A Closely-Knit Cluster

Pécsdevecser, along with the neighbouring settlements of Peterd and Kiskassa, form a closely-knit cluster positioned alongside the main country road linking Pécs and Villány. Despite being separate administrative entities, their close proximity and interconnectedness through public roads and dirt tracks requires a collective examination from an urban planning perspective. Viewing them as a cohesive unit is essential, as the uplift of one settlement can directly influence the well-being of its neighbours. This interconnectedness extends beyond the immediate vicinity, potentially impacting even larger and more populated neighbouring centres like Újpetre, where municipal governance is centralised.

In 5 min of walk, the entire residential area of Pécsdevecser can be reached, and the main residential part of villages Peterd and Kiskassa can be walked in, starting from the village centre. Considering Pécsdevecser as the centre point of the three neighbouring settlements, the border of Kiskassa and the fishing lake towards Peterd can be reached in 15 min by walking. By bicycle, in just 10 min, the whole residential area of Peterd and Kiskassa can be covered, and in a 15-min bike ride, the furthest end of even Újpetre can be reached Fig. 3.2.

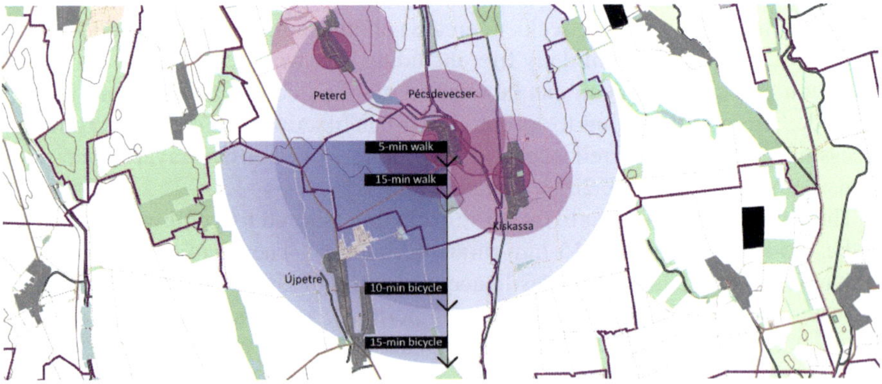

Fig. 3.2 5-, 10-, 15-min distance. *Source* The Authors

Table 3.3 Population decline

Population	1949	1960	1980	1990	2001	2011	2022
Pécsdevecser	366	262	186	145	115	92	90
Peterd	449	398	287	212	205	182	203
Kiskassa	738	591	403	355	298	249	216
Újpetre	1381	1206	1168	1036	1153	1029	959

Source KSH census data. Edit: The Authors

Table 3.4 Pécsdevecser: population, housing, employment

Settlement	Population	Natural reproduction	Domestic migration	Registered job seekers	Housing stock	Built apartment	Receiving pensions, benefits, annuities and other benefits
Pécsdevecser	90	−1	−3	4	49	–	35
Peterd	203	−1	−4	27	95	–	48
Kiskassa	216	−2	−6	29	112	–	75
Újpetre	959	−11	−40	87	387	–	301

Source KSH census data, 2022. Edit: The Authors

3.3.2.2 Population

Like many countries globally, Hungary also struggles with depopulation in its rural regions, with Baranya experiencing the most significant decline in natural reproduction rates. [KSH TIMEA] Pécsdevecser and neighbouring villages like Peterd and Kiskassa, situated just a few kilometres apart, have decreasing populations spanning through decades, as illustrated in Table 3.3. These areas face imminent threats from depopulation, a challenge even compounded in larger surrounding settlements like Újpetre.

Pécsdevecser and its environs predominantly house elderly residents, and the number of children in the villages is significantly low. Moreover, only half of the residents are employed and within working age, resulting in a pronounced labour shortage throughout the area (Table 3.4).

3.3.2.3 Mobility

While mobility between the settlements is mostly solved by bicycles or walking, as previously discussed, the transportation infrastructure in the region presents a mixed picture (Lieszkovszky 2023). While road connections to urban centres like Pécs, Villány, Siklós and Harkány are generally good, public transport options leave much to be desired. Although there is a bus service available, the absence of a rail connection poses a significant limitation. The nearest train station is 10.7 km away, in Vokány. The number of direct bus and train connections between municipalities and micro-regional centres within a working day is low. According to data collected through menetrendek.hu (accessed February 2024), the region has between 3 and 5 direct connections per day to Pécs, which meets the current passenger traffic needs, but to Villány it has only 3 per day, 2 during the morning and the latest in early afternoon at 2 pm. This highlights a need for improvements in public transit accessibility, since despite its geographical proximity to Pécs, the local community of Pécsdevecser finds itself far more intertwined with the Villány region. Through conversations with residents, it becomes apparent that Pécs is perceived as distant, while their ties to Villány—historically, economically, in terms of employment opportunities, and supply—are deeply ingrained, it also comes from Pécsdevecser being part of the Siklós microregion alongside Villány, and Pécs belonging to a separate microregion.

3.3.2.4 Services

The territorial division of Pécsdevecser consists of residential areas, vine cellars, vineyards and farmland. More than 60–70% of the agricultural land are vineyards (data sourced from THEIR—National Regional Development and Spatial Planning Information System). The village is located in a valley, the terrain slopes down from the vineyard hill in the direction of the stream. The plots running up from the main street are complemented by vineyards, while the plots running down are complemented by agricultural lands. Currently, only a very small part of the population is engaged in agriculture or viticulture.

There are two working businesses in the village, established in the recent years: a biodynamic winery run by a German couple and a permacultural vegetable farm run by a French-Hungarian couple. Besides that, the village only has a church and the village hall which contains the municipality office, the library and the cultural centre. Commercial services have almost completely disappeared from the village, such as the bakery, the convenience store or the pub (Fig. 3.3 and Table 3.5).

For the regional overview of the available service analysis, we combined the four close village's offers and categorized them into 4 categories that are essential for a quality life: Commerce, Healthcare, Education and Entertainment.

Újpetre, being a more densely populated village, offers the most extensive range of services in the vicinity. It is the sole settlement equipped with a primary school. While traditional stores may no longer operate in Pécsdevecser, the village is catered to by mobile shops that traverse the area on specific days and hours. These mobile shops, which are quite frequent, serve as grocery stores, meat markets, and feed suppliers, also ensuring a steady provision of fresh vegetables and fruits to the villages. Additionally, a weekly postal service addresses various personal needs. The area has two major employers: the Farmer's Association and the Butchery in Mohács. An organized minibus service transports the employees commuting to Mohács.

Fig. 3.3 Map of Pécsdevecser. *Source* The Authors

3.3.2.5 Problematization

The main challenges faced by the rural area is the low population density, which results in limited offer of local essential services, the distance from the urban centre combined with the poor mobility prevents access to services available in Pécs, Siklós or Villány. The lack of services results in lack of jobs. The primary focus is on fostering business opportunities capable of generating jobs and offsetting the distance barrier from urban hubs. Introducing a co-working space in the rural region emerges as a potential solution, as digital work minimizes the reliance, and reduces the role of physical space and distance. By embracing digital space, co-working initiatives offer a pathway for rural communities to access employment opportunities and essential services, transcending geographical limitations and fostering economic development Fig. 3.4.

3.3.3 Digital Nomad Hub of Pécsdevecser

3.3.3.1 Defining the Rural Co-Working Business Model in Pécsdevecser

Vibrant gathering spots are essential for fostering vital communities, and co-working spaces provide much more than just a place to work within rural settings. Multifunctional venues like New Village Centres serve as dynamic hubs within villages and communities, accommodating various activities such as meetings, markets, and events, with promising synergies for co-working environments. Co-working acts as a unifying force, bringing together diverse individuals, ideas, and objectives, while

Table 3.5 Available services

Category	Facilities	Spatial distribution				Total amount
		Pécsdevecser	Kiskassa	Peterd	Újpetre	
Healthcare facilities	Health centre	0	0	1	1	2
	Kindergartens, Nursery	0	0	1	1	2
	Elderly facility	0	0	0	0	0
Cultural and educational facilities	Elementary school	0	0	0	1	1
	Middle school	0	0	0	0	0
	Library	1	1	0	0	2
Convenient service and commerce facilities	Convenience Store	1	2	2	3	8
	Feed store	1	1	1	1	4
	Vegetable shop	1	1	1	1	4
	Butcher shop	1	1	1	1	4
	Drugstore	0	0	0	1	1
	Restaurant	0	0	0	0	0
	Municipal office	1	1	1	2	5
	Post office	1	1	1	2	5
	Police station	0	0	0	0	0
	Fire station	0	0	0	0	0
	Bank, ATM	0	0	0	2	2
	Church, chapel	1	2	2	1	6
	Open air market	0	0	0	1	1
	Gas station	0	0	0	1	1
	Guesthouse	0	1	0	1	2
Entertainment	Playground, soccer field	1	1	1	3	6
	Cultural facility	1	0	0	1	2
	Museum	0	0	0	1	1
Total		10	12	12	25	59

Source The Authors-on-site information gathering

Fig. 3.4 Problematization. *Source* The Authors

also laying the groundwork for attracting additional services and activities, including experimental retail concepts. The potential for such spaces is extensive, encompassing a range of amenities from cafes and seminar rooms to essential services like post offices, copy shops, bistros, medical centres, and libraries etc.

Successful models of New Village Centres already exist in countries like Germany, Norway, and Switzerland, offering diverse operating models for co-working spaces. Examples such as Prädikow Farm and Viehbrook Farm in Germany, which integrate a kindergarten, corner shop, and co-working space, demonstrate the transformative impact of rural co-working environments when integrated into community life. These spaces serve as catalysts for the digitalization of rural areas and the creation of vibrant village communities, attracting new residents seeking a blend of rural peace and urban amenities (Kremkau et al. 2021; Bertelsmann 2020).

Although rural co-working brings social advantages, attaining economic sustainability remains an ongoing challenge. However, the New Village Centre model presents a promising long-term vision by integrating various functions that sustain both the local economy and community. This model aligns well with the context of Pécsdevecser, offering a feasible and suitable approach to fostering rural co-working initiatives.

3.3.3.2 Szezon Färm, Pécsdevecser—A Place that Feeds and Connects

Rural coworking spaces serve as vibrant hubs where individuals gather to engage in discussions, celebrate, learn, and foster a sense of community. Given their potential to drive economic development, promoting these spaces should be a priority for public authorities. However, a significant challenge in Hungary is the lack of understanding of the concept from the municipality, which can and should be one of the key partners. Without a private initiative, co-working places do not get established. In the case of Pécsdevecser, the founders are the owners of Szezon Färm, an organic garden, where plenty of fresh vegetables grow. The eco-farm is located in the centre of the village. Its 7500 m^2 plot runs down to the stream. The farm established in 2021, operates under chemical-free, cultivation-based permaculture principles (Fig. 3.5).

Fig. 3.5 Szezon Färm. *Source* The Founders of Szezon Färm, Google Maps Streetview

At the centre of Szezon Färm's ideology lies a profound commitment to the community. Their produce finds its home primarily within the close-knit basket community of Pécs and local restaurants. Yet, their vision extends beyond mere transactions, central to their philosophy is the elevation of a communal spirit. Szezon Färm is a place that feeds and connects, where in addition to the literal meaning, it also means spiritual feeding, which is embodied in their initiatives, such as the weekly pop-up dinners where strangers gather to savour farm-fresh products and connect.

3.3.3.3 Applying the New Village Centre Concept at the Szezon Färm

The ideology centred on vegetable cultivation and community development is enriched by the integration of co-working spaces and innovative services such as co-living arrangements, pop-up restaurant, market, and various programs. This symbiotic approach fosters the essential elements for a high-quality rural lifestyle within the community, including aspects like housing, employment, provision of goods, leisure, and cultural enrichment. It becomes a hub that not only caters to these fundamental needs but also serves for connection and collaboration, nurturing a thriving ecosystem where individuals can flourish and contribute to the collective well-being. It's a space designed to feed and connect, catering to two primary target groups: digital nomads seeking for a balance between work and lifestyle, and the local community drawn to the sense of community and opportunities for relationship building.

Table 3.6 Identification of the descriptive characteristics of the Pécsdevecser Digital Nomad Hub

Szezon Färm	A place that feeds and connects	
Space	Within the project, the co-working spaces provide a dynamic environment for digital nomads and local professionals alike. The old barn is transformed into comfortable and fully equipped workspaces, offering a blend of traditional but also open-air, multi-use spaces and semi-open options within the garden	•••○○
Resilience	The Pécsdevecser project focuses on two distinct target groups: digital nomads, who are looking for a balance between work and lifestyle, and local professionals, who are attracted by flexible workspace solutions and networking opportunities The multifaceted digital nomad hub encompasses an array of amenities, including a co-living house, apartments, a pop-up restaurant, fitness area, farmers market, event space, community areas, etc	•••○○
Community	In the ideology of the farm, community-centred thinking and design was always a key factor, in case of the co-working this attitude is kept and even strengthened It places great emphasis on the development of other domestic partnerships in addition to the community of the region. Through digital nomads, not only domestic but also international relations can be formed	••••○
Catchment	It is important to emphasize, that the co-working hub being established here, not only addresses people of the region, but mainly digital nomads from across Hungary and abroad	••••○
Region	It can be assumed that if the village is filled with life, development processes will also start in the surrounding villages, which will have a positive effect on the economy and the community Markets, which were a base to village community, have now completely disappeared from these villages, and it needs to be restored. The farm is aiming to cultivate bonds with fellow producers (wine, meat, truffle etc.) to create a local market in the future, to provide for Pécsdevecser and also the close surrounding Emphasis is placed on ethical, sustainable sourcing, where most of the vegetables come from the garden or involving local producers and service providers, fostering an ethical provider model with a short supply chain, minimizing long-distance procurement to only essential raw materials not readily available within Hungary	••••○

Source The Authors

Based on the categorization established by BertelsmannStiftung and CoWork-Land, the distinguishing features of Szezon Farm's co-working model are the following (Table 3.6):

3.3.3.4 Design Guidelines

Two residential buildings, a barn, and several agricultural buildings are located on the site, which are being repurposed for the establishment of the digital nomad hub.

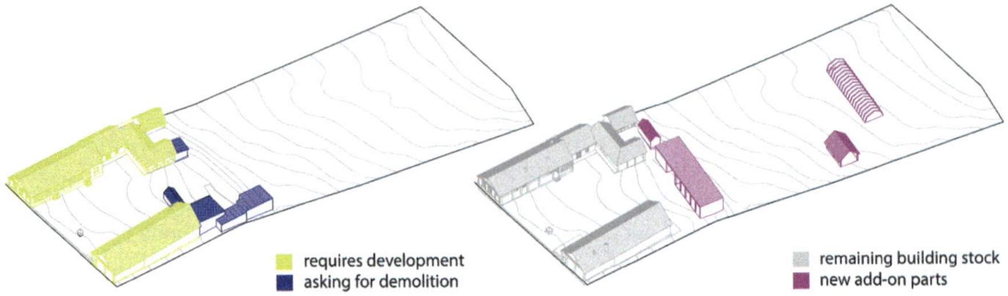

Fig. 3.6 Building stock utilization and development. *Source* The Authors

While the main buildings will be retained, they necessitate development and adaptation. The buildings dating back to the 1940s, bear distinct architectural elements from the 1960s and 1970s. They are now being removed, for example, the porch of the main residential building, which was glazed in is being reopened. Additionally, outbuildings are being repurposed. The preservation of the charming barn is a priority. Low-quality and architecturally insignificant light constructions will be demolished. Furthermore, new additional building components will also be integrated into the plot Fig. 3.6.

The accommodations are established through the conversion of the Swabian farmhouse, originally constructed in 1940, into a co-living house, while other farm buildings, such as the pig sty, are repurposed into apartments. The old barn, now transformed into a co-working space, offers comfortable and fully equipped work environments: open-plan shared offices, communal areas on the ground floor, including an incubator for collaborative group projects, a meeting room, and additional shared spaces on the upper level, along with private booths for online meetings and calls.

A newly constructed multi-use building serves as a semi-open shared space for co-working activities, as well as a venue for community events such as concerts, theatre performances, music events, producer markets, and more.

The garden functions as the establishment's pantry, while the old summer home houses a pop-up restaurant, retaining its original purpose with the restoration of the traditional open-fire kitchen.

One of the main new additions to the project is a community greenhouse. The greenhouse not only serves as an educational and demonstrative space but also hosts community-building activities, workshops, and culinary courses, fostering connections alongside cultivation.

At the rear of the plot, a forest garden is established alongside a wastewater stabilization pond, while two rainwater collecting channels on either side of the site are utilized for watering the garden Fig. 3.7.

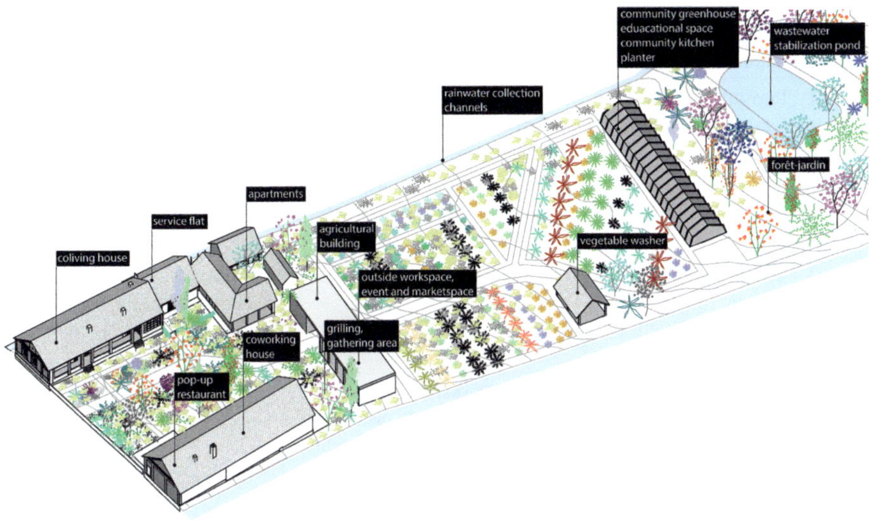

Fig. 3.7 Functional distribution of the DN Hub. *Source* The Authors

3.3.3.5 Integration of Workspaces into the Life of the Garden

The area can be divided into 5 zones. The triad of the vegetable garden, community and workspaces weaves through the entire area and forms a symbiosis with each other Fig. 3.8.

The transition between the three distinct categories within the zones manifests in varying forms. Commencing from the lively areas where daily life intertwines with leisure activities, serving as hubs for work and community interactions, these

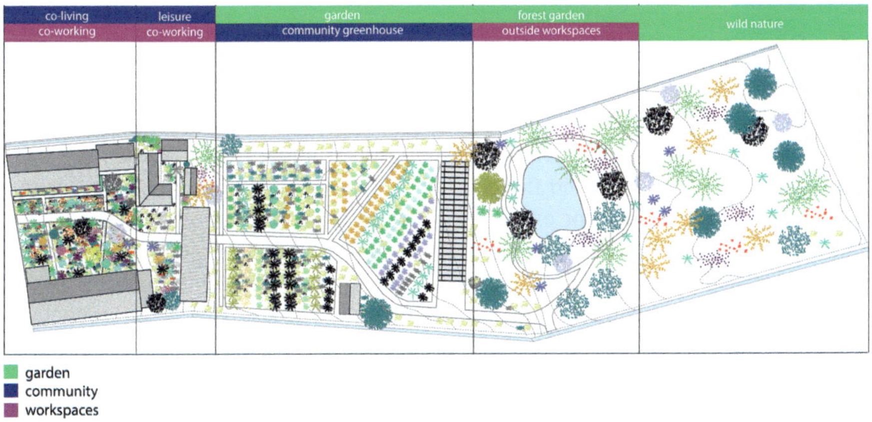

Fig. 3.8 Symbiosis between the garden, community and workspaces. *Source* The Authors

Table 3.7 Integration of workspaces into the life of the garden

Zones	Garden	Workspaces	Community
Zone 1	Pop-up restaurant: fresh food from the garden	Inside workspaces: the main co-working house is located in this front garden	Co-living house: communal housing for the digital nomads
Zone 2		Semi-open workspaces: the multipurpose building provides place for shared workspaces	Leisure area: grilling, gathering, campfires, events etc.
Zone 3	The vegetable garden: the heart of the establishment		Community greenhouse: planter, education, but also community kitchen, cooking classes, events
Zone 4	Forest garden: leisure area, walkway around the pond, resting areas	Outdoor workspaces: self-sufficient workspaces, solar-powered, with electricity and internet connection	
Zone 5	Wild nature: untouched nature, fifth zone in permaculture		

Source The Authors

vibrant spaces gradually merge into the wilderness, embodying the essence of permacultural principles. Positioned between these two ends lies the garden and community greenhouse, serving as a vital link. Workspaces span across three of the five zones, each adapting to the evolving surrounding. As the zones transition, so do the workspaces, from interior spaces through semi-open areas, to outdoor workspaces immersed within the natural landscape Table 3.7.

3.4 Discussion and Conclusions

Pécsdevecser presents several strengths that could position it as an attractive rural tourism destination. Benefits from its proximity to many tourist attractions, such as the cultural capital Pécs and the popular Villány wine region, providing visitors with several opportunities for exploration and cultural enrichment. Szezon Färm's unique product range, created by the symbiosis between the vegetable garden, the workspaces, and the communal spirit can attract domestic and international digital nomads. The availability of various services, including access to an organic vegetable garden and the pop-up restaurant, adds to the allure of the destination, offering visitors a taste of authentic, locally sourced cuisine (Kovács and Horeczki 2023).

Despite its great location and emerging possibilities, there are a lot of shortcomings and weaknesses that need to be addressed.

Pécsdevecser needs to enhance digital infrastructure to establish the digital nomad residence. Internet connectivity and digital facilities are essential for narrowing the digital divide and fostering remote work opportunities. Furthermore, both the farm and the village lack the necessary capacity to accommodate a substantial influx of people. The farm's ability to host digital nomads is limited, while the village lacks the essential housing infrastructure to handle a significant population surge, prompting the need for future development and expansion concepts.

The example of Vászoly, another small village where the coworking space was established in 2021, offers a useful point of comparison. In contrast to the situation in Pécsdevecser, PortusHome in Vászoly was able to leverage the area's pre-existing infrastructure, particularly its proximity to Lake Balaton. The area's dormant infrastructure, impacted by depopulation and migration, experienced a revival following the launch of the co-working space, and led to the resurgence of accommodation options, and hospitality services. Conversely, the scenario is different for Szezon Farm, as the surrounding area lacks pre-existing infrastructure.

The lack of recognition of potential by municipalities, as well as the absence of social acceptance and understanding of co-working spaces within Hungarian society, presents a significant barrier. In contrast, the extensive presence of rural co-working in Germany indicates societal acceptance and necessity. This disparity stems from cultural differences. In Germany, an already existing lifestyle and philosophy was adapted to the digital nomad lifestyle, building upon an existing socially desired system. Communal living is ingrained in the culture, fostering a sense of community from an early age, reinforced through practices like communal housing during university years. This norm provides a solid foundation for co-working spaces to seamlessly integrate into society. Conversely, Hungary's individualistic society lacks such traditions, making it challenging for co-working spaces to survive, particularly in rural areas.

In conclusion, the increasing number of mobile workers, both domestically in Hungary and globally, offers an opportunity to draw individuals seeking remote work-friendly destinations and experiences. A digital nomad establishment in Pécsdevecser could serve as a new guiding example for rural co-working initiatives in Hungary. Although Pécsdevecser holds substantial potential as a new rural hub, addressing its weaknesses and capitalizing on its strengths will be key to the success of realization.

3.4.1 Future Directions

Co-working and the community garden can revitalize the neighbourhood, but the already mentioned absence of infrastructure, especially the lack of accommodation, requires a solution. It seems necessary to prepare a comprehensive value inventory of the entire area, thereby defining a longer-term settlement planning approach that reviews and offers possible expansions for services and accommodation opportunities.

Acknowledgements We extend our heartfelt gratitude to the founding partners of Pécsdevecser for their dedication and valuable contributions. We are deeply appreciative of their commitment, belief, and collaboration.

References

Almeida M, Souza J, Correia A, Schneider D (2023) Post-Covid-19 digital nomadism: beyond work from (almost) anywhere 4605–4611

Bertelsmann S (2020) Coworking im ländlichen Raum: Menschen, Modelle, Trends

Bosworth G, Whalley J, Fuzi A, Merrell I, Chapman P, Russell E (2023) Rural co-working: new network spaces and new opportunities for a smart countryside. J Rural Stud 97:550–559

Foertsch C (2022) 2021–22 Coworking Space Trends. Deskmag.com, https://www.deskmag.com/en/coworking-spaces/2022-results-of-the-coworking-space-business-trends-survey

Foertsch C (2023) Global Coworking survey: 2023 coworking space business trends. Deskmag.com, https://www.deskmag.com/en/coworking-spaces/2023-coworking-space-business-trends-results-of-the-global-coworking-survey

Füzi A (2015) Coworking spaces for promoting entrepreneurship in sparse regions: the case of South Wales. Reg Stud Reg Sci 2(1):462–469

Görmar F (2021) Collaborative workspaces in small towns and rural areas. The COVID-19 crisis as driver of new work models and an opportunity for sustainable regional development?

Hannonen O (2020) In search of a digital nomad: defining the phenomenon. Inf Technol Tourism 22:335–353

Horeczki R (2022) Új típusú munkavégzési terek a Covid-19 árnyékában Magyarország vidéki térségeiben 18

Horeczki R, Egyed I (2021) Small town development in peripheral areas. Deturope 13(2):52–65

Horeczki R, Póla P (2023) Fejlesztési lehetőségek a periférián egy Baranya megyei felmérés tükrében. Tér És Társadalom 37(3):77–97

Jiwasiddi A, Schlagwein D, Leong C (2022) Assessing the impacts of digital nomadism on local communities: the case of Chiang Mai, Thailand

Kovács D, Horeczki R (2023) ÚJ JELENSÉGEK A BARANYAI GASZTRONÓMIÁBAN. SZÜLETŐBEN A MECSEK–VÖLGYSÉG–HEGYHÁT GASZTROTÉRSÉG?

Krasilnikova N (2024) Re-thinking the role of municipalities in mobility transitions: CO-working spaces in suburban and rural areas as a "new normal"., Cities 145

Kremkau T, Schmied A, Voll J (2021) Rural coworking in Europe: Status Quo, as far as we know: green paper & country reports

Lieszkovszky JP (2023) A vidéki térségek térszerkezetének elemzése a közforgalmú közlekedési kínálat alapján, 2018–2020. Területi Statisztika 63(1):38–66

Logan M, Hobbs MH, Conrow LC, Reid NL, Young RA, Anderson MJ (2022) The x-minute city: measuring the 10, 15, 20-minute city and an evaluation of its use for sustainable urban design. Cities 1

Nagel L (2020) The influence of the COVID-19 pandemic on the digital transformation of work. Int J Sociol Soc Policy (ahead-of-print)

Ohnmacht T, Z'Rotz J, Dang L (2020) Relationships between coworking spaces and CO_2 emissions in work-related commuting: first empirical insights for the case of Switzerland with regard to urban-rural differences. Environ Res Commun 2(12):125004

Papadopoulos E, Sdoukopoulos A, Politis I (2023) Measuring compliance with the 15-minute city concept: state-of-the-art, major components and further requirements. Sustain Cities Soc 99

Policies to improve rural areas' innovation systems by professionalising networking activities and use of innovation tools: establishing co-working spaces in rural areas, P-IRIS 2021

Smercina V (2019) Living on the move: the digital nomad mobile phenomenon identity and practice. UNLV theses, dissertations, professional papers, and capstones
Sutherland W, Jarrahi MH (2017) The gig economy and information infrastructure: the case of the digital nomad community. In: Proceedings of the ACM on human-computer interaction (1)
Sutton-Parker J (2021) Determining commuting greenhouse gas emissions abatement achieved by in-formation technology enabled remote working. Procedia Comput Sci 191:296–303
Wang S, Cecez-Kecmanovic C (2019) Digital nomadism and the market economy: resistance and compliance
Woodcock J, Graham M (2019) The gig economy: a critical introduction, pp 8–13
Yuen Thompson B (2018) Digital nomads: employment in the online gig economy. Glocalism

Part II
Macro Scale: Design with Nature and a More-Than-Human Design Culture

Chapter 4
Bases for a Mapping of Granada City with an Ecosystem Approach: Foundational Features, Urban Fabrics, Green Infrastructure and Urban Livability

Juan L. Rivas-Navarro and Belén Bravo-Rodríguez

Abstract The formation of heat islands, poor air quality, loss of biodiversity and high motorisation, among others, have been aggravating pollution levels in the city of Granada for a long time. Meanwhile, the government, the regional and local administrations, and most of civil society have been disconnected from the problem. For this reason, it was decided to carry out a study on these problems and on how certain cities, both nationally and internationally, have adopted ecosystem mechanisms that contribute to the fight against climate change. In this way, an inventory of urban elements that make the green infrastructures of cities more resilient to the urban heat island phenomenon and function as a refuge for biodiversity is obtained. After several tables have been compiled, we systematise eight initial criteria for categorising spaces with ecosystem functions in three groups of urban facts, corresponding to three cartographic moments and an extra chapter: foundational trait, urban fabric, green infrastructure, and livability. With preliminary results presented, the focus would be mainly on understanding the special and urban capacity of the city to strengthen itself eco-systemically and prepare for real policies of decarbonisation and improvement of urban health and well-being.

Keywords Urban ecosystem · Mapping · Urban form · Urban planning · Climate change

J. L. Rivas-Navarro (✉) · B. Bravo-Rodríguez
Department of Urban and Regional Planning, University of Granada, Granada, Spain
e-mail: juanluisrivas@ugr.es

M. Vukmirović et al. (eds.), *Digital Dreams, Urban Realities*, The Urban Book Series,
https://doi.org/10.1007/978-3-031-76353-3_4

4.1 Introduction: The Current Urgent Situation

Located in Andalusia, the city of Granada benefits from a privileged settlement in this semi-desert territory in southern Spain. The historical functional domestication of the Genil, Darro, Beiro and Monachil rivers by the network of ditches allowed the flourishing of great civilisations. However, the morphology of its territory is a catalyst for contamination in our current situation. According to the report of the Observatory of Sustainability 2020, "The orography of the Metropolitan Area or the burning of biomass in the Vega, the agricultural area of the geographical depression, aggravates the effects of the emission of polluting gases", placing Granada in second place in the ranking of the most polluted cities in Spain (Polo 2021).

The city is in a space of transition and meeting point between the ecosystem of the Sierra Nevada and that of the agricultural plain of Granada. The urbanised hillside on which Granada and other centres of the metropolitan area are located fulfil an important ecosystem function. Besides, water has always been established as an important landscape-shaping factor, as a modelling agent of relief, responsible for the presence of fertile soils, hydrophilic vegetation, irrigated crops, and the configuration of an original distribution system.

4.1.1 The Loss of Urban Resilience

Problematic situations caused by climate change have overlapped with the disconnection of administration and its inability to overcome conceptually outdated planning. There are only some worthy segments of civil society fully aware of the urgency of reversing the loss of urban resilience to fight or mitigate the many effects of climate change, among which we would highlight the heat island effect, decreased humidity, poor air quality, decrease in rain, biodiversity loss or excessive motorisation.

As we can observe in Fig. 4.1, there could be a decrease in the rain and desertification in the Iberian Peninsula. The average annual rainfall could decrease by up to 42% by the end of this century because of climate change. In the evolution of the climate model in the projections for 2040–2070 and 2071–2100 were confirmed, the desertification trend would be clearly upward, with the area affected at the regional level, going from 9.57% in 2003 to 13.32% in 2010 (Argüeso et al. 2012).

Considering the problem of excessive motorisation rate, number of trips, etc., private transport by car, which is one of the most polluting means, accounted for almost 50% of total trips in October 2019 in the metropolitan area of Granada. On the contrary, public transport accounted for 9.46%, cycling for 3.44% and walking for 38.65% (Junta de Andalucía 2021). In addition to the above, more than a third of the vehicles registered in Granada are highly polluting (Morales 2022), and it has been proven that one of the main sources of respiratory diseases in the urban area of Granada is poor air quality (Abuín 2019; Rodríguez-Acosta 2022).

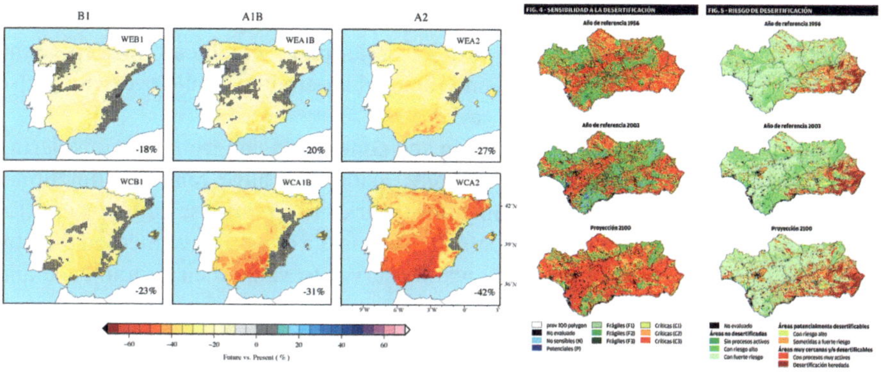

Fig. 4.1 (Left) Projected changes in the annual mean precipitation (2070–2099 vs. 1970–1999) for the different WRF simulations. Areas with black dots indicate that changes are not significant using a two-sided Student's t-test at the 95% confidence level. The spatially averaged changes over Spain are shown in the bottom right corner of each panel: (Right) Desertification sensitivity and desertification risk. *Source* (Left) From Argüeso et al. (2012: 6); (Right) From Moreira and Rodríguez (2008: 20)

4.2 Objectives and Research Methodology

4.2.1 How Can Green Infrastructures Be More Resilient in the Face of the Urban Heat Island Phenomenon and Refuge for Biodiversity?

The objective of this research aims to identify the elements of the city and its surroundings that would make up, at the same time, the ecosystem structure of Granada. The research work is based on the prospective vision of the metropolis, considering public and private spaces as functional elements of a complex system of overlapping scales, focusing on improving the quality of life and the health of its citizens. The research methodology includes a previous regulation analysis and a definition of basis and elements, followed by an assessment of the laboratory of certain cities, to subsequently establish a criterion for ecosystem mapping, which permits us to cartography Granada's Ecosystem. This methodology is based on various phases, which are:

4.2.1.1 Phase I: Analysis of Primary Sources

Primary Source I—Theoretical Approach. The first phase consists of a theoretical approach to the ecosystem city as a project based on seven categories of urban elements with ecosystem potential.

"The city, as an object of architecture, is always a rewriting of a previous city [...] At another level, we are also faced with the urban plan, which can be seen as the ground on which traces are engraved and retained indefinitely while everything else changes" (Gandelsonas 1999).

The disconnection between urban form and context, arising from not considering the conditions of the supporting geography, leads to a huge range of functional and formal deficiencies. Urban planning always tries to recompose complex entities through the analysis of specific elements. It often uses a constructive method that, by observing parts or components of the whole, serves to value their prominence and generate hierarchies.

Recognising urban identity as an essence that is renewed and constructed precisely through human decisions, it is worth asking about the urban elements or concepts that give meaning to urban identity (Rivas-Navarro 2009). This approach would be based on a morphological analysis, a typological classification of elements and their relationship, and spatial and temporal intertwining between them. Without the intention of looking for a common trunk that turns the parts into a structure, the following seven categories of components are identified as relevant in the open process of building an ecosystem urban identity. This new identity would be understood as a new organisation, characterised, in the words of Carlo Aymonino, by "a multiplicity of hierarchical, interdependent and intensely selected relationships" (Aymonino 1981):

(1) The meaning of cities
(2) The importance of topography and geography
(3) Collective structures
(4) The participation of individual forms
(5) Urban and territorial links
(6) Essential corners
(7) Synthesis of essential forms
To which we add a supplementary category:
(8) Citizen initiatives as a qualitative criterion.

Primary sources II—Study of real cases. The analysis of primary sources is followed by the study of cities that show in their urban planning an ecosystem commitment with the logic of their existing fabric. In the spatial framework of cities, this combination of opportunities and challenges necessitates the integration of strategies that achieve a friendly environment, sustainable transportation, reduced air pollution and increased biodiversity, greater proximity to green spaces, more productive and connected neighbourhoods, etc. (Dijkstra 2017).

The primary capital for improving urban ecosystems and green infrastructures is open urban space, particularly public space. To acquire such regenerative powers, it is vital to grasp its existing position as well as its increasingly various aspects (Rivas-Navarro and Bravo-Rodríguez 2024). Recent perspectives stressing the green infrastructure approach provide a more socioeconomic perspective, having the possibility to address the economic value system of spatial development (Seiwert and Rößler 2020). Evidence for this is his view of the interrelated components that guide the integration of sustainable urban development into the New Urban Agenda (United

Nations 2017) and the central role that public spaces play in social and economic development (Mehaffy et al. 2019).

Understanding the ecological function of urban open space, particularly the components of potential green infrastructure, necessitates a search for interconnection that may be integrated into networks. In the definition and interpretation of the term green infrastructure, substantial emphasis is given to the excellent quality of these pieces and their participation as a vital part of a linked network (Seiwert and Rößler 2020; European Commission 2022). This network structure provides systemic capability to the varied group of previously separated territorial places and resources (Feria et al. 2020). We will discuss a very extensive catalogue of elements that will include almost all the unbuilt areas and the biological matrix of open spaces, which must be viewed as a single interconnected system (Rivas-Navarro and Bravo-Rodríguez 2024).

Based on various urban and environmental reconsiderations that other cities have been facing, through fieldwork, ecological reading of the support territory, unoccupied space analysis, cartographic interpretation, morphology of the urban present, history, and current urban dynamics.

Phase II: Building of the cartographic's legend and Urban Purpose. With the aim of obtaining a new image meaningful to the city, territorial resources, current ecosystem services and potential green infrastructure are carefully identified and categorised in that second phase, considering elements from the city and its surroundings. The ecosystem structure for Granada is revealed using two complementary perspectives. On the one hand, the registration of urban-ecological elements belonging to the founding of the original settlement. On the other hand, future urban projects and urban policy logic should be considered to identify possible systems and urban patterns according to sustainable development.

Study of real cases of implementation of ecosystem policies. To compile information about ecosystem policies applied to real cities, it's fundamental to establish a criterion for ecosystem mapping. To meet with this research phase, it was to study real cases of implementation in various NATIONAL AND INTERNATIONAL CITIES:

(1) **Lleida.** ***Municipal Plan for Urban and Territorial Planning 1995–2015.*** The Municipal Plan of Lleida goes beyond the simple urban land planning, aspiring to be a sustainable model of city that puts in the center of attention the treatment of the Segre River as an articulating element, on a territorial scale, of the urban and periurban landscape, focusing the attention on the free spaces as a plural system.
(2) **Barcelona.** ***Green Plan and Biodiversity of Barcelona 2020.*** The city of Barcelona, the largest port in the Mediterranean, has positioned itself at the forefront of contemporary ecosystemic challenges of revitalisation and/or recovery of urban land and natural heritage on its edges and streets. By turning its city into a veritable laboratory, the Catalan city council has demonstrated an ecological awareness in its urban development policy. After several studies on the quality of life of the citizens (air quality, sound quality, environmental temperature,

mobility), a great work of cataloguing and theorising concepts for the development of a green plan in the image of a protocol was undertaken. In doing so, greater flexibility is sought in the interaction between nature and urbanisation, considering its very characteristic urban fabric inherited from the Cerdá plan, its seafront and its reliefs, habitat and refuge for biodiversity, which frame the growth of the city (Fig. 4.2, left).

(3) **Vitoria.** ***Green Belt of Vitoria-Gasteiz, 1993.*** Established for ten years in the urban and peri-urban landscape, Vitoria's green ring meets the criteria of sustainability, livability and efficiency in terms of mobility and energy. The urban green infrastructure strategy has demonstrated the increased benefits of operating as a connected network and has been rewarded for this by the title of *European Green Capital 2012*.

(4) **Granada.** ***Paseo del Salón Rehabilitation Competition, Genil Central proposal*** (Rivas-Navarro and Cabrera-Manzano 2013). The *Genil Central* project relocates its position within the city and questions the functionality of this river corridor. It also proposes the reconquest of the denied riverbank, taking advantage of the transversal axes to enhance the continuity of the routes. Considering the riverbank as a margin of opportunity, it widens the river for a progressive approach to the *water*.

(5) **Madrid.** ***Madrid Río Project, 2004; Metropolitan Forest Project, 2020.*** After the liberation of the land near the Manzanares River due to the burying of the M-30 highway, the project for the recovery of the riverbanks was imposed as a socio-environmental necessity. The river corridor becomes a key backbone element in the reconstitution of a fragmented and interrupted territory. For its part, with the ambition of becoming a major ecosystem reference, the Madrid metropolitan forest is characterised by a pre-project assessment methodology. For this, the collaboration of multiple stakeholders leads to a complex reflection on the benefits and risks of assessing the sustainable potential of the project. This contrast is essential for the creation of values with a simple objective: to achieve climate neutrality in the next ten *years*.

(6) **Girona.** ***Las veras de Girona. Design and management laboratory for urban green infrastructure in Girona., EMF Arq, 2014.*** Aware of the potential of its outskirts, Girona focused on the valuation of its natural spaces, turning them into the largest facility in the city. It is a multifunctional green infrastructure on a metropolitan scale: a total of 600 hectares divided into 30 proximity parks available to its inhabitants. Its affordability favours social appropriation in the low meadow areas, always focusing the project on ecology and cultural *activities* (Fig. 4.2, right).

Other cases studied:

Salamanca. Special Plan for the Protection of the green infrastructure and biodiversity of Salamanca 2019.

- New York. City Street Trees by Species.
- Frankfurt. Integrated Urban Development Plan for Frankfurt 2030, 2017.

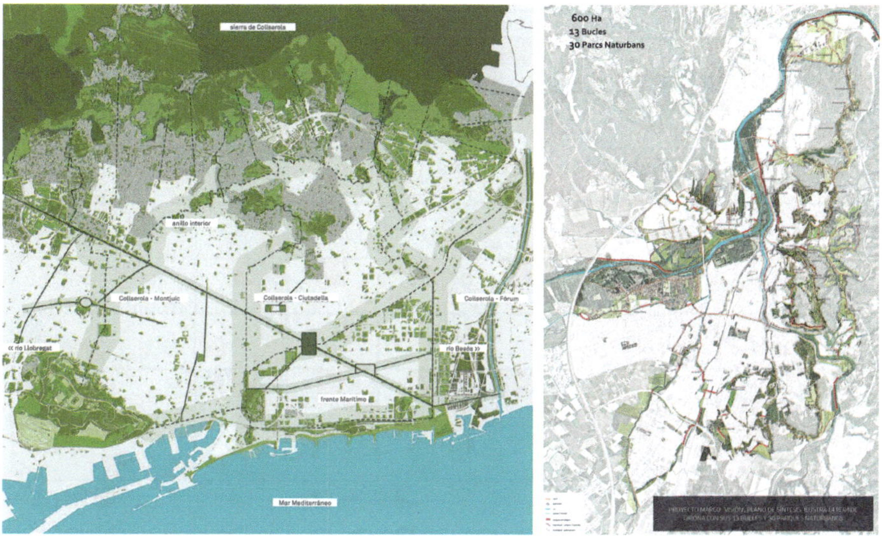

Fig. 4.2 (Left) Urban green corridors, *Green Plan and Biodiversity of Barcelona 2020*; (right) 'Vision' Synthesis Plane, *Las veras de Girona. Source* (Left) (Ayuntamiento de Barcelona 2013: 67); (Right) (EMF Arquitectura del Paisaje 2020: 11)

- Lyon. 2021–2026: Relance écologique et solidaire. La Métropole du Grand Lyon (Fig. 4.2).

4.3 Results

4.3.1 Assessment of the Laboratory of Cities According to the Eight Basis and Elements of the Ecosystem Identity

Once the cities mentioned are known and studied, we proceed to make a comparative analysis that consolidates all the information obtained. The ensuing synthetic information table tries to exhibit the qualities of the national cities we've just mentioned regarding the seven categories of urban elements with ecosystem potential listed above. Each criterion is broken down into several tangible urban elements that can be assessed according to their values: ecosystemic, structuring, aesthetic, functional, identity, singularity, sociocultural, historical, health, comfort, geopolitical, economic, and educational (See Fig. 4.3). This allows us to understand why certain cities became pioneers in the fight against climate change and what tools they have been using for this purpose.

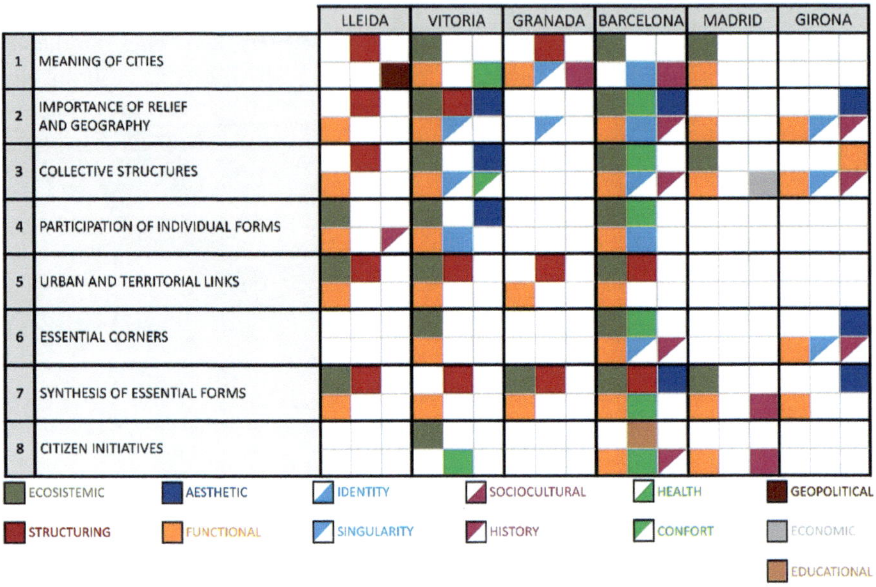

Fig. 4.3 Multidimensional assessment of the "Cities laboratory" according to the eight bases to address the Ecosystem Identity

(1) *The meaning of cities.* Preservation of urban parks with a historical and artistic character. 30% of the currently public green spaces are former private gardens of artistic quality, 27 of which are included in the Barcelona city council's catalog of architectural heritage.
(2) *The importance of relief and geography.* The hill and the river (territorial elements within the city) are used as free spaces and green areas.
(3) *Collective structures.* Increased water infiltration to reduce runoff and flooding: permeable pavements; creation of rain gardens as runoff water filters; and recovery and ecological management of mediaeval pipes.
(4) *The participation of individual forms*
(5) *Urban and territorial links*
(6) *The essential corners*
(7) *Synthesis of essential forms.* The integration of rivers previously limited the growth of the historic city in the dynamics of metropolitan Granada. Conversion of these corridors into axes of interest and/or central parks.
(8) *Citizen initiatives.* Work in multidisciplinary teams to maximise environmental and social services. The forest is a place of opportunity to favor pedagogical, cultural, technological, and digital innovation.

4.3.2 Mapping of Granada Through an Ecosystem Approach: Proposal in a Transversal Framing

Spatial categories with ecosystem function. After the corresponding tables in relation to a group of elements with ecosystem potential have been compiled, we have systematised the eight initial criteria for spatial categories spaces with ecosystem functions. We propose three groups of urban facts related to three cartographic moments: foundational traits (past), urban fabric (present), and green infrastructure (future). The experimental drawing appeals to the urban landscape that has been travelled, inhabited, legislated, and politicised within the chronological process of the territory transformation. The proposal of ecosystemic components is presented in detail below and divided into the blocks described above. Those components that are not yet represented in the exemplary framework of the urban territory initially studied have been placed in italics.

I. FOUNDATIONAL TRAITS. Domesticating and sewing. Identifying primary elements of natural origin with ecosystem potential:

(a) Riverscapes
   (i) **Fluvial axis**: The actual riverbed along which the rivers run.
   (ii) **Nearby extension**: Space adjacent to the river that is understood as part of the riverbank.
   (iii) **River frontage**: Frontages of buildings that have a direct view of the river at close or medium distance.
   (iv) **Riverbank blocks**: Those which are understood to be part of the river fabric and which have clearly been configured under its influence.
   (v) **Tributaries Street**: Streets perpendicular to the river that direct the flow of movement towards it.
   (vi) **Urban basin of the crossing**: Space adjacent to the river course to which it is intended to amplify its influence, as it is considered as an extension of it, whether for historical, urbanistic or ecosystemic reasons.
   (vii) **Confluence nodes**: The point where two river axes meet.
   (viii) **Bridges**: Elements connecting both banks of the river axis.

(b) Water culture
   (i) **Acequia**: Uncovered or buried channels that carry water, historically for supplying populations or irrigation.
   (ii) **Fountain**.
   (iii) **Runoff**: Land through which rainwater circulates freely and naturally.
   (iv) **Private swimming pool**.
   (v) Others:

   - *Washhouse:* A space used for washing clothes.

- *Water reservoir:* A hydraulic structure is used for the storage of water.

II. URBAN FABRIC. Settling down and defining the legacy. Built elements that speak of the objective transformation of territory from the foundation of the settlement to the present:

(a) The fabric of the major connectors
   (i) **Historic roads**: Axes of the city that predate all others and whose layout has been preserved over time. Normally, these same axes were the backbone of the existing plain fabric.
   (ii) **Efficient axes of traffic flow**: Efficient penetration of road traffic into the urban fabric, axes of great power to accommodate road flows.
   (iii) **Pedestrian parallel axes**.
   (iv) **Historical gates and walls**: remains of walls built during the Nasrid period in Granada.
(b) The structuring axes and their infrastructure
   (i) *Railway and railway station:* Railway line and its stops.
   (ii) **Granada's Beltway**: roadway system transition to the Vega, motorway that cuts the connection between the city of Granada and its Vega.
(c) Structure of public land
   (i) **Block**
   (ii) **Plot**
   (iii) **Sidewalk**
   (iv) **Driveway**
   (v) **Road or mixed street**
   (vi) **Pedestrian street**
   (vii) **Linear walkways**
   (viii) **Plaza**
(d) Building
   (i) **Residential building**
   (ii) **Urban** facility
(e) *Free space attached with potentiality*

III. GREEN INFRASTRUCTURE. Suture and reproduction. Prospective of the spatial potential of the Granada through an Ecosystem approach considering its territory as a place of opportunities, including:

(a) Free walkable space
   (i) **Public green** space
(b) Free space non-transitable
   (i) **Riverside vegetation**: vegetation present on the banks of rivers, linked to the passage of river axes.
   (ii) **Public space vegetation**: vegetation present in and shaping public open spaces.

   (iii) **Vertebral axes vegetation**: vegetation linked to the passage of important road or pedestrian axes.
- (c) The production spaces.
   (i) **Agricultural plot and nursery**: Different plots that make up the Vega of Granada.
- (d) Green mobility
   (i) **Pedestrian path**
   (ii) **Linear/river walkway**
   (iii) **Cycling** route
- (e) *Biodiversity*
- (f) *Free space attached with potentiality*

IV. [extra] HABITABILITY. Realistic sustainability. Evaluation of qualitative performance of the proposal, whose fundamental values would be:

- (a) *Realistic sustainability*
- (b) *Short-term changed behavior*

Transversal framing of Granada with ecosystem approach, experimental cartography. As computer code, red, green, and blue are considered as complementary colors of synthesis of white one. Being represented together and mixed, three colors codes -red, green, and blue-, are speaking to us of the need to make diverse and plural the look at the three times of the city: the legacy of the past, the concerns of the present, and the planning future (See Fig. 4.4).

The cartographic attempt has been developed in a central frame of the city of Granada that represents a 1200 m by 4500 m plan section. The selected framework includes the historical limit of the city's growth towards the south of the Genil riverbed

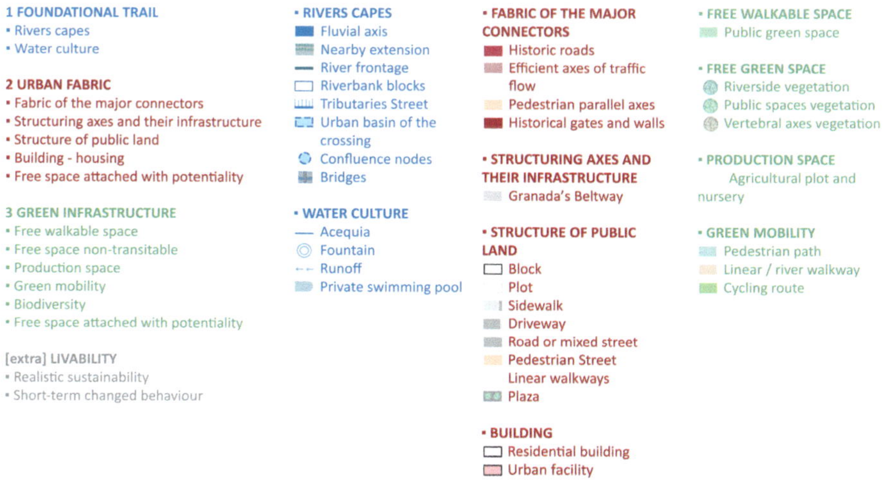

Fig. 4.4 Legend of the laboratory-framing of the mapping of Granada with ecosystem approach according to thematic blocks or times of the territory. *Source* Own elaboration

and includes some of the main urban landmarks of the city's urban planning from a typological perspective (See Figs. 4.5 and 4.6): the Armilla Road, the southern edge of the historic center, the Darro river as it passes through the city, the city-countryside contact of the valley at the confluence of the Genil and Monachil rivers, etc. This "laboratory window" tests the ecosystemic structure proposed above in a position that is selected as very useful for these initial research results (See Fig. 4.7).

If we make a more careful observation of the representation of the ecosystemic components in a central framing of the window initially addressed, we can draw specific conclusions from the result (See Fig. 4.8). In this central area of the city -in its southern limit with the historical periphery behind the Genil river-, the complete system of the riverbed is the absolute protagonist, which implies, in addition to considering its immediate banks and associated streets, the reconsideration of the urban fabrics annexed and involved in the "fluvial environment" of the surroundings, the calming of traffic in its road and the increase of pedestrianism of the environment,

Fig. 4.5 Identification of green areas in Granada and situation of the "laboratory window" for the attempt of ecosystem mapping. *Source* Own elaboration

Fig. 4.6 View of the Genil riverbed from west to east in its central section (historic center on the left). *Source* Own elaboration

of its environmental conditions and urban comfort; the enhancement of heritage elements associated with water, as in this context the Acequia Gorda, as well as the space involved in the meeting between the Darro and Genil rivers, and the recovery of the spatial and environmental quality of the banks of the Darro river itself.

It is remarkable the spatial relevance that the greater cohesion of the surrounding public spaces acquires when they are agglutinated as a true system from this perspective, thus reinforcing the fluvial condition of the area. It would be necessary to add to the considerations on this framework the importance that private open spaces could acquire, in terms of the ecosystemic commitments that they could assume. This would be valid both for the singular residential uses -relevant properties of the Paseo del Salón and Bomba facades- and for the unoccupied spaces of educational or other types of facilities, both public and private, the permeability of their soil and the increase and strengthening of vegetation in these private spaces.

If we move westward in the initial window approached, we can focus our gaze on the contact between city and central Vega, placing the analysis of results in the territorial position of the exit of the fluvial system of the Genil river from the urban area to the agricultural area of the Vega (See Fig. 4.9). In this closer framework, both the development of the riverbed and the traditional relationship between the city of Granada and the town of Armilla and other later ones that go deeper and deeper into the Vega de Granada are established in an east–west direction. Both east–west linear systems cross the north–south system of streets and avenues that have in the Camino de Ronda the main axis, the street-city of Granada, as pointed out by Rivas (Rivas-Navarro 2015). Here again, the improvement of the spatial, environmental and ecological conditions of riverside spaces must transcend to the axes parallel to the river system, as well as to the systems of public spaces that, from a topological perspective, are easier to establish relations of proximity and continuity with the river area itself. The pedestrianism that could arise from this "calm and environmentally improved environment" of the Genil, would also be key when reconsidering certain axes that connect the banks with more inland positions (especially if these streets can also connect with public facilities of all kinds).

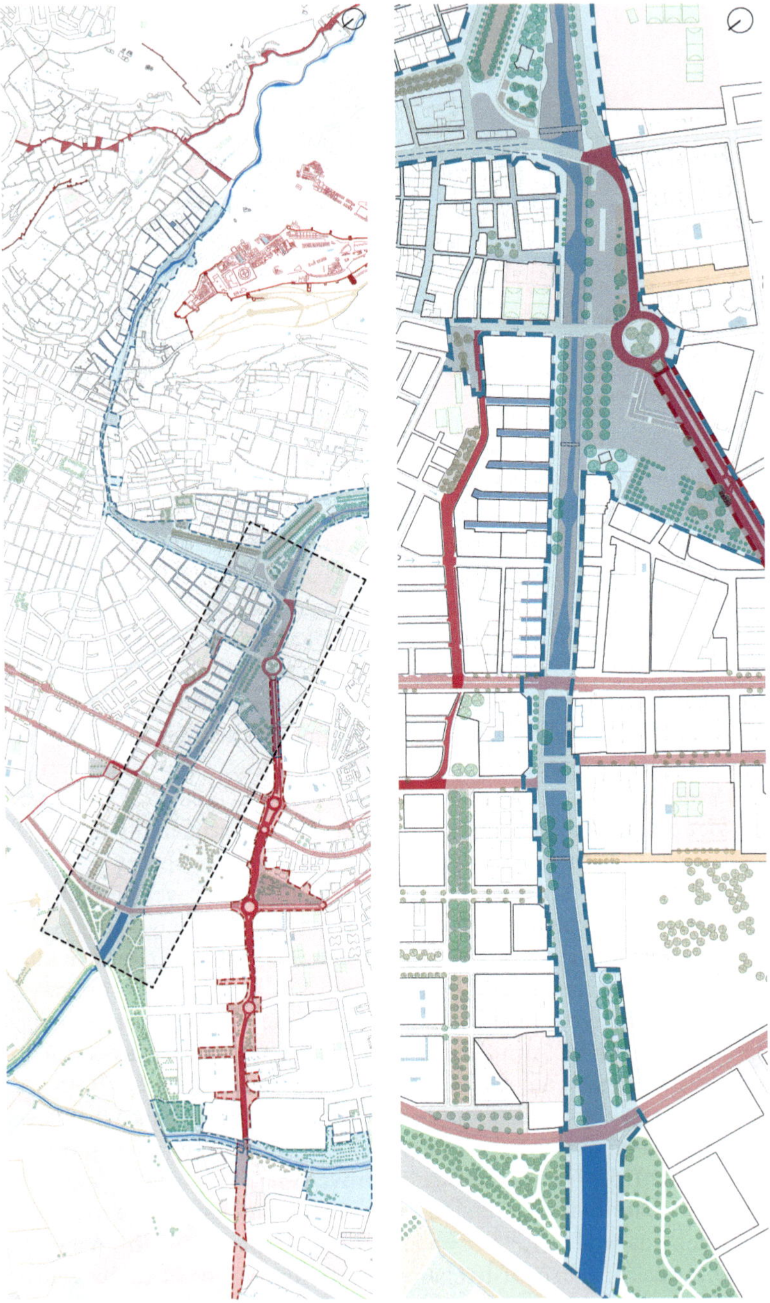

Fig. 4.7 (Left) "Laboratory-framing" of the mapping of Granada with ecosystem approach; (Right) Zoom applied to the central urban area around the Genil River. *Source* Own elaboration

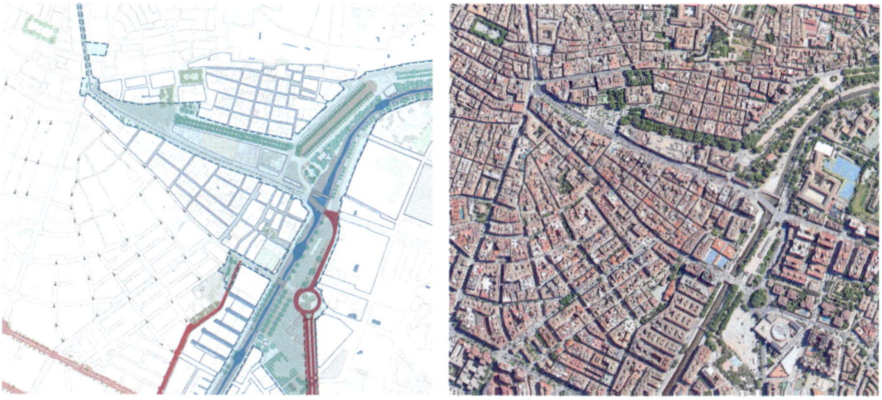

Fig. 4.8 Central framework of the laboratory framing addressed for the mapping of Granada from an ecosystem approach. *Source* Own elaboration

Fig. 4.9 West framework of the laboratory framing addressed for the mapping of Granada from an ecosystem approach. *Source* Own elaboration

The value of space in the large streets and arteries of vehicular traffic in this western frame of the initial window addressed is also fundamental. Large sidewalks and their continuity are a first level spatial resource for tree planting and the improvement of temperature, humidity and biodiversity connectivity conditions, especially in this context in which the streets cross a space that we could consider an "ecosystem node" of mapping such as the urban environment of the urban river environment (Fig. 4.7).

4.4 Conclusions

This research is being developed through: (1) referential national and international case studies; (2) a local field work in depth; (3) an ecological reading of territory, considering an analysis of unoccupied space; and (4) a multiscale and multidimensional cartographic interpretation of the city of Granada, its urban history, and current urban dynamics. The point would be transforming and mobilizing awareness of the time we are facing, with a huge climatic crisis with locally palpable effects. The aim is to prepare future reflections and interventions involved in a real possibility to tackle the serious environmental and health threats. The focus would be mainly to understand the spatial and urban ability of the city to strengthen itself eco-systemically, preparing for real policies of decarbonization and improvement of urban health and well-being of people.

The fight against climate change takes place in the urban scenarios of the city, and the spatial dimension of physical resources plays a fundamental role as dissipator, enhancer, blocker, or enabler of urban strategies of more positive response of cities to local effects of the climate emergency.

The approach of mapping of the Granada with ecosystem approach is just a starting point. By the moment, that vision of Granada only exists as a project. As a foundational gesture, the research is setting the basis for further development, but it is mainly rescuing historical values that may be part of a much better relationship between the city and its geographical environment. The possibility that it is opened would be to establish an eco-structure that implies a lot of urban fabrics and the assignment of different roles to urban facts (streets, blocks, public space, collective facilities, etc.)

Another outcome would be to balance inequalities, increasing the ties of this environmental armour with respect to the area for which it is responsible. It would also mean increasing the collective and public space dimensions of urban space for geographical support in a more resilient way and creating a positive network to guide future steps.

References

Abuín, E (2019) La Granada metropolitana respira el aire más contaminado de España. [Trad.: Metropolitan Granada breathes the most polluted air in Spain]. GranadaHoy, https://www.granadahoy.com/granada/Granada-metropolitana-respira-aire-contaminado-peor-calidad-Espana_0_1368763403.html. Accessed 13 June 2023

Argüeso D, Hidalgo-Muñoz JM, Gámiz-Fortis SR, Esteban-Parra MJ, Castro-Díez Y (2012) High-resolution projections of mean and extreme precipitation over Spain using the WRF model (2070–2099 versus 1970–1999). J Geophys Res Atmos 117, https://doi.org/10.1029/2011JD017399

Aymonino C (1981) El significado de las ciudades [Trad.: The meaning of cities]. Blume, Madrid

Ayuntamiento de Barcelona (2013). Plan del Verde y de la Biodiversidad de Barcelona 2020 [Trad.: Barcelona's Green and Biodiversity Plan 2020]. Medi Ambient i Serveis Urbans—Hàbitat Urbà,

Ajuntament de Barcelona, Barcelona. https://ajuntament.barcelona.cat/ecologiaurbana/es/que-hacemos-y-porque/ciudad-verde-y-biodiversidad/plan-verde-y-la-biodiversidad
Dijkstra L (ed) (2017) 7th Report on economic, social and territorial cohesion. Publications Office of the European Union, Luxemburgo. https://ec.europa.eu/regional_policy/information-sources/publications/reports/2017/7th-report-on-economic-social-and-territorial-cohesion_en. Accessed 20 March 2024
EMF Arquitectura del Paisaje (2020) Las Veras de Girona. Equipamiento y Servicios Municipales 193:106–117
European Commission (2022) Guidance on a strategic framework for further supporting the deployment of EU-level green and blue infrastructure. commission staff working document. Available online: https://environment.ec.europa.eu/topics/nature-and-biodiversity/green-infrastructure_en. Accessed 20 March 2024
Feria JM, Santiago J, Iglesias R, Andújar A, Hurtado C, Gómez F, Gutiérrez J (2020) Ciudades inteligentes y sostenibles. Infraestructura verde y hábitats urbanos integrados. [Trad.: Smart and sustainable cities. Green infrastructure and integrated urban habitats]. Centro de Estudios Andaluces, Sevilla
Gandelsonas M (1999) X-Urbanism: architecture and the American city. Princeton Architectural Press, New York
Junta de Andalucía (2021) Plan de Transporte Metropolitano del Área de Granada. Diagnóstico preliminar. [Trad.: Metropolitan Transport Plan for the Granada Area. preliminary diagnosis]. Consejería de Fomento, Infraestructuras y Ordenación del Territorio, Junta de Andalucía, Sevilla, https://www.juntadeandalucia.es/organismos/fomentoarticulaciondelterritorioyvivienda/areas/infraestructuras-movilidad/planes-movilidad-sostenible/paginas/plan-movilidad-granada.html. Accessed 13 June 2023
Mehaffy MW, Elmlund P, Farrell K (2019) Implementing the new urban agenda: the central role of public space. Urban Des Int 24(1):4–6. https://doi.org/10.1057/s41289-018-0063-2
Morales J (2022) Un tercio de los vehículos censados en Granada capital están entre los más contaminantes [Trad.: One third of the vehicles registered in Granada city are among the most polluting ones]. Ideal, https://www.ideal.es/granada/cuatro-vehiculos-empadronados-20220517184053-nt.html. Accessed 13 June 2023
Moreira JM, Rodríguez M (2008) Escenarios climáticos y desertificación en Andalucía. Medioambiente 62:16–21, chrome-extension://efaidnbmnnnibpcajpcglclefindmkaj/https://www.juntadeandalucia.es/medioambiente/web/Bloques_Tematicos/Publicaciones_Divulgacion_Y_Noticias/Publicaciones_Periodicas/revista_medio_ambiente/revista_ma_62/escenarios_clima.pdf. Accessed 13 June 2023
Polo JC (2021) La contaminación en Granada, un problema que no solo está en la capital. [Trad: Pollution in Granada, a problem that is not only in the capital]. Granada Digital, https://www.granadadigital.es/granada-ya-es-segunda-ciudad-mas-contaminada-espana-tras-superar-barcelona/. Accessed 13 June 2023
Rivas-Navarro JL (2009) Fundación de la tesis: la travesía más transparente. La visión de Córdoba, Málaga y Granada desde su calle ciudad [Trad: Foundation of the thesis: The most transparent traverse. The vision of Córdoba, Málaga and Granada from its city street]. Universidad de Granada, Granada, https://digibug.ugr.es/handle/10481/19760
Rivas-Navarro JL (2015) Un urbanismo de la observación: Metodologías prospectivas en torno a la idea de calle ciudad en tres capitales andaluzas (Córdoba, Málaga y Granada) [Trad.: An urbanism of observation: prospective methodologies around the idea of the city street in three Andalusian capitals (Cordoba, Malaga and Granada)] EURE (Santiago) 41(123):131–158. https://doi.org/10.4067/S0250-71612015000300006
Rivas-Navarro JL, Bravo-Rodríguez B (2024) Beyond multidimensional vulnerability approach: a triple network notion for urban cohesion in at-risk neighborhoods of Jaen's historic center. Sustainability 16(10)
Rivas-Navarro JL, Cabrera-Manzano D (2013) Genil Central: de salón barroco periférico a distribuidor central contemporáneo. Concurso de ideas para la rehabilitación del Paseo del Salón

[Trad.: Genil Central: from a peripheral baroque hall to a contemporary central distributor. Ideas competition for the rehabilitation of the Paseo del Salón]. Universidad de Granada, Granada, http://hdl.handle.net/10481/23309

Rodríguez-Acosta S (2022) La mala calidad del aire de Granada, una fuente de enfermedades respiratorias [Trad.: Grenada's poor air quality, a source of respiratory diseases] Granada Digital, https://www.granadadigital.es/mala-calidad-aire-granada-fuente-enfermedades-respiratorias/. Accessed 13 June 2023

Seiwert A, Rößler S (2020) Understanding the term green infrastructure: origins, rationales, semantic content and purposes as well as its relevance for application in spatial planning. Land Use Policy 97:104785. https://doi.org/10.1016/j.landusepol.2020.104785

United Nations (2017) New urban agenda. United Nations conference on housing and sustainable urban development, Habitat III Secretariat. United Nations, Quito, https://habitat3.org/the-new-urban-agenda/

Chapter 5
Geodesign Framework in Green Infrastructure Planning: Prioritizing Uncertainty in the Future of Urban Ecosystem Services

Boris Radić, Suzana Gavrilović, and Siniša Polovina

Abstract Contemporary urban landscapes face increasing pressure on natural and semi-natural elements, threatening vital ecosystem services. The proliferation of sealed surfaces precipitates a decline in crucial ecosystem services, pivotal for enriching urban environments and augmenting economic and cultural well-being. Recognised as a sustainable development approach, green infrastructure emerges as a multifunctional and strategically planned to counterbalance these pressures. The amalgamation of rapid urbanisation and climate variability paints an enigmatic future for urban landscapes, necessitating methodologies conducive to more apparent scenario analyses. A Geodesign framework emerges as a systematic instrument, furnishing a nexus of interconnected facts that foster discerning deliberation and facilitate optimal solutions for improving urban landscape resilience. Research endeavours pivot towards applying Geodesign's Change and Impact models to green infrastructure, focusing on evaluating primary ecosystem services within the urban matrix. The foundational blueprint for Belgrade's green infrastructure is delineated within the General Regulation Plan of the Green Area System, further refined by the draft version of the Master Plan. The current state of the urban landscape structure, coupled with planned enhancements through integrating new green infrastructure elements at the watershed level, represents the main spatial context. An array of geostatistical models furnishes profound insights into the dynamics of natural processes, elucidating ecosystem services at the landscape level.

Keywords Green infrastructure · Geodesign · Ecosystem services · Urban landscape

B. Radić (✉) · S. Gavrilović · S. Polovina
University of Belgrade—Faculty of Forestry, Belgrade, Serbia
e-mail: boris.radic@sfb.bg.ac.rs

S. Gavrilović
e-mail: suzana.gavrilovic@sfb.bg.ac.rs

S. Polovina
e-mail: sinisa.polovina@sfb.bg.ac.rs

M. Vukmirović et al. (eds.), *Digital Dreams, Urban Realities*, The Urban Book Series,
https://doi.org/10.1007/978-3-031-76353-3_5

5.1 Introduction

In recent decades, the rise in urban population has frequently been termed as "population ante portas", underscoring the milestone established in 2007 when it was confirmed that over half of the world's population lives in cities. Urbanisation is a global trend, and more and more people are moving to cities and towns in search of better economic opportunities, education, healthcare, and quality of life (Zhang 2016). However, population growth is only one side of the coin. Conversely, the simultaneous expansion of the intricate urban "organism" and its identifiable structures, including continuous fabric, roads, industrial zones, commercial areas, public spaces, and private units, is a pressing concern (Zhang 2016). Urban infrastructure, predominantly comprised of impervious surfaces, leads to soil sealing, widely acknowledged as the primary catalyst for the deterioration of ecological integrity and disruption of natural processes discernible at a landscape scale. (Tobias et al. 2018; Radić Gavrilović 2021). Due to the projection of population change at the global level, which forecasts that by 2050, 70% of the world's population will live in cities, environmental problems are expected to be even more emphasised (Gu et al. 2021).

The quality of the environment in urban landscapes can be gauged by the extent to which ecosystem services are realised (Breuste et al. 2013; de Groot et al. 2002). Ecosystem services encompass the diverse benefits and resources that cities, operating as spatial systems on a landscape level, obtain from natural ecosystems (McPhearson et al. 2014). Ecosystem services are the benefits humans receive from ecosystems, such as clean air and water, fertile soil, pollination, climate regulation, and different non-material benefits from ecosystems, such as recreation and aesthetic experience. In urban landscapes, where natural ecosystems are often fragmented or altered, the level of realisation of these services can indicate how well the environment supports human well-being and sustainability. (Mitchell et al. 2015). An increasingly effective strategy in recent decades has been applying the concept of green infrastructure. This approach entails integrating natural elements and processes into urban and landscape planning and design to yield multiple benefits, particularly those associated with ecosystem services (Wang 2014; Calderón-Contreras 2017). The green infrastructure concept is significantly acknowledged within the strategies and directives of the European Union, actively advocated for, and implemented to bolster environmental sustainability, resilience, and urban and regional quality of life (Mell 2015; Lafortezza et al. 2017). Implemented green infrastructure initiatives serve various purposes, including conserving biodiversity and providing ecosystem services such as climate regulation, air and water purification, flood protection, and cultural and recreational amenities. (De Groot et al. 2010). Green infrastructure enhances connectivity between existing natural elements, mitigating habitat fragmentation and loss while increasing ecological resilience. Nevertheless, it's essential to acknowledge that different urban landscape types necessitate diverse green infrastructure and design approaches to realise various ecosystem functions effectively. The landscapes of urban outskirts, in particular, are identified as fragile zones where the interests of multiple stakeholders and unique environmental conditions intersect. (Radić et al. 2023). These zones represent isolated enclaves characterised by high-quality natural soils, which serve as conduits for mitigating the impacts of torrential floods (Ristić et al. 2012; Gkiatas et al. 2021). Additionally, natural and semi-natural patches within urban outskirts contribute to creating diverse, compositionally and configurationally rich landscapes with significant levels of ecosystem diversity (Asikainen 2009; Cornelis 2004). Ultimately, landscape units situated at the forefront of rapid urban sprawl, albeit temporarily, often serve as the final refuge for outdoor recreation opportunities for residents (Pietilä et al. 2015; Pyky et al. 2019). Considering these factors, urban outskirts are highly regarded for their intrinsic value, underscoring the need for careful planning and design efforts to preserve their ecological integrity and ensure their rightful integration into the inevitable urban grid.

Amidst forthcoming shifts in climate parameters, the relatively preserved level of naturalness, and the partially unpredictable expansion of urban fabric, understanding potential development scenarios becomes paramount in the planning process of modern cities and their outskirts. In this context, generating scenarios does not just transcend strategic foresight; it involves creative endeavours to forge optimal connections between the structure and function of green infrastructure, ecosystem services, and sustainable development. Due to their value in times of uncertainty and complexity, scenario planning techniques are garnering more attention in landscape and urban planning.

The Steinitz framework has been instrumental in generating alternative land use scenarios at the landscape scale, from implementing various choices for over half a century. These scenarios have been used in planning and urban development practices (Steinitz 2014). With advancements in spatial and informational technologies, a distinct approach known as Geodesign has emerged from this framework. Geodesign integrates concepts from geography, urban planning, and design to address complex landscape challenges. Using Geographic Information Systems (GIS), data, spatial analysis, and visualisation tools, Geodesign facilitates more sustainable, informed, and innovative decisions concerning the built and natural environment. In the context of Geodesign, scenarios denote diverse future possibilities or alternatives that can be created and assessed using geospatial data and analysis techniques. These scenarios are typically used to determine the potential impacts of different landscape decisions or interventions in the context of natural processes to provide the best choices about planning, design, and development. In summary, Geodesign and its scenarios play a pivotal role in tackling intricate spatial challenges in a data-driven and informed manner. They empower landscape planners, designers, and decision-makers to examine multiple options, evaluate potential impacts, and ultimately make more sustainable and resilient choices, particularly in fragile landscapes.

5.2 Materials and Methods

The broader research area encompasses the territory of Belgrade, the capital of the Republic of Serbia, boasting a populace exceeding 1.6 million inhabitants, according to the latest census. At the pivotal juncture of two great rivers, the Danube and the Sava, Belgrade holds a strategically significant position, serving as a bustling nexus for transportation networks linking Europe and Asia. A particularly valuable but also challenging aspect of Belgrade's territory is the pronounced heterogeneity of the environmental factors: extensive alluvial plains of large rivers, narrow beds of smaller watercourses, hilly and mountainous hinterland with very different geological formations that form a diverse soil substrate, and, accordingly, different natural values. All these factors resulted in highly diverse landscape types: the highly dense urban fabric of historical and cultural centres, thought-out suburb zones with industrial heritage, remnants of natural patches, agro-forestry landscapes with satellite estates, and nature-protected areas. The level of achievement of ecosystem services at the level of landscape type is highly variable and defines various degrees of landscape sensitivity to different impacts (Vasiljević et al. 2021).

The territory of Belgrade has a long and significant history of urban planning. The first signs of city planning with the philosophy that it is necessary to prescribe urban rules and direct the city's further development began in 1867 under the leadership of the first Serbian city planner, Emilijan Josimović (1823–1897). Belgrade's first formal urban plan was adopted in 1923, presenting an ambitious vision of the city that unfortunately could not be realised in a country impoverished by war in the previous decade (Vuksanović-Macura 2014). In the following half-century, numerous plans shaped Belgrade's dynamic scenography to create an operational and healthy stage for more than a million inhabitants. After years of isolation during the nineties, a modern urban plan was established in 2003 with greater flexibility in light of political support for ongoing socio-economic changes. The plan introduced numerous innovations, and in the context of green infrastructure and ecosystem services, the use of the principle under the metaphorical name "In agreement with nature: green spaces system" (Macura et al. 2019). The system of green areas is defined as a singular system of the city, consisting of individual elements under different ownership and whose land use is subordinated to the category of land cover. The organisation of the green spaces system is based on principles of connectivity, multifunctionality, scalability, integration, and multiscale, which for that period were "avant-garde approach" and today are recognised as an indispensable part of the green infrastructure concept (Cvejić 2010; Monteiro et al. 2020). Simultaneously with the adoption of the Urban Plan 2003, the city administration initiated the realisation of the "Green Regulative of Belgrade", which, in addition to other aspects related to the preservation and maintenance of existing green areas, provides a conceptual guideline for defining the methodology, building a database, and establishing a plan for the city's green areas system. Due to the spatial comprehensiveness, the pioneering activities of creating a plan of a new character, and the need for coordination with the planning solutions of other plans, in 2019, after ten years of working processes, the Plan of General Regulation of the Green Area System of Belgrade (PGR) was adopted (OGCB 2019). The plan significantly enhanced the system of green areas, leveraging green infrastructure principles to create a novel integrative value intertwining natural and built elements. This initiative not only improved Belgrade's landscape identity and character but also fostered a more resilient and sustainable environment.

5.2.1 Scenario Design

In this research, the objective is to elucidate the sub-performance of Geodesign through spatial interventions and their associated scenarios, focusing on their impact on the provision of ecosystem services. As part of this endeavour, a portion of the planned land cover outlined in the Plan of General Regulation will be integrated into the intervention phases: the Change, Impact, and Decision models. (Steinitz 2012) (Fig. 5.1). The PGR has the character of the physical planning result, so elements of the assessment phase in the Geodesign framework are unnecessary in this case and are not part of our research. In this framework, the Change model represents a landscape design solution, designated as Scenario B, while Scenario A mirrors the current condition. The Impact model will delineate the disparities in delivering selected ecosystem services between scenarios A and B. Ultimately, insights from Decision models will furnish comprehensive information, aiding in identifying key stakeholders and streamlining the implementation process of green infrastructure at lower spatial scales through design solutions.

Regarding the selected ecosystem services that will be a criterion subject to applying the modelling method through the impact model phases of Geodesign, it is necessary to define methodological procedures and the essential databases

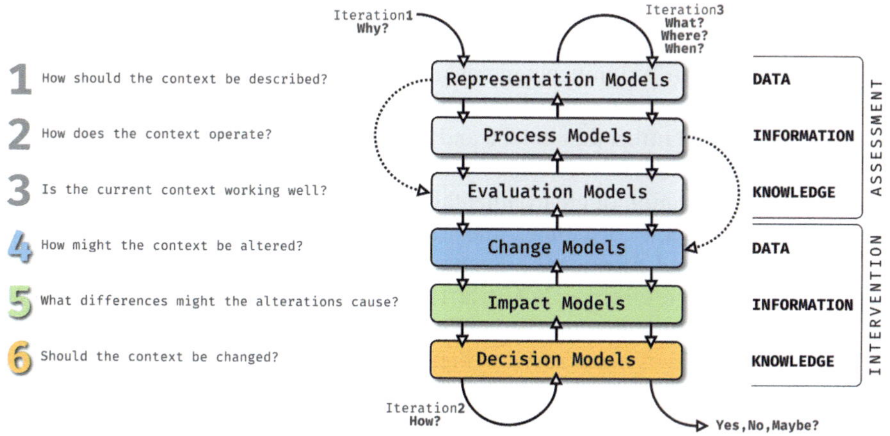

Fig. 5.1 Selected phases of the Geodesign framework (modified Steinitz 2012)

Table 5.1 Ecosystem services and elements of their operational procedures related to impact models

Criteria	Indicator	Method	Data set
Regulating services	Soil erosion control	Revised universal soil loss equation (RUSLE)	Land cover (landscape level), rainfall, soil erodibility, slope and length, conservation measures
	Surface runoff regulation	Flash flood potential index (FFPI)	Land cover (landscape level), slope, soil structure and texture, rainfall
Supporting services	Connectivity of natural and semi-natural elements	Graph theory	Land cover (class level of natural and natural and semi-natural elements)
Cultural services	Recreation opportunities	Distance analysis	Land cover (class level of recreational elements)

(Table 5.1). Of course, as certain aspects of ecosystem services are related to natural processes that depend on climate change, both scenarios related to changes in climate parameters (RCP4.5 and RCP8.5) were used.

5.2.2 *Spatial Domain of Geodesign Models*

The specific domain of application for the Geodesign framework encompasses the territory on the outskirts of Belgrade, characterised as a peripheral zone within the General Urban Plan (OGCB 2016). In this area, spatially independent neighbourhoods dominate, with sparse agricultural and natural elements. The zone is recognised as an area where a significant increase in the number of inhabitants is expected. Concerning capacity, this zone has a minimal number of developed recreational functions. The inefficient utilisation of land and inadequate transportation and communal infrastructure characterise the fundamental attribute of the peripheral zone. This results in the degradation of environmental elements.

On the other hand, PGR defines this area as a "green infrastructure outer ring", which has spatial and functional potential that derives from a high percentage of surfaces with sponge effects (OGCB 2019). The precise research area is defined by the complex of urban watersheds (Fig. 5.2). Since part of the impact model follows changes in ecosystem services that are part of natural processes, the choice of the watershed boundaries is a logical choice because these areas unite all biological, physical, and social phenomena (Steiner 2008). The researched area represents the watershed system made of tributaries of a more significant watercourse, the

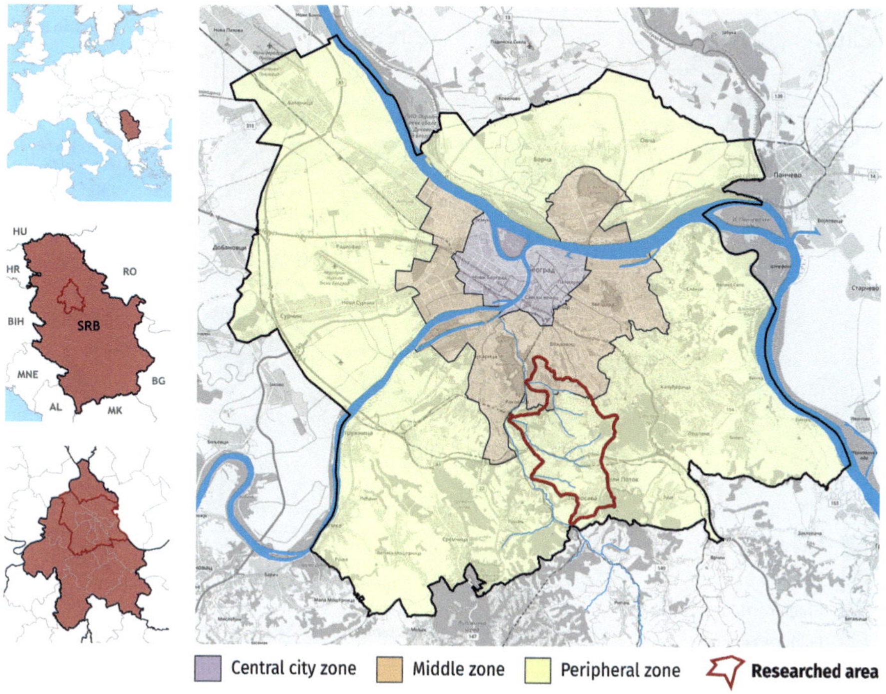

Fig. 5.2 Location of investigated watersheds in the urban context

Topčiderska River, that flows through all the city zones—peripheral, middle, and flows into the Sava River in the central city zone.

5.2.3 Land Cover Data

As previously stated, Scenario A depicts the current state of the land cover (Fig. 5.3 left). This scenario draws upon data from the Urban Atlas database for 2018 (EEA 2002), an integral part of the European Union's Copernicus Earth Observation satellite program. In this terminology, 27 land cover classes are represented, grouped under artificial surfaces, agricultural areas, semi-natural areas, wetlands, forests, and water surfaces. Except for artificial surfaces, the minimum mapping unit is 1 hectare. This database allows very detailed spatial analysis since granulation for urbanised elements (artificial surfaces) is 0.25 hectares, and this leading group is stratified into 17 classes. Scenario B presupposes the land cover resulting from the maximal implementation of green infrastructure as outlined in the PGR, supplanting the current land cover (Fig. 5.3 right).

The class "Green corridors and protective belts" was particularly problematic for identification in **Scenario A** because the Urban Atlas nomenclature does not

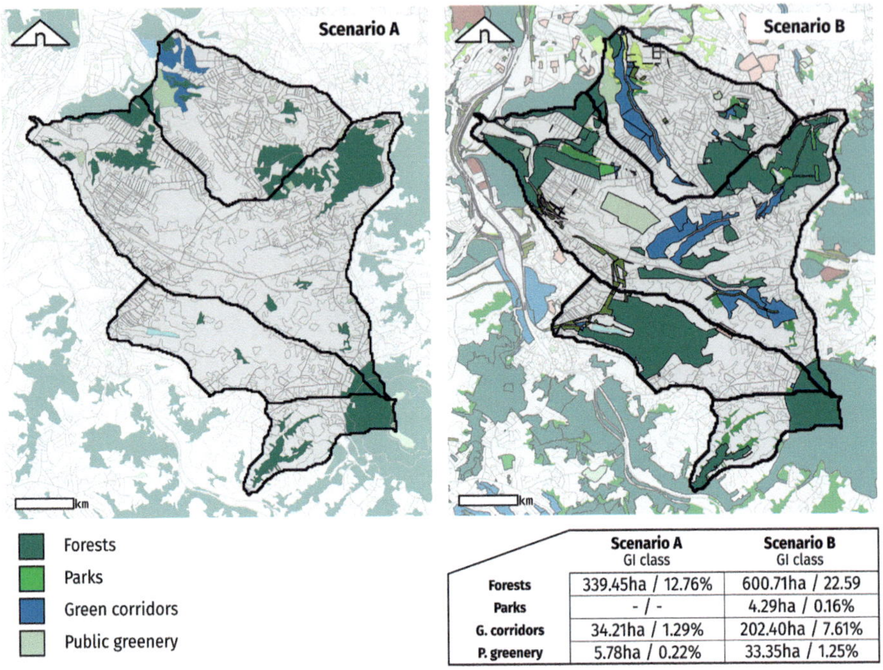

	Scenario A GI class	Scenario B GI class
Forests	339.45ha / 12.76%	600.71ha / 22.59
Parks	- / -	4.29ha / 0.16%
G. corridors	34.21ha / 1.29%	202.40ha / 7.61%
P. greenery	5.78ha / 0.22%	33.35ha / 1.25%

Fig. 5.3 Landscape pattern and green infrastructure elements in Scenarios A and B

recognise it. However, the overlay method determined that the class "Herbaceous vegetation associations" is part of the ruderal vegetation of riparian zones and areas under the direct influence of urbanisation. Hence, it found its analogy in **Scenario B** with the mentioned class (Green corridors and protective belts).

5.2.4 Soil Erosion Control

Soil erosion is a complex natural process and spatio-temporal phenomenon (Fistikoglu 2002; Hoyos 2005; Pandey 2009) which is conditioned by the physical and chemical characteristics of the soil, topography, land cover and human activities. Erosion processes, as one of the factors causing torrential floods, act on 75% of the territory of Serbia (Ristić 2007), with an average annual production of erosive material of 30 million m3, of which about 8 million m3 reaches river and stream beds (cause occurrence of enormous amounts of mud and stones in flooded places and cities) (Ristić et al. 2016). Many models are available today to estimate soil losses and erosion risk levels. One of the most frequently applied empirical models in RUSLE is the Revised Universal Soil Loss Equation (Renard et al. 1997; Panagos et al. 2015), which is used in this work. The concept of the RUSLE formula is based on the same form as the USLE, with several improvements in the determining factors. The model calculates soil loss per unit area by considering various factors, including climatic factors (rainfall—erosivity factor), pedological characteristics (soil erodibility), topographic (slope and length), vegetation and land use factors, and anthropogenic (conservation measures factor). RUSLE is an erosion model for predicting average annual losses from agricultural soils under surface and furrow erosion conditions and specific cultivation and management systems of farming crops. Wide application in these conditions has proven its convenience and value. In addition, it was also applied to other non-agricultural conditions as one of the instruments in the spatial planning process.

5.2.5 *Surface Runoff Regulation*

Flash floods are a hydrological phenomenon, the most frequent and widespread natural disaster, which, in recent decades, represents an ecological and economic challenge throughout the world and causes disastrous consequences for human health and existence. To identify the occurrence of flash floods, author Smith (2003) developed the Flash Flood Potential Index (FFPI) method for determining the territory's predisposition to flash floods. The FFPI was developed to quantitatively describe an area's flash flood risk and susceptibility based on its inherent and static characteristics. According to this, the FFPI can be classified as a new generation "tool" for flash flood risk assessment (Lazarević et al. 2023; Smith 2005; Duong et al. 2020; Costache, 2017). Input parameters in the calculation of this method are terrain slope, land cover, land use method, soil infiltration-retention capacity and rainfall. The slope of the terrain determines the speed and concentration of surface runoff. The physical characteristics of the soil (soil structure and texture) are the properties that determine water retention and infiltration. Vegetation cover greatly influences the runoff process, which depends on the type of vegetation, structure, and age, which manifests itself through interception and infiltration-retention characteristics of the soil. Land use, especially urbanisation, has a significant role in water infiltration, reduction of the porous surface and frequent occurrences of intense surface runoff (after heavy rains, snowmelt, and its coincidence). Rain intensity and spatial distribution of precipitation are key components of flash floods. The FFPI method holds great significance and potential through the application of modern technology and Geographic Information Systems. The process involves developing a raster dataset for each parameter influencing the predisposition to flash flood occurrence.

5.2.6 *Connectivity of Natural and Semi-Natural Elements*

Measuring urban biodiversity is critical in determining the availability and efficacy of ecosystem services. Cities occupy strategically advantageous territories, usually in fertile land areas, river deltas, or forested regions historically rich in natural habitats (Farinha-Marques et al. 2017). In urbanisation, the remaining fragments of natural elements are the remaining exponents of biodiversity, and the need for their conservation and enhancement represents a significant task for urban development. Given the scarcity of naturally succeeding biodiversity carriers in urban areas, assessing biodiversity levels presents a difficult task. In some instances, the measure of spatial connectivity between natural and semi-natural elements is used as one of the indicators of the state and potential of biodiversity (Vasiljevic et al. 2018; Lookingbill et al. 2022). The use of graph theory as a platform to define relationships and connections between spatial entities has become increasingly prevalent in evaluating the connectedness of habitat networks and other aspects of naturalness, such as species flows (Thornhill et al. 2018). The landscape graph-based principle is widely adopted for assessing the degree of spatial connectivity attained (Minor 2008). This approach facilitates the identification of spatial gaps within the ecological continuum. It enables the prioritisation of elements crucial for the flow of material, energy, and organisms—key components essential for biodiversity preservation (Xun et al. 2014). For this research, the Graphab 2.8 application was used (Foltete et al. 2021), in which the local current flow metric was applied to existing green infrastructure elements (Scenario A) and those planned by PGR (Scenario B). The advantage of the graph-based approach lies in using electrical circuit theory, which measures connectivity levels through achieved conductivity levels (Koh et al. 2013). This proves to be useful in assessing the connectivity of green infrastructure elements.

5.2.7 Recreational Opportunities

Assessing the value of cultural ecosystem services encompasses diverse methods and techniques. The main problem in evaluating this group of services is that only some approaches can avoid direct contact with the users (using surveys, interviews, focus groups, etc.) that can give sufficiently good results. On the other hand, such research requires significant time and personnel capacity and is often coloured by mutual non-objectivity. In this research, the ecosystem's cultural service value will be assessed through its potential to provide a recreational function to the urban population. The accessibility of green infrastructure elements is expressed through the degree of physical accessibility at the local, city, and regional levels. As it is the peripheral territory of the city, positioned in the peripheral zone, its green infrastructure elements provide local services. This implies daily recreation, which is why elements of green infrastructure of a public nature must be located at a walking distance of no more than 10–15 min, corresponding to an aerial (actual) distance of about 300–400 m (Koppen et al. 2014). The distance analysis method is a cornerstone in most GIS applications, providing insights into spatial relationships. At its core, distance represents the separation between two locations, typically measured along a straight line, which offers the shortest path between them.

5.3 Results and Discussion

The simulation of climate parameters was conducted based on the reference (baseline) period (1986–2005), using the EURO-CORDEX database (Jacob et al. 2014), for the middle of the twenty-first century (2046–2065). Future periods were analysed according to two selected RCP (Representative Concentration Pathways) emission scenarios for greenhouse gases: RCP4.5 (stabilisation scenario, which predicts emission stabilisation from the year 2040) and RCP8.5 (constant increase scenario) (IPCC 2014). The resulting soil losses and areas exposed to flash floods are presented in a raster database at a resolution of 30 m for both observed scenarios.

Modelling the erosion process using authoritative and detailed databases, such as the applied RUSLE model, enables precise identification of the soil erosion intensity category, and using a climate model adds a dimension of reality to their prediction.

In terms of the spatial distribution of categories of erosion processes, according to the presented scenarios, significant reductions of erosion processes were identified in the categories from excessive to weak erosion in Scenario B, while under the category of very weak erosion, it remained unchanged. The weak erosion category experienced a substantial reduction in surface area in Scenario B (after land cover modification), especially in analyses with the RCP4.5 climate scenario. Comparing the results obtained under the moderate erosion category, significant reductions in surface area were noted in the RCP8.5 climate scenario. Concerning the strong erosion category, a substantial reduction in surface area was observed in the RCP4.5 climate scenario. The excessive erosion category underwent a significant reduction in Scenario B (after land cover modification), where a higher percentage reduction in surface area was noted in the RCP8.5 climate scenario. Figures 5.4 and 5.5 present the surface area differences between both scenarios.

The modification of urban landscapes can lead to the development of erosion processes of different categories, which would become catalysts for the occurrence of flash floods under conditions of changed climatic parameters (Kusky 2010). This work applied the FFPI method to identify the areas exposed to flash floods. The results were obtained in raster format with a resolution of 30 m, using GIS analysis, and classified into three categories (low risk for flash flood, medium risk for flash flood, and high risk for flash flood). A comparison was made between the two scenarios based on the results obtained. According to the presented scenarios, it can be stated that there has been a significant reduction in the areas identified for the occurrence of flash floods according to Scenario B (in the scenario where land cover modification has occurred). As per the outlined scenario, there has been a notable decrease in the areas categorised as very high-risk and medium-risk for flash floods. The percentage changes in risk categories obtained by the FFPI method for both climatic scenarios are shown in Figs. 5.6 and 5.7. A significant reduction in the surface area under the high-risk category of flash flood occurrence was noted under the climate scenario RCP4.5. Looking at the category representing a moderate risk of flash flood occurrence, the percentage decrease in surface area is similar in both observed climate scenarios. A significant increase in the surface area under the low-risk category was observed in

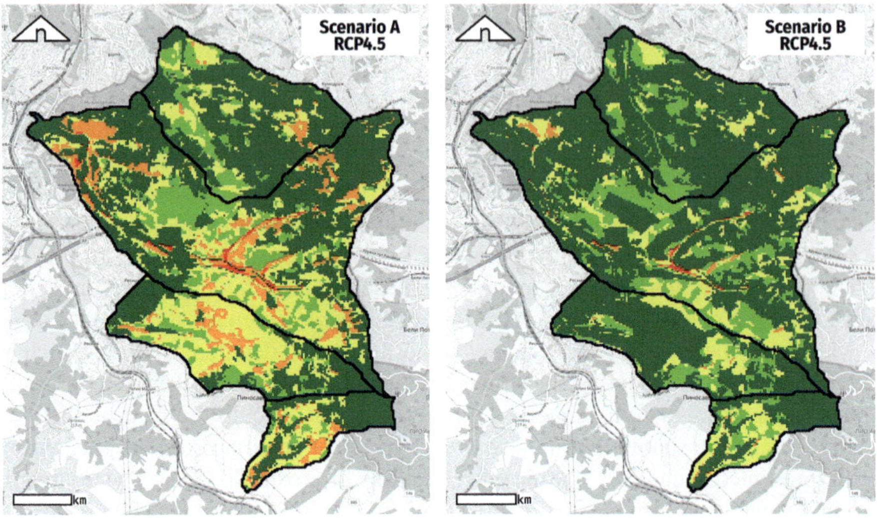

	Scenario A % of erosion category	Scenario B % of erosion category
Excessive	0.71	0.27
Strong	9.04	1.12
Medium	13.99	13.74
Weak	25.66	16.45
Very weak	50.60	68.41

Fig. 5.4 Soil erosion control as indicator in both scenarios due to RCP4.5

the climate scenario RCP4.5 compared to the climate scenario RCP8.5. Considering these changes in both climate scenarios, RCP4.5 represents a stabilisation scenario, which predicts a stabilisation of greenhouse gas emissions from 2040. According to anomalies in mean annual precipitation, a decrease of 5% in precipitation is expected for the middle of the twenty-first century in the researched area. Concerning the climate scenario RCP8.5 for the middle of the twenty-first century, precipitation increased between 5 and 10% compared to the simulated reference (baseline) period.

A significantly higher percentage of areas was recorded under the low risk of flash floods. According to the presented results, it can be concluded that forest vegetation and porous surfaces significantly reduce the amount and speed of surface water and the intensity of erosion processes, helping soil development and infiltration-retention capacity (Ristić et al. 2011). Reducing deforestation and forest destruction and stopping the conversion of agricultural areas into urban areas in the basin would lead to a favourable hydrological regime (Carlson 2000; Arthur-Hartranft et al. 2003) and a reduction in the conditions for the occurrence of rapid surface runoff and the occurrence of flash floods. (Corbet et al. 1997; Murdock et al. 2004; Ferreira et al. 2012).

Graph-based modelling of scenarios reveals an enhancement in landscape connectivity in Scenario B due to the intricate spatial relationships between natural and close-to-natural elements, exemplified by green infrastructure. (Fig. 5.8). According

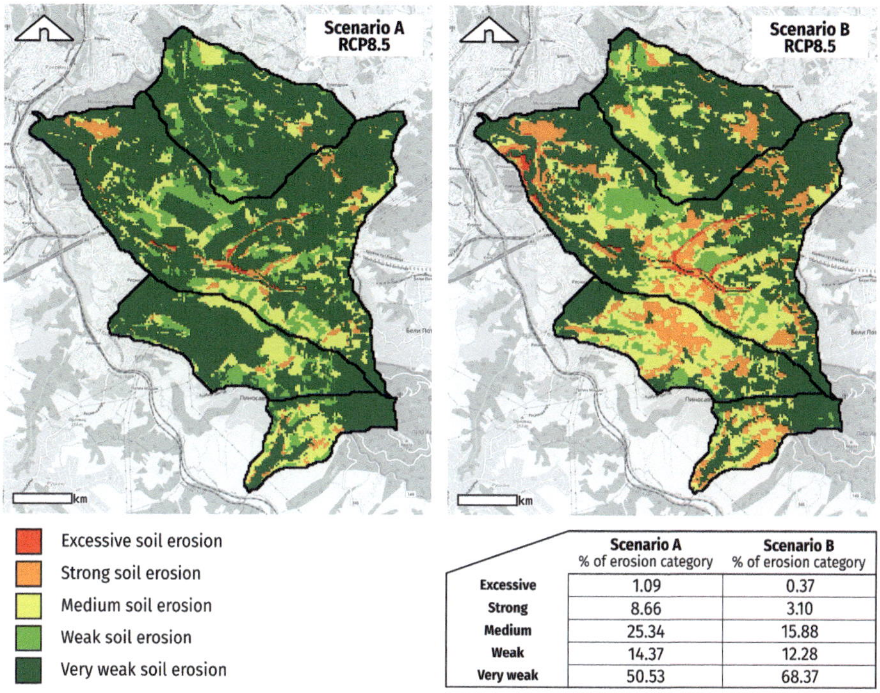

	Scenario A % of erosion category	Scenario B % of erosion category
Excessive	1.09	0.37
Strong	8.66	3.10
Medium	25.34	15.88
Weak	14.37	12.28
Very weak	50.53	68.37

Fig. 5.5 Soil erosion control as indicator in both scenarios due to RCP8.5

to the approach that is verified in other urban landscapes in Serbia (Bajić et al. 2022), the level of importance of separate elements of green infrastructure in a landscape urban mosaic can be classified as hubs, nodes, and gateways. The focal points of ecosystem services, such as green infrastructure hubs, are the most important components of green infrastructure. Green infrastructure nodes play a significant role in the hub-to-gateway connection matrix. These components display a high degree of naturalness and unquestionably enable a certain level of ecological services. Gateways of green infrastructure are singular pockets of nature in the city. They can be crucial for providing services on a local level. A particular aspect of connectivity improvements is the development of an "ecological spine" through the geometrical centroid of watersheds that correlates with the hydrographic network; these zones are most vulnerable to the genesis of flash floods and the deposition of pollutants.

The application of distance analysis methods to the data of both land covers that belong to the public land use of green infrastructure showed significant changes in accessibility for users (Fig. 5.9). The substantial shift in accessibility levels in Scenario B can be attributed to large-scale interventions such as afforestation of terrain prone to soil erosion, establishment of substantial protection belts around populated enclaves, and the creation of a new park area.

Decision models allow the prioritisation of proposed green infrastructure elements based on categorical values of ecosystem services registered during impact models.

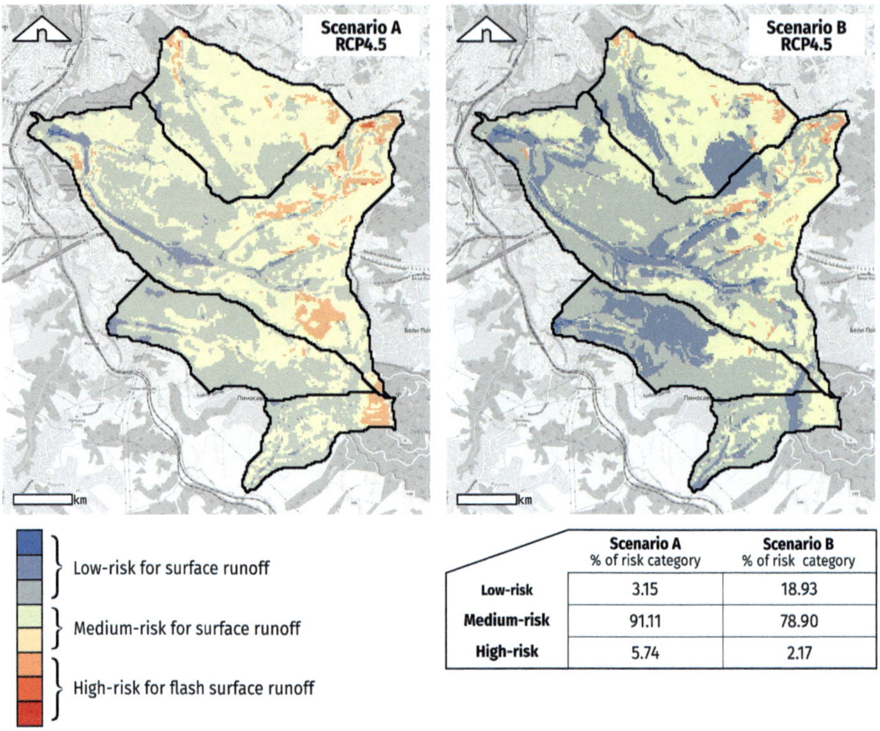

	Scenario A % of risk category	Scenario B % of risk category
Low-risk	3.15	18.93
Medium-risk	91.11	78.90
High-risk	5.74	2.17

Fig. 5.6 Surface runoff regulation as indicator in both scenarios due to RCP4.5

This final stage of the Geodesign framework provides rational observations and makes choices of the best paths for landscape change that will highlight levels of functional importance of green infrastructure. Suppose the time frame of the European Union until 2050 should improve the continent's resilience to the effects of environmental and climatic changes. In that case, applying the Decision model allows for the phased prioritisation of green infrastructure to be carried out in the following period (Fig. 5.10). This phase of Geodesign enables the spatial and functional classification of ecosystem services of existing and planned elements of green infrastructure to form new elements at reasonable intervals up to 2050 and more efficiently maintain the existing ones.

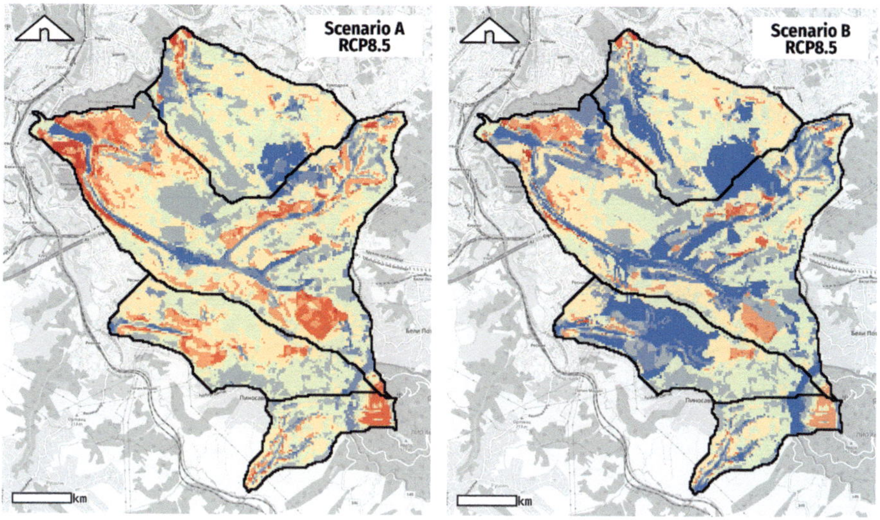

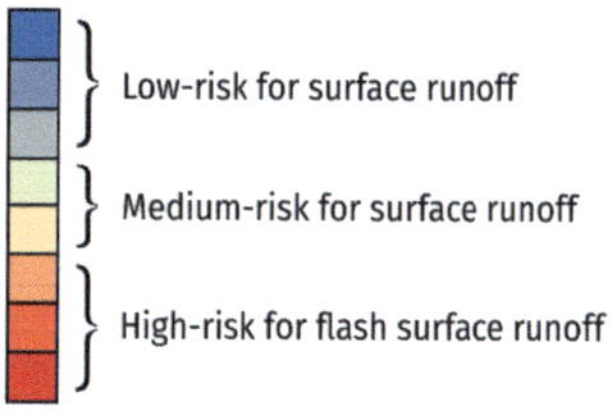

	Scenario A % of risk category	Scenario B % of risk category
Low-risk	25.57	42.77
Medium-risk	57.48	49.83
High-risk	15.95	7.41

Fig. 5.7 Surface runoff regulation as indicator in both scenarios due to RCP8.5

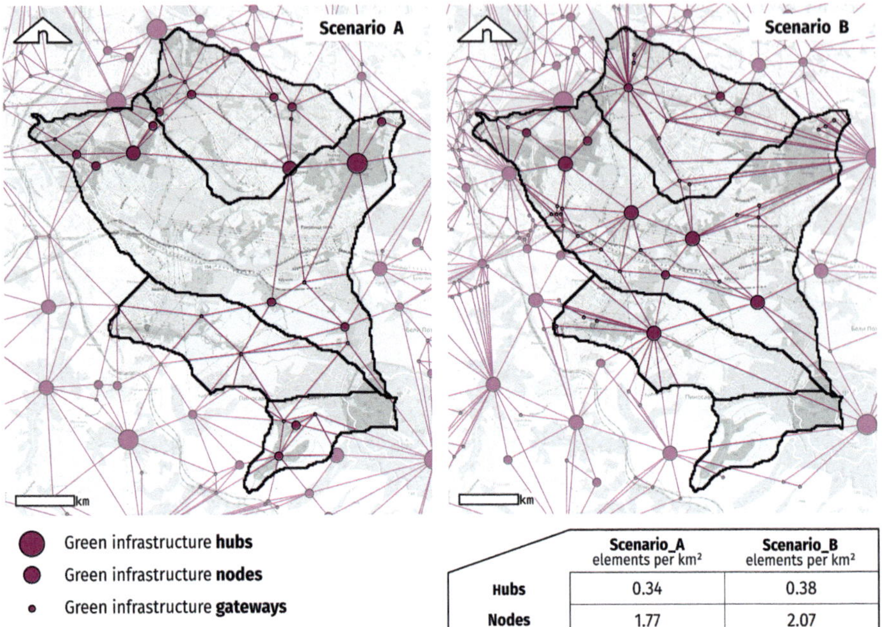

	Scenario_A elements per km²	Scenario_B elements per km²
Hubs	0.34	0.38
Nodes	1.77	2.07
Gateways	5.94	11.36

Fig. 5.8 Connectivity of natural and semi-natural elements as indicator in both scenarios

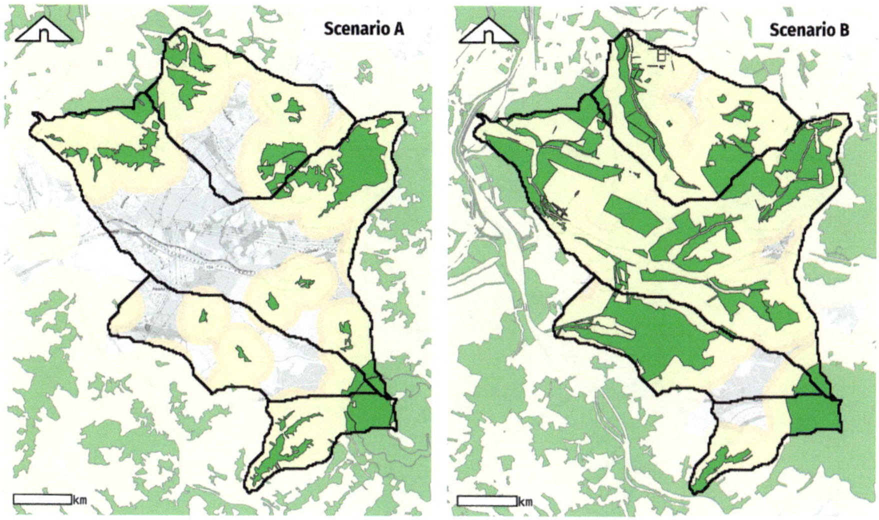

	Scenario A % of accessible area	Scenario B % of accessible area
High	56.31	89.13
Medium	12.52	6.11
Low (no)	31.17	4.76

Fig. 5.9 Recreation opportunities as indicator in both scenarios

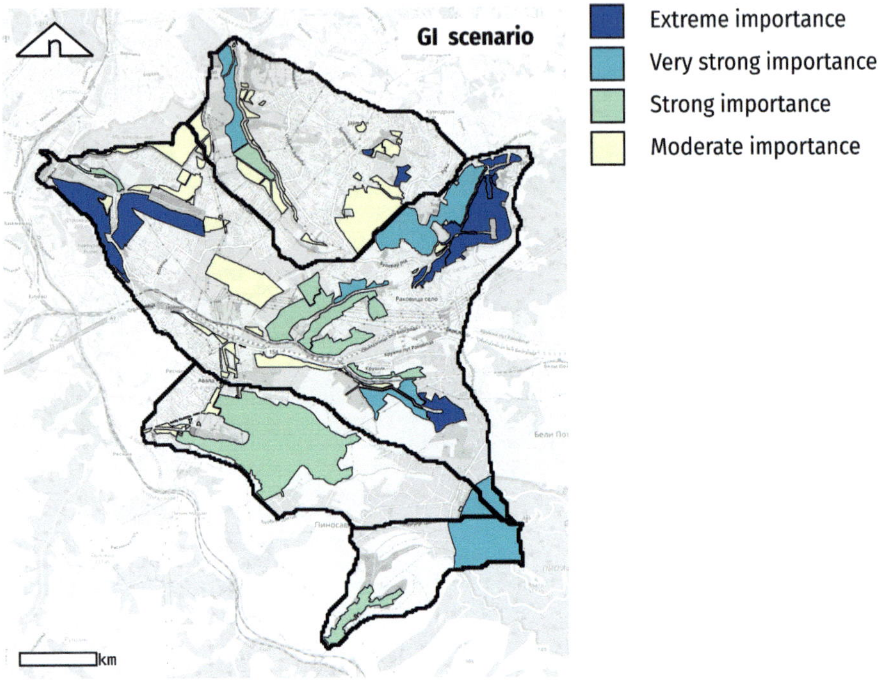

Fig. 5.10 Prioritization of green infrastructure elements due to their spatial and functional role in ecosystem services providing

5.4 Conclusion

Contemporary cities generate urban landscapes which are carriers of complex natural and artificial processes—the takeover of "unused" areas materialises requirements for the fulfilment of essential living functions, resulting in ecological conflicts that reduce safety and quality of life. Given the projections of population growth and climate change, urban areas are expected to become even more under pressure, making cities an open-air laboratory that enables the testing of sustainable solutions. As Serbia's capital and largest city, Belgrade is inherently influenced by these dynamics and future forecasts. Consequently, there has been a sustained devotion to formulating an urban plan and legislative framework to guide the city's growth that promotes pragmatic expansion while safeguarding the environment, preserving natural spaces, and protecting ecologically sensitive areas.

Concerning recognised environmental and societal threats, the European Commission has defined ambitious goals that point to the year 2050 as a time tag for forming a climate-neutral continent. In this context, numerous strategic documents have recognised the importance of green infrastructure as a connected network of natural and semi-natural elements of the city that provides the realisation of ecosystem services and enables the progressive functioning of ecological systems while providing various benefits to city residents. In this sense, the City of Belgrade has adopted the Plan for the General Regulation of the System of Green Areas, in which existing areas are prioritised. In contrast, the planning and creation of new areas based on the principles of connectivity, multi-functionality, accessibility, and landscape character improvement are positioned in the focal plane.

Geodesign is a relatively new approach that significantly contributes to synthesising the Geographic Information System as a comprehensive analytical tool and landscape design as a creative process. This approach also emphasises the role of decision-makers and other stakeholders in the final phase of the planning and design process so that it articulately brings them different development scenarios. Given that the General Regulation Plan was adopted relatively recently (2019), applying the Geodesign framework can enable its further implementation process and realising goals at local spatial levels.

Change models are usually referred to as the core of a Geodesign study. The goal of the data used in this phase is to accurately present possible future scenarios. In our research, the Change model is defined by two scenarios: the first is a potential future of land cover without implementing the vision of the Plan of General Regulation of the Green Area System of Belgrade. At the same time, the second is the maximal realisation of spatial guidelines through elements of green infrastructure provided by the Plan.

The impact models "breathe life" into change models and assign measurable process quality parameters that result from different scenarios. In our research, the impact models materialise functional aspects of green infrastructure elements categorised through ecosystem service achievement. Selected criteria are recognised as the functional domain of ecosystem service realisation and are estimated to be relevant for the context of the outskirts of Belgrade. Of course, the availability of databases required to implement modelling methods and techniques affects the choice of indicators. The impact model is used to assess the achievement of the ecosystem regulatory service by analysing the level of soil erosion control and surface runoff regulation using the Revised universal soil loss equation and Flash flood potential index. Soil erosion and the risk of flash floods in the landscape of urban outskirts are recognised as the main threats to urban safety.

On the other hand, supporting ecosystem services provided by biodiversity has the potential to humanise these parts of the city landscape, which are usually burdened with low environmental quality. The level of biodiversity is estimated based on the particular principle of landscape ecology related to the importance of spatial connectivity between natural and semi-natural elements. A straightforward method for assessing the level of connectivity between habitats is based on the principles of graph theory. Another critical aspect of the quality of life on the city's edge is the degree of development of the ecosystem's cultural services, which is evaluated through the prism of spatial possibilities for the realisation of recreational functions and their availability.

The Decision models are an essential phase of the Geodesign process, primarily for city authorities but also for other relevant stakeholders. With much more understanding, they can accept the importance of and support creating new links in the complex interaction between green infrastructure and ecosystem services through strategies and plans at lower design levels. The obtained results indicate that the implementation of green infrastructure principles through the legal and operational framework of the General Regulation Plan will improve the future of Belgrade's urban landscapes and provide meaningful efforts to create an ecologically resilient system that has the potential to meet the challenges of global and climate change.

References

Arthur-Hartranft ST, Carlson TN, Clarke KC (2003) Satellite and ground-based microclimate and hydrologic analyses coupled with a regional urban growth model. Rem Sens Environ 86:385–400

Asikainen E, Jokinen A (2009) Future natures in the making: Implementing biodiversity in suburban land-use planning. Plan Theory Pract 10(3):351–368

Bajić L, Vasiljević N, Čavlović D, Radić B, Gavrilović S (2022) A green infrastructure planning approach: improving territorial cohesion through urban-rural landscape in Vojvodina. Serbia Land 11(9):1550

Breuste J, Haase D, Elmqvist T (2013) Urban landscapes and ecosystem services. Ecosyst Serv Agric Urban Landscapes, 83–104

Calderón-Contreras R, Quiroz-Rosas LE (2017) Analysing scale, quality and diversity of green infrastructure and the provision of urban ecosystem services: a case from Mexico City. Ecosyst Serv 23:127–137

Carlson TN, Arthur ST (2000) The impact of land use—land cover changes due to urbanisation on surface microclimate and hydrology: a satellite perspective. Glob Planet Change 25:49–65

Corbet CW, Wahl M, Porter DE, Edwards D, Moise C (1997) Nonpoint source runoff modeling: a comparison of a forested watershed and an urban watershed on the South Carolina coast. J J Exp Marine Biol Ecol 213:133–149

Cornelis J, Hermy M (2004) Biodiversity relationships in urban and suburban parks in Flanders. Landsc Urban Plan 69(4):385–401

Costache R (2017) Assessment of building infrastructure vulnerability to flash floods in Panatau River Basin, Romania. Analele Univ Din Oradea Ser Geogr 27:26–36

Cvejić J, Teofilović A (2010) Concept of the green spaces system—Belgrade case study. Proceed Fábos Conf Landscape Greenway Plan 3(1):64

De Groot RS, Wilson MA, Boumans RM (2002) A typology for the classification, description and valuation of ecosystem functions, goods and services. Ecol Econ 41(3):393–408

De Groot RS, Alkemade R, Braat L, Hein L, Willemen L (2010) Challenges in integrating the concept of ecosystem services and values in landscape planning, management and decision making. Ecol Complex 7(3):260–272

Duong Thi L, Van Do T, Le Van H (2020) Detection of flash-flood potential areas using watershed characteristics: application to Cau River watershed in Vietnam. J Earth Syst Sci 129:120

EEA—European Environmental Agency (2002) Towards an urban atlas: assessment of spatial data on 25 European cities and urban areas, environmental issue report 30; European Environmental Agency: Copenhagen, Denmark

Farinha-Marques P, Fernandes C, Guilherme F, Lameiras JM, Alves P, Bunce RG (2017) Urban habitats biodiversity assessment (UrHBA): a standardized procedure for recording biodiversity and its spatial distribution in urban environments. Landscape Ecol 32:1753–1770

Ferreira CSS, Ferreira AJD, Pato RL, Magalhães CM, Coelho CO, Santos C (2012) Rainfall-runoff-erosion relationships study for different land uses, in a sub-urban area. Zeitschrift Für Geomorphol 56(3):005–020. https://doi.org/10.1127/0372-8854/2012/S-00101

Fistikoglu O, Harmancioglu NB (2002) Integration of GIS with USLE in assessment of soil erosion. Water Resour Manage 16:447–467
Foltete JC, Vuidel G, Savary P, Clauzel C, Sahraoui Y, Girardet X, Bourgeois M (2021) Graphab: an application for modeling and managing ecological habitat networks. Softw Impacts 8:100065
Gkiatas G, Kasapidis I, Koutalakis P, Iakovoglou V, Savvopoulou A, Germantzidis I, Zaimes GN (2021) Enhancing urban and sub-urban riparian areas through ecosystem services and ecotourism activities. Water Supply 21(6):2974–2988
Gu D, Andreev K, Dupre ME (2021) Major trends in population growth around the world. China CDC Weekly 3(28):604
Hoyos N (2005) Spatial modeling of soil erosion potential in a tropical watershed of the Colombian Andes. CATENA 63(1):85–108
IPCC (Intergovernmental Panel on Climate Change) (2014) Climate change 2014: synthesis report. Contribution of working groups I, II and III to the Fifth assessment report of the intergovernmental panel on climate change [Core Writing Team, R.K. Pachauri and L.A. Meyer (Eds.)]. IPCC, Geneva, Switzerland, p 151
Jacob D, Petersen J, Eggert B et al (2014) EURO-CORDEX: new high-resolution climate change projections for European impact research. Reg Environ Change 14:563–578. https://doi.org/10.1007/s10113-013-0499-2
Koh I, Rowe HI, Holland JD (2013) Graph and circuit theory connectivity models of conservation biological control agents. Ecol Appl 23(7):1554–1573
Koppen G, Ode Sang Å, Tveit MS (2014) Managing the potential for outdoor recreation: adequate mapping and measuring of accessibility to urban recreational landscapes. Urban Forestry Urban Greening 13(1):71–83
Kusky MT (2010) Urbanization and flash flooding. Encyclopedia of earth and space science (pg. 282). Facts on File Inc., Infobase Publishing, New York, p 916
Lafortezza R, Davies C, Sanesi G, Konijnendijk CC (2013) Green Infrastructure as a tool to support spatial planning in European urban regions. iForest-Biogeosci Forestry 6(3):102
Lazarević K, Todosijević M, Vulević T, Polovina S, Momirović N, Caković M (2023) Determination of flash flood hazard areas in the Likodra watershed. Water 15:2698. https://doi.org/10.3390/w15152698
Lookingbill TR, Minor ES, Mullis CS, Nunez-Mir GC, Johnson P (2022) Connectivity in the urban landscape (2015–2020): who? Where? What? When? Why? And how? Curr Landscape Ecol Rep 7(1):1–14
Macura V, Stojkov B, Cvejić J, Đorđević-Ciganović J, Mihaljević V, Ferenčak M, Kurtović-Folić N, Stojanović B, Vuksanović-Macura Z, Milanović D, Milašinović-Marić D (2019) Urban innovations from the 2003 master plan: preserving Belgrade's identity. Archit Urbanism Printed Serbian Lang 49:40–59
McPhearson T, Hamstead ZA, Kremer P (2014) Urban ecosystem services for resilience planning and management in New York City. Ambio 43:502–515
Mell I C (2015) Green infrastructure planning: policy and objectives. Handbook on green infrastructure: planning, design and implementation, 105–123
Minor ES, Urban DL (2008) A graph-theory framework for evaluating landscape connectivity and conservation planning. Conserv Biol 22(2):297–307
Mitchell MGE, Suarez-Castro AF, Martinez-Harms M, Maron M, McAlpine C, Gaston KJ, Johansen K, Rhodes JR (2015) Reframing landscape fragmentation's effects on ecosystem services. Trends Ecol Evol 30(4):190–198
Monteiro R, Ferreira JC, Antunes P (2020) Green infrastructure planning principles: an integrated literature review. Land 9(12):525
Murdock J, Roelke D, Gelwcik F (2004) Interactions between flow, periphyton, and nutrients in a heavily impacted urban stream: Implications for stream restoration effectiveness. Ecol Eng 22:197–207
OGCB—Official gazette of the city of Belgrade (2016) Master Plan of Belgrade. LX (11)

OGCB—Official gazette of the city of Belgrade (2019) Plan of general regulation of the green area system of Belgrade. LXIII (110)
Panagos P, Borrelli P, Poesen J, Ballabio C, Lugato E, Meusburger K, Montanarella L, Alewell C (2015) The new assessment of soil loss by water erosion in Europe. Environ Sci Policy 54:438–447. https://doi.org/10.1016/j.envsci.2015.08.012
Pandey A, Mathur A, Mishra SK, Mal BC (2009) Soil erosion modeling of a Himalayan watershed using RS and GIS. Environ Earth Sci 59(2):399–410
Pietilä M, Neuvonen M, Borodulin K, Korpela K, Sievänen T, Tyrväinen L (2015) Relationships between exposure to urban green spaces, physical activity and self-rated health. J Outdoor Recreat Tour 10:44–54
Pyky R, Neuvonen M, Kangas K, Ojala A, Lanki T, Borodulin K, Tyrväinen L (2019) Individual and environmental factors associated with green exercise in urban and suburban areas. Health Place 55:20–28
Radić B, Gavrilović S (2021) Natural habitat loss: causes and implications of structural and functional changes. In: Leal Filho W, Azul A M, Brandli L, Lange Salvia A, Wall T. (eds) Life on Land. Encyclopedia of the UN Sustainable Development Goals. Springer, Cham. https://doi.org/10.1007/978-3-319-95981-8_6
Radić B, Gavrilović S, Stevanović I, Polovina S, Vukmirović M (2023) Leave no land scape left behind: creating and connecting the landscape values of urban outskirts. ECLAS 2023, 10–13. September 2023, Brno—Lednice—Czech Republic
Renard KG, Foster GR, Weesies GA, McCool DK, Yoder DC (1997) Predicting soil erosion by water: a guide to conservation planning with the revised universal soil loss equation (RUSLE). Agriculture Handbook, US Department of Agriculture, Washington, DC, Vol 703. pp 1–251
Ristić R, Radić B, Vasiljević N, Nikić Z (2011) Land use change for flood protection—a prospective study for the restoration of the river Jelašnica watershed. Bull Fac Forest 103:115–130
Ristić R, Kostadinov S, Abolmasov B, Dragićević S, Trivan G, Radić B, Trifunović M, Radosavljević Z (2012) Torrential floods and town and country planning in Serbia. Nat Hazard 12(1):23–35
Ristić R, Nikić Z (2007) Sustainability of the water supply systems in Serbia from the point of view of land erosion hazards, Vodoprivreda No. 225–227(ISSN: 0350–0519), p 47–57, Belgrade. (In Serbian)
Ristić R, Milčanović V, Malušević I, Polovina S (2016) Torrential floods ad erosion as dominant factors of land degradation in serbia—concept of prevention and protection, degradation and soil protection, thematic collection, ISBN 978-86-7299-242-7 (In Serbian)
Smith G (2003) Flash flood potential: determining the hydrologic response of FFMP basins to heavy rain by analyzing their physiographic characteristics. 2003. Available online: https://www.cbrfc.noaa.gov/papers/ffp_wpap.pdf, last accessed 2023/10
Smith G (2005) Developing a relative flash flood potential index to assist in the flash flood warning decision making process. In: Proceedings of the NWS workshop on hydrologic forecasting, Prague, Czech Republic, pp 20–24
Steiner FR (2008) The living landscape. Island press (496)
Steinitz C (2012) A framework for geodesign: changing geography by design. Esri Press, Redlands, CA
Steinitz C (2014) The beginnings of geographical information systems: a personal historical perspective. Plan Perspect 29(2):239–254
Thornhill I, Batty L, Hewitt M, Friberg NR, Ledger ME (2018) The application of graph theory and percolation analysis for assessing change in the spatial configuration of pond networks. Urban Ecosyst 21:213–225
Tobias S, Conen F, Duss A, Wenzel LM, Buser C, Alewell C (2018) Soil sealing and unsealing: State of the art and examples. Land Degrad Dev 29(6):2015–2024
Vasiljević N, Radić B, Gavrilović S, Šljukić B, Medarević M, Ristić R (2018) The concept of green infrastructure and urban landscape planning: a challenge for urban forestry planning in Belgrade Serbia. Iforest 11:491–498

Vasiljević N, Radić B, Matić A, Medojević E, Gavrilović S, Tutundžić A, Krč M, Ćorović D, Galečić N, Mitrović S, Pešikan D, Mićić S, Elčić I (2021) Atlas of landscape character types of Belgrade. City of Belgrade—city administration, secretariat for environmental protection, Belgrade (ISBN 978–86–7299–334–9 COBISS.SR-ID—50710793)

Vuksanović-Macura Z (2014) Belgrade's general plan 1923: a comparison of planned to completed. University of Belgrade Faculty of Architecture (Doctoral Dissertation), Belgrade

Wang Y, Bakker F, De Groot R, Wörtche H (2014) Effect of ecosystem services provided by urban green infrastructure on indoor environment: a literature review. Build Environ 77:88–100

Xun B, Yu D, Liu Y (2014) Habitat connectivity analysis for conservation implications in an urban area. Acta Ecol Sin 34(1):44–52

Zhang XQ (2016) The trends, promises and challenges of urbanisation in the world. Habitat Int 54:241–252

Chapter 6
Revitalising the Bank of the Danube-Tisa-Danube (DTD) Canal: A Participatory Design Approach Rooted in Local Identity for Urban Transformation

Melisa Alcan Musović, Lejla Zećirović, Enis Hasanbegović, Danilo Dragović, and Branko Slavković

Abstract The waterfront, an integral concept in the realm of urban planning and design, stands as a linchpin for the development of dynamic urban structures and diverse city forms. In accordance with the unique characteristics of the surrounding environment, this research seeks to identify and investigate the elements impacting the development of urban space along rivers and their activation. This study analyses what content and activities residents lack in their daily lives based on a survey that included people of various categories (ages, professions, genders, etc.). Potential locations for the creation of active urban space were also mapped in accordance with the data that was acquired. Therefore, it is necessary to consider people's demands in order to maintain an active and high-quality urban public space. As previously mentioned, people's needs vary based on the context and habits they have in a certain area. The setting up of guidelines for promoting and utilising the riverfront's urban area as well as for activating it reflects this paper's contribution. This will highlight the significance of the waterfront as a bustling urban public area that promotes human interaction and the development of an authentic visual identity.

Keywords River bank · Urban transformation · Participation · DTD canal

6.1 Introduction

As urban landscapes continue to evolve, the need for inclusive and sustainable urban transformations becomes paramount. Urban design of public spaces implies a process of action, not just a response to problems—it provides a framework for activities.

M. Alcan Musović (✉) · L. Zećirović · E. Hasanbegović · D. Dragović · B. Slavković
State University of Novi Pazar, Novi Pazar, Serbia
e-mail: malcan@np.ac.rs

M. Vukmirović et al. (eds.), *Digital Dreams, Urban Realities*, The Urban Book Series,
https://doi.org/10.1007/978-3-031-76353-3_6

Participative implementation can contribute to the development of the knowledge and abilities of actors by encouraging joint reasoning, organising around common problems, and formulating ideas on how to improve the area (Čolić and Čolić 2018). By prioritising the unique cultural, historical, and social attributes of the community, the participatory design process aims to create spaces that resonate with the lived experiences and aspirations of the people who inhabit them., while the choice of approach in planning and space organisation and content provides integration of city and water. This paper aims to research and define the factors that influence the activation of urban space along the river and its development, depending on the specificity of the local environment. Based on a survey in which residents of different profiles (ages, professions, genders…) who live in this area were included, this paper analyses what contents and activities they lack in everyday life. This imposes the hypothesis that to ensure an active and quality urban public space, the needs of people should be considered, depending on the local environment and habits they own. The contribution of this paper is reflected in the creation of guidelines for activating the urban space on the riverfront. Ultimately, this study contributes to the discourse on urban design methodologies by advocating for a participatory approach that places local identity at its core. It underscores the importance of recognising the unique attributes of each community, fostering a more inclusive and sustainable urban future that reflects the collective vision of its residents.

Transformability is the capacity to develop a fundamentally new system when ecological, economic, political, or social conditions make the existing system untenable. The capability can also be considered a strategy to react to changing urban paradigms concerning the concept, which is inherently rooted in flexibility (Jun 2023). Revitalisation is a key driver of urban transformation, influencing various aspects of a city's development. Economically, it attracts investment, businesses, and tourists, fostering increased economic activity and job creation. Community engagement is a crucial element, as involvement in revitalisation projects instils a sense of ownership and pride among residents. The creation of safer, cleaner, and aesthetically pleasing urban environments contributes to an enhanced quality of life for residents.

The needs of contemporary society have led to significant changes in the use and design of city riverbanks in the modern era. An increasing number of new opportunities are being created for the use and structure of coastal space with the goal of enhancing the overall quality of the city.

The redefining of the general significance of city riverbanks in the context of these spaces' potential for the socio-cultural, economic, and ecological revitalization of cities is becoming a growing trend in a number of studies and modern strategic plans for the development of cities worldwide.

6.1.1 Methodology

There are a few basic methods for research that can be applied in participatory processes to gather information from the public and give specialists sufficient details. The most popular method for gathering data on people's overall sentiments is through survey questionnaires. A survey is popular because it is a reasonably cheap way to get information from a lot of respondents. To ensure maximum validity of the data gathered, it is imperative to integrate these procedures with those more appropriate for professionals, such as architects and urban planners, to enhance overall process effectiveness. The methodology employed in this study consists of a dual-pronged approach, wherein surveys serve as the principal instrument for garnering insights from a demographically diverse populace. The survey instrument, carefully constructed to capture different perspectives, aims to discern community needs. In conjunction with this, a comprehensive examination of the current state of the location is conducted, entailing a meticulous analysis of the existing urban area. This multi-faceted analysis extends to the mapping of locales exhibiting untapped potential for achieving the ambitious goals of coastline revitalisation.

6.1.2 Problem Statement

One of the most notable issues is that Novi Sad has failed to successfully disperse its population from the city centre to the periphery despite the abundance of potential and opportunities that Canal DTD has. The chosen area for research and elaboration is close to the outskirts of the city, but it is also close to the centre, offering the opportunity to swiftly deal with the congestion issue in the city's central urban core.

6.1.3 Overall Goal of the Paper

Determining the significance of rivers as vital natural features in urban environments and researching the variables that shaped the establishment of riverine cities and how they can evolve in dependence on local conditions are two of the research's objectives. In addition, it is important to come to a conclusion about the activities that are carried out on the coast and whether this affects the change in the attitude of the population towards the same, as well as what are the methodologies for the improvement and development of the coastal area. The DTD canal's banks appear to have a lot of potential, but for unjustified reasons, Novi Sad's residents continually disregard them. The objective was to investigate the possibility of altering the coastline and connecting it with Klisa's settlement because of its great position. Is it possible to create an appropriate setting for an intervention that would encourage the development and revitalisation of this area? The discussion will conclude with a practical plan that will be put forth in the form of a conceptual solution for a number of urban areas that will effectively improve and transform the Danube-Tisa-Danube canal banks while also connecting the Klisa with it.

6.1.4 Specific Objectives for the Participatory Design Approach

Current methods of planning the way to use the coast point to the necessity of creating a strategic plan that will take into consideration the primary elements of the urban space—that is, its residents and their needs—during the entire process of choosing the activities and content that will be implemented in the coastal zone. Along with the local populace, collaboration with experts is crucial because they will oversee the entire revitalisation process of the coast, guided by the findings of their analysis of the present situation, their survey, and the perceived benefits and drawbacks of the site. It is assumed that in order to transform and revitalise the topic of the study area, three expected objectives need to be achieved.

6.2 Affirmation of the Contact Between the City and the Water and the Selection of Content in Accordance with the Information of the Participatory Process

Determining which amenities and activities will be located in the coastal zone according to the degree of integration with water, that is, the intensity of dependence and connection with water, is necessary in order to affirm and improve the quality of contact between the city and the river. This will benefit the city's population on a daily basis. This study combined location analysis and observation techniques in addition to the most commonly utilised methodology for participation processes, which is a survey.

6.3 Increased Public Usage and Access to the River (Development of a Traffic Network from the Centre, Pathways for Bicyclists and Pedestrians, Etc.)

In modern planning, one of the principal goals of coastal regeneration is to prioritise public access and riverbank utilisation. An important lesson from practical instances of revitalisation efforts is the need to achieve a range of uses along the coastline while also making sure that there are strong connections to the inland and, most importantly, to the city core. Given that from the previous steps where the existing situation was analysed, it was observed that the lack of traffic connectivity is one of the main shortcomings that hinder people's access to the coast; it was singled out as a segment of the problem to be solved.

6.4 Tourist Attractiveness of the River Banks

The successful promotion of the coast as a tourist attraction and the genuine urban form, which allows the riverbanks to serve as the city's new representative boundaries, are the foundations for initiatives aimed at making cities more appealing to tourists. The following planning and design initiatives underpin city coastlines' tourist appeal:

- Tourist destinations (parks, marinas, sports fields, museums, sports and recreation facilities, cultural centres, vistas, etc.) should be placed with recreational, cultural, and educational content in the coastal zone.
- Creating a continuous network of public open spaces that guarantees citywide access to the river and the coast's constant utilisation for leisure, culture, and education.

6.5 Danube-Tisa-Danube (DTD) Canal

6.5.1 Brief History of the Bank of the Danube-Tisa-Danube (DTD) Canal

The network of canals that links the rivers Danube and Tisa is known as the Danube-Tisa-Danube hydro system. It symbolises a system for transportation, irrigation, flood control, forestry, tourism, fishing, and internal water drainage. Located in Vojvodina, the DTD hydro system is one of the biggest streams in Europe today. The hydro system DTD is not only an amazing project but also a premier urban, community, and spatial phenomenon (GradskeInfo 2022). New roles and information that were not initially possible to incorporate into the project were added to the system through additional research and development. Open water throughout the settlement directs urban planners and designers to focus on and prioritise water for particularly valuable and appealing areas and contents of the settlement, enhancing the environment's visual appeal. The canal's surroundings were made greener, the network's topography was examined, and suggestions for the programs and contents of certain locations were developed for a variety of uses. Based on the investigation, it was determined that the area along the canal has a lot of potential for use, and this may significantly boost and extend Vojvodina's tourism industry.

6.5.2 *The Current State of the DTD Canal*

This area of the coastline never quite managed to come to life, even though the canal is positioned far away from major roads and the bustle of the city. The canal's surrounds are disorganised and unutilised. The majority of its users are local fishermen who occasionally host small-scale fishing tournaments. The paved dirt roads that the locals have created as walking trails run the entire length of the canal. Four significant traffic routes that connect the DTD canal to Novi Sad's downtown were identified during the analysis of the chosen covering of the canal's banks. Temerin Road and St. Andrew's Road are two significant routes that go through Klisa and link to the e-75 motorway. Furthermore, the Boulevard of Europe and Partizanska, two of the main roads in the area, are connected to the highway. This results in a lot of people commuting in this direction via road transportation, which presents an opportunity for the area to be developed from a number of angles, including an economic and touristic side. Visual representation of the aforementioned roadways is given in Graphic Attachment Fig. 6. 1. Of particular importance is the fact that four bridges, whose purpose is to facilitate vehicular traffic, cross the subjected DTD canal. One obvious disadvantage is the lack of bike and pedestrian routes the whole way along the coast, as well as a big, overgrown green area that detracts from the coastline's overall visual appeal.

The graphic attachment Fig. 6.2 serves as a visual representation of sites earmarked for the integration of novel content and activities to address prevailing community needs. The selection criteria for these sites were underpinned by the absence of extant built structures, the existence of undefined green spaces, and a notable attribute common to all—their direct adjacency to the river, thereby enhancing their intrinsic appeal and accessibility.

An industrial zone was noticed during the analysis of the areas as important for the DTD canal bank. Apart from the industrial zone, the Republic of Serbia derives significant economic resources from the port of Novi Sad, which is located in close proximity. There are also residential areas close to the coastline, most of which are made up of single-family homes (Fig. 6.3). Two notable commercial objects are the Nylon market and the BIG shopping centre. In the graphic attachment, Fig. 6.4, the locations of the aforementioned uses in relation to the DTD canal are illustrated.

6.6 Participatory Design Process

Participatory design is an approach that actively involves different stakeholders in the design process to create environments that are responsive to their diverse needs. In the context of urban design, participation is usually related to the inhabitants and future users of space and is considered as one approach to place-making. It encompasses various forms of public involvement in project design processes (public surveys, presentations, workshops, design charrettes, etc.) in order to develop or test design options with users. The appropriate type of community participation depends on

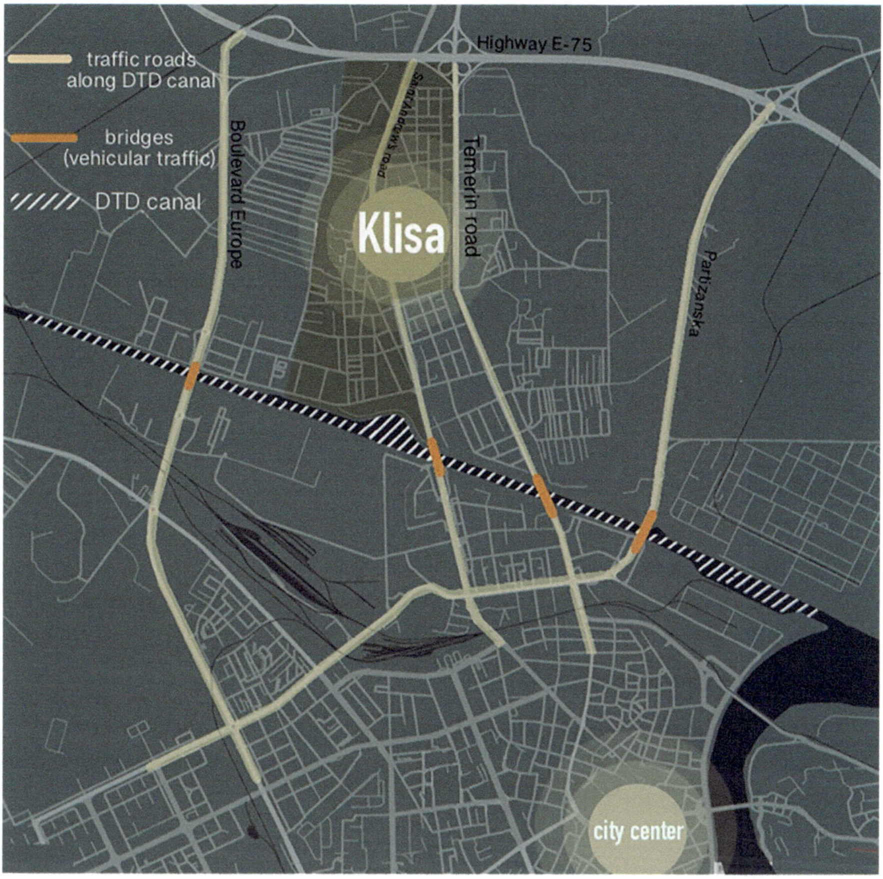

Fig. 6.1 Traffic analysis of the selected scope Of DTD canal

the scale and nature of the project, as well as on the social and political context of development. (Dukanovic et al. 2021). Participatory processes play many different roles, but at their core, they serve to guarantee that the public interest in a certain region is acknowledged and taken into consideration during the planning and development stages. This secures the public interest by emphasising the citizens as space users and ensuring their participation in the planning procedures. A well-executed participative method enhances the success of the implementation and use of space by allowing individuals and the public to identify with the project. The process's stakeholders value mutual interaction, which connects the expert with actual-life scenarios and the user while also fostering the user's ability to comprehend professional principles. This bridges the gap between theory and practice or between professional principles and contemporary needs. Aside from identifying and comprehending the

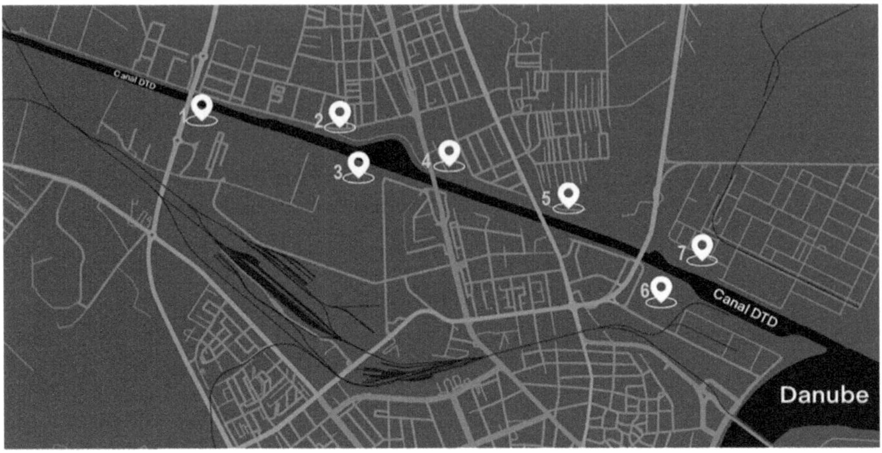

Fig. 6.2 Locations of current state on the site

public interest, the primary challenge we have in participatory processes is communicating with a large number of users who are frequently unavailable or disinterested. Several approaches must be used for the participative process to be successful.

Community Engagement:

- Engage with the local community to understand their needs and aspirations.
- Data sources—workshops, interviews, surveys and exhibitions to involve residents in the design process.
- Data analysis.

Inclusivity and Accessibility:

- Design spaces that are accessible to people of all ages and abilities.
- Prioritise inclusive and participatory design processes, involving the community in decision-making.
- Ensure that public spaces are welcoming and accommodate diverse user needs.
- Design pedestrian-friendly zones with parks and recreational areas along the canal banks.

Collaboration and Cross-Disciplinary Approach:

- Foster collaboration between architects, urban planners, environmental experts, other stakeholders and residents.
- Proposal of conceptual designs (Fig. 6.5).

The questions were composed in accordance with the potential problems and potentials noticed during the analysis and observation of the existing situation and data. As previously said, the goal of this research is to revitalise the DTD Canal's coast and produce advantages for the local populace, especially the people who inhabit Klisa. To ensure that the majority of respondents were locals, the survey was

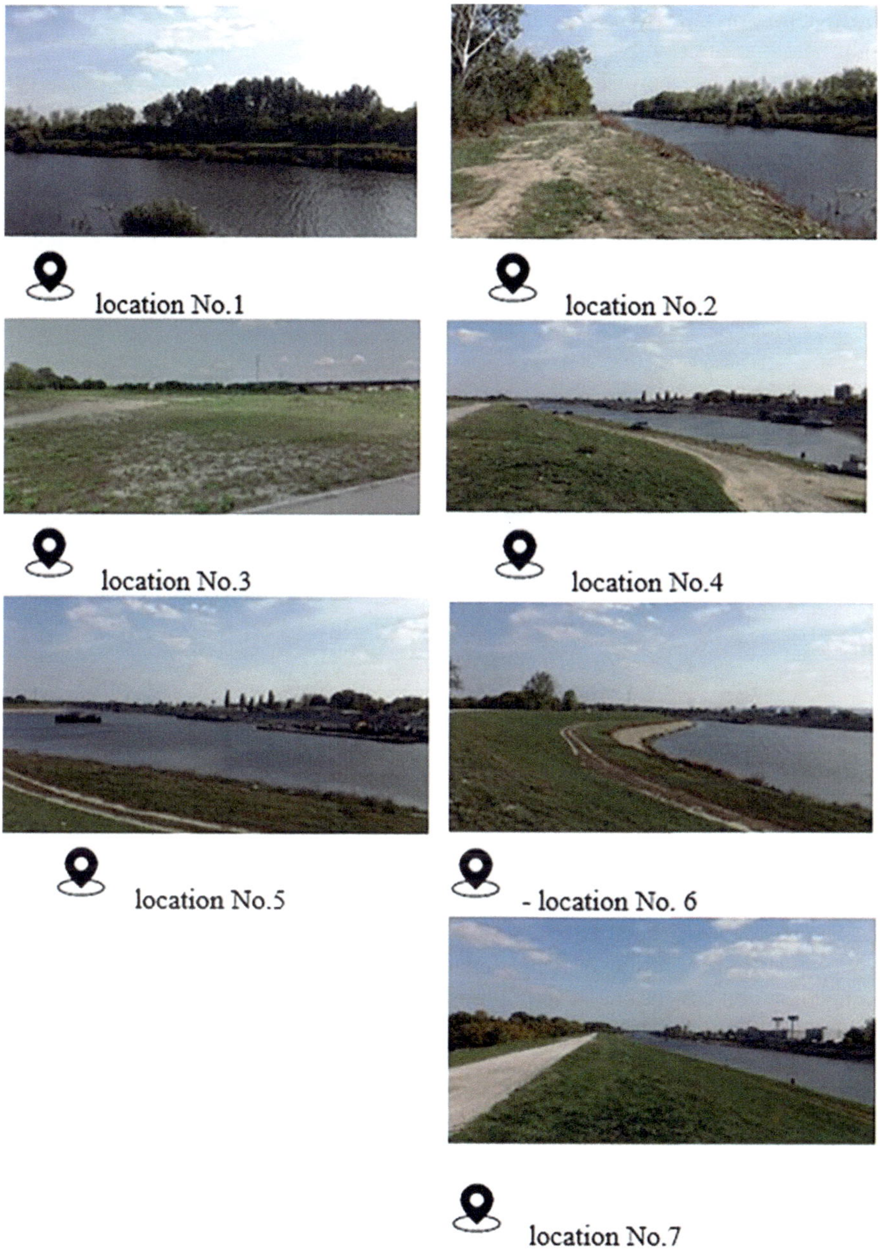

Fig. 6.3 Photo documentation of current state

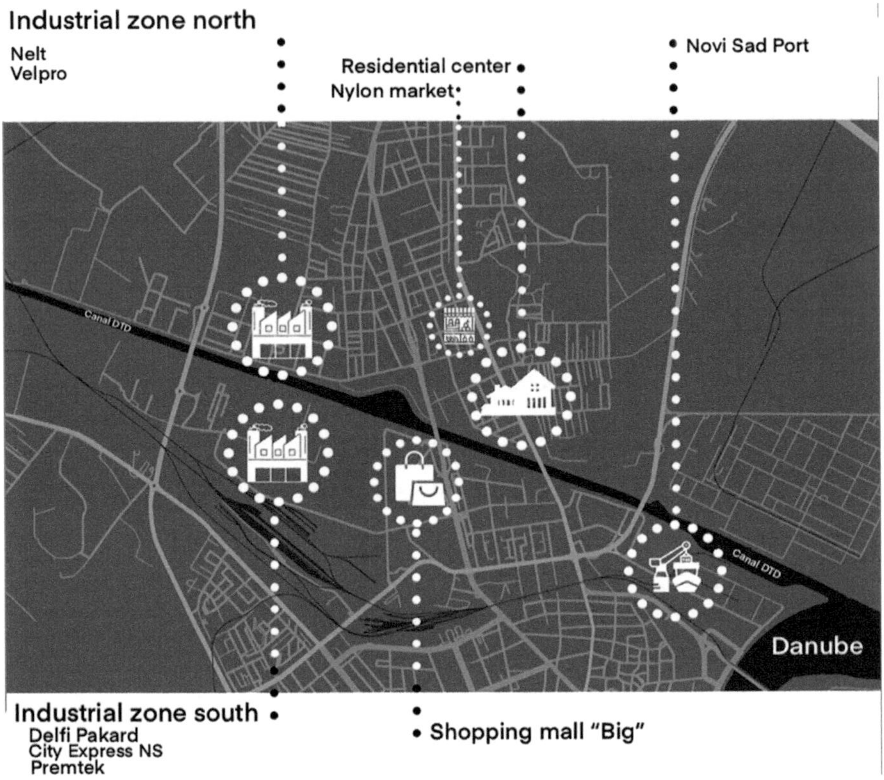

Fig. 6.4 Surface purpose—current state

therefore limited to the area of the aforementioned community (streets, BIG shopping centre and Nylon market). The purpose of the survey's random participant selection process was to incorporate as many characteristics as possible related to age, gender, occupation, and education. Despite the vivid personas, the survey's findings clearly highlighted the deficiencies, missing components, and socialisation requirements. The information gathered showed that people have varying needs when using public places on a regular basis, and these needs fall into distinct categories. These categories cover needs for sports and leisure, socialising and meetings, maintaining green spaces and nature in urban settings, and issues with cleaning, traffic, and public order. The current style of living has created a discernible need for greater citizen mobility. The stated need for environmentally friendly alternatives to driving (like cycling) reflects this. Establishing detours for traffic and strengthening the links between the city's different parts and parks. Thanks to the collected data, three phases and guidelines were proposed, which are based on achieving a successfully revitalised coast (Fig. 6.6).

SURVEY

PERSONAL INFORMATION :

Date of Birth::

Job: Education: Marital Status:

Gender: Male Female Prefer Not to Answer

Address: Are you resident of Klisa? Street Name

1.On a scale of 1 to 10, how important is access to green spaces in your daily life?
2.How often do you visit riverbanks or waterfront areas in your city?
3.What improvements would you like to see in your local riverbank areas to enhance your daily experience?
4.Are there specific amenities or features you believe are lacking on your local riverbanks?
5.Do you currently use riverbanks for recreational activities, such as walking, jogging, or cycling?
6.How would you rate the cleanliness and maintenance of riverbank areas in your city?
7.Would you be more likely to visit riverbanks if there were more cultural or community events and activities held in those areas?
8.Would you prefer riverbanks with a focus on natural habitats and biodiversity, or more urbanized and developed spaces?
9.What transportation options would make it easier for you to access and enjoy riverbank areas?
10.Are you aware of any existing riverbank development projects in your city? If yes, please share your thoughts.
11.Are there any specific events or festivals you would like to see organized on riverbanks in your city?
12.What role should technology play in enhancing the overall experience on riverbanks? (e.g., smart benches, Wi-Fi zones)
13.How often do you participate in or attend events at local gathering spaces, such as community centers or parks?
14.How important is it for riverbank transformations to reflect the cultural and historical identity of the local community?
15.How often do you use public transportation, specifically buses, in your daily routine?
16.What improvements would you like to see in the design and amenities of bus stops in your area?
17.How can local gathering spaces be designed to foster a sense of community and inclusivity?
18.Should local gathering spaces be adaptable for various purposes, such as markets, festivals, or outdoor performances?
19.What safety measures do you think are essential for playgrounds, and how can they be improved?
20.In your opinion, how can local gathering spaces contribute to building a stronger sense of community and social connection?

Fig. 6.5 Survey form defined for the purpose of this research

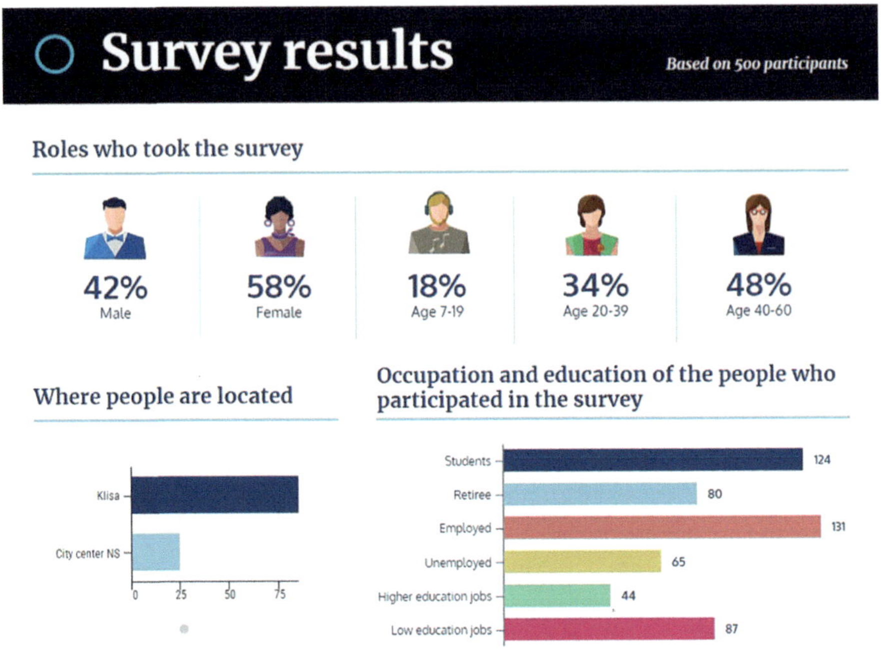

Fig. 6.6 Survey results—graph presentation

6.6.1 Guidelines for Revitalization and Transformation of Bank

The most effective actions that would result in the revitalisation of the banks of the DTD and Klisa canals were identified by analysing the current situation. The goal of the proposed interventions is to make this area of the city better so that locals are able to access it easily. The creation of bicycle and pedestrian paths, which are lacking in this area, will create the framework for the development of new content that will benefit tourism, the economy, and society. Therefore, the interventions that will be carried out include the remodelling of traffic infrastructure, the revitalisation of unused and unorganised green spaces, and the introduction of new content and public spaces that will affect better communication between people. These are the requirements that must be met for this coast to become the spotlight in socialisation. It should be emphasised that the main idea is to show the importance of the coast as an urban space, which can be said to be one of the most important, whose primary role is to serve residents for rest, leisure and entertainment. Even though the current situation analysis reveals that structures have been built around the canal bank, this is insufficient to support its existence. Instead, citizens who actively participate in the activities of the city are essential. Users and coastlines and the link between users and activities (content) on both sides will be covered in more detail in the text that

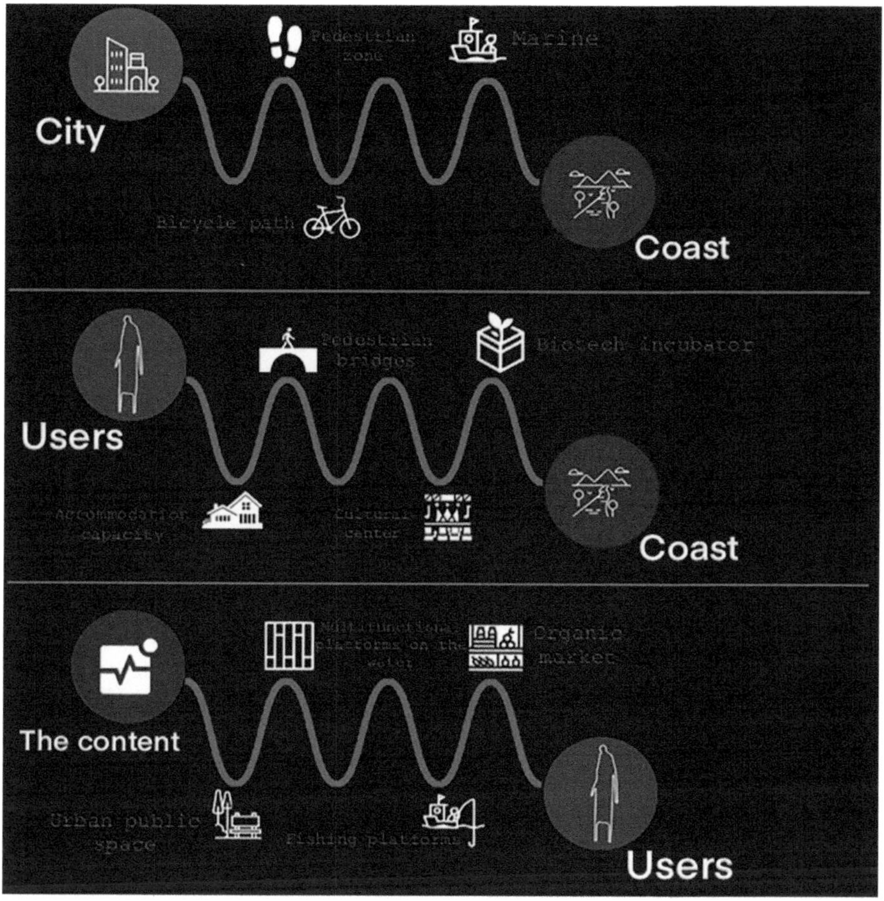

Fig. 6.7 Conceptual diagram of transformation phases

follows, along with a graphic presentation as an additional attachment. Based on these relationships, the revitalisation process can be solved in a couple of phases (Fig. 6.7).

6.7 First Phase (Traffic Solution)

A significant intervention is planned for the repurposing of the traffic infrastructure. Novi Sad is known as a city that pays great attention to bicycle traffic. However, this type of traffic is not provided adequately from a part of Novi Sad to Klisa, i.e. St. Andrew's road, as well as in the most important streets within Klisa itself. Traffic must be converted into mixed traffic in order to provide simpler access from the city

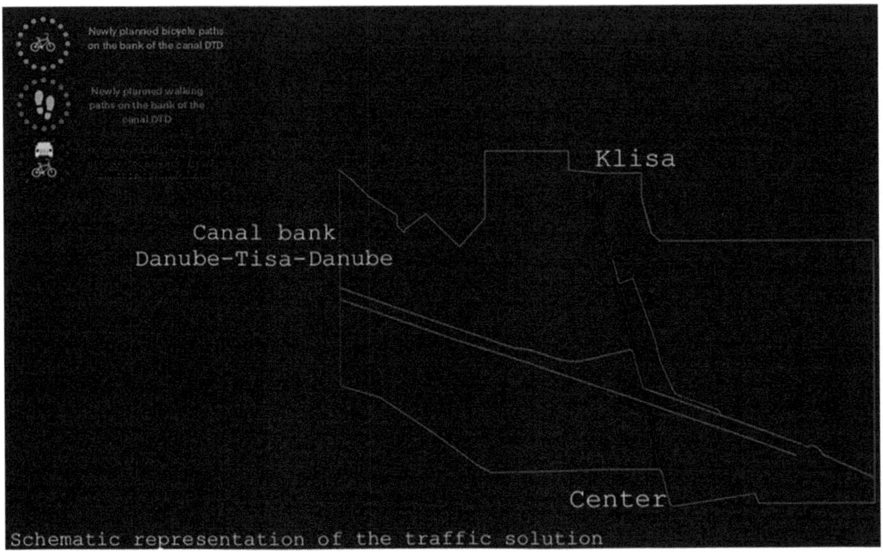

Fig. 6.8 Schematic representation of the traffic solution

to Klisa or from Klisa to the seaside. The term "mixed traffic zone" refers to a set of roads where automobile, bicycle, and pedestrian traffic is permitted. In regard to the streets inside the Klise community, Čenejska Street stood out as the busiest street that could be claimed to symbolise the core of that area of the city, containing all the most significant public facilities. Since there isn't an open area for socialising on Klisa, the plan is to connect it to the canal bank. With its newly designed contents, the bank would be able to fulfil all the requirements needed to take centre stage for communication. Additionally, the plan is to link it to the bank and Slana Bara, the sports complex located along Čenejska Street. So far, the most exciting attraction drawing visitors to this area of the city is the Nylon market, which is situated right next to Čenejska Street and the coast. Bicycle and pedestrian lanes must be established so that market visitors have clear and convenient access to the coast. These actions create opportunities for improved socialising and life on the coastline by increasing the frequency of people and making the coast simpler to access (Figs. 6.8 and 6.9).

6.8 Second Phase (Surface Repurpose)

The transformation of surfaces is a necessary step towards the development of the Danube-Tisa-Danube canal banks as a live urban structure. A few areas were chosen where an unhindered intervention may be possible based on the location analysis. Large areas of uncontrolled vegetation are to be transformed into open public places with varying links to the water. One such interaction will be achieved by cascading

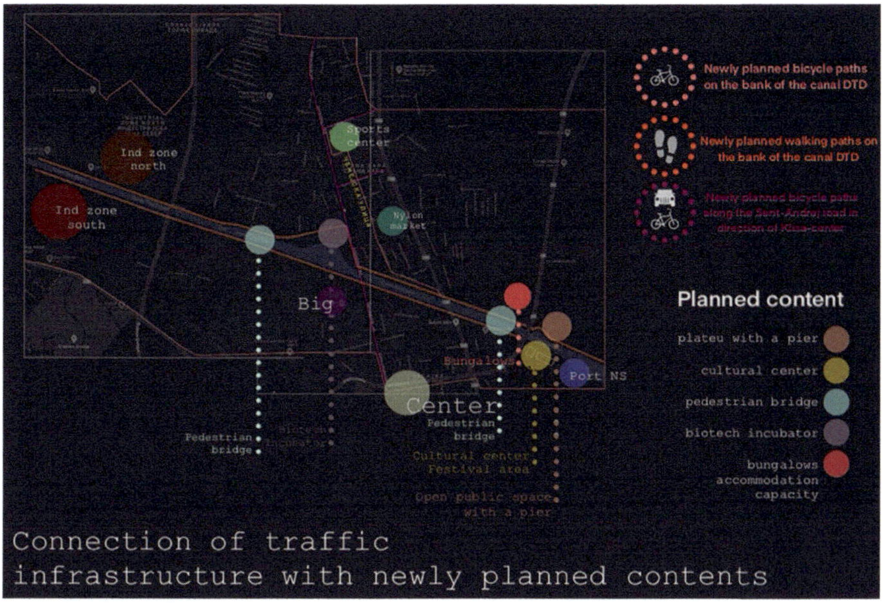

Fig. 6.9 Connection of traffic infrastructure with newly planned contents

stairs that descend towards the water's edge, which brings people and the water closer together. There will be walking lanes, trim tracks, sitting and lying seats, and rest areas in these open spaces. Agriculture stands out as the greatest potential of Vojvodina, and Klisa is one of the few where the majority of the population is still engaged in agriculture. As part of the solution, a conceptual solution for a biotech incubator will be proposed, which will be designed to include the residents of Klisa—farmers and students from Novi Sad. The idea is to introduce urban agriculture with the use of modern technology, as well as ephemeral structures in which the organic products resulting from this cooperation will be sold. In this way, conditions are created for the active participation and income of the inhabitants of Klisa. In addition to the construction of new buildings, it is also possible to repurpose the functions of existing structures. Given that Klisa is considered a remote part of Novi Sad due to the lack of cultural content, the idea is to use the abandoned Agrohem facility. The idea is to repurpose one part into a cultural centre and the other into a multifunctional space where festivals can be held (Figs. 6.10 and 6.11).

6.9 Third Phase (New Content and Activities)

One of the key characteristics of any area is the diversity of activities and content. The primary drawback of the coast's current condition, as was already indicated, is the absence of any pedestrian amenities. Authentic spaces for socialising will be

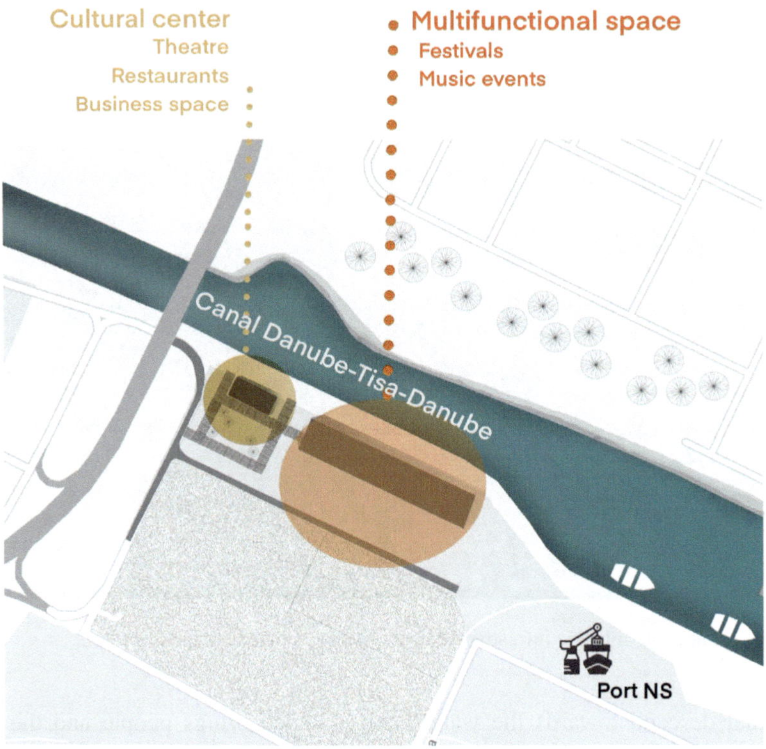

Fig. 6.10 Newly planned purpose for abandoned agrohem facility

Fig. 6.11 Current state of agrohem building (left)—design proposal of facade revitalization on newly planned cultural center (right)

created by building pedestrian bridges that principally connect the two sides of the DTD canal bank. It became clear from analysing Klisa residents' daily lives that platforms and piers for fishermen were necessary. A promenade area with a planted greenbelt, restaurants, and a recreation area are just a few of the features that should be considered when developing the pier. New urban furniture can be installed along the promenade's freshly designed promenade.

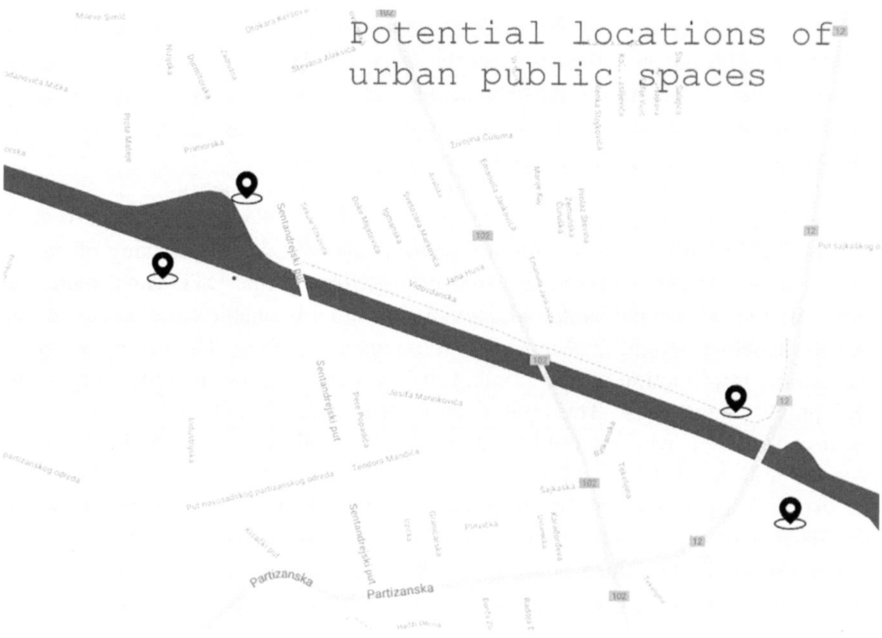

Fig. 6.12 Potential locations for transformation into designed urban public spaces

In addition, a national food festival and fishing competitions are proposed to be hosted alongside the markets featuring locally produced organic goods. By introducing all of the predicted activities and contents, the coast will be enhanced from multiple angles, creating a favourable environment for the provision of accommodation capacity that would be used by a variety of users, including visitors to festivals and competitions, students and experts who will support the people of Klisa in the development of their agricultural, and participants residing at the cultural centre. Given that there are two football fields and the Slana Bara sports facility, athletes are another group of potential consumers. Bungalow-style accommodations would consist of lightweight materials, compact design, and reasonably priced rates. Situated close to the water, they offer consumers a unique ambience and create an overall setting that satisfies all the requirements of a high-quality coastal lifestyle (Fig. 6.12).

6.10 Conclusion

By researching and creating a study on possible transformations on the coast of the Danube-Tisa-Danube canal, a strategy was arrived at that would help in reviving this coast, as well as the area of Klisa. One of the goals of this research is to define the importance of the river as an important natural element in the urban environment and to consider the factors that are necessary for a living river bank and its development depending on the specifics of the local environment. In addition, it is important to come to a conclusion about the activities that are carried out on the coast and whether this affects the change in the attitude of the population towards it. This indicates the importance of the coast itself as an active urban public space that leads to the revitalisation of the space and the socialisation of people, as well as the creation of an authentic visual identity. With that in mind, a series of transformations were proposed that would lead to the development of this part of Novi Sad. The revitalisation strategy was processed in three phases. The first phase deals with the repurposing of traffic, which is focused in two directions: connecting Klisa and the DTD canal bank and connecting the DTD canal bank with Novi Sad in order to enable easier access to the coast itself, which would lead to higher frequency of people. The second phase is based on the introduction of new content that would also economically support the development of the coast. The revitalisation of the abandoned Agrohem premises was also proposed, which would be used as a cultural centre and multifunctional space for holding festivals, which would give Klisa cultural facilities for the first time. Apart from agriculture, fishing is the only activity that manages to live on this coast. Therefore, a plateau with landscaped green areas and platforms with shallow pools is proposed. The third phase of work is devoted to the introduction of new urban public spaces. The planned interventions would take place in a couple of locations that are overgrown in unorganised green areas. Small transformations that lead to the socialisation of people are very important and include the introduction of furniture, reading areas, and children's playgrounds. In the creation of the strategy for the transformation of the DTD canal bank, the needs of the inhabitants and their active inclusion in all the proposed contents were of great importance. The above-mentioned transformation proposals would achieve the economic, social and cultural development of this area, which definitely contains enormous potential for the creation of a modern riverfront.

References

Čolić N, Čolić R (2018) Dijaloški i javni prostor—participativni pristup urbanom dizajnu. Arhitektura i Urbanizam 47:37–45. https://doi.org/10.5937/a-u0-17445

Dukanovic Z, Živkovic J, Radosavljevic U, Lalovic K, Jovanovic P (2021) Participatory urban design for touristic presentation of cultural heritage sites: the case of Negotinske Pivnice (wine cellars) in Serbia. Sustainability 13:39. https://doi.org/10.3390/su131810039

GradskeInfo.2022 https://gradskeinfo.rs/koji-je-znacaj-kanala-dtd-i-kako-se-odrzava-najveca-kanalska-mreza-u-srbiji-01-09-2022/ Accessed 14 Jan 2024

Jun J (2023) Towards sustainable urban riverfront redevelopment: adaptability as a design strategy for the Hangang Riverfront in Seoul. Sustainability 15:9207. https://doi.org/10.3390/su15129207

Part III
Macro Scale: Local and Regional Communities as Innovators in Our Society

Chapter 7
Beautiful and Inclusive Transitions to Resilient, Climate-Neutral Cities and Communities, Driven by Radical Collaboration

Annemie Wyckmans

Abstract During a recent local workshop, participants highlighted a significant challenge: despite efforts to ensure co-creation projects are inclusive, many people still perceive a high threshold for required competence and capacity. How can we improve project designs to facilitate radical collaboration among diverse stakeholders, addressing complex societal challenges? This chapter explores key challenges and strategies using examples from + CityxChange (2018–2023), CrAFt (2022–2025), NEB-STAR (2022–2025), and Re-Value (2023–2026) projects coordinated by NTNU. Funded under the European Mission on Climate-Neutral and Smart Cities and the New European Bauhaus initiative, these projects prioritise multi-actor cooperation "for, by, and with citizens." In + CityxChange, supported by Horizon 2020, a consortium of 32 partners adopted an Open Innovation 2.0 framework, fostering active involvement and collaboration among public and private sectors, NGOs, and academia. This approach connected building owners, public authorities, local businesses, residents, and more, enhancing collective learning and successful strategy replication Drawing on + CityxChange insights, subsequent projects aim to enhance radical collaboration further, boosting societal readiness through mission-oriented approaches. The CrAFt, NEB-STAR, and Re-Value projects embrace inclusivity, transparency, and comprehensive stakeholder engagement. These initiatives promote systems thinking, inclusiveness, and experiential learning, ensuring that science and innovation are integral to daily life and community well-being. By fostering these collaborative frameworks, we pave the way for sustainable, inclusive, and resilient urban environments.

A. Wyckmans (✉)
Department of Architecture and Planning, Faculty of Architecture and Design, Norwegian University of Science and Technology, Trondheim, Norway
e-mail: annemie.wyckmans@ntnu.no

M. Vukmirović et al. (eds.), *Digital Dreams, Urban Realities*, The Urban Book Series,
https://doi.org/10.1007/978-3-031-76353-3_7

7.1 Introduction

Who can afford to participate in co-creation?

During a recent local workshop, a participant highlighted this significant challenge. Despite efforts to ensure that co-creation projects and activities are inclusive and accessible to a diverse range of participants, many people still perceive a high threshold in terms of required capacity and competence.

How can we improve the design of projects to facilitate radical collaboration among diverse stakeholders, thereby addressing complex societal challenges?

In this chapter, we explore key challenges and strategies for addressing them, drawing on examples from the + CityxChange (2018, 2023), CrAFt (2022, 2025), NEB-STAR (2022, 2025), and Re-Value (2023, 2026) projects coordinated by the Norwegian University of Science and Technology (NTNU). These projects are funded under the European Mission on Climate-Neutral and Smart Cities (2021) and the New European Bauhaus initiative (2021).

Both of these programmes prioritise multi-actor cooperation, respectively "for, by and with citizens" and "Beautiful. Sustainable. Together".

7.2 Multi-Actor Collaboration Using an Open Innovation 2.0 Framework

In the + CityxChange project, supported by the Horizon2020 (2014, 2020) research and innovation funding programme, we had a consortium of 32 partners spanning public and private sectors, non-governmental organisations, and academia.

In order to make this cooperation work, we adopted an Open Innovation 2.0 framework (Curley and Salmelin 2018). This framework is grounded in principles such as integrated collaboration, co-creation of shared value, nurturing innovation ecosystems, leveraging exponential technologies, and fostering rapid adoption through network effects. By embracing this model, we facilitated the active involvement and collaborative contributions of partners and stakeholders from diverse sectors. Each participant brought their unique perspectives and resources to the table, resulting in a unified and comprehensive project approach.

Beyond the typical stakeholders such as building owners, architects, contractors, and occupants, our + CityxChange network also included public authorities, housing associations, cooperatives, utility companies, network operators, regulatory bodies, various investors, residents' associations, local businesses, commercial associations, voluntary and community groups, as well as local politicians. Supporting the interaction among these groups gave us unique insights into their experiences and learnings throughout the project and made it easier for all of us to find successful strategies for replication among the partner cities and stakeholders.

Moreover, extracting insights from the successes and failures of other cities empowered our municipalities to adopt solutions that have already proven effective elsewhere, either through direct replication or as sources of inspiration. In order to make this work in practice, we implemented a range of collaborative initiatives within and between our cities. These included local workshops and events for inhabitants and professionals (see amongst others García et al. 2020; Haugslett 2021; Hourigan et al. 2022; Mee 2023).

Additionally, we organised study tours among the seven participating cities, along with learning workshops that facilitated peer-to-peer interaction with local collaborators and stakeholders (for details see + CityxChange, all deliverables related to work package 9). These initiatives provided opportunities for hands-on learning from local demonstration experiences in the cities and other relevant locations. During these visits, cities and solution providers could showcase practical examples, pose inquiries, and gain insights from their peers.

Furthermore, each city established a localised Innovation Playground (Grabinsky 2022; Mee et al. 2020; Mee 2023; Walsh et al. 2022) to prototype experiences and enhance collective learning efforts. Within these Innovation Playgrounds, residents, local businesses, and various stakeholders were supported to engage in networking, idea generation, and the creation and testing of urban prototypes.

In addition to reports per topic as indicated above, these collaborative experiences and learnings are summarised in the "How to PED" Cookbook (Ahlers et al. 2023).

7.3 Radical, Mission-Oriented Collaboration to Create Societal Readiness

Expanding upon the knowledge and insights gained from the + CityxChange project, our next objective was to explore ways to enhance collaboration even further by co-designing solutions and strategies together with different types of stakeholders and partners. For this purpose, we aimed to adopt a more mission-oriented approach (Mazzucato 2021), with a strong emphasis on mutual accountability, to effectively use resources. Our goal was to develop projects that would facilitate cross-cultural dialogue, understanding, collaboration, and creative confidence among stakeholders while also accelerating the learning process and refining outcomes across the entire value chain.

We recognised a very interesting environment to pursue this objective further when two new European programmes were launched:

- The European Mission on Climate-Neutral and Smart Cities (Cities Mission) aims to establish one hundred climate-neutral and smart cities by 2030 and facilitate the adoption of similar practices by all European cities by 2050. To this end, it engages local authorities, citizens, businesses, investors, as well as regional and national authorities in a cross-sectoral and demand-driven approach.

- The New European Bauhaus (NEB) initiative encourages all Europeans to envision and construct a future that is both sustainable and inclusive, one that captivates our senses, intellect, and spirit through collective effort and creativity.

While + CityxChange and other projects in the Smart Cities and Communities Lighthouse programme had a main technical focus, in the Cities Mission and New European Bauhaus programmes, multi-actor collaboration is strongly at the forefront.

Within these programmes, we are experimenting with radical collaboration in three projects:

- The CrAFt Cities Coordination and Support Action (2022, 2025) aims to support cities and communities in becoming climate-positive, i.e. working towards climate neutrality and resilience in a manner that makes places more beautiful, inclusive and sustainable. The project includes eight partners and more than 70 CrAFt Cities.
- The NEB-STAR New European Bauhaus Lighthouse project (2022, 2025) aims to demonstrate how Stavanger, working in collaboration with the twin cities of Prague (Czech Republic) and Utrecht (Netherlands), can offer valuable experimentation, demonstrations, testbeds, and guidance on the principles of the New European Bauhaus (NEB) in urban transitions. The project includes sixteen partners from public and private sectors, academia, arts and culture, youth and social entrepreneurship.
- The Re-Value Innovation Project (2023, 2026) works with nine waterfront cities to re-value their urban planning and design in the context of climate change. Taking cues from the aesthetic sustainability principles of the New European Bauhaus and aligning with the EU Mission for Climate-Neutral and Smart Cities, the project adopts a comprehensive approach that encompasses not only the physical infrastructure but also prioritises the well-being of communities, paving the way towards achieving climate neutrality in urban environments. The project has 26 partners, including cities, local solution providers, and transversal partners from universities, research and non-governmental organisations.

Within the CrAFt, NEB-STAR and Re-Value projects, we aim to improve societal readiness (NewHoRRizon 2021) through radical collaboration, by practising the following principles together with the participating cities and stakeholders.

7.4 Our project activities and structures

- Are guided by the needs, values, and expectations of various social groups, considering factors such as gender, age, socioeconomic status, geography, and vulnerability
- Comprehensively grasp different viewpoints and interests
- Nurture inclusive partnerships among various stakeholders to drive change
- Embrace inclusivity and transparency throughout the processes and outcomes

- Proactively identify, mitigate, and prevent negative impacts
- Foster positive social, environmental, and economic outcomes

In the following section, we explain how we created a learning environment for this work.

7.5 A Learning Environment for Radical Collaboration

The intention to incorporate radical collaboration into our projects, deeply embedded within the formal project structures and informal activities, prompts us to explore why we invite others to collaborate and co-create—really. While cooperation is very useful in complementing our individual strengths and offset limitations, it requires continuous refinement and strengthening. We are often not fully aware of even our own expectations of others and vice versa, and this may hinder effective cooperation and co-creation. Sennett (2012) underscores the significance of understanding and adapting collaborative formats to others' perspectives, particularly when their interests diverge from our own. Moreover, he highlights the imperative of establishing common ground to facilitate collaboration, even amidst varying interests and priorities.

To foster radical collaboration in the CrAFt, NEB-STAR, and Re-Value projects, we are therefore integrating essential aspects of experiential learning, a method championed since the early twentieth century by Dewey (1998). Experiential learning prioritises the development of decision-making and problem-solving abilities through active engagement, analysis, and reflection, diverging from passive instructional approaches such as lectures.

Bloom, another influential figure in 20th-century pedagogy, devised a taxonomy to catalogue and enhance learning progression, encompassing six cognitive processes (Anderson 2021):

- Remembering: Recalling information, facts, or concepts from memory
- Understanding: Comprehension of the meaning and significance of the information remembered. Ability to explain concepts using own words and demonstrate understanding through examples or applications.
- Applying: Apply understanding of concepts to new situations or contexts
- Analysing: Breaking down information into its constituent parts, examining relationships, and identifying patterns or connections
- Evaluating: Critically assess the quality, validity, or effectiveness of information, arguments, or solutions
- Creating: Synthesising and integrating knowledge to generate new ideas, products, or solutions

Anderson (2021) later expanded on this taxonomy by adding a second dimension that includes four types of knowledge:

- Factual knowledge: Basic information or facts that learners need to know and remember
- Conceptual knowledge: Understanding the underlying principles, theories, or conceptual frameworks that organise and structure information
- Procedural knowledge: Knowing how to do something, such as performing a task or solving a problem
- Metacognitive knowledge: Awareness and understanding of one's own cognitive processes and strategies for learning and problem-solving

If we really want to harness the collective intelligence and diverse perspectives of individuals, organisations, and communities in our projects, we need to activate all of these learning dimensions. In the next paragraphs, we offer some examples of the formats we use in the projects to enable this.

7.6 Radical Collaboration for Societal Readiness

7.6.1 Sharing Individual Experiential Knowledge

Pooling our resources, expertise, and insights has the potential to yield innovative solutions and strategies that are more robust and effective than those conceived in isolation. Collaborations spanning various sectors, disciplines, and stakeholders empower us to develop more holistic responses to societal challenges. For this reason, it is important for us to cultivate an environment supportive of collective experimentation and innovation in which we can explore novel ideas and approaches, even if we need several iterations before these are successful (Sennett 2012).

A significant hurdle in establishing such an environment is the difficulty of communicating and learning from individual experiential knowledge. Such knowledge is often acquired within specific contexts and may not easily resonate with others. People are often not aware of the full knowledge they possess—as captured by the meta-cognitive knowledge component in the previous section. In order to convey this knowledge to others, therefore, it is imperative not only to gather and disseminate project outcomes but also to document the decision-making processes and priorities that guided the entire design journey. This documentation should encompass the context in which decisions were made, who made them and how, and what the ensuing repercussions (in other words, the procedural learning component explained in the previous section). By sharing the narratives behind project development, we can start to construct a shared repository of environmental learning.

In the first edition of the CrAFt Cookbook (Wyckmans Laschet 2024), we have started to document such stories from cities and communities across Europe. We are gathering additional input for second and third editions to show how approaches can be localised in different contexts and cultures with minor adjustments, always firmly anchored in the local ecosystem of stakeholders.

7.6.2 Systems Thinking

Radical collaboration also helps us to promote systems thinking, which involves understanding the co-benefits that arise in cooperation between socio-cultural, economic, health, environmental, and governance sectors. Co-benefits are additional, intended or unintended, positive outcomes that result from taking action to pursue a specific goal. These benefits may not be the primary focus of the action but are welcome side effects that contribute to overall well-being and sustainability. By taking a holistic view of complex systems, we can identify such co-benefits and use them as leverage points and intervention opportunities to promote societal readiness.

This approach is explained in our New European Bauhaus-inspired Impact Models (Vandevyvere Wyckmans 2023; Wyckmans Vandevyvere 2023), and integrated in the CrAFt, NEB-STAR and Re-Value projects.

7.6.3 Inclusiveness and Diversity

We use radical collaboration to deepen social awareness and celebrate a diversity of perspectives and experiences. When we encounter other project partners and local stakeholders from different cultures, languages, professional backgrounds, and more, the differences may prompt us to reflect on the values we typically take for granted. This, in turn, leads us to a deeper appreciation of our understanding of shared—or diverging—values and norms (Sennett 2012). It also helps us to develop and support inclusive attitudes and behaviours that promote respect, empathy, and cooperation across the many diverse backgrounds and perspectives in our projects.

For this reason, CrAFt, NEB-STAR and Re-Value include management tasks on inclusiveness and diversity to promote societal readiness throughout our cities, stakeholders and project structures. Inclusiveness and diversity form part of the overall project management and coordination in order to balance the formal project structures with informal processes that nurture participation and co-creation among the project partners and local stakeholders.

In addition, story-building (Carlsen et al. 2012) activities within the cities and across the partners and stakeholders encourage us to challenge the assumptions we often carry with us, think critically, and explore innovative solutions to the complex, systemic challenges we are trying to address in the projects and cities. We organise lateral thinking exercises and activities that encourage the participants to explore unconventional ideas, make unexpected connections, and challenge long-held assumptions. These exercises can help stimulate creativity, spark innovation, and push participants to think outside the box. In this manner, we work with our partners and stakeholders to tackle challenges creatively and adaptively.

7.6.4 In-Person Interaction

As explained earlier in this chapter, learning goes beyond sharing facts and figures as information; it involves exchanging context, emotions, and subtle nuances that add depth to the message. This includes showing empathy, actively listening, and being attuned to non-verbal cues, all of which help forge genuine connections and relationships. When context is overlooked or oversimplified, the true essence of learning is diluted, reducing the opportunity for sharing meta-cognitive knowledge, increasing the risk of misunderstandings or misinterpretations, and limiting the chance for others to learn how to recreate this experience in their own environments. Face-to-face interactions often offer different—sometimes serendipitous—types of opportunities for spontaneous brainstorming, creativity, and problem-solving that are harder to replicate in virtual or asynchronous environments.

For this reason, the CrAFt, NEB-STAR and Re-Value projects include city visits, peer-to-peer workshops, study tours by bike, story-building activities, open-ended discussion sessions and other types of interactive formats that provide opportunities for skill-building, team-building, and relationship-building. Our in-person meetings are set in physical surroundings that have a strong connection with the challenges the project aims to address. We facilitate the events and workshops in a manner that promotes active listening, constructive feedback, and recognition of diverse perspectives. By doing this, we aim to enhance trust and collaboration within and between the different participant groups. We are always testing new formats that can support this creative in-person interaction—rather than replicating online meeting formats in a physical setting.

In addition, we organise regular evaluation, feedback, and reflection on the effectiveness of workshops, meetings, and other forms of social exchange, to identify areas for improvement and optimising cooperation over time. These in-person formats are complemented with asynchronous and virtual collaboration tools, and participatory online meetings and platforms to provide additional opportunities for social exchange and cooperation.

7.6.5 From Place-Based Experiment to Long-Term Structure

In this chapter, we have described various formats and structures that are put in place to provide support for radical collaboration in our projects. As time passes, we see that common values and standards start to develop among the participants. These shared values gradually become part of our group's identity and influence how we work together to prepare our cities and communities for various challenges. However, projects are restricted in time (typically 3–5 years). For this reason, we also include measures to integrate what we learn into long-term structures.

In each project, we aim to make our experiences and lessons a permanent part of how things are done. This means integrating them into official policies, rules, educational programmes, and other initiatives that aim to improve society's readiness and encourage larger cultural changes.

In time, we aim for these changes to also include shifts in people's attitudes, behaviours, and societal norms, all geared towards making our cities and communities better prepared and more resilient.

Acknowledgements +CityxChange (Grant Agreement No. 824260) has received funding from the European Union's Horizon2020 research and innovation funding program under the Smart Cities and Communities Lighthouse program. Craft (Grant Agreement No.101056946), NEB-STAR (Grant Agreement No. 101079952) and Re-Value (Grant Agreement No. 101096943) have received funding from the European Union's Horizon Europe research and innovation funding programme, as part of the work programme of the Mission on Climate-Neutral and Smart Cities.

All projects are the result of an extensive collaborative effort by the project partners and stakeholders—without their continuous effort and engagement, these results would not have been possible to achieve.

References

+CityxChange (2018–2023) Positive city ExChange. https://craft-cities.eu/. Accessed 14 May 2024

Ahlers D et al (2023) How to PED. The +CityxChange cookbook: experiences and guidelines on positive energy districts (eds). https://cityxchange.eu/knowledge-base/how-to-ped-cookbook/. Accessed 14 May 2024

Anderson LW, Krathwohl, DR (2001) A taxonomy for learning, teaching and assessing (eds). A Revision of Bloom's Taxonomy of Educational Objectives. Longman, New York

Carlsen A, Clegg S, Gjersvik R (2012) Idea work: lessons of the extraordinary in everyday creativity. Cappelen Damm, Oslo

CrAFt (2022–2025) Creating actionable futures. https://craft-cities.eu/. Accessed 14 May 2024

Curley M, Salmelin B (2018) Open innovation 2.0. The new mode of digital innovation for prosperity and sustainability. innovation, technology, and knowledge management. Springer International Publishing, Cham, Switzerland

Dewey J (1998). Experience and education. West Lafayette, Ind: Kappa Delta Pi. 1st edition in 1938

European Mission on Climate-Neutral and Smart Cities (2021) https://research-and-innovation.ec.europa.eu/funding/funding-opportunities/funding-programmes-and-open-calls/horizon-europe/eu-missions-horizon-europe/climate-neutral-and-smart-cities_en. Accessed 14 May 2024

García JB, Mora MS et al. (2020) Delivery of the citizen participation playbook. Deliverable 3.2. https://cityxchange.eu/knowledge-base/delivery-of-the-citizen-participation-playbook/. + CityxChange. Accessed on 14 May 2024

Grabinsky C et al (2022) Trondheim innovation lab solutions catalogue. Deliverable 5.10. https://cityxchange.eu/knowledge-base/d5-10-trondheim-innovation-lab-solutions-catalogue/. +CityxChange. Accessed 14 May 2024

Haugslett A et al (2021) +Trondheim citizen observatory. Deliverable 5.8. https://cityxchange.eu/knowledge-base/d5-8-trondheim-citizen-observatory/. +CityxChange. Accessed 14 May 2024

Horizon2020 research and innovation funding programme (2014–2020) https://research-and-innovation.ec.europa.eu/funding/funding-opportunities/funding-programmes-and-open-calls/horizon-2020_en. Accessed 14 May 2024

Hourigan S, Tierney J, Fitzgerald H (2022) Limerick citizen observatory. Deliverable 4.8. https://cityxchange.eu/knowledge-base/d4-8-limerick-citizen-observatory/. +CityxChange. Accessed on 14 May 2024

Mazzucato M (2021) Mission Economy: A Moonshot Guide to Changing Capitalism. Harper Business, London

Mee A, Crowe P et al (2020) Framework for innovation playgrounds. Deliverable 3.3. https://cityxchange.eu/knowledge-base/d3-3-framework-for-innovation-playgrounds/. +CityxChange. Accessed 14 May 2024

Mee A et al (2023) Report on community participation and playground results. Deliverable 6.5. https://cityxchange.eu/knowledge-base/d6-5-report-on-community-participation-and-playground-results/. +CityxChange. Accessed 14 May 2024

NEB-STAR (2022–2025) New European Bauhaus Stavanger https://nebstar.eu/. Accessed 14 May 2024

New European Bauhaus (2021) https://new-european-bauhaus.europa.eu/index_en. Accessed 14 May 2024

NewHoRRizon (2021) Societal Readiness Thinking Tool https://newhorrizon.eu/thinking-tool/. Accessed 14 May 2024

Re-Value (2023–2026) Re-valuing urban quality & climate neutrality in European waterfront cities. https://re-value-cities.eu/. Accessed 14 May 2024

Sennett R (2012) Together. The rituals, pleasures & politics of cooperation. Allen Lane, London

Vandevyvere H, Wyckmans A (2023) NEB Impact Model (initial version) https://craft-cities.eu/wp-content/uploads/2023/07/D1.1-CrAFt-NEB-Impact-Model-updated.pdf. CrAFt. Accessed 14 May 2024

Walsh G et al (2022) Limerick innovation lab solutions catalogue 2. Deliverable 4.10. https://cityxchange.eu/knowledge-base/d4-10-limerick-innovation-lab-solutions-catalogue-2/. +CityxChange. Accessed 14 May 2024

Wyckmans A Laschet M (2024) The CrAFt Cookbook. Local collaboration for climate-positive cities & communities (eds). https://craft-cities.eu/craft-cookbook/. Accessed 14 May 2024

Wyckmans A, Vandevyvere H (2023) Re-value Impact Model (initial version) https://re-value-cities.eu/documents/re-value-impact-model-initial-version. Re-value. Accessed 14 May 2024

Chapter 8
Integration Factors and Social Cohesion of Urban Facilities and Urban Design Criteria in the North and South Peripheries of Granada

Belén Bravo-Rodríguez, Juan L. Rivas-Navarro, Cecilia Hita-Alonso, and Pilar Martos-Fernández

Abstract The results of the study of the urban facilities of the northern and southern districts of the city of Granada are presented based on their spatial location conditions. A methodology is proposed to evaluate the way in which these conditions influence the capacity of each urban facility to generate social cohesion in the urban environment. Urban facilities have a social function as collective urban nodes, becoming places of meeting and linking the collective sphere. Currently, the social and spatial fragmentation of cities is increasing, which deepens the social fragmentation of public spaces. Thus, it is necessary to strengthen social cohesion using the available public resources: public spaces and the set of urban facilities. An adequate location of public buildings in the urban fabric promotes social interaction and the strengthening of the social capital of the neighbourhoods. The research establishes a study at two levels, morphological and social. A comparable set of multifunctional urban facilities (civic centres, sports centres and health centres) is addressed. The results allow us to conclude with a series of patterns and criteria for the location and urban design of facilities in the city to promote the economic and social activation of urban areas and strengthen the social fabric of these vulnerable neighbourhoods.

Keywords Urban facilities · Public space · Social cohesion · Urban design · Periphery · Granada

B. Bravo-Rodríguez (✉) · J. L. Rivas-Navarro
Department of Urban and Regional Planning, University of Granada, Granada, Spain
e-mail: bbravo@ugr.es

C. Hita-Alonso · P. Martos-Fernández
Department of Sociology, University of Granada, Granada, Spain

M. Vukmirović et al. (eds.), *Digital Dreams, Urban Realities*, The Urban Book Series,
https://doi.org/10.1007/978-3-031-76353-3_8

8.1 Introduction

Urban facilities, regardless of their specific function, have a social function as collective urban nodes, as meeting places and links of the collective sphere (Ledrut 1976). Nowadays, cities are increasingly prone to social and spatial fragmentation, aggravated by conditions arising from emergencies such as the health crisis and the economic crisis. It is necessary to strengthen social cohesion using public resources from those that are arranged, which at a spatial level are public spaces and urban facilities. Such socioeconomic and spatial changes push us to sustain a more leisurely, reflective, and valuing look at cities. It is time, perhaps, for an urban planning that (Solà-Morales and Parcerisa 1987) called for, which proposes the city and its urban fabrics "between topography and layout (…) as a project for the promotion and configuration of public and collective spaces". From previous studies carried out, it has been observed and verified in nearby urban contexts that a correct position of public urban functions favours social interaction and the strengthening of the social capital of the neighbourhoods. It is also valid for their architectures and the design of their own spaces related to the urban fabric in terms of accessibility, proximity to residential areas, urban access quality, etc.

Research focused on urban facilities is proposed from the view of their spatial conditions of urban location and of urban design of public space around them. The aim is to assess how these conditions can influence the generation of social cohesion in urban environments, mainly in peripheral neighbourhoods. Specifically, the study is developed regarding sports equipment, health, and social services, as these types usually attract a very diverse population range and mean an increasingly greater demand by citizens.

8.2 Dimensional Bases of Urban Facilities to Promote Greater Social Cohesion

The concentration of poverty in vulnerable urban neighbourhoods is now one of the main problems of the European Union. This is pointed out by much evidence, such as the 'Seventh Report on Economic, Social and Territorial Cohesion in the EU' (Dijkstra 2017), which puts the risk of poverty or social exclusion in cities in countries that joined the Union before 2004 at higher levels than before the crisis (Dijkstra 2017).

The concept of vulnerability can be defined as a state of high exposure to certain risks and uncertainties, in combination with a reduced ability to protect or defend oneself against those risks and uncertainties and cope with their negative consequences.

Understood in this way, vulnerability is the prelude to social exclusion and the breeding ground for the decay of urban residential areas, bringing together the spatial dimension and the social structure dimension, which are directionally linked (Castells 1997).

The research carried out proposes to respond to one of the greatest social emergencies within the urban context, that is, urban cohesion and the identification of shared identities. The mechanism on which we rely is the urban exploitation of public or shared resources.

The current logic of the production of urban space frequently results in spatial segregation. Many principles are the basis for such segregation: the isolation of the area or neighbourhood due to the lack of public transport connection; the deficit in urbanization; from the absence of basic services to a chronic shortage of equipment and public space; the reduction of urban functions in the neighbourhood (Pinto and Remesar 2012).

To counteract the above, urban science and technique also need to be linked to the knowledge generated about the social dynamics of appropriation, filling the urban space with meaning with the practice of citizenship (Certeau 2000; Harvey 2003). In this sense, it would be crucial to understand the true transformative dimension of public space, which is an inducer of collective spaces in the city. In the words of Solà-Morales (Solà-Morales and Parcerisa 1987), "All the places where collective life takes place are represented and remembered".

Shared use of public spaces can also create a degree of mutual recognition and feelings of comfort based on public familiarity: recognising others and being recognised in local spaces (Blokland 2014). Informal encounters favour class and ethnic encounters, while formal encounters maintain stereotypes and class ideas (Ahmadi 2018). According to "the contact hypothesis" (Allport 1958), under the right circumstances, interpersonal contact between majority and minority groups has the potential to reduce prejudice. This is especially relevant when working to improve social cohesion in vulnerable peripheries such as the areas studied in the research.

The way of understanding the design of public space must incorporate the condition of being flexible and evolutionary in order to allow appropriation strategies according to unforeseen actions. Public space must reflect an evolving society and support different forms of collective expression, especially in neighbourhoods where diverse cultural groups coexist. For this reason, in this research, public space is understood as an infrastructure, as defined by Allen (Cannon 2018), flexible and anticipatory support which specifies what should be fixed and what is subject to change, assuming the evolving condition of public space and landscape. To incorporate these premises in the methodology of evaluation and design of public space, the Urban Indicators methodology is used, which establishes intervals of validity for the different criteria and combinations open to each case study.

In many definitions of social sustainability, the physical environment leads to social coexistence and promoting the well-being of diverse groups (Ali et al. 2019). On one hand, Jabareen (2006) associates social sustainability outcomes with seven design concepts or typologies, including compactness, sustainable transportation, density, mixed land uses, diversity, passive solar design, and greening. Leyden (Leyden 2003), on the other hand, highlights the role of urban form design in improving social ties among residents.

It is widely found that the accessibility and usefulness of the facilities improve with their networked operation, although they require the study of the local spatiality of their environment (Taleai et al. 2014). It is considered of great relevance to assess the conditions of the facilities both in their insertion in the urban fabric, the design conditions of the urban environment, and their socialisation and actual use.

In accordance with these dimensional bases of urban facilities and their contribution to social cohesion, the methodological approach presented in the following section is developed. It is intended to study the conditions of their implantation, their urban synergies and the existence and design of the surrounding public spaces.

8.3 Methodological Approach and Selection of Case Studies

The case studies are selected among urban facilities located in the peripheral area of the city of Granada, in two areas with a high vulnerability index: on the one hand, the Zaidín neighbourhood, located in the southern district; on the other hand, the Cartuja neighbourhood, which is in the northern district of Granada. The two peripheral areas improved their urban fabric with the urban planning and development of new urban facilities due to the master plans of the late s. XX. However, urban dynamics in recent decades have led to the fragmentation and dissolution of traditional neighbourhood-community ties. Among them, we can mention the expulsion of the population towards the metropolitan periphery due to housing prices, the arrival of the migrant population and changes in the social structure, or the polarisation of the work-residence relationship. Therefore, it is essential to establish contemporary urban mechanisms that help strengthen the social urban fabric. To this end, the urban facilities selected should constitute a key variable on which daily life and the possibilities of meeting and building community networks are based. Urban facilities have been chosen for analysis, as they have enough development and time from their foundation. Thus, the urban facilities have established, in a better or worse way, a level of relationship with the urban environment to have generated the positive or negative effects and appropriation dynamics that could now be recognised and assessed (Martínez 2014; Peimbert 2019).

The methodological approach designed combines the morphological study of the urban structure of the urban facilities -the local scope of the facility and its collective functions-with the study of the sociological variable. Regarding the last one, the forms of use and appropriation of public space linked to the urban facility are analysed (Fig. 8.2). This approach allows an assessment of the facilities based on their capacity to build the collective and revitalise urban environments. A more friendly and linked urban environment around the urban facility is partially the consequence of various characteristics of their design and their implementation in the neighbourhood fabric.

Research is proposed in 2 levels, MORPHOLOGICAL and SOCIAL, both considering a triple methodological approach related to each other: CARTOGRAPHIC, QUALITATIVE and QUANTITATIVE. A multidisciplinary research team was set up so that each of the research dimensions could be formulated and developed

Fig. 8.1 General views of the accesses to the selected urban facilities in the north of Granada (from left to right: sports centre, Civic centre, and Health centre). *Source* Authors

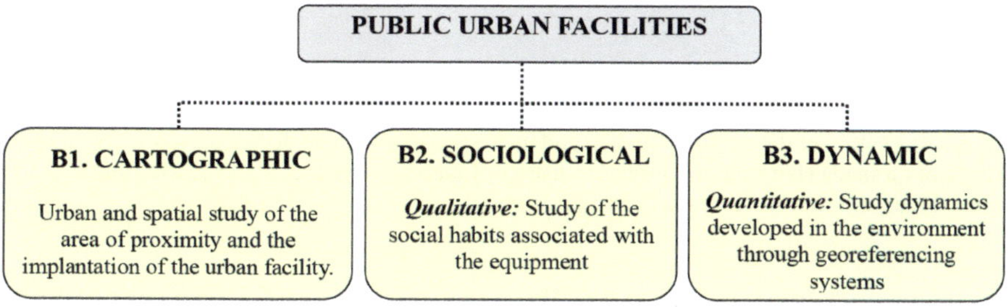

Fig. 8.2 Multidisciplinary work areas for the development of research. *Source* Authors

from expert knowledge. Furthermore, the research itinerary had to intrinsically contain the verification of contextual knowledge produced through urban analysis and fieldwork, participant observation and interviews. Three work areas are organised: B1. CARTOGRAPHIC—B2. SOCIOLOGICAL—B3. DYNAMIC (Fig. 8.3). This methodological system allows for better disciplinary development in each area and better management of researchers involved in each set of tasks.

The main objective of the research is to find out how spatial characteristics of urban facilities -free spaces of the facility, the relationship between indoor and outdoor, and public spaces around the facility, among others- influence their greater or lesser capacity to generate complementary and casual activities in their urban environment of influence. Therefore, the design criteria that allow us to take advantage of urban facilities' potential as nodes of urban activity and in the configuration of the social urban fabric are revealed.

8.4 Results

Firstly, the health service facilities (Health Centers), social services (Civic Centers) and multi-sports centers have been selected to be studied, one of each type in each district (Fig. 8.4). The aim would be to have a comparative vision but diverse. After

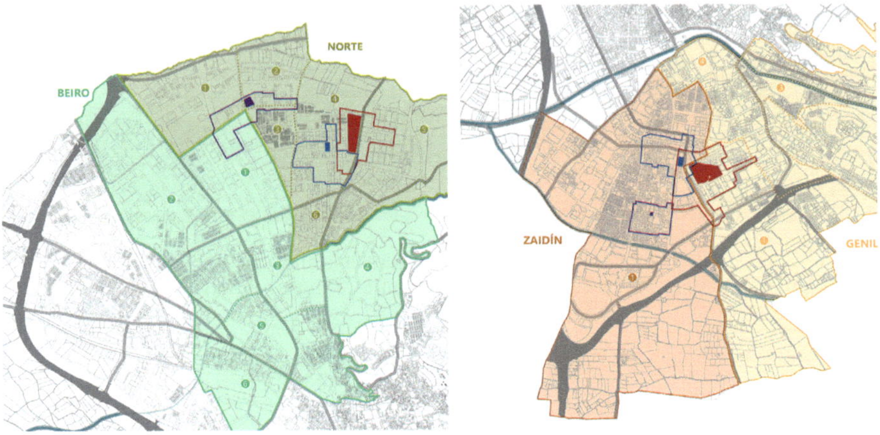

Fig. 8.3 Location of the facilities studied and their area of influence (distance 5 min). The northern district of Granada (left) and the southern district of Granada (right). Health centre (purple), civic centre (blue) and sports centre (red). *Source* Authors

the analysis of urban facility types, the target population of each facility and the intensity of use in a daily routine of the neighborhood are also addressed.

A characterisation of each urban facility environment has been developed. In this sense, its influence on sociability in the public space and the synergies with retail and other facilities located in the nearest surroundings have been observed. The concept of space attached to the urban facility (Bravo and Rivas 2021; Low et al. 1992) has been used to identify the area of influence of urban facilities. The consideration of the "space attached" has been combined with the measure of a 5-min walking distance, affected by the conditions of the urban fabric. This is the way to deduce an "accessibility contour" to the urban facility. Both information layers are contrasted with direct observation, which has allowed us to understand the forms of space used by the population. A differentiation between mandatory, optional, and resulting social activities (Gehl 2011) has been applied.

The methodology implies a development according to the following work areas:

8.4.1 Cartographic Study

An accurate cartography of the current features of the public space related to each urban facility has been prepared (Fig. 8.5). The urban analysis has meant dealing with the following tasks: the arrangement of urban furniture, the existing vegetation in the open space, night lighting, pedestrian crossing, housing typology and access, bus stops, level of urbanisation, etc. Besides, the retail and service sectors located on the ground floor, differentiated according to commercial profiles in the sphere of influence of each urban facility, have been identified. The urban and architectural

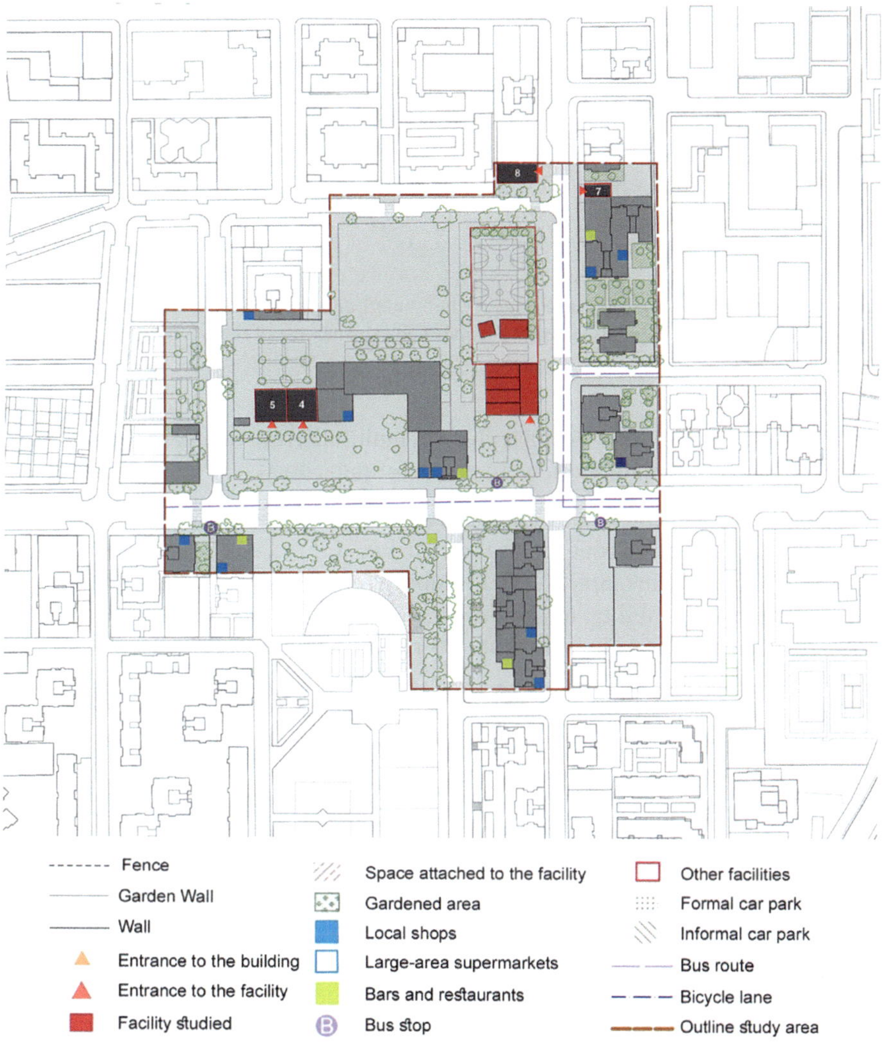

Fig. 8.4 Area of influence of the Civic Center of the Northern District of Granada. *Source* Authors

design of the limits of each urban facility is relevant to analysing the relationship existing throughout its perimeter. This is why the different types of current walls, fences, or plant filters in the boundaries of public lots for the urban facility have been pointed out.

The conditions of public space around the facility have been assessed: (1) setting a wide range of indicators to measure them; (2) analysing normative sources and related research in relation to the strategy for urban location; and (3) interpreting the influence of the adjacent urban fabrics and the design of the proximity public space of each urban facility. All this is considering its capacity to consolidate itself

Fig. 8.5 Mapping the 5-min isochron, the 10 and the 15 ones referenced the Zaidín Sur Health Center in relation to the population density (centre) and the heights of the building (right). *Source* Authors

as a social infrastructure. In the final phase, a multi-criteria evaluation is going to be carried out according to the following groups of indicators:

- Urban mobility: pedestrian traffic, distance to public transport stops, car parks and ease of pedestrian route to and from them.
- Accessibility: architectural barriers, visibility of accesses, etc.
- Urban comfort: livability, materials, benches and other urban furniture, lighting, safety, vegetation, relationship with other uses, width of sidewalks, existence of playgrounds, etc.
- Implementation in the urban fabric: relationship with the residential urban fabric, relationship with public spaces, connectivity with the main urban network, distance to other facilities, etc.

Social cohesion: population analysis, social relations, cultural activities, population scope of urban facilities, etc.

8.4.2 *Socio-Demographic Study*

The socio-demographic study of social habits associated with urban facilities has been developed through three methodological components:

(a) **Socio-demographic profile of users** of each of the urban facilities chosen.
(b) **Stratified survey** (age and sex), which collects the most relevant information on frequency of use, reasons, and motivations for the use of the facilities, way of access and the most favorable and unfavorable aspects of the services (Martos et al. 2014).
(c) **In-depth interviews with professionals and discussion groups among users** to ascertain the existence and intensity of the "social construction" of social habits beyond those that are purely functional. The interviews and meetings have been performed considering the specific characteristics of the urban space

observed and the possible emergence of "finite domains of meaning" (Schutz 1970).

The information obtained has been spatially interpreted and transferred to representative cartography of diverse forms of appropriation of the local environment of each urban facility.

8.4.3 *Study of Urban Dynamics*

On the one hand, descriptive data of the population profile and residential buildings have been analysed, and they have been related to the urban location of each urban facility through the isochrones (Fig. 8.6). These isochrones have been drawn to describe the distance to the urban facility based on the time spent walking and taking the facility as a starting point (Molero 2020). On the other hand, the social dynamics developed in the local environment are studied through georeferencing systems. Such systems have used data mining from GIS and scraping techniques on different sources of mobile information such as social networks, browsers, games, or applications associated with public transport. All the data have been collected and scanned at appropriate intervals at different times of the day and week. The point would be to obtain information on the variability of the use of public space and facilities and, with it, their influence on the local urban environment.

By obtaining isochrones for each facility, three areas of influence of the facility on the urban environment have been identified:

- AI. Influence area (10 min)
- EPA. Attached public space (5 min)
- EI. Immediate space (2 min)

In each of them, the multivariable characterisation described above has been carried out, and the indicator methodology has been used to characterise each of the environments according to certain criteria. An urban indicator is an urban variable or estimate that provides aggregated, synthetic information on a phenomenon beyond its own capacity for representation, i.e. it is exogenously endowed with an added meaning (Palenzuela 1999). An urban indicator is thus a variable that has been given added meaning in order to synthetically reflect a social concern with regard to the urban environment and to insert it coherently into the decision-making process (Leva 2005).

Urban indicators play a key role in assessing the quality of public space and its ability to promote community life. This method is an objective tool to measure and understand how an urban space contributes to the well-being of citizens and facilitates urban cohesion. Therefore, in this research, its use is relevant both in the analysis of the environment of the facilities and the study of the special conditions of the implementation of each one of them, as well as in the subsequent definition of design criteria to promote social cohesion in vulnerable neighbourhoods.

EVALUATION INDICATORS (R2)			
Goals	Scope of influence (AI)	Attached public space (EPA)	Immediate space (EI)
MOBILITY (M) Evaluate travel for facility users	• Areas and streets with pedestrian priority (M.AI.1) • Parking spaces (formal and informal) (M.AI.3)(M.AI.4)	• Proximity to public spaces from the facility (M.EPA.5) • Access to the bicycle network (M.EPA.6) • Proximity of the facilities to a bus or metro stop (M.EPA.7)	• Bicycle parking (M.EI.8) • Public transport stops with visibility (M.EI.10)
ACCESSIBILITY(A) Evaluate the ease of pedestrian traffic in the surroundings of the facility	• Accessible public transport stops (A.AI.2)	• Pedestrian accessibility (A.EPA.3)(A.EPA.4)	• Ease of access to the facility (A.EI.6) (A.EI.8) • Visibility of access to the facility (A.EI.7)
URBAN COMFORT (CU) Evaluate the quality of public space to achieve maximum well-being	• Maintenance of green areas and public use spaces close to the facility (CU.AI.4)	• Shadow in urban environments of interest close to the facility (CU.EPA.8)	• Free waiting space (CU. EI.13)(CU.EI.12)
URBAN FABRIC IMPLEMENTATION (IT) Evaluate the conditions of the existing urban fabric in relation to the location of the facility	• Other facilities (IT.AI.5) (IT.AI.6)	• Shops (IT.EPA.8) • Public space linked to the activity of the facility (IT.EPA.9)	• Service linked to the main use of the facility (IT. EI.12)
SOCIAL COHESION (CS) Evaluate the degree of integration of citizens	• Main population characteristics (CS.AI.1) (CS.AI.2) • Population density (CS. AI.4)	• Relationship spaces. Location of social gatherings (CS.EPA.6)	• Population coverage of the facility (CS.EI.8) Activities offer (CS.EI.9) CS.EI.10)

Fig. 8.6 Urban indicators were selected and categorised according to urban area for evaluation and applied in the Equisocial research facilities. *Source* Authors

Since Habitat II and the 'Urban Indicators Programme', the aim has been to define indicators that make it possible to evaluate the state of urban quality of life at a global level, in a trend towards the definition of local indicators that reflect the specificities of each territory. In the research carried out, filtering of different national and international reference documents was carried out, selecting those that are directly related to the subject of study, to conclude a proposal of indicators classified according to four dimensions:

- Quality of urban life
- Sustainable urban development
- Urban regeneration
- Gender perspective

Thus, the indicators are grouped by five urban qualities: mobility, accessibility, urban comfort, implantation in the urban fabric and social cohesion. An evaluation of these socio-urban indicators has been carried out by calculating and graphically representing them (Table 8.1).

Table 8.1 Relevant set of indicators for the assessment of the urban infrastructure of the facility. Source Authors

	Urban qualities	Criteria
Indicators of the urban infrastructure of facilities	*Mobility	Pedestrian friendly Distance to public transport stops Existence of car parks Ease of pedestrian route
	*Accessibility	Architectural barriers Visibility of accesses
	*Urban comfort	Watertightness Materialities Presence of benches Lighting Security Vegetation Relationship with other uses Width of pavements Existence of playground equipment
	*Implantation in the urban fabric	Relationship with residential fabrics Relationship with public spaces Connectivity with the main urban network Distance to other facilities
	*Social cohesion	Population analysis Social relations Cultural activities Population coverage of facilities

8.4.4 Evaluation Process

The application of these urban indicators has been carried out individually and at the appropriate scale for each of the facilities established in order to create patterns of relationship between them. A relationship is established between the indicators and the different areas of urban synergy of the facility in order to find the scale that best suits each parameter, differentiating between the sphere of influence, adjacent public space and immediate space.

The speed of movement and the distance to stimuli influence the relationship we generate with our environment and the links we establish with others. Gehl (Gehl 2011) defines the importance of distances through the 'social field of vision', whose maximum limit is 100 m, the point at which it is still possible to observe the movement of people, and the minimum limit is 25 m at which we can speak of a turning point, from which we begin to decode emotions and facial expressions. The connection between distance, intensity, closeness and warmth in contact situations has an interesting parallel when it comes to decoding and experiencing cities and urban space. In turn, this distance depends on the activities with which one wants to generate synergy since the relationship with nearby commercial activities is conditioned by the loading situation that reduces the walking distance and, therefore, requires more proximity between these activities (Pozueta et al. 2013). For this reason, the influence of the facility must be assessed in different time radii, understanding that in each environment different relationships or synergies are established with urban elements, whether they are shops, transport stops, parks, etc.

A set of indicators is proposed, categorised by the thematic objectives and by the scale of the urban area where they should be applied (Fig. 8.6).

For the evaluation of the indicators, a 30×30 m grid has been established in the area of the Adjacent Public Space (EPA), and a selection of indicators have been calculated that are considered to be priorities for the social activation of the space around the facility. Figure 8.7. Shows the calculation of the pedestrian accessibility indicator in the two civic centres, showing the best accessibility in the North Civic Centre (left), as there is a square at the entrance to the facility, and the block is surrounded by pedestrian space. However, in the Zaidín Civic Centre, the pedestrian space is fragmented and disconnected from the facility, which has a very low pedestrian accessibility index at the entrances to it.

8.5 Discussion of Results and Conclusions

Regarding the research focus, it has been revealed that the local environment of influence of the urban facility is largely consistent with a 5-min walk distance. In this area, there are retail on the ground floor and labor activities that generate synergies with the type of service offered by the public urban facility. In addition to this, the public space attached to the urban facility acts as a space for sociability, mainly if it participates in the pedestrian flows of entrances, waiting, or visibility of the own activity. This is why the remarkable public space in relation to the building must have conditions in terms of urban design: urban comfort conditions, spatiality, and physical and visual connection with the interior of the urban facility, among others. The comparative vision between neighbourhoods and between urban facilities case studies have allowed us to deduce several conclusions. Firstly, the urban facilities located in central positions of the neighbouring urban fabric develop their social function under better conditions, integrating into the daily dynamics of the local population in a better way.

Secondly, through analysing the urban facilities in vulnerable areas in peripheral districts of Granada city, it is shown that social factors (the feeling of security, isolation, or mistrust, among others) are very relevant when it comes to establishing neighbourhood networks. Thus, strengthening urban networks is a fundamental base for facilitating the use of urban facilities. Moreover, configuring its local environment as a safe space is a key factor, even more so, in helping socialisation in these areas and

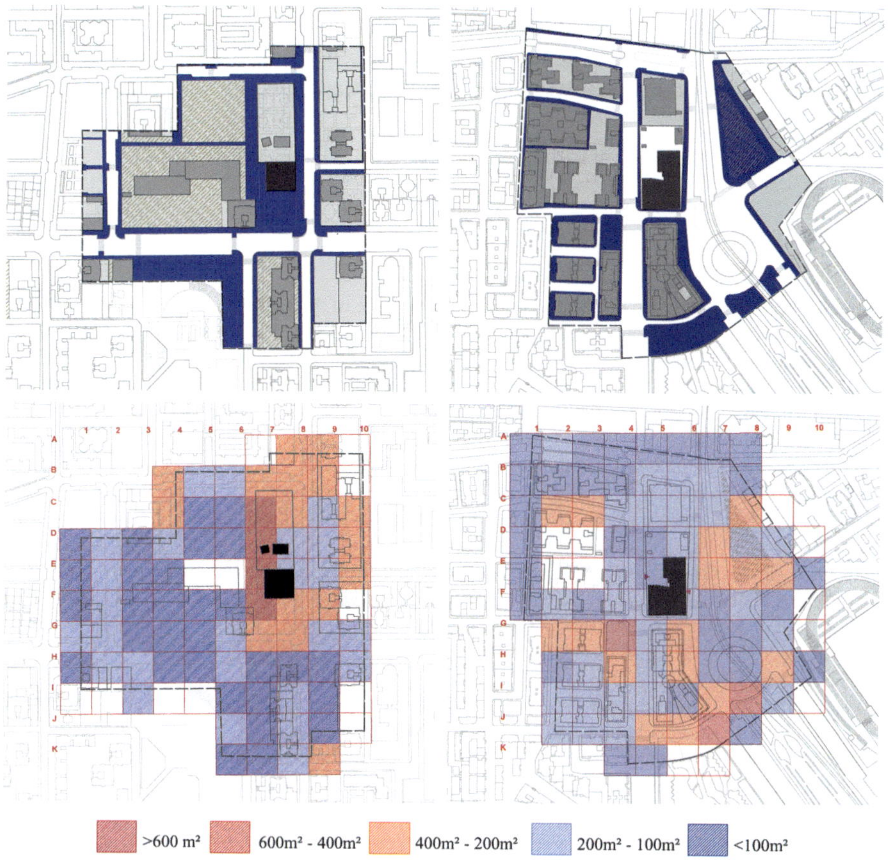

Fig. 8.7 Evaluation of the Indicator "Pedestrian Accessibility" in the Civic Centres Norte (left) and Zaidín (right). *Source* Authors

generating bonds of trust. Several studies reveal the influence of the care of public space on citizens' feelings of safety and security, or conversely, how walking past an abandoned void produces a reaction of fear and insecurity (Klinenberg 2021). It is, therefore, especially important in vulnerable neighbourhoods to pay attention to the conditions of public space and to provide comfort and safety in order to encourage neighbourhood participation in public space and to build trusting relationships.

Thirdly, a multidisciplinary methodology produces more accurate results when it comes to an understanding of the urban behaviour of the facilities studied. In addition of that, the results are more adjusted to the reality of the social use of space. The creation of social infrastructures, linking the arrangement of urban functions and public space, is the way to give physical support to the forms of neighbourhood relations and urban sociability.

Finally, interesting conclusions have been reached from the methodological point of view on the ways of interacting between specific procedures of sociology and architecture and criteria related to urban location. For this, it is crucial to the ability to understand and map the urban space involved, figuring out the urban relations from and to each urban facility, which can be extrapolated to further research and other contexts.

References

Ahmadi D (2018) Diversity and social cohesion: the case of Jane-Finch, a highly diverse lower-income Toronto neighbourhood. Urban Res Pract 11(2):139–158. https://doi.org/10.1080/17535069.2017.1312509

Ali H, Al-Betawi Y, Al-Qudah H (2019) Effects of urban form on social sustainability—a case study of Irbid, Jordan. Int J Urban Sustain Dev 11(2):203–222. https://doi.org/10.1080/19463138.2019.1590367

Allport G (1958) The nature of prejudice. Doubleday, N.Y.

Blokland N (2014) From public familiarity to comfort zone: the relevance of absent ties for belonging in berlin's mixed neighbourhoods. Int J Urban Reg Res 38(4):1142–1159

Bravo B, Rivas J (2021) Planificación y proyecto para el fortalecimiento del tejido dotacional: el distrito sur de Granada [Trad: Planning and project to strengthen the endowment fabric: the southern district of Granada]. Ciudad y Territorio Estudios Territoriales (CyTET) LIII (210): 1007–1030. https://doi.org/10.37230/CyTET.2021.210.06

Cannon B (2018) Staging urban landscapes: the activation and curation of flexible public spaces. Birkhauser, Basel

Castells M (1997) El futuro del estado del bienestar en la sociedad informacional. In: Giner S, Sarasa S (eds) Buen gobierno y política social. Ariel, Barcelona

De Certeau M (2000) La invención de lo cotidiano. Universidad Iberoamericana, Mexico

Dijkstra L (2017) 7th report on economic. Publications Office of the European Union, Luxembourg, Social and Territorial Cohesion

Gehl J (2011) Life between buildings: using public space. Island Press, London

Harvey D (2003) The right to the city. Int J Urban Reg Res 27(4):939–941. https://doi.org/10.1111/j.0309-1317.2003.00492.x

Jabareen Y (2006) Sustainable urban forms: Their typologies, models, and concepts. J Plan Educ Res 26:38–52

Klinenberg E (2021) Palacios del pueblo. Políticas para una sociedad más igualitaria. Capitán Swing

Ledrut R (1976) Sociología urbana. Instituto de Estudios de Administración Local, Madrid

Leva G (2005) Indicadores de calidad de vida urbana. Universidad Nacional de Quilmes, Teoría y metodología

Leyden K (2003) Social capital and the built environment: the importance of walkable neighborhoods. Am J Public Health 93(9):1546–1551. https://doi.org/10.2105/ajph.93.9.1546

Low S, Altman I (1992) Place attachment: a conceptual inquiry. In: Attachment P (ed) Altman I, Low S. Plenum Press, New York, pp 1–12

Martínez E (2014) Configuración urbana, hábitat y apropiación del espacio [Trad: Urban configuration, habitat and appropriation of space]. Scripta Nova, XVIII, 493(33)

Martos P, Hita C, Romero A (2014) Biosaludando. Los parques biosaludables en Granada: Interacción social y actividad físico-deportiva [Trad: Biosaluting. Biohealthy parks in Granada: Social interaction and physical-sports activity]. In: Llopis R (eds) Crisis, cambio social y deporte, Nau Llibres, Valencia. pp 149–160

Molero E (2020) SIG para cartografiar la actividad urbana: una aplicación en la ciudad de Almería [Trad: GIS to map urban activity: an application in the city of Almería]. In: Temes-Cordovez R (eds) SIG Revolution. Ordenación del territorio, urbanismo y paisaje, Síntesis, Valencia. pp 178–192

Palenzuela SR (1999) Modelos de indicadores para ciudades más sostenibles. In: Documento del Taller sobre Indicadores de Huella y Calidad Ambiental Urbana de la Fundació Fòrum Ambiental. Departament de Medi Ambient de la Generalitat de Catalunya

Peimbert A (2019) Tensiones sincopadas y narrativas paradójicas sobre el espacio público en una ciudad fronteriza [Trad: Syncopated tensions and paradoxical narratives about public space in a border city]. EURE 45(136):255–276. https://doi.org/10.4067/S0250-71612019000300255

Pinto AJ, Remesar A (2012) Urban cohesion: a guiding concept for new urban realities. In: Dias Coelho C (eds) Ambivalent Landscapes sorting out the present by designing the future. FAUTL, Lisbon

Pozueta J, Lamíquiz FJ, Porto M (2013) La ciudad paseable. Cedex, Marid

Schutz A (1970) On phenomenology and social relations. University of Chicago Press, Chicago

Solà-Morales M, Parcerisa J (1987) El urbanismo urbano. Forma urbana y planeamiento urbanístico en siete capitales españolas [Trad: Urban planning. Urban form and urban planning in seven Spanish capitals]. Estud Territ 24:33–51

Taleai M, Sliuzas R, Flacke J (2014) An integrated framework to evaluate the equity of urban public facilities using spatial multi-criteria analysis. Cities 40:56–69

Chapter 9
The Eibenthal Initiative: Sustainable Development Through Responsible Tourism

Ştefana Bădescu, Ana Branea, and Mihai Danciu

Abstract In recent years, the issue of sustainable development through responsible tourism has been at the forefront of global trends, with both theoreticians and practitioners looking for solutions to integrate the opportunities offered by tourism development in a sustainable way. Looking for examples of good practice, one seeks to identify models of intervention in terms of both specific means of implementation as well as regulations or possible directions for development. One such example is the Czech village of Eibenthal, a small settlement in the Almăj Mountains (Banat, Romania), which has become, in recent years, a reference for sustainable development through responsible tourism. Established during the nineteenth century by Czech settlers as part of a network of Czech villages in the Banat Mountains, Eibenthal and its people were deeply affected by the closure of the nearby coal mine in 2006. However, in recent years, Eibenthal has been going through a revival phase, with a number of projects currently being developed by the local community in order to attract tourists, enhance a resilient economy, and promote local values and heritage. In doing so, the Eibenthal Initiative adheres to the principles of the New European Bauhaus—with surprising results, as well as innovative functioning mechanisms.

Keywords Responsible tourism · Community involvement · Heritage valorisation

Ş. Bădescu (✉) · A. Branea · M. Danciu
Urban Planning and Architecture Research Centre, Politehnica University of Timişoara, Timişoara, Romania
e-mail: stefana.badescu@upt.ro

A. Branea
e-mail: stefana.badescu@upt.ro

Ş. Bădescu · M. Danciu
Non-Formal Studio for Spatial Planning, Timişoara, Romania

M. Vukmirović et al. (eds.), *Digital Dreams, Urban Realities*, The Urban Book Series,
https://doi.org/10.1007/978-3-031-76353-3_9

9.1 Introduction

Increasingly, theorists and practitioners alike are searching for ways to integrate the opportunities presented by tourism development in a sustainable way, ensuring that the economic impact does not disrupt the social and environmental balance. This is because the issue of sustainable development through responsible tourism has been at the forefront of global trends in recent years.

Seeking models of intervention—both in terms of particular methods of implementation and rules or potential avenues for development—one looks for instances of exemplary practice. One such instance is the Czech village of Eibenthal, a small community in Banat, Romania's steep Almăj Mountains, that has emerged in recent years as a model for sustainable development via ethical tourism.

Originally founded in the nineteenth century by Czech settlers, Eibenthal and its residents suffered greatly when the neighbouring coal mine closed in 2006. Nevertheless, the settlement is presently experiencing a renaissance, with several initiatives being established by the locals to draw tourists, strengthen a thriving local economy, and promote the area's values and legacy.

The informal initiatives demonstrate the local community's dedication to maintaining the settlement's unique identity. Moreover, to integrate the villagers in all these operations, the Eibenthal community has a private organisation that handles tourist coordination—by providing information, marketing lodging in the village, and selling local goods.

The goal of the Eibenthal Initiative is thus to achieve environmental, social, and economic sustainability. The community revitalization project is attracting increasing interest due to its adherence to the New European Bauhaus ideals in all its operations—all the way from the project's working mechanisms to its unanticipated results.

9.2 Sustainable Development of Peripheral Communities Through Responsible Tourism

Sustainable tourism is seen as an important means of promoting community development while minimising negative impacts on the environment and local cultures (Richards 2011). It is important to involve local communities in tourism planning and decision-making processes to ensure that tourism development aligns with their needs, aspirations, and cultural values. Also, relevant principles such as minimising environmental degradation, respecting local cultures, and promoting socio-economic benefits for peripheral communities are introduced as responsible tourism practices. Case studies on sustainable community development illustrate successful sustainable tourism initiatives in peripheral communities around the world, highlighting various approaches to responsible tourism and their impacts on community development. The most important challenges faced by these communities in implementing sustainable tourism initiatives are limited resources, infrastructure constraints or socio-cultural conflicts, although some of the most important opportunities for overcoming these challenges present innovative strategies and partnerships. (Buckley 2012) proposes that sustainable tourism serves as a vital means of attaining the equitable development of marginalised communities around the globe.

Peripheral communities, (Gato et al. 2022; Wanhill 1997) understood as communities residing in urban or rural areas affected by phenomena such as isolation, economic decline, depopulation and subsequent loss of relevance, represent one of the most fragile environments in the present-day context of territorial development.

At the same time, peripheral situations often occur in areas rich in valuable heritage elements—be it natural or cultural. (Mayer and Baumgartner 2014) As such:

The isolation from main mobility and economic development corridors usually indicates an area of natural beauty (be it up in the mountains, along water courses or in protected areas for biodiversity);

The economic decline usually leaves behind traces of an industrial past, manifested in the form of a network of industrial heritage (with both architectural and symbolical value);

The depopulation of a settlement sometimes allows for the involuntary preservation of domestic architecture (which is no longer transformed to fit the changing needs of the inhabitants).

Peripheral communities may leverage the potential advantages of tourism (Salvatore et al. 2018) while reducing its negative effects by implementing responsible tourism practices, such as community participation, environmental protection, and cultural preservation. But to fully realise this potential, all parties involved—governments, enterprises, local communities, and even visitors themselves—must work together. Planning and managing tourism should prioritise sustainability principles in order to promote equitable growth, preserve cultural and natural heritage, and improve the long-term well-being of outlying communities. In the United Nations World Tourism Organization's (UNWTO) report "Tourism and the Sustainable Development Goals—Journey to 2030," (Tourism and the Sustainable Development Goals 2024) the UNWTO highlights how important responsible tourism is in supporting the sustainable development of peripheral communities. The paper emphasises that in these frequently neglected communities, tourism can be a potent instrument for promoting economic growth, conserving cultural heritage, and safeguarding natural resources, provided it is handled appropriately. Peripheral communities may increase social cohesiveness, provide new economic possibilities, and improve the standard of living for citizens by enabling local communities through sustainable tourism initiatives, including capacity-building programs and community-based tourist businesses. In addition, the UNWTO emphasises how critical it is to match the growth of the tourism industry with the sustainability ideals expressed in the Sustainable Development Goals (SDGs) of the United Nations. This will guarantee that tourism supports larger initiatives aimed at reducing poverty, protecting the environment, and promoting inclusive growth. By means of strategic alliances, regulatory frameworks, and stakeholder involvement, responsible tourism manifests as a transformative agent, propelling the sustainable advancement of marginalised communities in the direction of greater prosperity and fairness.

Peripheral communities, therefore, usually benefit from a wide patrimonial infrastructure that can—and should—be valorised in order to help these communities regain a certain degree of economic resilience and thus ensure the sustainable development of the territory. (Dinis 2004, 2006) As such, the practice of responsible tourism often constitutes a suitable solution for the revitalisation of peripheric communities (Munoz 2003; Virkkala 2007), building on the local values and valorising both the natural and cultural heritage of the area.

The sustainable development of a territory through responsible tourism to attract visitors, strengthen the local economy and promote local values and heritage is a significant driver for both urban and rural growth. As such, it has the potential to:

Prevent residents from leaving the area;

Promote everyday economic activities, customs and products as essential components for tourism and sustainable development;

Preserve site-specific building techniques as mediators between the built and natural environment in a cultural landscape rich in potential.

In order to facilitate the sustainable development of responsible tourism, communities and authorities alike (Gato et al. 2022; Hospers 2003) need to work together to better understand and strengthen the following issues: promotion and conservation of natural and cultural resources, economic, social and environmental sustainability, tourism development through value chain integration, good governance for the promotion of tourism, good infrastructure and connectivity, and safety and security.

However, tourism development can only provide sustainability if the services and products delivered meet certain quality standards. (Pezzi et al. 2020) This applies to accommodation, catering and tourist orientation services, as well as tangible and intangible products specific to the area. This criterion can be objectively measured on three distinct levels of assessment. Firstly—the existence of the necessary approvals, which guarantee a minimum of quality for each service and product; second—increased visibility (such as participation in fairs online presence) and, subsequently, sales; and finally—the award of quality labels (such as designation of controlled origin / organic products / local origin seals / other quality seals).

9.3 Responsible Tourism Through Community Involvement

9.3.1 The Network of Czech Settlements in the Almăj Mountains. Evolution and Distribution

The Czech villages in the Almăj Mountains (Fig. 9.1) were founded during the nineteenth century as part of the process of colonisation of these territories with Central European populations, initiated by the Habsburg imperial administration in

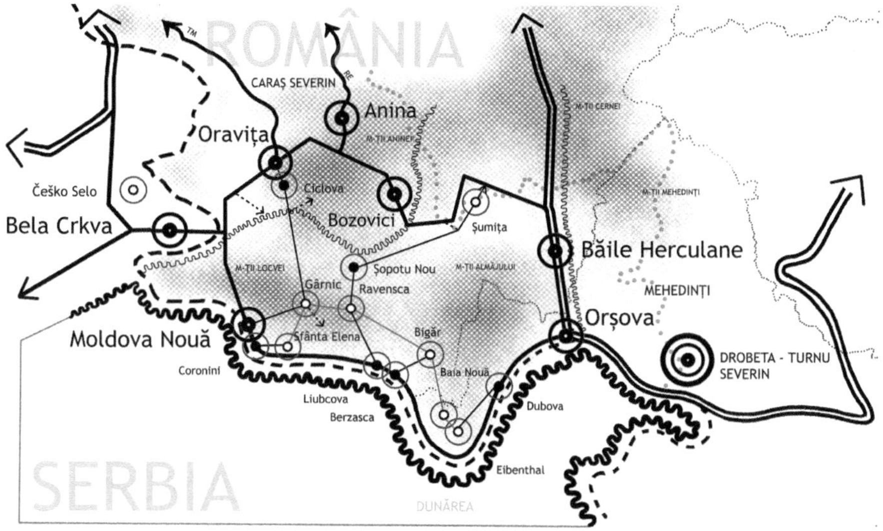

Fig. 9.1 The network of Czech settlements in the Almăj Mountains. Territorial chorema

the eighteenth century (and continued by the Austro-Hungarian administration until the end of the nineteenth century).

The purpose of these efforts was twofold. On the one hand, the aim was to exploit existing resources in the area (forestry, coal deposits and other subsoil resources) in an organized manner by communities with experience in these areas. On the other hand, the colonization of this region was also aimed at consolidating the borders of the empire—the Iron Gates being a critical area of direct contact with the territories under Ottoman occupation (or influence).

Known as "semi" (a name reflecting the area of origin of the settlers—i.e. Bohemia), the Czechs in the Almăj Mountains still have seven rural settlements today, forming a territory-wide network. It is also worth mentioning the Czech settlements that are still preserved today on the territory of Serbia, having a similar historical background—but a much weaker representation in the contemporary context.

Typically developed in regions of rugged terrain near traditionally exploited resources, the Czech settlements in the Almăj Mountains are remote from today's major mobility corridors and are generally accessible by narrow, winding roads. The reduced accessibility, as well as the gradual decline of the economic activities traditionally carried out by these communities, contribute to the peripheral character of the Czech settlements—strongly affected by phenomena such as depopulation or exclusion.

Today, however, these settlements are undergoing a phase of gradual revitalisation, with local communities currently developing a series of projects aimed at promoting local values and heritage in order to increase the level of tourist attraction and strengthen a resilient economy.

The network of Czech settlements in the Almăj Mountains (Fig. 9.1) includes:

Three settlements located in Caras-Severin county, accessible from the Danube area: Gârnic I Gerník, Sfânta Elena I Svatá Helena and Bigăr I Bigr.

Two settlements in Caras-Severin county, accessible from the Almăj region: Şumiţa I Šumice and Ravensca I Rovensko;

Two settlements in Mehedinţi county, accessible from the Danube area: Eibenthal I Tisové Údolí and Baia Nouă I Nové Doly.

This network is completed by the Czech settlements on the current territory of Serbia, the most important of which is Cesko Selo (near Bela Crkva).

9.3.2 *The Village of Eibenthal. Challenges and Opportunities*

The main challenges of the village of Eibenthal. The main challenges of the village of Eibenthal (Fig. 9.2), as well as of the other Czech settlements in the Almăj Mountains, generally refer to:

Low accessibility, caused by the limited quality of access roads; however, this situation prevents the development of mass tourism—which can actually be considered an opportunity;

The massive depopulation of recent years, coupled with the ageing of the local residents—resulting in the loss of valuable members of the community (through migration or by natural means);

The low impact of economic activities, other than tourism, on the well-being of the inhabitants.

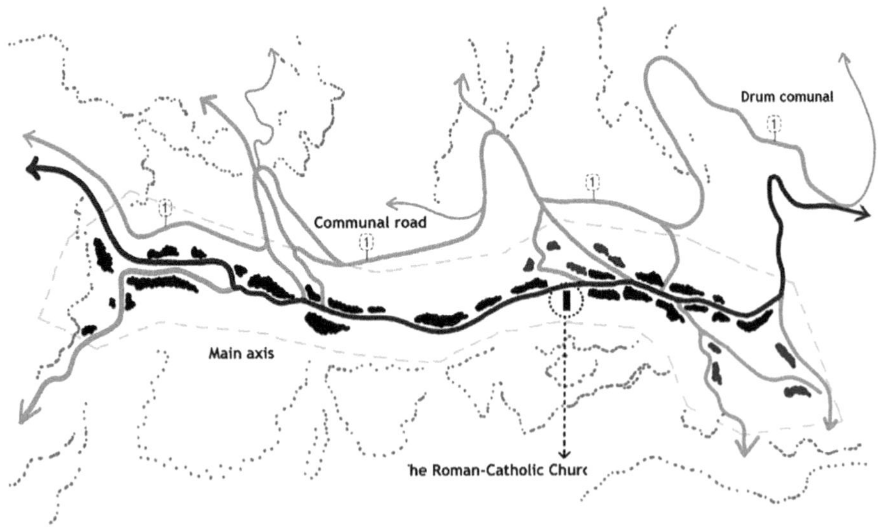

Fig. 9.2 The village of Eibenthal

The main opportunities are in the village of Eibenthal. The main opportunities of the village of Eibenthal (Fig. 9.2) refer to the cultural and natural heritage of the area.

The village of Eibenthal thus preserves a number of cultural resources, such as:

Tangible heritage: The Catholic Church "St. Joan Nepomuk" in Eibenthal, built in 1912 as part of a program of ecclesiastical architecture, implemented within the Czech communities between 1858 and 1923;

Tangible heritage: remnants of built heritage assets reminiscent of mining activity, especially in Baia Nouă: dwellings for miners, managers of mining units or industrial heritage sites;

Tangible heritage: traditional residential architecture, in the form of entire households or individual buildings that preserve valuable characteristics regarding the physical organisation of the household, the spatial and functional configuration of the buildings themselves, as well as the general aesthetics of the facades and of the property limits; such examples (most of which are relatively well preserved) can be seen all throughout the Czech settlements in the Almăj Mountains, contributing to a coherent ambience within the rural environment;

Tangible heritage: traditional functional architecture, in the form of isolated households for agricultural activities, situated outside the villages—some of which still preserve their original form and function;

Intangible heritage: traditional Czech dishes served in the Eibenthal Restaurant (such as garlic soup—"Česneková polévka"), traditional Czech beer (the "Eibenthal" beer, produced in the Czech Republic, but using a local recipe);

Intangible heritage: traditional Czech folk performers and traditional Czech folk costumes.

At the same time, the village of Eibenthal is part of the Iron Gates Natural Park—a protected area established by Law no.5 / 2000 on the approval of the National Spatial Planning Plan—Section III—Protected Areas. This is a territory in which the remarkable landscapes and biological diversity can be capitalised in the conditions of preserving unaltered traditions, where the quality of life of communities is to be the result of the economic activities of the inhabitants, carried out in harmony with nature. Holder of many recognised natural resources, the protected area corresponds to IUCN category V: "Protected landscape: protected area managed mainly for landscape conservation and recreation".

Although the village of Eibenthal does not contain officially recognised cultural heritage objectives, considered valuable at national or international level, some European-funded projects currently seek to recognise the quality of the tourist services offered in Eibenthal, as well as their impact on the economic development of the village and of the larger area.

The specifics of the region, its stories, and the current initiatives developed locally by members of the community demonstrate the capacity of the settlement for economic recovery through tourism. At the same time, the capacity of tourism entrepreneurs in the area to understand the importance of—as well as the opportunities which derive from—the preservation of the local specificity represents a strong argument in favour of awarding a quality seal that internationally recognises the village of Eibenthal, as well as other initiatives developed by the Czech communities all throughout the Almăj Mountains, as good practices.

9.3.3 *The Eibenthal Initiative*

Established during the nineteenth century by Czech settlers as part of a network of seven Czech settlements developed in the Almăj Mountains (present-day Romania) under Austro-Hungarian rule, the village of Eibenthal and its people suffered immensely following the closure of the nearby coal mine in 2006. However, in recent years, Eibenthal has been leading a revival process impacting the entire region, with a number of projects currently developed in order to attract tourists, enhance the local economy and promote the local values and heritage. The main objectives of Eibenthal's tourism development, as a driver for the growth of the wider rural area, comprise:

- Preventing inhabitants from leaving the area and thus limiting the negative impact of depopulation and shrinkage—two phenomena that usually affect rural areas but can prove particularly challenging for communities such as the one in Eibenthal, which is undergoing a recovery process following the 2006 disaster that led to the closure of Baia Nouă coal mine;
- Promoting day-to-day manufacturing activities, customs and products as essential components for sustainable tourism and development in the area;
- Preserving site-specific building techniques as mediators between built and natural environment in a cultural landscape rich in potential.

The most important opportunities brought by tourism to the village of Eibenthal, as well as to the other Czech settlements in the Almăj Mountains, refer to:

Higher employment rates—the growth of the tourism sector provides more jobs, both directly (in tourism businesses) as well as indirectly (in adjacent fields, either supporting the tourism sector or developed as opportunities determined by the economic growth); these refer to the administration of the public domain, which is done locally in Eibenthal, as well as to the development of individual initiatives, ideally by each family in the community;

Better access to information—the continuous influx of tourists brings not only people but also knowledge to the village; as such, the local community of Eibenthal developed its informal policies regarding ecology, health, economics and even tourism development through the direct interactions with foreign tourists;

A diverse and extended network of individuals who work towards the development of the region—as many of those who come to the area as visitors become advocates for its further growth.

In order to promote tourism in the village of Eibenthal, as well as in the wider area, the local entrepreneurs work closely with communities from the Czech Republic and organise, mainly for them, an annual festival: Banat Festival. This event provides dedicated transportation (from Prague to Orşova and, further on, to Eibenthal) that allows Czech tourists to take part in this festival. Of course, the event is open to everyone, but the connection with the country of origin helps to preserve the specificity of the place.

Furthermore, the Banat Festival creates the premises for local producers to promote their services while also receiving informal confirmation regarding the quality of their products from foreign customers. Last but not least, during the Banat Festival and in other circumstances, participants are encouraged to explore the wider area using the pedestrian routes that connect Eibenthal with the other Czech settlements in the Almăj Mountains.

The success of organising a large international event that contributes to the sustainable development of the village, as well as of the wider area, is exceptional, as a natural accentuation of its usual rhythms, as well as a demonstration that the rural context can live through tourism. This opportunity is further developed through the organisation of a series of additional events (such as the Beer Festival), aiming to cover the entire tourist season.

Another representative component is the combination of the opportunities offered for mass, cultural and event tourism, each of which contributes to the economic sustainability of the context.

At the same time, the practice of responsible tourism developed by the village of Eibenthal impacts the entire network of Czech settlements in the Almăj Mountains, which share a common history—while also preserving specific characteristics that contribute to a rich tourist offer.

Last but not least, the transfer of knowledge from tourism entrepreneurs to villagers, through examples of good practice, encourages a certain local specificity in terms of building and managing households.

9.3.4 *The Commitment of the Local Community for the Promotion and Conservation of Cultural Resources*

The local community from the village of Eibenthal supports the conservation and promotion of cultural resources through the utilisation of construction methods specific to the area, using local materials and techniques. Specific crafts are also promoted through the regular organisation of fairs and the presentation of products at outdoor events. Another relevant aspect is the cooperation with the only academic institution in the region (Universitatea Politehnica Timişoara) that has the capacity to produce strategic planning documents through scientific research. Thus, a guide for preserving the local specificity within the Czech settlements in the Almăj Mountains has recently been elaborated—and will be put at the disposal of all those who build in the region (RuRAl 2024; RuRAl 2024).

The commitment of the local community, a recurring theme of local rural entrepreneurism (Kneafsey 2000; North and Smallbone 2006), in the village of Eibenthal to preserve the specific character of the settlement can be seen in all the new tourist developments that have appeared here in the last 15 years (such as the bar "U medvěda", the Eibenthal guesthouse, the Eibenthal restaurant or the Manufaktura production cooperative). Moreover, nowadays, similar initiatives appear in the village, following the good example of the former.

At the same time, it is remarkable the way in which the local community reacts to certain projects, such as the recent rehabilitation of Eibenthal's parish church, having noticed that during the construction works, some of the original details of the church have been damaged, the locals interfered and demanded the immediate correction of this situation—thus ensuring the preservation of one of the most important buildings within the village.

The Eibenthal community manages a private structure that coordinates tourists called U Medvěda. It is a multifunctional space that has the role of informing tourists, offering local products and promoting accommodation in various structures in the village. More than a tourist information centre, it aims to involve the community in all types of activities carried out.

In the village of Eibenthal, one can find several accommodation options that reflect rural and local values. Adhering to the principle of using site-specific construction techniques and local materials, accommodation spaces such as the Eibenthal Pension, as well as other private dwellings open for tourism, promote the specifics of the area and thus provide authentic experiences to tourists while preserving the character of the settlement itself.

Eibenthal Restaurant is another family business in the village of Eibenthal, which offers tourists authentic experiences inspired by traditional cuisine. As such, the menu comprises local dishes (such as garlic soup or slow-cooked pork chops), which are still prepared according to traditional recipes and procedures.

In addition, the Eibenthal community has also established a cooperative for the production of culinary goods (such as sausages, cheese, canned food, syrups and herbal teas), which continues the generational practices and produces enough to cover the daily needs of the restaurant (selling the surplus to tourists).

Within the village of Eibenthal, local farmers and producers are included in the tourism activities. As such, local produce and other culinary goods are utilised within Eibenthal Restaurant and even sold as such to tourists. Moreover, local fairs are organised periodically (during the Banat Festival, the Beer Festival and other occasions), during which local craftsmen are invited to present and sell the manufactured goods they produce (textiles and woven toys, wooden objects and toys, canned goods and so on). These events also create an opportunity for producers from other Czech settlements in the Almăj Mountains to promote their products.

At the moment, the recognition of quality standards regarding tourism-oriented businesses within the village of Eibenthal, as well as within the other Czech settlements in the Almăj Mountains, is being considered at the European level through a Quality Seal currently developed by DIA—Danube International Association. At the same time, the local brand of "Eibenthal" is gradually being promoted, with specific products being sold in a dedicated shop for tourists, inaugurated in Eibenthal in April 2023.

Finally, in order to promote tourism in the village of Eibenthal, as well as in the neighbouring area, local entrepreneurs work closely with communities in the Czech Republic and organise, mainly for them, an annual festival: the Banat Festival. The relation between small-scale, bottom-up initiatives such as festivals organised by rural communities and their revitalisation strategies is well documented. (Anwar McHenry 2011; Black 2016; Borrup 2016; Cwiertka and Machotka 2018; Qu and Cheer 2021; Richards 2011) This event also offers dedicated transport (from Prague to Orsova and, further, to Eibenthal), which allows Czech tourists to take part in this festival. Of course, the event is open to all, but the connection with the country of origin helps to preserve the specifics of the place. Moreover, the Festival creates the premises for local producers to promote their services while receiving an informal confirmation of the quality of their products from foreign customers. (Borrup 2016) Thus, the village of Eibenthal contributes to the promotion of local values and practices in the area through the sale (and production) of specific Czech food and beverages, household goods and other products.

By adhering to the principles of the New European Bauhaus in all its activities, the Eibenthal Initiative thus seeks to reach economic, social and environmental sustainability.

9.3.5 The Economic Sustainability

The local community from the village of Eibenthal is working towards obtaining financing at different levels, in order to improve the existing tourism infrastructure. As such, the village recently developed a new access road, a new lightning system and promotes financing opportunities to the locals in areas such as education, accessibility and tourism.

The economic activities that have already succeeded here (the "U medvěda" bar, the Eibenthal Restaurant, the Eibenthal Pension accommodation or the Manufaktura production cooperative) are the living proof of a business development conducive environment within the village of Eibenthal. Moreover, their success encourages others to follow the same pattern and thus contribute to a growth in the tourism capacity of the settlement as a whole (through accommodation, services, products and so on). Following Eibenthal's example, similar initiatives are being developed by all the other Czech settlements in the Almăj Mountains, demonstrating a business-conducive environment throughout the region.

Finally, as part of the "Oltenia Danubius" Microregion, the village of Eibenthal is included in the Integrated Territorial Development Strategy of this area, which focuses on measures to increase the economic sustainability of the Dubova Municipality (including Eibenthal). According to this strategy's action plan, investments concern water supply and sewer rehabilitation, modernisation of communal roads, an extension of the public lighting network, construction of a cultural centre, the establishment of a tourist information centre and modernisation of forest and agricultural roads.

Moreover, at the initiative of the Association of Czech People in Banat, a series of joint projects are being developed in collaboration with the Czech government, as well as with other Czech public authorities, aiming to increase the opportunities for tourism development in the entire region.

Similar measures are being prioritised for all the other Czech settlements in the Almăj Mountains—some of which are included in the neighbouring microregions.

9.3.6 The Social Sustainability

Although there are no official policies that foster employment in the tourism sector within the village of Eibenthal, the examples of good practices already implemented here represent informal models for growth. Usually developed as family endeavours, the local tourism businesses employ all the members of the family unit, thus ensuring a uniform distribution of efforts and resources alike.

This model is gradually being reproduced in all the other Czech settlements in the Almăj Mountains, with family businesses being developed in the tourism sector all throughout the region.

In the village of Eibenthal, all children have access to school enrolment. As such, the numbers indicate:

Kindergarten: 7 children (2 boys, 5 girls).
Primary school: 12 children (7 boys, 5 girls).
Secondary school:
5th grade—0 children,
6th grade—3 children (1 boy, 2 girls),
7th grade—3 children (2 boys, 1 girl),
8th—0 children.

These numbers indicate a constant growth in the number of school children in the village of Eibenthal—with similar situations in the other Czech settlements in the Almăj Mountains.

Eibenthal Village seeks to promote informal policies of integration and education, regardless of gender or age. Non-formal and informal educational activities are based on transversal principles, such as respect towards the environment, social equity and gender equality; these are considered essential for the sustainable development of a responsible tourist destination. As family businesses, the economic activities developed by the inhabitants include all the members of a household (including women) while also having a significant impact on the education of the children.

Thus, the village of Eibenthal promotes education on two levels:

Formal education—in the local school, where, with the help of the teachers, the curricula are updated to reflect the contemporary realities and thus provide the relevant skills;

Non-formal and informal education—within families / in the work environment, where the passing of knowledge and skills regarding traditional crafts and customs happens naturally and the educational activities are based on transversal principles, such as respect towards the environment, social equity and gender equality.

More recently, the Czech settlements in the Almăj Mountains began hosting Erasmus students from European countries, invited to work with local teachers in order to organize both educational and leisure activities for local children.

Vulnerable populations in the village of Eibenthal, as well as in other Czech settlements in the Almăj Mountains, include:

- Czech ethics (the majority of the village's population, but a minority in the country); the entire touristic development of the settlement is built around the Czech specific and the ethnic minority;
- Children (vulnerable age group) benefit from both formal education programs and their inclusion in informal practices connected to tourism;
- Elders (a vulnerable age group) benefit directly from tourist development through the promotion of their main activities (accommodation services, manufacturing products and goods) as the core of touristic experiences.

The latest project for modernisation and rehabilitation of the communal roads in Dubova commune—villages of Eibenthal and Baia Nouă, improve mobility within the settlements but also increase accessibility for disabled travellers. Since Eibenthal is a rather quiet village, with one narrow main road, without any public transport (for reasons of economic sustainability), and a single way for communication, the investment produced a major impact on the local community by improving the accessibility between households. Moreover, locals offer access services for disabled persons from the main national road upon prior request.

Working towards recovery (following the 2006 disaster that led to the closure of the Baia Nouă coal mine), the village of Eibenthal is currently gradually changing its profile—from that of a mining community to a settlement promoting sustainable tourism development. Through this process, both economic activities and mentalities are being challenged and transformed. At the same time, the gradual success of the Eibenthal Initiative and its positive impact on the socio-economic development of the local community is currently inspiring other Czech settlements in the Almăj Mountains, thus leading a revival process all throughout the region through responsible tourism.

9.3.7 The Environmental Sustainability

Through the projects it implements, the Eibenthal community puts into practice the policy of using local resources. Moreover, the educational programs carried out at the village school focus on respect and protection for the environment.

Thus, the village of Eibenthal focuses on the subject of climate change, by putting this aspect at the core of the education programs for local children and thus emphasizing the need to understand this concern (including the impact it has on local development). Also, the local community periodically organises workshops and presentations on the subject of ecology and healthy living for adults.

The Banat Festival is also centered around activities that raise awareness regarding the quality of the natural environment, which are dedicated to both Romanian participants, as well as to those who come from the Czech Republic. At the same time, the transport made available to the participants is a sustainable one: the access from the Czech Republic is facilitated through a long-distance charter train, entitled "Banat Express".

Last but not least, the location of the village in the Iron Gates Natural Park allows for the implementation of joint projects to inform the local community and tourists about the environmental values that need protection. The tourist activities in the village of Eibenthal are therefore built around the motto "without plastic", with the products and drinks being served directly on the spot, in reusable glass or ceramic containers. Also, the largest event organised in the village (the Banat Festival) promotes the reuse of containers by limiting the possibility of purchasing products in plastic containers or packaging.

For the village of Eibenthal, the most important infrastructure objective (following the modernisation of the access road) is the implementation of a water supply and sewage system. Dubova Municipality already initiated the formal procedures in this regard. Until then, however, traditional methods of water supply and wastewater are utilised, teaching young generations to use them rationally in order to preserve the available resources. Also, the local community periodically organises workshops and presentations on the subject of waste management and healthy living.

With such a small and closely-knit community, the policies implemented in the village of Eibenthal are primarily informal, being kept as a custom of a growing settlement in search of long-term sustainability. However, given the development of tourism in the village in recent years, the implementation of such written policies has become mandatory in order to cope with the expected pressure of tourists in the coming years. To this extent, it is worth mentioning the Action Plans for a more performant Danube Region, developed in 2022 within an InterReg-funded international cooperation project along the Danube (DANUrB +) (DANUrB+—Interreg Danube 2024) which tackle the situation of the Czech settlements in the Almăj Mountains.

9.4 Conclusions. A Community-Based Model

Part of a network of seven Czech settlements developed in the Almăj Mountains (present-day Romania) under Austro-Hungarian rule, the village of Eibenthal is currently leading a revival process impacting the entire region. By adhering to the principles of the New European Bauhaus in all its activities, the Eibenthal Initiative stands out through both its informal management processes and its determination to influence the entire network of Czech settlements in the Almăj Mountains for the better.

The approach promoted by the Eibenthal Initiative was that of creating a community-based model for sustainable growth through responsible tourism. (Gato et al. 2022) This refers to a gradual process that develops organically, building upon previous experiences in order to constantly improve the existing situation and adapt to new stimuli.

At the same time, the entire process is built around the community's needs and rhythms; as such, the desirable end result is not that of achieving unconditional economic growth but rather that of creating a safe and inclusive environment in which people can live their lives freely. Eibenthal Village thus seeks to promote informal policies of integration and education, regardless of gender or age. As family businesses, the economic activities developed by the inhabitants include all the members of a household (including women) while also having a significant impact on the children.

Eibenthal also stands out through its informal processes of management in tourism, specific to small communities, in which the institutionalisation and implementation of rigid policies could affect the existing relationships in the local community and the specifics of a place that is attractive precisely because of its naturalness. The mechanisms developed by the Eibenthal community can be recognised as exceptional precisely because of their bottom-up, informal character. Therefore, the development of tourism in the village of Eibenthal is, in fact, the result of entrepreneurial initiatives developed around community involvement regarding the common interest, in which the support of local authorities is limited to infrastructure.

Moreover, environmental preservation and cultural preservation are given priority in Eibenthal's sustainable development initiatives in addition to economic growth. The Almăj Mountains' natural beauty and the distinct cultural legacy of the Czech communities are valued assets that the community understands the significance of preserving. The settlement strives to preserve its natural and cultural resources for future generations by encouraging eco-friendly lodging, putting in place waste management systems, and planning cultural events that honour regional customs. These are examples of responsible tourism activities. The community works to achieve a balance between economic success and environmental stewardship by incorporating sustainability into all facets of tourist development, guaranteeing the long-term survival of its tourism business.

Furthermore, other outlying towns confronting comparable possibilities and problems should draw inspiration from Eibenthal's successful sustainable tourism approach. It serves as an example of how responsible tourism may spur positive change on a larger scale by disseminating exemplary practices and experiences to other towns. Peripheral communities have the opportunity to learn from each other's experiences, modify tried-and-true methods to fit their particular circumstances, and work together to create a more sustainable future by means of cooperative networks and knowledge-sharing platforms. In addition to helping its own citizens, Eibenthal's leadership in sustainable tourism advances the resilience and general development of the surrounding area.

To sum up, the instance of Eibenthal serves as an example of how responsible tourism may significantly impact the sustainable development of outlying villages. Through a focus on community empowerment, environmental protection, and cultural preservation, the settlement has established a booming tourism destination that protects the natural and cultural resources that are distinctive to the town and improves the well-being of its citizens. Its creative and inclusive approach to tourist development makes it a global source of hope for marginalised people, proving that ethical tourism can be an effective means of promoting social justice, economic growth, and environmental sustainability. As the globe moves closer to sustainable development, Eibenthal serves as an example of how community-driven solutions and grassroots efforts may make a beneficial impact.

References

Anwar McHenry J (2011) Rural empowerment through the arts: the role of the arts in civic and social participation in the Mid West region of Western Australia. J Rural Stud 27:245–253. https://doi.org/10.1016/j.jrurstud.2011.03.001

Black N (2016) Festival connections: how consistent and innovative connections enable small-scale rural festivals to contribute to socially sustainable communities. Int J Event Festiv Manag 7:172–187. https://doi.org/10.1108/IJEFM-04-2016-0026

Borrup T (2016) Creative placemaking: arts and culture as a partner in community revitalization

Buckley R, Buckley R (2012) Sustainable tourism: research and reality. Ann Tour Res 39:528–546. https://doi.org/10.1016/j.annals.2012.02.003

Cwiertka KJ, Machotka E (2018) Consuming life in post-bubble Japan. Amsterdam University Press

DANUrB+—Interreg Danube. https://www.interreg-danube.eu/approved-projects/danurb_plus. Accessed 19 Mar 2024

Dinis A (2004) Territorial marketing: a useful tool for the competitiveness of rural and peripheral areas. Eur Reg Sci Assoc ERSA Conf Pap

Dinis A (2006) Marketing and innovation: useful tools for competitiveness in rural and peripheral areas. Eur Plan Stud 14:9–22. https://doi.org/10.1080/09654310500339083

Gato MA, Costa P, Cruz AR, Perestrelo M (2022) Creative tourism as boosting tool for placemaking strategies in peripheral areas: insights from Portugal. J Hosp Tour Res 46:1500–1518. https://doi.org/10.1177/1096348020934045

Hospers GJ (2003) Localization in Europe's periphery: tourism development in Sardinia. Eur Plan Stud 11:629–645. https://doi.org/10.1080/0965431032000108369

Kneafsey M (2000) Tourism, place identities and social relations in the European Rural Periphery. Eur Urban Reg Stud 7:35–50. https://doi.org/10.1177/096977640000700103

Mayer H, Baumgartner D (2014) The role of entrepreneurship and innovation in peripheral regions. DisP Plan Rev 50:16–23. https://doi.org/10.1080/02513625.2014.926720
Munoz F (2003) Lock living: urban sprawl in Mediterranean cities. Cities 20:381–385. https://doi.org/10.1016/j.cities.2003.08.003
North D, Smallbone D (2006) Developing entrepreneurship and enterprise in Europe's peripheral rural areas: some issues facing policy-makers. Eur Plan Stud 14:41–60. https://doi.org/10.1080/09654310500339125
Pezzi MG, Faggian A, Reid N (2020) Agritourism, wine tourism, and craft beer tourism: local responses to peripherality through tourism niches. Routledge
Qu M, Cheer JM (2021) Community art festivals and sustainable rural revitalisation. J Sustain Tour 29:1756–1775. https://doi.org/10.1080/09669582.2020.1856858
Richards G (2011) Creativity and tourism: the state of the art. Ann Tour Res 38:1225–1253. https://doi.org/10.1016/j.annals.2011.07.008
RuRAl—Rural Relevant în Actualitate (2024) Așezările de cehi din Munții Almăjului: ghidul de intervenție by Atelierul NonFormal de Planificare Spațială Timișoara—Issuu. https://issuu.com/stefana.badescu/docs/20231125_brosura_pemeleaguri_-_etapa_de_ghid. Accessed 19 Mar 2024
RuRAl—Rural Relevant în Actualitate (2024) Așezările de cehi din Munții Almăjului: analiza by Atelierul NonFormal de Planificare Spațială Timișoara—Issuu. https://issuu.com/stefana.badescu/docs/20231116_brosura_pemeleaguri_-_analiza. Accessed 19 Mar 2024
Salvatore R, Chiodo E, Fantini A (2018) Tourism transition in peripheral rural areas: theories, issues and strategies. Ann Tour Res 68:41–51. https://doi.org/10.1016/j.annals.2017.11.003
Tourism and the Sustainable Development Goals—Journey to 2030. In: UNDP. https://www.undp.org/publications/tourism-and-sustainable-development-goals-journey-2030. Accessed 19 Mar 2024
Virkkala S (2007) Innovation and networking in peripheral areas—a case study of emergence and change in rural manufacturing. Eur Plan Stud 15:511–529. https://doi.org/10.1080/09654310601133948
Wanhill S (1997) Peripheral area tourism: a European perspective. Prog Tour Hosp Res 3:47–70. https://doi.org/10.1002/(SICI)1099-1603(199703)3:1%3c47::AID-PTH38%3e3.0.CO;2-F

Part IV
Mezzo Scale: Holistic Perspective and Land Use

Mezzo Scale: Bridging Vision and Detail

In the intricate tapestry of urban design and planning, the mezzo scale is a vital link that bridges the expansive vision of macro-scale strategies with detailed, human-centered interventions at the micro-scale. The mezzo scale focuses on the planning and organization of neighborhoods, districts, and specific urban areas within a city or region. On this scale, the broad goals of sustainability, inclusivity, and beauty take a tangible form in the daily lives of urban residents.

The mezzo scale is where ambitious strategies laid out at the macro-level are translated into functional, livable, and cohesive urban environments. This scale is critical for creating urban areas that function efficiently and resonate with the social and cultural identities of the communities they serve. By carefully considering the distribution of land uses, the design of public spaces, and the integration of green infrastructure, planners and designers at the mezzo scale craft the environments that shape our everyday experiences in the city (Jacobs 1961; Lynch 1960).

A central theme of the New European Bauhaus initiative is creating spaces that are sustainable, inclusive, beautiful, and enriching. The mezzo scale is particularly suited to this endeavor, as it involves the detailed planning of neighborhoods and districts where these values can be most effectively realized. Carmona highlights the importance of public spaces at this scale, emphasizing their role in enhancing social cohesion and cultural identity (Carmona 2021). At this scale, we observe the integration of diverse housing options, the design of streetscapes that invite social interaction, and the development of public spaces that foster a sense of community and belonging (Gehl 1971; Whyte 2001). By focusing on the mezzo scale, urban designers can ensure that the ideals of the New European Bauhaus permeate the fabric of everyday urban life.

The mezzo scale also plays a crucial role in promoting environmental sustainability in urban areas. By integrating green infrastructure such as parks, urban forests, and sustainable drainage systems, planners can create neighborhoods that not only support biodiversity but also contribute to the city's overall resilience (Hagan 2015; Rueda-Palenzuela 2019). Sustainable mobility options, including public transit, cycling, and walking, are also prioritized at this scale, helping reduce the environmental footprint of urban areas while enhancing accessibility and quality of life for residents (Cervero 1998; Lehmann 2010).

Social and cultural integration is another key focus of the mezzo scale. Urban designers at this level work to ensure that neighborhoods and districts reflect and celebrate the cultural heritage of their communities. This intervention might involve preserving historic buildings, creating spaces for cultural events, or incorporating public art that conveys the story of the area's unique identity (Alexander et al. 1974). Additionally, the provision of community facilities, such as schools, libraries, and healthcare centers, ensures that social infrastructure is in place to support the diverse needs of the population (Hawkes 2001).

As we explore the mezzo scale in the following chapters, we delve into the various elements that contribute to creating vibrant, sustainable, and inclusive urban environments. From the planning of mixed-use developments that promote walkability and reduce car dependency to the design of public spaces that enhance social cohesion, the mezzo scale offers a rich canvas for implementing the core principles of the New European Bauhaus (Talen 2019; Dovey 2010). At this scale, the grand visions of urban planners are brought to life, creating neighborhoods and districts that form the backbone of resilient, inclusive, and beautiful cities.

Concerning all the mentioned, the mezzo scale section of the book explores the holistic perspectives on land use, as seen in the TExTOUR Cultural Tourism Lab in Trebinje, where co-creation processes play a critical role in sustainable urban renewal. The implementation of New European Bauhaus values is further examined through the Nyhavna, Trondheim case study, illustrating how mezzo scale interventions can reflect these ideals in real urban settings.

Chapters on carbon-neutral buildings, spaces, and infrastructures underscore the importance of environmental sustainability at the mezzo scale, with detailed analyses of climate impacts on building energy demand in Greece and green urban renewal strategies. The preservation of heritage buildings through sustainable restoration and adaptive reuse demonstrates how cultural identity can be maintained while advancing sustainability goals; a theme echoed in UNESCO's *Culture: Urban Future* (UNESCO 2016).

The mezzo scale, therefore, serves as a critical bridge between strategic, large-scale planning at the macro-scale and detailed, human-focused design work at the micro-scale. By ensuring that these different levels of spatial design are interconnected and aligned with broader urban development goals, we can craft urban areas that truly reflect the aspirations of the New European Bauhaus—cities that are not only sustainable and inclusive but also places of beauty, culture, and community (Montgomery 2013;Lynch 1981).

References

Alexander C, Ishikawa S, Silverstein M (1974) A pattern language. Oxford University Press, New York

Carmona M (2021) Public places urban spaces. The dimensions of urban design. Routledge, UK

Cervero R (1998) The transit metropolis. Island Press, London

Dovey K (2010) Becoming places : urbanism/architecture/identity/power. Routledge, UK

Gehl J (1971) Life between buildings: using public space. Danish Architectural Press, Copenhagen

Hagan S (2015) Ecological urbanism: the nature of the city. Routledge, UK

Hawkes J (2001) The fourth pillar of sustainability: culture's essential role in public planning. Common Ground Publishing

Jacobs J (1961) The death and life of great American cities. Random House, New York

Lehmann S (2010) The principles of green urbanism: regenerating the post-industrial city. Earthscan

Lynch K (1960) The image of the city. MIT Press, USA

Lynch K (1981) A theory of good city form. MIT Press, USA

Montgomery C (2013) Happy city: transforming our lives through urban design. Farrar, Straus and Giroux

Rueda-Palenzuela S (2019) "Vista de El Urbanismo Ecosistémico," Cituidad y Territorio. Estudos Territoriales, vol LI, no 202, pp 723–752

Talen E (2019) Neighborhood. Oxford University Press, New york

UNESCO (2016) Culture: urban future; global report on culture for sustainable urban development; summary. Accessed: Aug 28, 2024. Available: https://unesdoc.unesco.org/ark:/48223/pf0000246291 (Online)

Whyte WH (2001) The social life of small urban spaces. Project for Public Spaces, New York

Chapter 10
Relevance of Co-creation Process for Sustainable Urban Renewal in the Local Context: TExTOUR Cultural Tourism Lab in Trebinje

Isidora Karan, Igor Kuvač, and Dorotea Ottaviani

Abstract Cultural tourism development deeply involving local communities fosters sustainable and smart cultural heritage management, as well as broader community-led urban development. Experimentation with co-creation methods within Living Labs supports this development. H2020 TExTOUR, an EU-funded project, pioneers sustainable cultural tourism strategies, including in Trebinje, Bosnia and Herzegovina, renowned for its cultural heritage. This paper examines Trebinje Cultural Lab's co-creative process using the TExTOUR project's methodology. It analyses challenges, benefits, and areas for improvement, focusing on urban renewal through cultural tourism. Findings contribute to understanding local co-creation processes and offer enhancement guidelines.

Keywords Co-creation · Cultural tourism · City lab · Trebinje · EU projects

10.1 Integrating Cultural Tourism Development, Smart Heritage Management, and Co-Creative Processes for Sustainable Urban Renewal

In the realm of urban research, the fusion of cultural tourism (CT) development, smart heritage management, and co-creative processes stands as a cornerstone for achieving sustainable urban renewal. CT thrives on the rich tapestry of cultural,

---

I. Karan (✉)
Centre for Spatial Research, Banja Luka, Bosnia and Herzegovina
e-mail: icprostor@gmail.com

I. Kuvač
Faculty of Architecture, Civil Engineering and Geodesy, University of Banja Luka, Banja Luka, Bosnia and Herzegovina

D. Ottaviani
Department of Architecture, Alma Mater Studiorum—University of Bologna, Bologna, Italy

M. Vukmirović et al. (eds.), *Digital Dreams, Urban Realities*, The Urban Book Series,
https://doi.org/10.1007/978-3-031-76353-3_10

social, and environmental assets present within specific regions and sites. Yet, its success hinges not only on these intrinsic factors but also on effective management strategies and community involvement. By embracing a community-based approach, CT development transcends mere economic gains to foster authentic, sustainable, and responsible tourism experiences (Dangi and Jamal 2016). Involving local communities ensures that cultural heritage is preserved, both tangible and intangible, while also accurately portraying these experiences.

Smart cultural heritage management serves as a linchpin in the preservation and sustainable utilisation of cultural resources, essential for the vitality of cultural tourism. It entails leveraging technological advancements and innovative strategies to safeguard heritage assets while concurrently harnessing their economic potential (Okazaki 2008). This approach not only safeguards the integrity of heritage sites but also augments the tourist economy, thus creating a symbiotic relationship between cultural preservation and economic prosperity. In the context of community-led city development, the integration of cultural tourism development and smart heritage management emerges as imperative. These complementary approaches not only bolster a city's identity and tourist appeal but also engender sustainable economic growth. Moreover, they play a pivotal role in preserving cultural heritage and enhancing the quality of life for residents, thereby fostering social cohesion and community pride (Urošević 2012).

Central to the discourse on sustainable urban renewal is the concept of co-creation, which champions collaborative decision-making and stakeholder engagement. By involving diverse voices in the urban planning process, co-creation ensures that renewal strategies are contextually relevant and socially inclusive (Gallagher et al. 2020). Moreover, it catalyses the emergence of innovative solutions to complex urban challenges, thereby fostering resilience and adaptability within urban communities (Healey 2016). This chapter elucidates the intertwined nature of cultural tourism development, smart heritage management, and co-creative processes in fostering sustainable urban renewal. Drawing upon insights from academia and practice, it underscores the importance of community engagement, technological innovation, and participatory governance in shaping the cities of tomorrow. As cities continue to evolve and adapt to changing socio-economic and environmental dynamics, embracing a holistic approach that integrates cultural, economic, and social dimensions is essential for fostering resilience, inclusivity, and vibrancy within urban communities (Evans-Cowley and Hollander 2010).

10.2 TExTOUR Cultural Tourism Lab

In trying to achieve the integration of cultural tourism development and smart heritage management in processes of sustainable urban development and urban renewal, the TExTOUR, an EU-funded project, developed and implemented co-design pioneering and sustainable cultural tourism strategies to improve deprived areas in Europe and beyond. Its goal was to integrate cultural heritage, tourism activities and local communities—sustainably expecting the realisation of the vision for a greener environment, a healthier economy, and happier people. To do this, it sets up Cultural Tourism Labs at eight pilot locations in less-known regions to invent and design collaborative work methods for the development of CT policies and strategies. The project applied ICT and innovative instruments focusing on the positive impact of such strategies and policies on socioeconomic territorial development. Integrating cultural heritage and tourism sustainably, the TExTOUR project works with local communities to create social, environmental, and economic benefits for all stakeholders.

There isn't a formal definition for CT-Lab beyond a pilot that involves all necessary stakeholders collaborating to co-implement CT actions. These labs aim to gather strategic information on CT and assemble knowledge, research, and key players to develop a practical methodology for promoting the pilots in a sustainable manner (Fig. 10.1). Each TExTOUR pilot operates as a living lab centred on CT, with eight CT-Labs actively involved in co-designing and implementing tailored sustainable CT strategies. These labs collaborate synergistically, covering various types of CT and aligning with UNESCO's guidance to provide customised capacity-building activities based on specific needs and contexts. CT-Labs facilitate co-creation between users and producers within a secure, open ecosystem, fostering social innovation, inclusion, and sustainability while also assessing the impacts of the actions (economic, social, cultural, and environmental), identifying beneficiaries, and considering ICT integration.

This paper examines the co-creative activities in the TExTOUR Trebinje CT- Lab, both the results and the process itself, focusing on the activities oriented towards urban renewal strategies. Those activities, developed within the unique TExTOUR methodology, are being analysed considering the local context characteristics, particularly from the aspect of participative urban development practices and social engagement. The aim of this paper is to provide a deeper understanding of the co-creative processes in the urban renewal field in Trebinje and the context of Bosnia and Herzegovina and provide tentative guidelines for their enhancement.

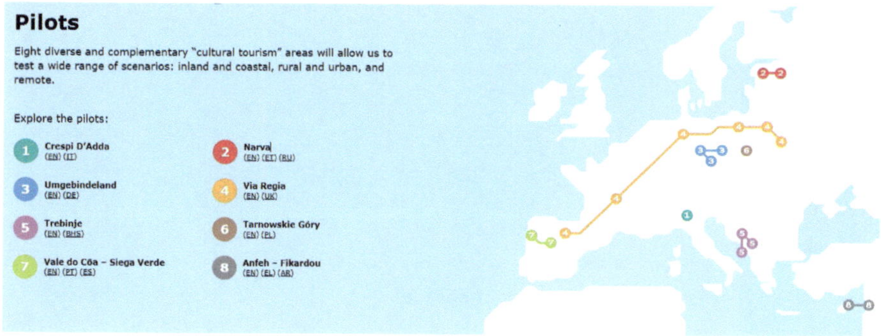

Fig. 10.1 TExTOUR CT-Labs. *Source* https://textour-project.eu/

10.3 TExTOUR Methodology

The TExTOUR project has developed a specific and unique integrated methodology for ad hoc strategies and an appropriate set of cross-thematic key performance indicators (KPIs). The methodology provided context-free indicators that served to assess the current state and monitor changes to estimate the impact of cultural tourism strategies. It also identified key success factors and replicability conditions for strategy development or improvement. The methodology approached the evaluation of the impact from a general perspective that facilitated its replication, but at the same time, it allowed adaptation to specific areas with less attractiveness for tourism. The main question the methodology tends to provide is: How can sustainable cultural tourism support inclusive and sustainable development in remote and peripheral areas? Ottaviani et al. (2023) proposed a specific participatory methodology for cultural tourism developed and tested in eight case studies across Europe and beyond. The methodology and tools presented are based on three replicable steps that aim to enhance cultural tourism in a sustainable, diverse, inclusive, and innovative way (Fig. 10.2).

Within the overall project methodology, a set of co-creative methodological steps was developed to co-create, together with local stakeholders and local community members, ad hoc strategies for cultural tourism development. The co-creative process was based on principles of inclusiveness and diversity and aimed to engage local stakeholders in the tourism sector but also local community members, including minorities, who are not sufficiently represented in the current tourist flows, to trigger social innovation and community-led cultural heritage management. The co-creative process was designed in a way that recognised that public participation alone is not sufficient to ensure sustainable cultural tourism planning. Instead, cultural tourism must embrace the four pillars of sustainable development: focusing on economic, cultural, environmental, and social sustainability (Ottaviani et al. 2023). The strategy was developed during the three activities specifically designed for participatory workshops.

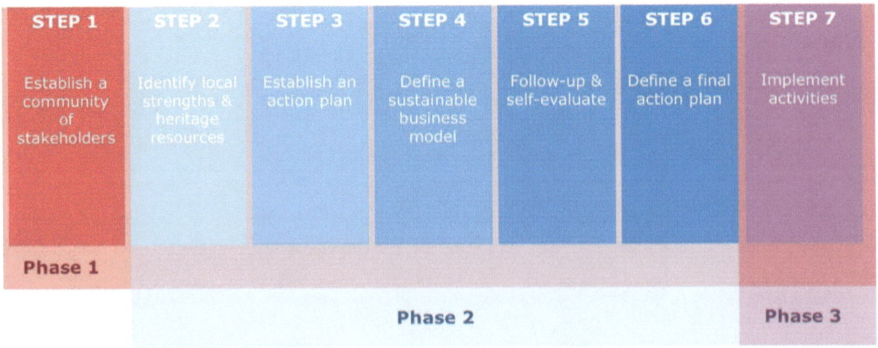

Fig. 10.2 A seven-step methodology developed in the Textour Project. *Source* https://icons.pageflow.io/

Workshop 1- the "co-mapping" exercise aimed at engaging the community, asking them to identify existing local heritage resources and the hopes, values, and goals that the participants have, as a community, regarding their area. The workshop consisted of two different parts, activities "From Individual to Shared Goals" and Co-Mapping Local Heritage Resources". Workshop 2- "Actions" was inspired by good practices in heritage-based tourism. The participants were asked to select and/or create a set of actions that would be based on the goals expressed in the previous workshop.

The Workshop 3 "implementation" was built upon the traditional Business Model Canvas (BMC) tool, expanding the scope of the BMC to introduce the concepts of inclusion, ICT, social innovation, impact, and sustainability. Participants broke down and analyzed the specific aspects of the selected Actions Business Model Canvas using the TExTOUR BCM.

10.4 Trebinje Cultural Lab is Putting on the Scene

Trebinje is a border city in Bosnia-Herzegovina, located less than 35 km from Dubrovnik (Croatia) and 80 km from Kotor (Montenegro), both UNESCO sites. Since ancient times, Trebinje has always been on the frontier between political powers and at the intersection of trade routes, which has marked its urban structure, ethics, and co-living culture. It is one of only a few cities in Bosnia and Herzegovina with preserved urban fabric that originates from different historical and cultural periods. All these layers of urban life, together with the prominent natural landscape of the river, the mountains, and the flatlands, have created an authentic and attractive place to visit and live in. However, the cultural diversity in the first two decades of the post-renewal and reconciliation was not recognised by the local community (Karan 2015) as a value, and it hasn't been used either to enhance tourist offers or to bring people together. Despite efforts to promote the city as a tourist destination, Trebinje still has significantly less cultural tourism compared to its neighbours, Dubrovnik and Kotor. However, following the rise of national and regional tourism in recent years, the city centre of Trebinje, with its monuments and cultural heritage sites, has experienced overcrowding and even construction pressure, which requires sustainable strategic planning and management together with the process of urban renewal and overall urban development (Tourism Development Strategy of the City of Trebinje 2018–2027).

Trebinje's urban development and urban renewal activities correlate with investment levels and political stability. Despite efforts to modernise the urban legislation for increased flexibility and participation (Milojević 2013), institutional weakening has diminished planning's efficacy and relevance. Local governments lacked effective management mechanisms, leading to politically driven development rather than expert planning, while public participation was reduced to formality. Corruption proliferated due to inexperience and power imbalances, privileging private interests over public spaces. The market economy reshaped economic activities and land usage, favouring investor-driven transformations over holistic urban planning.

Thus, urban development is faced with many challenges to protect the city's identity and inherited urban matrix (Cvijić 2016; Guzijan 2009). Unfortunately, social reality and new planning practices have led to many challenges in the sustainable urban renewal of Trebinje as well as a rapidly and radically changing urban image characterised by uncontrolled informal urban fabric, the expansion of the construction of printing and business facilities, the imbalance of the size of the historic city against new settlements, the unbalanced relationship between private and public interests, a lack of public investment, and ineffective participation. Private investors now dominate, with housing construction as the primary investment (Milojević and Kuvač 2021), while the city is getting a "summer house" status. This phenomenon, which has been raised with the COVID-19 pandemic, has been influencing not only the authentic image of the city but also the local people's everyday life as well as their ability to afford a property and/or living space. Moreover, Trebinje might be grappling with challenges such as infrastructure, ecology, and socio-economic disparities. If this trend continues, the possible result could be that the relatively small urban core of Trebinje would be significantly threatened and that it would not be able to offer a pleasant stay experience with its generous capacities (Fig. 10.3).

According to the Commission to Preserve National Monuments, there are 90 official cultural heritage sites in the territory defined by the city of Trebinje administrative borders, including ancient and prehistoric sites, and more than 80 located outside the urban area, in villages and hilltops surrounding the city, which make them difficult to manage and protect. On the other hand, many cultural heritage areas are not under protection, and some of them have lost some value due to inappropriate acting. Thus, the urban core of the city, assessed as a functional and aesthetically rounded entity, requires careful consideration and treatment in further planning processes, especially in terms of the integral urban renewal and expanding its capital capacities towards nowadays neglected areas such as Krš and many other remote points of interest. Addressing these issues would require comprehensive and integrated planning efforts involving collaborations with regional and national authorities, as well as international organisations.

In these terms, Trebinje has been undergoing various initiatives reflective of broader trends in contemporary urban planning and design, encompassing

Fig. 10.3 The urban landscape of the city of Trebinje. Photo by: Ninoslav Ilić

elements such as infrastructure improvement, heritage preservation, tourism promotion, and sustainable development practices (Урбанистички план града Требиња 2002–2016). In terms of contemporary urban development approaches, Trebinje may be incorporating principles of sustainability, resilience, and community participation into its planning processes. This could involve initiatives such as green infrastructure development, pedestrian-friendly urban design, and participatory planning strategies aimed at engaging residents and stakeholders in decision-making.

10.4.1 Trebinje Cultural Tourism Lab- Co-development Workshops and Activities

During the TExTOUR project stakeholders mapping, consisting of interviews with different institutional representatives and focus groups with local community members, the Trebinje pilot site had identified different goals as enhancing the involvement of young people in the cultural tourism sector and diversifying the tourist offer, redirecting tourist flows, boosting cultural tourism while decongesting overloaded areas within a critical carrying capacity condition—and collaborating beyond borders and historical disagreements; raising awareness on the common and universal values of cultural heritage and on the multiple roles they may play; training local inhabitants and stakeholders and collecting useful data to help local stakeholders, and cultural and tourism managers in the decision-making processes (https://textour-project.eu/pilots/trebinje/). Those goals were later discussed and reconsidered within the three co-creative workshops: "co-mapping", "actions", and "implementation". The workshop participants were mostly from the public sector (administration, culture, and education), then individuals and young people (students and independent experts), and finally, SMEs and NGOs (small food producers, women's associations, etc.)

The number of participants varied in each of the workshops (from 12 to 19). Even though the methodology envisaged the same participants to be involved in each of the workshops, only 60% of the participants were involved in the co-creative activities from the start until the end. More than 80% of the participants were from the active part of the population, from 20 to 50 years, and were targeted during the stakeholders' mapping exercise. It is important to mention that almost 70% of the participants were female.

The workshop participants who represented different types of public institutions and agencies mostly were not in the power of decision-making within their representative institutions, which has been reflected in the lack of institutional coordination when it comes to the actions implementation and required extra effort from the part of the project team. This implies that cultural heritage and its sustainable and collaborative management are still not recognised as a top-level topic and a resource that can be linked to direct economic benefit (for the difference of services in the tourism sector, for example). On the other hand, even though the participants coming

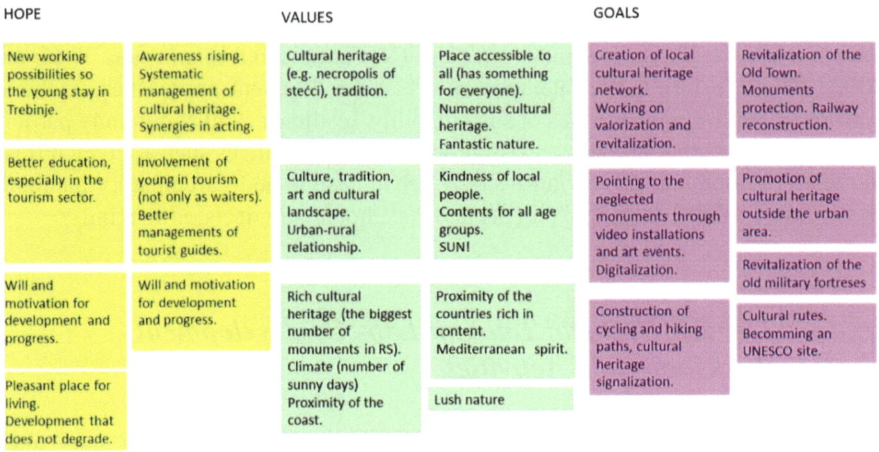

Fig. 10.4 "From Individual to Shared Goals" cumulative board produced as the result of Workshop 1

from SMEs, especially small food producers, were interested in participating in and taking a more active role in the action's implementation phase, they found it difficult to follow all the workshops due to the lack of time and resources. This leads to the conclusion that the co-creative methodology applied should be more locally adjusted to meet the needs of groups that are mostly excluded from the current tourism and cultural heritage flows but where their involvement could have a great potential for sustainable cultural tourism development.

The feedback survey that was shared among participants after each of the workshops showed that the level of the workshop's usefulness and quality was assessed as good (5 on the scale from 1 to 6). Even though the co-creative process intention was to bring different actors around the same table, the survey results indicated that participants mostly knew each other, which is also the condition of small environments as Trebinje is (Fig. 10.4).

During Workshop 1, The participants were very willing to share their opinions, positions and ideas and a fruitful discussion was developed. They also showed a high level of understanding of cultural heritage types and cultural heritage site mapping and showed interest in future workshop participation. Workshop 1 identified that the biggest hope for Trebinje is that young people will stop migrating to bigger cities in the region and worldwide, putting the social dimension of tourism development in the first place. This is related to job possibilities, diversification, and better education in the cultural tourism sector. Besides, they stressed the importance of developing in a more sustainable way, loving and caring for the city, raising awareness, and protecting the city space values and genius loci. Based on the values identified, the participants co-mapped more than 40 tangible and intangible cultural heritage.[1]

[1] It may be observed within three main groups based on the heritage types defined in the Textour methodology: 1. First group includes Archaeological Sites, Monuments and Historic City Centers;

In the results of Workshop 1, 16 action examples were prepared by the Textour project team and shared with the participants during Workshop 2 with the aim of selecting the actions they consider the most relevant for Trebinje. They were also asked to propose their own Actions by themself. In total, 11 actions were proposed, but only a few of them were presented in a structured way, and the aim and the impacts of the action, the potential challenges, and the roles that different actors could take in the action implementation were clearly stated. The workshop results indicated that most of the participants did not have a clear role that they could possibly take in the Actions. This implies that the participants are mostly aware of the significance and importance of cultural heritage but do not have direct experience in sustainable tourism development, especially from the aspect of social integration and innovation. However, this also could be interpreted in a way that the participants were not prepared to take responsibility and initiate the change.

The previous hypothesis was confirmed during Workshop 3 when the participants were asked to develop BMC for the 3 Actions selected after Workshop 2. Participants mostly have difficulties further developing ideas presented in the Actions in the BMC form provided. The BMC created during Workshop 3 could be assessed as descriptive, even vague, and mostly with no concrete propositions in terms of activities-partner in charge and budget-activities relationships. One of the reasons is also that participants mostly have no experience in business plan development nor in the actual costs of the planned activities (Figs. 10.5 and 10.6).

Implemented methodology in Trebinje pilot site has shown different results, challenges, advantages, and points of improvements of the co-creation process focusing on the urban renewal aspect of cultural tourism, described on the topic of StartArt! Festival action. From the 3 Actions selected in the Workshop 2 and further developed in the Workshop 3, one of them The StartArt! The Festival, among its goals, marked the urban renewal of neglected and underused cultural heritage sites in the inner city of Trebinje.

10.4.2 StartArt! Festival

The action was about strengthening the local and regional art scene and creative sector through collaborative interventions and events in public spaces open to both locals and tourists. The action consisted of co-designing and implementing some of the following activities: temporary artistic and spatial installations, and pop-up events. The activities took place in an underused urban pocket and marginalized cultural heritage site in the inner-city center (Krš neighborhood) pointing to the values and potentials of architectural heritage and with the aim of "spotting the light" on their revitalization.

2. Second group refers to the natural heritage and includes and 3. Third group includes Living Culture, Gastronomy, Arts / crafts, literature, and music.

Fig. 10.5 As the resources of the pilot, and results of co-mapping, 13 local heritage resources were identified in total

Number or code	Action Name	Heritage resources	Heritage-based tourism type
1	Small School of Gastronomy	Gastronomy, living cultures, living heritage	Culinary tourism, village tourism, eco-tourism
2	Natural heritage, living heritage, monuments	Natural heritage, living heritage, monuments	Hiking tourism, village tourism, eco-tourism
3	Start!Art Festival	Art/Crafts, Literature and Music, Historic City Centre	Cultural events and Creative Tourism, City tourism

Fig. 10.6 The results of the workshop 2

The first StartArt! The festival was held in the Krš neighbourhood in Trebinje in the period from July 3rd to 7th. The festival program was divided into two parts, "Building the Krš" and "Inhabiting the Krš", including several activities such as a discussion about the project, a musical evening, the experience of building Krš for tourists and citizens, and the opening of two pop-up exhibitions. The main part of the festival was the instalment of a temporary spatial installation in the public space of the Krš neighbourhood, which marked the place of interaction and all events at the festival (Fig. 10.7).

The project tried to emphasise the potential of the Krš neighbourhood towards the development of cultural tourism through the activation of open public spaces. The strategic selection of the intervention site is the result of an analytical and creative process at various spatial scales, as well as an understanding of the complex social and cultural structure of the area. Considering the criteria for the development of small-scale (Karan et al. 2017) urban acupuncture projects (Lerner 2005), such as public land ownership, small plot size, accessibility, visibility, centrality, site intensity, function, and others, the main square of the Krš neighbourhood has been chosen as the intervention and festival site (Fig. 10.8).

Fig. 10.7 Posters announcing StartArt! Festival

Fig. 10.8 Krš neighborhood, 2023. Photo by Igor Kuvač

In the specific context,[2] the urban design project represents just one new interpretation of the rich spatial layers and its subtle enhancement through adding a new one. The temporary spatial installation[3] is a pergola with a traditional mušebek motif

[2] The Krš neighborhood (known as Omanovića-mahala) is one of the oldest urban units in the city of Trebinje, and it is part of the national monument "Historical Urban Area of Trebinje." The mahala developed during the 17th century on steep slopes north above the Old Town. It is characterized by narrow and steep streets with small stone houses for living. The document on protection by the Commission for the Preservation of National Monuments of Bosnia and Herzegovina states the following: "The buildings are densely clustered, modest in size, single-story or one-story, with sloping roofs covered with stone slabs. The facades are stone, made of broken stone laid in irregular rows. By construction method and materials, the architecture is characterized as folk, typical for Herzegovina."

[3] The installation was conceived, designed, and realized by the students of the sixth semester of the architecture study program, within the scope of the small-scale design course, under the mentorship of assistant professor Igor Kuvač, Ph.D. Arch., at the Faculty of Architecture, Civil Engineering

(shades on the windows of old houses in Bosnia and Herzegovina) and a white canvas element fluttering in the wind, both serving to create shade and protection from the sun during hot summer days under the intense Herzegovinian sun.

In addition to characterising this specific place, such a thoughtfully designed structure allows for the formation of different ambiences and uses during various periods of the day, as well as longer periods of use equally throughout each day or during the season. The process of constructing installation was an event, simulating the possibilities of activating public space, which can become an exhibition space, a venue for concerts, film screenings, lectures, gatherings, socialising, relaxation, and anything else that a quality open public urban space actually enables.

Its setting represents an experimental model of action in sensitive areas of architectural and cultural heritage. It also stimulates the possibility of activating numerous other public spaces in the neighbourhood by using different spatial tactics and engaging local artists, artisans, and creatives. In this co-creative process, carefully selected locations are temporarily revived in cooperation with the local community and a wide group of interested actors from different fields (Fig. 10.9).

According to the model of the *small SCALE* (a project of the Center for Spatial Research, the work involved identifying and responding to real problems and needs of public space and the inhabitants of the Krš neighbourhood, using different tactics and innovative tools, such as co-creation and collaboration during the several days of the festival. It was this tactic of joint learning through work that was key to the success of the activity.

The project and the festival team in the Krš settlement were warmly welcomed by the small community of residents in this marginalised area. The neighbourhood fully embraced the group of young, creative individuals who showed interest in the values of their settlement, its problems, and needs, as well as an ear for their desires and visions of preservation and development. Throughout the duration of the festival, neighbours were active participants in the process, expressing their gratitude in various ways. They allowed participants to literally enter their homes and yards while working, took care of the location, and with constant presence, safeguarded materials, tools for construction (installation), and finished parts of the installation. They also ensured the comfortable stay of participants, helped during work, brought refreshments during work, and encouraged action. Finally, the residents of Krš showed great respect for the festival program itself, becoming an integral part of it and being its audience.

The festival brought together 14 directly involved participants in the program and approximately 15 indirectly involved participants in the organisation and logistics of the festival. Over the course of the five-day festival, around 80 participants took part in the activities. Given the location of the installation in the public space of the Krš settlement, it remained a point of interest for residents, passersby, and tourists,

---

and Geodesy, University of Banja Luka. The authors are Nemanja Ćuso, Miloš Marčeta, Nataša Jotić, Marijana Makarić, Tamara Novaković, Fatima Abazović, Branka Jakovljević and Dragoljub Kopranović. In the work with the students, a significant contribution was made by the mentors from the Center for Spatial Research, Assoc. prof. Isidora Karan, Ph.D. Arch., Saša Zečević, MA Arch. and Dejan Puhalo, B.Sc. spatial planner.

Fig. 10.9 Pop-Up Pergola, a small-scale urban intervention in the open public space in Krš Neighborhood in Trebinje, developed through the co-creation process. Photo by Igor Kuvač

serving as a meeting place and hosting several events even after the festival ended. Therefore, it is estimated that the number of visitors after the festival was much higher, totalling around 300. This led to extensive coverage of the event in various media outlets, contributing to the promotion of the project among both professionals and the public (Fig. 10.10).

The described collaboration is the key advantage of the project, its greatest and added value, as confirmed by festival participants. Students and tourists highlighted the unique experience of the "Building Krš" program, through which they were able to experience the spirit of the place and coexistence in the settlement in a special way. On the other hand, residents of the settlement gained trust in the tested methodology of work (collaboration and co-creation) in their community and saw real results in the improvement of public space for the first time in decades. They affirmed a sense of pride and a feeling that changes are possible and that their settlement can indeed undergo urban renewal as a part of the old town of Trebinje.

Similar reactions were expressed by representatives of the local administration, who highlighted the good dynamics of work on the project, active participation, and finally, concrete results that are not common (which they encountered for the first time) in international projects. They expressed readiness to continue participating in similar activities with more available resources, as the involvement of the local administration in this project was sporadic and limited to only a few representatives who followed the process from the beginning. The local administration did not make additional efforts or did not have enough resources to involve a larger number of participants or systematically integrate them into the process in various scales and directions, which could directly address their needs and be better utilised in accordance with local development policies.

Collaboration with the local administration was also one of the greatest challenges, as the project was led from a different location (Banja Luka, Krajina region) by a group of independent experts (Center for Spatial Research), addressing activities (strategising cultural tourism in the city through activities in the fields of culture, tourism, and heritage, which are part of the urban renewal process and urban planning

Fig. 10.10 Neighbors in Krš participated both in constructing the installation and pop-up events. Photo by Igor Kuvač

projects) within the jurisdiction of the local government unit (City of Trebinje and its institutions). In addition to implementing activities, the coordinator and project leader, the Center for Spatial Research, had the role of balancing between critical analysis of the state of cultural tourism, the state of cultural heritage, and the urban renewal process on one hand, and the legal role of the local government unit in the heritage protection process and the development of cultural tourism activities and responsibility for urban development methods on the other hand, as well as establishing and maintaining partnerships with the City of Trebinje as an inevitable stakeholder in this process.

The implementation of the action faced numerous other challenges. (1) The isolation and remoteness of the settlement from the usual tourist routes in Trebinje necessitated additional efforts in promoting activities and attracting audiences. (2) Although the action was conceived by the citizens of Trebinje during three joint workshops, it seems that the idea was not adequately communicated to the public, as the workshops involved isolated representatives of stakeholder groups as individuals who were not sufficiently motivated to promote the strategy within their own circles and wider networks. (3) Certain mapped and invited stakeholders, such as the Trebinje Academy of Arts, Youth Center Trebinje, Radio Trebinje, and similar institutions, besides declarative support, did not actively participate in the action, reducing the number of participants, narrowing the audience, and diminishing the media promotion of the project. (4) The local community of artists, artisans, and creatives is not organised and cohesive, limiting opportunities for their engagement in any activities. Thus, it was not possible to form new partnerships and initiate new collaborations. (5) Despite being emancipated to some extent, small communities typically exhibit specific characteristics of resistance to new, different, and foreign ideas, requiring much more effort and time for the acceptance of new ideas (Fig. 10.11).

Fig. 10.11 StartArt! Festival team mapping the Krš neighborhood and presenting the event at the main square in Trebinje during March 2023. Photo by Igor Kuvač

10.5 Guidelines for Co-creation Process Enhancement in the Local Context

The analysis of the process reveals both strengths and weaknesses of the applied methodology, offering critical insights to enhance its efficacy for future testing and application across various contexts. Success in co-creation primarily hinges on cultural, social, and economic dynamics, as well as the level of experience and tradition in participatory processes. While these factors may not be readily changeable, they can still be influenced to some extent by the process. To further elucidate the key principles and strategies for effective co-creation in urban renewal, the following points are highlighted:

Urban renewal is a multifaceted endeavour where planners, architects, and institutional representatives must collaborate with the end-users who will utilise, adapt, and maintain the spaces. Thus, the development of a co-creation methodology and framework for this process is imperative.

Engagement of citizens and the local community in the co-creative process of urban renewal should be strategically planned well in advance, initiating from the conception of the idea. Early negotiation with the community during the process design phase is crucial, with clear communication of benefits to garner acceptance and facilitate implementation.

Transparency, democracy, and inclusivity are fundamental principles of the co-creative process, necessitating promotion and encouragement at all stages among diverse participants. It should involve a wide array of stakeholders throughout development, with careful consideration to ensure accessibility for all interested parties.

Motivation alone is insufficient for participation; projects must be binding and aligned with participants' work hours. Strongly motivated stakeholders who recognise genuine interests should be directly involved in specific actions and spaces.

Continuity of participation from start to finish in all phases is essential for the co-creative process, with equal encouragement for all participants to continuously enhance their capacities.

The co-creative process should present understandable, clear, and concrete proposals for small experimental projects addressing real community problems and needs, thereby motivating engagement among members of small local communities.

It should be integrated into planning and development mechanisms or at least adopted as a best practice by local government units alongside institutional coordination mechanisms.

In addition to citizen involvement, economic entities should be included through mechanisms ensuring urban space improvement rather than its usurpation and degradation.

The co-creation process in urban renewal should embrace experimentation in open public spaces, demonstrating the potential for renewal and development through small-scale projects.

Ultimately, the successful implementation of the co-creation methodology demonstrates its relevance for sustainable urban renewal, contingent upon genuine interest, motivation, and engagement from the local community.

References

City of Trebinje Development Strategy for the Period 2018-2027 (2017) The city of Trebinje, Trebinje

Cvijić S, Guzijan J (2016) Urban regeneration as an instrument of identity preservation: a case study of Trebinje's Krš district. Arch Tech Sci 14(1):19–27

Dangi TB, Jamal T (2016) An integrated approach to sustainable community-based tourism. Sustainability 8(5):475. https://doi.org/10.3390/su8050475

De Lange M, Loorbach D, Frantzeskaki N (2020) Urban transition labs: co-creating transformative action for sustainable cities. J Clean Prod 269:122357

Evans-Cowley J, Hollander JB (2010) Reframing economic development: the relationship between social capital and local business climates for job creation. Econ Dev Q 24(1):5–1

Gallagher L, O'Neill E, McCormack J (2020) Co-creation for sustainable urban futures: new and critical perspectives. Urban Plann 5(3):1–3

Healey P (2016) Planning and complexity: what kind of knowledge for what kind of democratic politics? Plann Theory 15(2):186–203

Karan I (2015) Trebinje on the border between East and West: heritage and memory of Trebinje. Bitácora Urbano Territorial 24(2):31–41

Karan I, Stijak M, Kuvač, I (2017) The small-scale approach as a generator for urbanity increases in Banja Luka city. In: Proceedings of the conference places and technologies, Sarajevo, 8th-9th June 2017. Faculty of Architecture, University in Sarajevo

Lerner J (2005) Acupuntura urbana. Instituto de Arquitectura Avanzada de Cataluña, Barcelona

Milojević B (2013) Urban development and influential factors on the urban form of towns in Bosnia and Herzegovina in the period of socialism and transition: a case study of Banjaluka and Trebinje. Facta Univ Series Archit Civil Eng 11(3):237–249

Milojević B, Kuvač I (2021) Recognizing principles of integrated urban planning in historical development of the city: a case study of Banja Luka. J Urban Hist 49(2):290–308. https://doi.org/10.1177/00961442211030624

Okazaki E (2008) A community-based tourism model: its conception and use. J Sustain Tourism 16(5):511–529. https://doi.org/10.1080/09669580802159594

Ottaviani D, Demiröz M, Szemz"o H, De Luca C (2023) Adapting methods and tools for participatory heritage-based tourism planning to embrace the four pillars of sustainability. Sustainability 15(6):4741. https://doi.org/10.3390/su15064741

Urošević N (2012) Kulturni identitet i kulturni turizam—između lokalnog i globalnog (primer Pule u Hrvatskoj). Singidunum J Appl Sci 9(1):67–76. https://doi.org/10.5937/sjas1201067U

Гузијан J, Цвијић C (2009) The cultural and historical heritage of towns Trebinje and Jajce-a resource for growth of tourism. Архитектура и Урбанизам Београд 27:57–61

Урбанистички план града Требиња 2002–2016 (2002) Бања Лука, Урбанистички завод Републике Српске а.д

Chapter 11
Urban Realities for Implementation of New European Bauhaus (NEB) Values and Principles: Case Study Nyhavna, Trondheim

Slađana Lazarević and Alenka Temeljotov Salaj

Abstract Urban transformation is a very challenging process due to many factors, among them motivating citizens to invest financially in sustainable renovation. The New European Bauhaus movement seems to be ideal because it brings more people-oriented principles and adds stronger values and links, such as sustainability, aesthetics and inclusion. In the chapter, the main focus is on developing an assessment framework of NEB values and testing them through the students' work. For that, the main research aim was to see how a transformation area of Nyhavna in Trondheim could become a sustainable, inclusive and beautiful district and what New European Bauhaus values NTNU students from the year 2022 recognised and proposed. From the results, the social components, co-creation methods, and inclusive and participatory design for all are emphasised.

Keywords Urban transformation · New European Bauhaus · NEB

11.1 Introduction

Since its creation, the Bauhaus approach has provided a platform for the integration of various disciplines. Especially in today's age of technological advancement, where creative solutions are more sought after than ever, this idea is being embraced and often reinterpreted. The principles and values associated with the Bauhaus movement are experiencing a special kind of renaissance within the framework of the

S. Lazarević (✉)
Faculty of Architecture and Design, Norwegian University of Science and Technology, Trondheim, Norway
e-mail: sladjana.lazarevic@ntnu.no

A. T. Salaj
Department of Civil and Environmental Engineering, Norwegian University of Science and Technology, Trondheim, Norway
e-mail: alenka.temeljotov-salaj@ntnu.no

M. Vukmirović et al. (eds.), *Digital Dreams, Urban Realities*, The Urban Book Series,
https://doi.org/10.1007/978-3-031-76353-3_11

European Union's "The New European Bauhaus" (2021). EU policy aims for a people-centred and ecologically conscious future, seeking to achieve it through sustainability, aesthetics, and inclusivity as a transformative movement toward their integration into the fabric of urban environments. The NEB values and principles have been implemented in various urban settings through innovative projects and programs that emphasise co-creation, citizen engagement, and transdisciplinary approaches.

The NEB places an emphasis on co-creation and citizen engagement to enhance public spaces and address urban challenges (Rosana 2021). For instance, the Co-Create NEB program focuses on innovative solutions that improve public spaces in cities, peri-urban, and rural areas by integrating NEB values such as sustainability, aesthetics, and inclusion (Co-Create NEB 2023). This program encourages collaboration between diverse stakeholders and utilises design-led approaches to foster the development of projects that improve the quality of public spaces and contribute to positive social outcomes; it creates a platform for experimentation and connection and supports positive change by providing access to funding for beautiful, sustainable and inclusive projects. Moreover, an NEB Digital Toolkit (2022) has been created to increase the empowerment of the NEB movement by showing cases of over 200 digital solutions for the NEB ecosystem and connecting results from European research with all stakeholders. These digital tools enable architects and designers to craft living spaces that are more sustainable, inclusive, and visually appealing. Even though digital solutions are available to examine, address, or assist with citizen engagement and the creation of smart cities, they rarely include all three aspects of NEB: climate neutrality, beauty and inclusion (Takaoka et al. 2023). Alongside, inclusive solutions, local social sustainable approach, impactful participatory approach and cultural value should also be considered when designing digital solutions (Takaoka et al. 2023; Abdurahiman et al. 2024; Akberinejad et al. 2023; Senior et al. 2023).

The implementation of NEB values and principles in urban realities is a multifaceted endeavour that involves a wide range of activities, from co-creation and citizen engagement to business support, digitalisation, and education. Through various programs, calls for proposals, and financial instruments, the NEB initiative is actively transforming urban spaces to align with its core values, fostering a sustainable, inclusive, and aesthetically pleasing future for European cities and beyond. Within the first NEB project implemented by the Norwegian University of Science and Technology called CrAFt (Creating Actionable Future 2022), an NEB Impact Model is developed to support the implementation and follow-up of the NEB principles on the ground (NEB Impact Model 2023). The NEB Impact Model helps to identify essential leverage points for systemic change towards climate neutrality and resilience by including all aspectual layers of sustainability (ecological, infrastructural, social, cultural, economic, aesthetical, legal, etc.) into innovative models for local collaborative governance and value creation that optimally use the co-benefits between the different sectors and minimise potential conflicts. Some valuable examples of NEB space transformations have been presented in the CrAFt cookbook: A Recipe for Climate-Positive Cities and Communities (2023) and CrAFt A NEB Pocket Guide: Climate-Positive Cities and Communities (2024). It outlines seven stages for urban transformation inspired by the NEB, covering everything from conceptualisation, execution, and documentation to replication. To inspire the other city's initiatives, each stage highlights key principles and joint strategies through real-life examples from CrAFt cities.

Similar to the Bauhaus initiative in 1919 that fostered the innovative, collaborative and artistic community, the NEB utilises the spark to encourage artists, engineers, architects, and citizens to achieve the goals of the Green Deal. In accordance with Sadowski (2021), NEB responds to the most challenges of our time (climate crises) and brings systematic change with its own distinct aesthetics to match with sustainability, so the creative work based on the collaborative work of many specialists corresponds to the new kind of artisans promoted by Bauhaus. By implementing renewable energies, digital solutions, and intelligent materials, the NEB reinvigorates the design principles of the Bauhaus and transfers them into a modern, technologically advanced environment. The foundational teachings of the Bauhaus represent an exemplary model of utilising established knowledge sustainably. Rather than an obsessive pursuit of novelty, Walter Gropius (a founder of the Bauhaus initiative in 1919) embraced the principle of "back to the basics." It was essential to master the materials and tools before using them to create something innovative. This philosophy reinvigorated craftsmanship, blending it seamlessly with art, thereby achieving a clear harmony. A distinctive feature of this method was its receptiveness to novel and untested materials, along with a fascination with technology. This combination significantly contributes to the enduring vitality and relevance of the Bauhaus ethos.

Despite the NEB movement advantage, which gives the interested societal groups a multidisciplinary approach for researching, educating, governing, constructing, designing and implementing a better environment, and due to the very broad spectre of science and art for addressing it, there is still a need to develop and test the assessment framework for it. In this article, the New European Bauhaus assessment framework for urban realities for the implementation of NEB values and principles is discussed. It is presented from the focus of the Nyhavna case study, assessing the students' visions on "*How to co-create-Nyhavna as a sustainable, inclusive and beautiful district*". The student's research, with a focus on the challenges, questions, and solutions for future development of the brownfield area, was conducted by the students of the Urban Facilities Management course in 2021 and participants of Ursa Major Autumn School in 2022.

11.2 NEB as a New Normative for Urban Transformation

Even though the NEB initiative has been active for four years and that we have many projects financed through EU funds which are related to the NEB, there is still not many published scientific works dealing with the NEB concept, and especially the evaluation of urban transformations through the criteria of basic values and principles of NEB.

A fundamental tenet of the Bauhaus movement was the incorporation of industrial elements and methodologies into artistic and design processes (Jafee 2005). Proponents of Bauhaus favoured the adoption of mass production methods and innovative materials such as steel, glass, and concrete. Their objective was to design items that were not only straightforward and stylish but also practical, with the capability for widespread manufacturing. Those fundamental principles are translating into New European Bauhaus (Sadowski 2021), and many NEBs' successful initiatives have already been implemented in small pilot projects, e.g. from an urban local perspective Andrzejewska (2024), and Bilic et al. (2024). Moreover, Rosado-Garcia et al. (2021) see NEB as an ecosystem of innovation. Thus, academia should be able to ensure a transversal, experimental, and participative education in architecture, engineering, art, design, and the humanities, as well as provide a niche for disruptive innovation development. Hence, a true social shaping of technology should allow for social sustainability while embracing aesthetic and cultural dimensions (Rosado-Garcia et al. 2021). On the other hand, and with NEB in mind, Hu et al. (2022) recommend incorporating vernacular knowledge into courses to ensure that buildings are adapted to the local environment and able to withstand the impacts of climate change.

Since the NEB was created to initiate a new approach to sustainable challenges, NEB values and principles are an extremely valuable basis for the assessment of urban transformation projects. Among the first innovative assessment approaches developed in the framework of the funded CrAFt project is the CrAFt's NEB Impact Model, which has been created as an assessment and guidance tool geared at a whole system approach for use in complex urban interventions. The whole systems approach fully integrates the NEB triple bottom line of realising sustainability, beauty and inclusion. The Impact Model can be optimally put to work at the district and urban scale levels of intervention. Within CrAFt, an NEB Impact Model is developed to support the implementation and follow-up of the NEB principles on the ground in our cities. This way, a new basis for assessing urban transformations has been initiated.

Translating the fundamental Bauhaus principles, such are functionality, simplicity and aesthetic in corelation with NEB values (sustainability, aesthetic and inclusion) and NEB principles have become the basic for the assessment framework developed and presented in this article with the purpose of creating new normative for "NEBise" urban transformation and assessing the students design solutions.

By creating bridges between different backgrounds, cutting across disciplines, and building on participation at all levels, the New European Bauhaus inspires a movement to facilitate and steer the transformation of our societies along three inseparable values (NEB Compass, 2024) (Fig. 11.1):

- **Sustainability**, from climate goals to circularity, zero pollution, and biodiversity
- **Aesthetics**, quality of experience and style beyond functionality
- **Inclusion**, from valuing diversity to securing accessibility and affordability.

Fig. 11.1 NEB values and ambition levels *source* NEB Compass (2024)

The NEB value **"beautiful"** expresses the importance of connecting different places and people, fostering a sense of belonging through meaningful collective experiences and integrating new enduring **cultural and social values** through creation, "*creating a positive and enriching experience for people, beyond functionality.*"

The NEB value "**sustainable"** refers to climate goals, circularity, zero pollution, and biodiversity in its essential meaning under the NEB initiative. However, sustainability is achievable only through a collective effort by many actors. That means we can achieve just changes if we act "**together**".

The third NEB value, "together", refers to inclusion, from valuing diversity and equality for all to securing accessibility and affordability.

11.3 NEB Assessment Framework

The integration of the NEB values and principles was used as a methodological concept to develop it further as an assessment framework (Table 11.1). Diving deeply into NEB values led us to scientifically elaboration on assessing the values for urban transformation projects, formulating the first set of qualitative parameters to evaluate NEB values integration into students' design projects for our case study Nyhavna in Trondheim, Norway. The NEB values assessment tool was designed (Table 11.2) and compared with the students' perceptions of the NEB values and ambitions (Table 11.3). The main NEB principles have been applied in the student's work, so this study does not present the results of the applied principles.

Table 11.1 An assessment framework with NEB values is extracted into qualitative parameters for pre-evaluation/preliminary assessment of the urban transformation projects

NEB value	Pre-defined qualitative parameters for assessment of the urban transformation projects				
Beautiful	Parameter B1—Cultural value	Parameter B2—Social value	Parameter B3—Functionality		
Sustainable	Parameter S1—Zero pollution	Parameter S2—Circularity (regeneration)	Parameter S3—Biodiversity	Parameter S4—Economic models	Parameter S5—Governance structure
Together	Parameter T1—Diversity and equality	Parameter T2—Accessibility	Parameter T3—Affordability		

Table 11.2 Assessment framework with NEB values and ambitions extracted into qualitative parameters and sub-parameters for pre-evaluation/preliminary assessment of the urban transformation projects

NEB values and ambitions	Sub-parameters		Value—recognised sub-parameters in correlation with NEB ambitions	
Value: Beautiful Ambitions: To activate To connect To integrate	B1 Cultural value To Activate	B1.1. Aesthetic (protecting and activating)	B1.2. Collective experience and history	
	B2 Social value To connect	B2.1. A sense of belonging	B2.2. Co-creation design	
	B3 Functionality To integrate	B3.1. (re)activate the qualities of a given context	B3.2. Restructuring of space values	
Value: Sustainable Ambitions: To repurpose To close the loop To regenerate	S1 Zero pollution To close the loop	S1.1. Reduce CO_2 emission	S1.2. Closing the whole material cycle or waste transformation	
	S2 Circularity To repurpose To regenerate	S2.1. Preservation of the built environment	S2.2. Repair, re-use, renew of materials and building stocks (Carbon storing)	S2.3. Revalue public spaces for sustainable urban mobility
	S3 Biodiversity To regenerate	S3.1. Ability to preserve	S3.2. Ability to reconnect with nature	
	S4 Governance structure	S4.1. Digitalisation and smart systems	S4.2. New services for citizens	
	S5 Economic model	S5.1. Public -private partnerships		
Value: Together—Inclusive Ambitions: To include To consolidate To transform	T1 Equality To Include To Consolidate	T1.1. Social justice (Representation and social stability)	T1.2. Equal accessibility to services—Overcoming segregation	
	T2 Accessibility To Include	T2.1. Sharing resources and opportunities	T2.2. Accessibility of public transportation and other way of mobility	
	T3 Affordability To Include To Transform	T3.1. Affordable public housing and parks	T3.2. Transformative benefits in neighbourhood on a societal level	

Table 11.3 Assessment findings—NEB Values in the students' projects—recognised sub-parameters in correlation with NEB ambitions

NEB values and ambitions	NEB values in the students' projects—recognised sub-parameters in correlation with NEB ambitions
Value: Beautiful Ambitions: To activate To connect To integrate	B1.1. heritage "... to preserve and reuse existing buildings or building components." B1.1. and B1.2. Preserve cultural and housing heritage i.e., design housing that preserves, respects, and recognizes the unique historical and cultural characteristics of the Nyhavna neighbourhood and its future residents B1.2. old industry building to be a new industry museum B2.1 existing buildings to be used as identity markers of the area B2.1. people need to feel to belong to the place, to create an identity linked to that place B2.2. to connect with neighbourhood B2.2. social inclusion for more liveable public spaces B2.2. public participation for design co-creation of publica spaces and areas of interaction with a high potential of inclusion for places to meet and live B2.2. the development of the neighbourhood allows for the creation of neighbourhoods that are in line with their wishes and thus strengthen social inclusion B3.1. unique context of the area, industry B3.1. a more inclusive approach to urban design B3.2. new purpose of existing buildings

(continued)

Table 11.3 (continued)

NEB values and ambitions	NEB values in the students' projects—recognised sub-parameters in correlation with NEB ambitions
Value: sustainable Ambitions: To repurpose To close the loop To regenerate	S1.1. developing a climate adaptive district, to regenerate district, through the implementation of appropriate materials (permeable materials), increased share of greenery and adoption of nature-based solutions S1.1. preventive measure should be taken into action to protect the environment from the previous activities in the Nyhavna area, i.e., reducing the pollution of air, water, and soil in the area and providing a preventive measure in the future for sustainable development S1.2. in the public space grey water storage technologies can help mitigate increased levels of precipitation and reduce the impact of drought periods S1.1. and S1.2. management of chemical pollution can be improved by implementing an early-stage exploration of the 'solution space' (the set of potential activities that can be considered to protect or restore the water quality against hazards posed by chemical pollution) S2.1. reflection on all buildings and their impacts must be considered S2.2. increase adaptability and flexibility in sustainable housing development in the area S2.3. revaluing urban mobility for the implementation of the public transport network, ensuring that the entire neighbourhood is connected S2.3. developing a new bike route and pedestrian corridors, connecting two borders' neighbourhoods S2.3. evaluation of the road safety for the partial and long/term urban transformation of districts and neighbourhoods S3.2. decreasing the chance of runoffs and vulnerability to floods, reducing the level of pollution in contaminated areas, reducing wind speeds, reduce energy needs in buildings and increase the overall air-, soil- and water quality in an area through new vegetation and pervious (natural) materials regeneration S4.1. smart digital grid (a new network linking the future user of the infrastructures and the new local renewable energy sources) S4.1. use of social media can be a very useful tool to improve citizen participation in urban planning S4.1. development and use of new technologies for citizen participation that involve digital visioning techniques using game strategies S4.2. adapted educational and working environments S4.2. comprehensive health and leisure services S4.2. the neighbourhood must have and make accessible health services S4.3. developing energy management system at the neighbourhood level S5.1. avoiding gentrification of the district through private and public investments to avoid higher prices of residential and social housing S5.1. attracting external fundings for the community

(continued)

Table 11.3 (continued)

NEB values and ambitions	NEB values in the students' projects—recognised sub-parameters in correlation with NEB ambitions
Value: Together—inclusive Ambitions: To include To consolidate To transform	T1.1. and T3.1. participatory budgeting to give residents a permanent opportunity to help shape their neighbourhood and thus create their own identity T1.2. and S4.1. digital tools and posters with QR code to include citizens' voice T1.2. to have access to an active and civic life in the neighbourhood transformation and to participation, and decision making for the place they belong to T1.2. the neighbourhood must have and make a variety of leisure activities for its residents to make the neighbourhood a pleasant place to live T2.1. suitable housing accessible to all T3.1. avoiding gentrification of the district to reduce prices of residential and social housing and to offer equal possibilities for all T3.2. long-term strategical collaboration between citizens and institutions T3.2. public participation in the creation of new values for neighbourhood T3.2. advantages and disadvantages in mix land use in the partial urban transformation projects T3.2. safer and attractive public recreational areas, and healthy, affordable food outlets and support services, recognizing the relationship between occupant health and a safe, attractive, cohesive, and quality neighbourhood

The quality of our living environments is essential, and art and culture play a key role in improving it (ref. Europa. eu)). Therefore, NEB values have been built on existing frameworks like the Davos Baukulture Quality System. NEB Compass, which serves as a self-assessment tool setting out the key characteristics for exemplary NEB projects, has been developed to be applied to a wide range of typologies: buildings, products, services, education models, etc., but with limitations to large-scale urban transformation projects. Therefore, in this Article, apart from extracting NEB values for assessing the quality of the urban transformation project designs in relation to the New European Bauhaus, a set of sub-parameters have been proposed in Chap. 5 of this Article following three ambition levels for all three NEB values.

11.4 Nyhavna Case Study Results

11.4.1 Nyhavna Case—Background

Nyhavna is an industrial area with historic buildings next to the city centre (Fig. 11.1). From a historical perspective, two very interesting buildings are located there, one of which is the submarine bunkers constructed during World War II. As they were massively constructed, it wasn't possible to destroy them after the war without affecting the surrounding space, so they give the identity of Nyhavna today (Trondheim Kommune 2020). They are mostly used as headquarters for the city archive, offices to book, tech companies' offices, laboratories, and storage space. Present stakeholders in the area are commercial business (sport, gastronomy/brewery), culture (art), industry (salt, metal, stones/cement and maritime), and institutional stakeholders (museum, NGOs, and university). The executive stakeholders are the city council, landowner, municipality and Nyhavna Utvikling.

There is a long-term development plan (30 years) for changing the district from an industrial to a mixed-use district with a social mix, various businesses, artists, and historical aspects, where visitors want to go, and people want to live (Trondheim Kommune 2022). The municipality development plan for the district includes the requirements for including a population with great differences in terms of income, physical abilities, and age group. In May 2022, the city council adopted the quality program for Nyhavna, which describes the basic objectives and plans. The program also states that the participation of the urban community is crucial to creating an active and new city district. It calls for children and young people to participate in the design of public spaces. In addition, some other methods are mentioned as desired to help figure out the public interest: temporary uses and offers, experimentation and trial and error. As a part of temporary use, a café with saunas was established to host events, festivals, and day activities during the developing period. Visitors have a chance to gather more information and knowledge about the district, and public events are organised for discussions about the future.

While the program demands participation in many fields of action, there are no measures, direct goals or guidelines for that. This diversity should also be reflected through the participation process to obtain representative results, which means that besides the mentioned stakeholders, future users (visitors, artists, residents, workers) and neighbours (Svartalmon, Lade, Nedre Elvehavn) should be invited. As the goal is to achieve a social mix, low-income people, migrants, or the elderly must also be actively involved, especially as measures so far have not reached many people out of those groups (Spring Methods AS 2019).

11.4.2 Results from the Students' Field Works and Analyses

The students from the Urban Facility Management and participants from Ursa Major autumn school with different backgrounds (Construction Engineering, Architecture, Computer science, Geography, Facility management, Social science, etc.) took Nyhavna as a case study. The results are evaluated in Ch. 5. Here, only the results of the participants of the Ursa Major Autumnus School (UM) are presented based on the challenge of using NEB principles.

The UM group used the brainstorming method (Fig. 11.2) to find the real challenge based on the information and the meetings they had, and they came up with this map (Fig. 11.3) that they chose to focus on participation as the problem that is needed from the first stage of the development.

Fig. 11.2 Aerophoto of Nyhavna district in Trondheim

Fig. 11.3 Brainstorming during students' workshop

This implies that two (participatory process and transdisciplinary approach) of the three core NEB principles were introduced even at the beginning of the students' work on concrete case studies to understand the community's needs and challenges in the current urban reality settings.

The students used and developed several methods while conducting the fieldwork:

- Meeting with the public sector
- Interview with the private sector.
- Interview with citizens.
- Interview with stakeholders like artists.
- Survey and QR code.
- Designing and suggesting smart digital solutions like VR and AR

In conducting a comprehensive stakeholder engagement process (Fig. 11.4), the research team engaged in dialogues with a diverse array of local actors in Nyhavna, including artists, entrepreneurs such as those from Monkey Brew (local brewery production), and representatives from the metal recycling sector (Fig. 11.5).

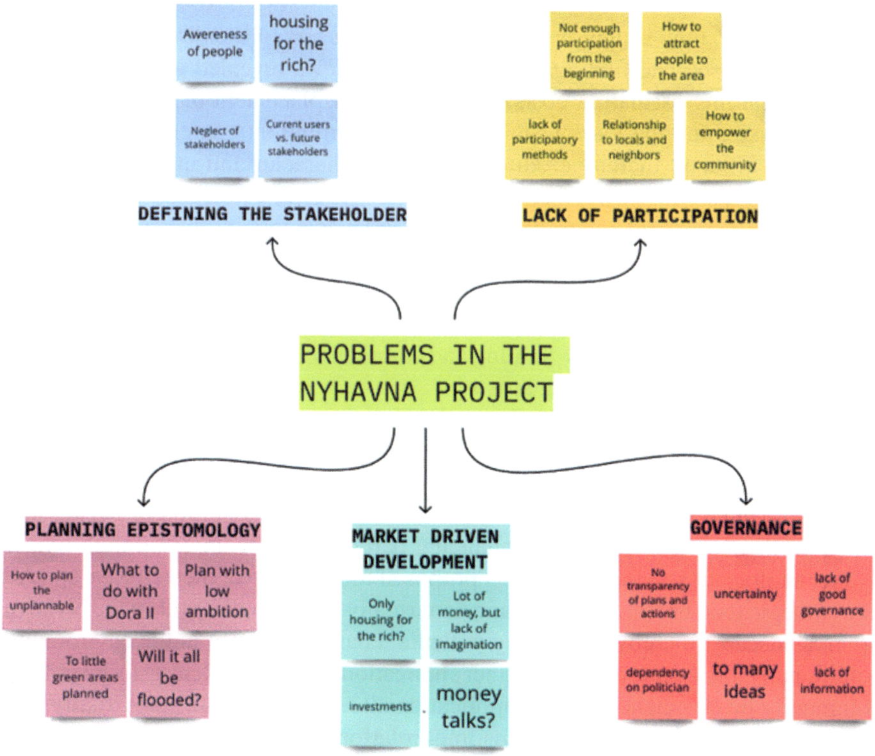

Fig. 11.4 Final mind map of the problem finding

Fig. 11.5 Interview with different stakeholders

Additionally, they sought insights from regular users of the district to gauge the community pulse. The findings from these interviews revealed a palpable apprehension regarding escalating costs, the potential displacement of current residents and businesses, and the fear of the district transforming into an enclave accessible only to the affluent, thereby eroding its original character and utility. Conversely, there was a palpable sense of optimism about Nyhavna's prospects. Stakeholders envisioned a transformed district that would serve as a vibrant hub for social interaction, featuring amenities such as cafes and eateries, spaces dedicated to sports and recreation, and cultural venues conducive to hosting concerts and musical events. Additionally, there was a strong desire for the future Nyhavna to include accessible waterfront areas suitable for swimming and communal spaces that cater to the diverse needs of stakeholders, including artists.

The second approach involved the development of a QR code system. This QR code facilitated access to various interactive platforms, including completing a survey, registering for a Virtual Reality (VR) experience at SINTEF, and engaging with Augmented Reality (AR) on mobile devices. This integration of digital tools offers numerous possibilities for enhancing participatory methods through innovative technological solutions. Students proposed the use of Augmented Reality (AR) as a strategic tool, enabling participants to virtually place objects within the Nyhavna urban landscape, thereby visualising potential changes directly in the environment. Additionally, they explored two distinct applications of Virtual Reality (VR) to foster community involvement. The first application allowed participants to virtually navigate the current state of the area, providing a platform for them to propose immediate modifications. The second application utilised VR to solicit feedback on future developmental plans, allowing participants to review proposed changes, experience them virtually, and contribute feedback on these visions.

Fig. 11.6 Smart digital solutions

The third strategy employed, and which is in line with the NEB's working principle—the participatory approach, was the administration of a survey designed to collect comprehensive insights from the community regarding their views and suggestions. This method ensures a structured and quantifiable means of gathering public opinion, which is crucial for informed decision-making in urban planning processes. Also, participants suggested the number of posters that could be implemented in different important locations to increase citizen participation and invite them more effectively to grab their attention. Some examples of the suggested locations were educational organisations, bus stops, tourist destinations, concert and art centres, healthcare, libraries, etc (Fig. 11.6).

Enhanced trust in future developments is anticipated through the implementation of a robust participation program. This way, community engagement could be bolstered further and foster a collaborative environment for the co-creative design of urban spaces. The proposed strategic directions came out through the transdisciplinary approach of the group of students, include following:

- Enhanced dialogue: Facilitate ongoing dialogues and debates among public entities, private sector representatives, and community members to disseminate knowledge about the health and well-being prospects of the area.
- Artistic engagement: Organize art workshops to capture diverse perspectives, emotions, and responses to the development plans, facilitating effective communication of these insights to public sector stakeholders involved in Nyhavna's transformation.
- Co-creation techniques: Employ co-creation methods as recommended by various stakeholders to enrich the comprehensive vision for the development phases.
- Student involvement: Focus on empowering students to take an active role in the development processes, enhancing their educational and practical experiences.
- Regular updates: Maintain and enhance trust through regular, clear communications about the development phases with all stakeholders.

- Community building activities: Introduce activities designed to foster a strong sense of community and belonging among potential future residents of Nyhavna.
- Addressing challenges: Work collaboratively with citizens to identify and overcome barriers to co-creation, ensuring inclusive and equitable development.
- Citizen-centric development: Prioritize citizen engagement to drive the creation of aesthetically pleasing, healthy, joyful, and sustainable urban environments.

These findings aim to not only enhance the urban fabric reality of Nyhavna but also to ensure that the redevelopment processes are inclusive, transparent, and beneficial to all stakeholders involved, what is in line with sustainability and inclusivity NEB values.

11.5 NEB Evaluation Findings

11.5.1 Evaluation of NEB Values and Principles

The presented long term development plan and Trondheim municipality's program which have already involved urban community in this long-term transformation process, could be assess as good basis for applying all NEB ambitions for its working principles (Fig. 11.7) defined in NEB Compass[1]:

- Participatory approach: ambition I—to consult and ambition II—to co-develop.
- Multi-level engagement: ambition I—to work locally and ambition II—to work across levels and
- Transdisciplinary approach: ambition I—to be multidisciplinary, ambition II—to be interdisciplinary and even ambition III—to be beyond disciplinary (involving non-academic partnership and public engagement).

Nyhavna case in Trondheim is selected to illustrate the NEB transformation parameters of sustainability, inclusiveness, and beauty. We draw from the NEB values and principles and present how NEB principles as a new dimension of urban development practices could be used and what NEB values are recognized in urban design projects for the concrete case.

To test our NEB values assessment framework in urban reality, Nyhavna case study is a good exemplary case, since two students' courses under which the students had opportunity to work together, showed that main NEB principles have been adopted in the work with students through defined methodological approach for the Nyhavna case study.

One of the key NEB principles such is transdisciplinary approach has been applied through the group work of students with the different educational background. Students with different backgrounds such as a Construction engineering,

[1] 405245f4-6859-4090-b145-1db88f91596d_en (europa.eu).

Fig. 11.7 NEB working principles and ambition levels connected with three principles: participatory process, multi/level engagement and transdisciplinary approach. *Source* page 6, NEB Compass, 405245f4-6859-4090-b145-1db88f91596d_en (europa.eu))

Architecture, Computer science, Geography, Facility management, Social science etc. worked on the Nyhavna case in two courses during the study year 2022.

The NEB principle which promotes participatory approach has been also applied, since students have been conducted different surveys and workshops to approach citizens and industry representatives of the Nyhavna case study. Assessing the principles of the work with students at the case study Nyhavna in Trondheim, authors of this paper have found that two working NEB principles have been integrated into working practice with students even those principles have not been explicitly predefined. In this chapter the results of our finding are presented and elaborated through the evaluation of NEB values and ambitions parameters and if the students projects on urban transformation of Nyhavna district are following NEB values.

To define the research question related to the school, the mentors arranged meetings with Trondheim Kommune and Nyhavna Utvikling in 2022. The aim of these meetings with the organization and the meeting of mentors was to define a real challenge for the district's future long-term development. The defined question and challenge were:

- Social sustainability
- Participation and co-creation
- Good urban space for community.

11.5.2 Urban Reality for NEB Values—Assessing the Students Works for Nyhavna Transformation District in Trondheim

To create an assessment framework that integrates the New European Bauhaus (NEB) values and ambitions into qualitative parameters and sub-parameters for the pre-evaluation or preliminary assessment of urban transformation projects, authors had to consider the core values and working principles of the NEB, as well as the specific ambitions associated with each value. This framework could help in assessing how well urban transformation projects align with NEB values before they are fully developed or implemented.

Apart of the identified sub-parameters defined in NEB Compass[2] and presented briefly in the Chap. 3 of this Article, authors of this paper have identified two more important set of parameters under "sustainability" value which are important for urban realities for preliminary evaluation of the urban transformation projects: governance structure (S4) and economic model (S5). Those two additional sets of parameters are presented with the simplified sub-parameters in Table 11.2, as part of assessment framework for NEB values and ambitions used for evaluation of alignment of NEB criteria in the students' work.

The students' findings are evaluated by the presented assessment framework for district level.

In total, 10 students' projects have been accessed in this Article, divided into three categories in line with NEB values for assessment of identified sub-parameters.

11.5.3 Urban Reality—Assessing NEB Values and its Ambition Levels

In this chapter, authors f the Article have presented findings on students' perception and interpretation of the NEB values: beautiful, sustainable, and together linked with three different ambition levels defined in the NEB Compass. In total, ten students' semestral projects have been evaluated. The assessment table in which three NEB values are evaluated through predefined sub-parameters presented in this chapter present a summarised fundings and results.

For the first NEB value "beautiful", authors of this Article have chosen a set of thee groups of sub-parameters to assess the students' visions on "How to co-create-Nyhavna as a sustainable, inclusive and beautiful district", as described in Table 11.2. The results of the assessment showed that students works have been referred "to activate" cultural and housing heritage and old industry to be used as an identity marker of the area (B1.1 and B1.2), "to connect" neighbourhoods with its citizens through social inclusion, public participation and co-creation activities

---

[2] *NEB Compass (accessed on 24.04.2024).*

(B2.1 and B2.2) and "to integrate" unique context of the area and existing buildings into inclusive approach to urban design (B3.1 and B3.2), thus reactivating the quality of context and restructuring of space values.

Since the NEB value "sustainability" is bringing many layers of complexity since its decades of developing and as an essence of the Agenda 2030, the authors of this Article proposed set of five groups of sub-parameters, connected with NEB ambitions, to assess sustainability concept applied in the students works answering the main research question "How to co-create-Nyhavna as a sustainable, inclusive and beautiful district'. The main terminology used in the students works and there understanding of the sustainability concept were linked with climate change impact, regeneration, increasing share of greenery, use of nature-based solutions, protection and restoration of environment (S1.1 and S1.2), thus reducing CO_2 emissions and closing the whole material cycle or waste transformation (linked with the NEB ambition "to close the loop"). When it comes to the NEB ambitions "to repurpose" and "to regenerate", sustainability value has been recognised by students in many different way linked with circularity and preservation of the built environment (reflection on all existing buildings, S2.1) through repair, re-use or renew of materials and building stocks (linked with carbon storing, S2.2.) and revalue public spaces for sustainable urban mobility for which a high numbers of sub-parameters have been recognised (S2.3). Under the group S3. Which relates to the biodiversity, the students found important "to regenerate" through ability to preserve and ability to reconnect with nature (S3.2). When it comes to the last two groups of set of sub-parameters which have been defined in the assessment framework as additional but essential segment of the sustainability concept (S4 governance and S5 economy), very positive surprise was to found many proposals for these two categories in the students works. From very innovative proposals for digitalisation and smart systems and the new services for citizens (S4.1, S4.2 and S4.3) all the way to the economic models such are public -private partnerships and attracting external fundings for the community (S5.1). The last, but not the least NEB value "together", students have shown good understanding of this concept assessing all three NEB ambitions under this value: "to include", "to consolidate" and "to transform". To include citizens' voice (T1.2 and S4.1), participatory budgeting (T1.1. and T3.1), to transform areas for the neighbourhood with a variety of leisure activities for its residents to make the neighbourhood a pleasant place to live (T1.2), for suitable housing accessible to all (T2.1), avoiding gentrification of the district to reduce prices of residential and social housing and to offer equal possibilities for all (T3.1) to many recognised actions under T3.2. Which refers to transformative benefits in neighbourhood on a societal level.

11.6 Conclusion

The aim of this research was to respond to the question "*How to co-create Nyhavna as a sustainable, inclusive and beautiful district*" and if NEB values could be already recognised in students' projects in 2022. To respond to this question both, NEB values and principles have been analysed under the different assessment criteria. For the NEB principles, authors of this Article recognised that two NEB working principles have been implemented during the co-creation process conducted by students and their mentors who were involved in the co-creation of the participatory process involving citizens, industry and different local institutions. Therefore, participatory approach has been recognised in this process. Since students have worked in groups which involved students and teachers from the different educational background, transdisciplinary approach has been applied.

The pronounced social components become evident through the creation of liveable spaces for all citizens. Inclusive and participatory design aims to ensure that all needs are considered, regardless of gender, age, or origin. This strongly reflects the social vision of the original Bauhaus movement, which aimed for a uniform quality of life for all. The relevance of this idea is therefore of great importance in today's context and urban reality in implementing NEB values and principles.

Acknowledgements The authors would like to thank all the master students who took part in the course 'Urban Facility Management' at NTNU Faculty of Civil ad Environmental Engineering in 2022 and participants of the Ursa Major Autumn School for their work and their innovative thinking which are the foundations of the case study part. This work includes results from the Ursa Major project funded by NFR 322317 and Craft project funded by Horizon Europe research and innovation programme under Grant Agreement 101056946

References

Abdurahiman S, Kasthurba AK, Nuzhat A (2024) Assessing the socio-cultural impact of urban revitalisation using relative positive impact index (RPII). Built Heritage. https://doi.org/10.1186/s43238-024-00118-3

Akbarinejad T, Salaj AT, Johansen A (2023) Implementing the integrated social sustainability assessment to Norway: a citizen-centric and expert-weighted approach. Sustainability 15:12107. https://doi.org/10.3390/su151612107

Andrzejewska AK (2024) Determining urban indicators in local plans—as one of the sustainable assumptions of the New European Bauhaus? Buildings 14:100. https://doi.org/10.3390/buildings14010100

Bilic B, Smit K (2024) Evaluation of the New European Bauhaus in urban plans by land use occurrence indicators: a case study in Rijeka Croatia. Buildings 14:1058. https://doi.org/10.3390/buildings14041058

Circular building adaptability in adaptive reuse: multiple case studies in the Netherlands: EMJEDT230052 1.23 (emerald.com)

Climate-Positive Cities and Communities (2024) https://craft-cities.eu/wp-content/uploads/2024/12/Neb-Pocket-Guide-Digital-Version.pdf

Co-create NEB (2023) https://www.eiturbanmobility.eu/co-create-neb-call-for-proposals-to-build-sustainable-beautiful-and-inclusive-public-space/

CrAFt Cookbook (2023) https://craft-cities.eu/wp-content/uploads/2024/02/D2.1-Craft-Cookbook.pdf

Creating Actionable Future (2022) https://craft-cities.eu/

Davos Baukulture Quality System 01-en_davos_baukultur_kurzbroschure_web_210428.pdf

The New European Bauhaus (2021) https://www.eitfood.eu/projects/the-new-european-bauhaus-designing-a-sustainable-and-inclusive-future

Hu M, Swierzawski J, Kleszcz J, Kmiecik P (2022) What are the concerns with New European Bauhaus initiative? Vernacular knowledge as the primary driver toward a sustainable future. Next Sustainability. https://doi.org/10.1016/j.nxsust.2023.100004

Jaffee B (2005) Before the new Bauhaus: from industrial drawing to art and design education in Chicago. Des Issue, 21(1). https://www.jstor.org/stable/25223979

Trondheim Kommune (2020) Kulturminneplan Nyhavna—Kulturhistorisk stedsanalyse av Dora ubåtanlegg.

Trondheim Kommune (2022) Kvalitetsprogram for Nyhavna—Ti strategiske virkemidler for utviklingen av en sentrumsbydel på Nyhavna

Spring Methods (2019) Innspill—hendelser på Nyhavna.

NEB (2021) Compass Ardeth #12: KEY WORDS. Beautiful, Sustainable, Together. 405245f4-6859-4090-b145-1db88f91596d_en (europa.eu)

NEB Digi toolkit (2022) The NEB Digital Toolkit: digitalisation empowering NEB (digineb.eu)

NEB impact model (2023) https://craft-cities.eu/wp-content/uploads/2023/06/CrAFt-NEB-Impact-Model-Summary.docx.pdf

Nunes JN (2024) The Bauhaus of the seas: a manifesto for the New European Bauhaus. Des Issues 40(2):90–97. https://doi.org/10.1162/desi_a_00758

Rosado-Garcia MJ, Kubus R, Arguelles-Bustillo R, Garica-Garcia MJ (2021) New European Bauhaus for a culture of transversality and sustainability. Sustainability 13:11844. https://doi.org/10.3390/su132111844

Rosana G (2021) The transformative power of the New European Bauhaus, Bringing the aspirations of the green deal closer to citizens and territories, https://www.iedonline.eu/download/green-deal/14_ROSANA_IED_Paper_2021.pdf

Sadowski K (2021) Implementation of the New European Bauhaus principles as a context for teaching sustainable architecture. Sustainability. https://doi.org/10.3390/su131910715

Senior C, Temeljotov Salaj A, Johansen A, Lohne J (2023) Evaluating the impact of public participation processes on participants in smart city development: a scoping review. Buildings 13:1484. https://doi.org/10.3390/buildings13061484

Sustainability | Free Full-Text | The relevance of the circular economy for climate change: an exploration through the theory of change Approach (mdpi.com)

Takaoka AJW, Ahlers D, Ådlandsvik FW, Dovland ES, Jaccheri L (2023) Towards understanding digital support contributing to climate neutral, inclusive, and beautiful cities: a systematic literature review. IEEE/ACM 7th International workshop on green and sustainable software. https://doi.org/10.1109/GREENS59328.2023.00012

Part V
Mezzo Scale: Carbon-Neutral Buildings, Spaces, and Infrastructures

Chapter 12
Assessing the Impact of Climate Change on the Energy Demand of Prefabricated and Conventional, Single-Family Buildings in Greece

Stella Tsoka, Katerina Tsikaloudaki, Theodoros Theodosiou, and Kondylia Velikou

Abstract This chapter aims to evaluate the energy performance of a prefabricated building in the context of climate change and compare the obtained results with the respective outcome for a similar conventional building, made of reinforced concrete and brick masonry. Dynamic energy performance simulations with the Energy Plus simulation tool are conducted for a one-story, family house, located in the city of Thessaloniki, Greece both for present and future climatic conditions. Different wall configurations and various insulation levels are examined. A future climate dataset for the period 2040–2060 is generated using dynamical downscaling approaches, while the RCP4.5 emission scenario is considered. Energy Plus simulations are then conducted for all the conventional and the prefabricated building scenarios for the current conditions and the future period, and the acquired simulation results are presented and discussed.

Keywords Climate change · Prefabricated buildings · Dynamic energy performance simulation

---

S. Tsoka (✉)
Department of Civil Engineering, University of Patras, Patras, Greece
e-mail: stsoka@upatras.gr

K. Tsikaloudaki · T. Theodosiou
Department of Civil Engineering, Aristotle University of Thessaloniki, Thessaloniki, Greece

K. Velikou
School of Geology, Aristotle University of Thessaloniki, Thessaloniki, Greece

M. Vukmirović et al. (eds.), *Digital Dreams, Urban Realities*, The Urban Book Series,
https://doi.org/10.1007/978-3-031-76353-3_12

12.1 Introduction

12.1.1 Climate Change and Built Environment

Climate change significantly affects the urban built environment not only regarding the higher probability of confronting extreme events such as heatwaves and flash floods but also with respect to the outdoor and indoor thermal comfort conditions and the buildings' energy demand. Especially for the buildings' energy performance, climate change may reduce the heating energy requirements, but it severely exacerbates the cooling energy needs, suggesting serious impacts on peak electricity demand, energy generation and energy supply (Berardi and Jafarpur 2020). In parallel, high indoor and outdoor air temperatures (Tair) due to climate change strongly aggravate the citizens' thermal comfort, constituting an important risk of their health and well-being. Given the severe social, economic and environmental extensions of the above-mentioned effects, it is of imperative importance to accurately evaluate the energy needs of buildings both under current and future climatic conditions so as to have a global perspective of the buildings, performance during their whole lifecycle and propose suitable design and adaption strategies.

Up to the present time, many existing studies aimed on the establishment of high accuracy methods to evaluate the magnitude of climate change along with its effect on the buildings' energy needs and the suggestion of suitable adaptation strategies. An overview of the methods that are used so as to generate future climate datasets for the evaluation of the buildings' future energy performance is given in Sect. 1.3

In the study of Shen (2017), the future heating and cooling energy demand for the period 2040–2069 of typical residential and office buildings has been evaluated. The analysis concerned 4 different cities and climatic zones across the USA, while 2 different emission scenarios have been considered. Firstly, the future climatic projections indicated an increase of the average, annual Tair value by 1.46 °C−2.76 °C and 2.3 °C−4.4 °C under the A2 and A1F emission scenario, respectively, with the highest discrepancies being reported for the A1F scenario in all the examined cities. The predicted hourly weather data were have been then used to evaluate the future energy performance of the typical residential and office buildings. The findings indicate an increase in monthly cooling energy consumption and a decrease in heating energy consumption across both office and residential buildings in all climate zones. In the warmest climatic zone, the highest changes in residential cooling energy usage reached 36.4% compared to the present-day simulation results, while for the office buildings, there is a 16.4% rise in cooling energy usage and a 23.7% reduction in heating energy usage.

Chan (2011) utilised the morphing technique to generate future hourly weather data, which were subsequently applied in assessing the energy performance of residential and office buildings in Hong Kong. Six sets of future weather data were generated involving two different future periods (i.e. 2046–2065 and 2080–2099) and 2 emission scenarios (i.e. SRA1B and SRB1). The study outcomes highlighted a significant rise in air-conditioning system energy consumption attributed to the anticipated higher Tair values. The increase ranged from 2.6% to 14.3% for office buildings and from 3.7% to 24% for residential units, with the highest values being found for the SRAIB scenario.

The influence of climate change on the energy demand of residential buildings in Valencia, Spain has been assessed by Perez-Andreu et al. (2018). Their analysis concerned two future periods (i.e. 2048–2052 and 2096–2100) and two different emission scenarios (i.e. RCP4.5 and RCP8.5) while different building envelope scenarios and insulation levels have been also considered. The authors indicated an increase on the average annual Tair under the RCP4.5 by 0.65 °C and 1.41 °C for the periods 2048–2052 and 2096–2100, respectively, with the corresponding values for the RCP8.5 being 1.33 °C and 3.44 °C. In terms of the buildings' energy performance, the analysis indicated that in all the examined scenarios, regardless of the insulation levels, the future heating energy demand is lower than the present day and the cooling energy needs are severely exacerbated. In the same context, Nik and Kalagasidis (2013) evaluated the impact of future higher Tair values on the energy demand of the Swedish building stock. Five different scenarios were considered for the future period 2081–2100. The authors suggested that the heating demand of typical buildings in Stockholm would be considerably lower in 2100 than the respective values in 2011, while cooling demand would increase in relatively small amounts, and overheating problems could be mitigated by natural ventilation measures.

Yet, as in the previously mentioned studies, most of the existing research assess the climate change impact on buildings that are constructed under conventional construction methods such as on-site processes, while relevant studies for prefabricated, modular buildings are far less. Indicatively, Zhang et al. (2023) evaluated the present day and future energy demand of prefabricated timber houses in China for the years 2050 and 2080. The Meteonorm weather generator and the Energy Plus dynamic simulation model have been applied for future climate projections and the energy performance simulations respectively. The results suggest an increase on the annual average temperature by 2.3 °C from 2020 to 2080, a rise that would lead to 13% increase in the total annual energy demand for terraced houses, a 10% increase for semi-detached houses, and a 6% increase for detached houses. The estimated increase on the annual energy demand, mainly attributed to the exacerbated cooling energy needs is also expected to lead to a respective increase in carbon emissions of 27%, 26%, and 23% for the terraced, semi-detached and detached houses respectively.

In the same vein, Ismail et al. (2021) estimated the future energy demand of a case study prefabricated building for the future years 2030, 2050, and 2080, considering two climate change emission scenarios (A1B and A1F1) and using the degree days method. Their analysis suggested that in 2030 and 2080, the anticipated energy requirements for heating in expected to decrease by approximately 12% and 34%, respectively, compared to the levels observed in 2017.

It has to be mentioned that, for many years, prefabricated buildings were regarded as of inferior quality or made for temporary constructions. But today, things have considerably changed as prefabricated buildings compete the conventional ones in every aspect of performance, i.e. structural, energy, hygrothermal, acoustic, fire, environmental, etc. Besides, among the advantages of this building system is the significant reduction in cost and time, the improved quality and accuracy in manufacture, the speed of installation on-site, the easy dismantle and reuse of its components (Navaratnam et al. 2019).

In this context and to assist the gap in the existing literature, the present study aims to evaluate, via dynamic simulations, the current and future heating and cooling energy demand of a prefabricated building and compare the obtained results to the respective energy demand of a conventional building with similar geometrical characteristics, made of reinforced concrete and brick masonry. Firstly, an overview of studies that assess the performance of prefabricated buildings is presented. The methodological approach for the assessment of the current and future energy demand is then described and finally, the main results of the study are summarized and discussed.

12.1.2 Overview of Studies Assessing the Performance of Prefabricated Buildings

Prefabricated buildings are made of components or units constructed in a factory setting, under controlled conditions and standardized procedures, which are then transported and assembled on-site to create a functional structure (Tsikaloudaki et al. 2019). On the basis of the prefabrication level, three major categories of building structures can be discerned (Gibb 1999) (i) the ones that have a framework made of "non-volumetric" prefabricated units, i.e. elements that do not enclose usable space (such as columns, beams, partitions, claddings, services, etc.)., (ii) the ones that comprise of "volumetric" prefabricated units, i.e. units that enclose usable space without constituting a whole building (such as toilets/washrooms, kitchens, lifts, etc.), and (iii) modular buildings, which refer to a complete building or parts of a building constructed off-site and assembled on-site.

A variety of materials, namely steel, wood, concrete, etc., or their combination, can be used for forming the structural framework and the envelope of the prefabricated buildings, depending on the prefabrication level, the design requirements and the performance standards to be achieved. Each material presents distinct benefits in terms of speed, durability, sustainability, and design flexibility. For example, steel is often preferred for its versatility, strength and speed of construction (Dalalbashi et al. 2024), wood has gained momentum for offering lightweight, earthquake-resistant, and sustainable-low carbon-constructions (Bhandari et al. 2023) concrete offers durability, fire resistance, and thermal mass (Kontoleon et al. 2022). The prefabrication industry has developed significantly in the last decades, due to the plethora of advantages offered by such a system. Among them, the most significant are:

- The reduced construction time.

The construction of prefabricated buildings comprises of a two-stage procedure: the off-site industrialized production of components or modules and their transportation, and the on-site assembly and finalization. The shift from the one stage on-site construction of conventional buildings towards the off-site manufacturing of the prefabricated components can shorten the construction time. The reduction ratios reported in the bibliography amount to 12% on the decreased construction time delays (Yuan et al. 2024) to an average of 34% in time savings, estimated for a significant number of case studies in (Auckland Shahzad et al. 2015) and even up to 40% through using modular technologies with respect to the conventional methods, as found by conducting a time-savings analysis on different building projects across the globe, namely UK, China, USA, Singapore and Myanmar (Tsz Wai et al. 2023). The speed is mostly associated with the shortened times of the on-site activities, as shown vividly by Tam et al. (2015), who analyzed two prefabricated high rise residential buildings in Hong Kong and presented very shortened installation times for the precast elements (facades, floors, partitions, etc.) and structural walls, as well as the precast modular units (bathrooms, lifts, stairs, lintels). However, it must be noted that some cases presented in the bibliography faced the problem of construction time overrun. This is attributed to various uncertainty factors, such as the delayed material delivery, the potential lack of work familiarity, the project complexity and poor weather conditions during the on-site assembly (Tsz Wai et al. 2023).

- The reduced cost.

In general, prefabrication is regarded as one of the most cost-effective construction methods. The total cost is shaped mainly by the material cost (30%–50%), the labor cost (14%–24%) and the transportation (6%–11%) (Hong et al. 2018). In a study conducted by Shahzad et al (2015), it is summarized that the marginal cost saving delivered through using prefabrication methods was found equal to 19% when compared to the conventional construction techniques. This can be attributed to several factors, such as the increased productivity and control quality, the decreased management needs, the decreased labor material and construction waste (Hong et al. 2018). Within this context, Tam et al. (2015) analyzed two prefabricated high rise residential buildings in Hong Kong and they stated that the reduction of the on-site construction can lead to a 30% decreased site manpower, 40%–75% reduced material usage, while 6% of the gross floor area was gained from the precast construction, leading to positive implications to the projects' cost. However, some studies indicated that the economic performance of the prefabricated buildings may be under question, as, in some cases, they presented higher unit costs with respect to conventional buildings (Jaillon and Poon 2008). The extra costs can be attributed to the need of highly skilled workers, design changes, initial investment (new machinery, fabricate molds, and factories), logistics and required components storage space (Hong et al. 2018).

- Enhanced safety and labor conditions.

The manufacturing of the prefabricated components in the factory, in a systemized and fully controlled environment, with repetitive tasks in a foreseen production process, minimizes safety risks. Furthermore, the decrease of on-site construction works encountered in prefabrication projects results in a lower demand for on-site workers and an increase of skilled ones among them -who are more familiar with safety risks-which in turn leads to less workplace accidents (Zhong and Zhang 2020). Accordingly, the accident ratios caused by workforce congestion, severe weather, dangerous activities, such as working at increased heights for a considerable time period, etc., which are often met in general construction, are regarded as lower (Shen 2022). Particularly for modular construction, the on-site reported accidents can be reduced to 80% with respect to the conventional construction field (Kamali and Hewage 2016). Although prefabrication and modularization are regarded safer than the conventional construction methods, accidents still occur; they are mostly associated with defective materials and machinery, as well as poor safety measures, including the personal protection equipment (Jeong et al. 2022). Beyond the construction accidents, it is essential to consider the occupational health risks under which the workers are exposed during the production of the prefabricated components and units, given that this stage is carried out in factories and the materials, machinery and energy use may lead to harmful environments (Cui et al. 2022).

- The increased quality.

Prefabricated buildings are regarded to present higher construction accuracy than conventional buildings (Yuan et al. 2023) due to the systematized operations and the use of automated machinery for the components manufacturing. Furthermore, the need for transporting the components to the construction site and lifting them for their assembly, necessitates increased strength, load-bearing capacity, durability and weather resistance, leading to standard and usually higher quality requirements (Kamali and Hewage 2016). Construction accuracy is among the determinants of the prefabricated structures' quality, and it is affected by several factors, namely the material properties, the equipment and the personnel quality through all stages of prefabrication, i.e. during the production, the transportation and the installation of prefabricated components (Yuan et al. 2023). New technologies, such as BIM, AI, IoT, etc., act as drivers for the quality assurance of prefabricated constructions, in terms of guaranteeing the geometry, accuracy and correctness of the units, scheduling, tracking and storage of components for maximization of construction productivity (Rangasamy and Yang 2024).

- Advanced environmental performance.

The potential of prefabricated buildings to reduce their environmental impacts is notable and addresses various aspects of environmental performance across their entire life cycle. The existing literature indicates that the embodied energy and carbon are generally lower for prefabricated buildings than the conventional ones, with the exception or prefabricated steel buildings, which that show similar or even higher impacts to the conventional ones (Tavares et al. 2019). Still, Aye et al. (2012) found that, even if prefabricated steel buildings resulted in a significant increase in embodied energy, the reusability of the material may counterbalance this impact due to the significant waste reduction and reduced requirements for additional virgin materials.

Contemporary prefabricated buildings demonstrate superior energy performance. The standardized components and construction processes enable better insulation and air sealing, reducing heating and cooling energy needs (Tsoka et al. 2020). Besides, the advanced building technologies employed for their construction foster the installation of smart HVAC and RES systems.

The transportation of the prefabricated components from the factory to the construction site results in fuel consumption and emissions from the transportation vehicles. However, this impact can be mitigated through efficient logistics planning and the use of low-emission transport methods. Furthermore, it must be noted that prefabricated buildings often exhibit higher resource efficiency compared to traditional construction, due to the controlled factory environment, which allows for precise material usage and minimization of waste (Aghasizadeh et al. 2022).

12.1.3 Models for the Prediction of Future Climate

General Circulation Models (GCMs) are numerical models that represent the physical procedures that occur in the atmosphere, the oceans, the cryosphere and the land surface (Grotch and MacCracken 1991). The definition of regional climates is achieved with the interaction between planetary and local scale forcing. The scientific community relies on these models for the simulation of climate and the evaluation of climate change in different temporal and spatial scales, and for the projection of future climate scenarios. GCMs are driven by various scenarios (Representative Concentration Pathways—RCPs) representing key drivers of future emissions, such as economic, technological, and population growth.

The Representative Concentration Pathways (RCPs) reflect the different emission scenarios available in literature from previous assessments that have similar characteristics and radiative forcing (RF).

RCPs must be compatible with a wide range of the available baseline stabilization, mitigation and emission scenarios and must contain information for factors such as long-term greenhouse gas emissions (GHG), air pollutant emissions, land use and socioeconomic conditions (Moss et al. 2008). There are four different RCPs (Table 12.1): a mitigation scenario that leads to very low radiative forcing Vuuren et al. (2011), two intermediate stabilization scenarios (Thomson et al. 2011; Masui et al. 2011) and a scenario with very high radiative forcing (Riahi et al. 2011).

To continue, the spatial resolution of the GCMs, driven by the various RCPs usually varies between 250 and 600 km. Consequently, they are rather coarse and not adequate for determining local climate in regions with complex topography. Their direct utilization of their output in dynamic building energy performance simulations is thus impractical, as these simulations demand climatic parameters at an hourly temporal resolution. To obtain climate data in a finer resolution, GCMs require downscaling.

Table 12.1 Summary of representative concentration pathways

RCP	Description
RCP8.5	Very high radiative forcing: Rise (8.5 W/m^2— ~ 1370 ppm CO_2 eq) by 2100
RCP6	Intermediate scenario: Stabilization without exceedance (6 W/m^2— ~ 850 ppm CO_2 eq) at stabilization after 2100
RCP4.5	Intermediate scenario: Stabilization without exceedance (4.5 W/m^2— ~ 650 ppm CO_2 eq) at stabilization after 2100
RCP2.6 (RCP3-PD)	Very low radiative forcing: Peak (~3 W/m^2— ~ 490 ppm CO_2 eq) before 2100 and then decline (declines to 2.6 W/m^2 by 2100)

PD peak and decline

There are two main approaches for this purpose: dynamical and statistical downscaling (Kostopoulou et al. 2007). Statistical downscaling methods establish statistical relationships between observed climatic variables and larger-scale variables from GCMs, employing techniques like regression analysis or stochastic methods. Statistical downscaling can also be integrated with bias correction for the improvement of the systematic errors of climate model outputs. On the other hand, GCMs can be dynamically downscaled in regional climate models (RCMs). RCMs are numerical climate prediction models that simulate the main climatic features that act on a local scale, where complex topographical characteristics are evident and mesoscale forcings are of great importance (Giorgi and Mearns 1991). They require specific lateral and ocean conditions derived from GCMs and reanalysis data. As a result, the dynamical extrapolation of the impacts of large-scale climate processes to regional or local scales is achieved. RCMs simulate the atmospheric and land surface procedures while considering the high-resolution topographic features of an area, the contrasts between land and sea, land surface characteristics and other Earth-system components.

Both approaches have been extensively employed in prior studies assessing the effects of climate change on building energy usage (Tsoka et al. 2021) with various studies analyzing the advantages and disadvantages of each method (Ekström et al. 2015). Within this framework, Tootkaboni et al. (2021) and Berardi et al. (2020) proposed that statistical downscaling methods may offer sufficient information for comparing long-term changes in building energy performance. Nonetheless, dynamical downscaling methods are considered more reliable due to their finer resolution. In the present study, dynamical downscaling approach was used as it was considered the most effective for this case.

12.2 Methodology

12.2.1 Generation of Current and Future Climatic Weather Datasets

To predict future energy needs within this study's context, dynamic downscaling methods are used as the basis for the generation of future weather files. More precisely, version 4.4.5.1 of the regional climate model, RegCM4 is used on a spatial model resolution of 25 × 25 km. The simulation is driven by HadGEM2 general circulation model (GCM) and for future projections the model is using the RCP4.5 scenario, which is an intermediate pathway with no exceedance of radiative forcing at a stabilization level of ~ 4.5 W/m^2. The work of Velikou et al. (2019) describes in further detail the main physics parameterizations of the model implemented in this study. For the evaluation of the energy performance for both the conventional and the prefabricated construction under current and future climatic conditions, two weather files are created according to the RegCM4 output: (a) an hourly weather file, corresponding to the present-day climatic conditions and issues by the reference simulation period 1981–2000 (i.e., Reference 1981–2000) and (b) an hourly weather file, reflecting the future climatic conditions for the period 2041–2060 (i.e., Future 2041–2060).

To this aim and for every simulation run of the RegCM4, five meteorological parameters are extracted. These parameters include the air temperature, the relative humidity, the wind speed, and global solar radiation. At the next step, RegCM4 extracted data for periods of 1981–2000 and 2041–2060 are organized and averaged for the 20-years period to generate a single weather file for each period. In parallel, given that the RegCM4 simulation output included 3-hourly temporal resolution data, linear interpolation was used to estimate the hourly values for each meteorological parameter for both periods of 1981–2000 and 2041–2060. This approach is similar to the one followed by Berardi and Jafarpur (2020). Finally, for both periods, the obtained 8760 hourly values of each meteorological parameter, have been introduced in the open-source platform for editing and creating custom EPW weather files, Elements which is compatible with the Energy Plus transient building energy modelling.

12.2.2 Description of the Case Study Building

The study concerns a small, single-family building, presented in Fig. 12.1. The 47.35 m^2, one-story, two-bedroom building features a simple, rectangular layout that is expanded along the south-north axis, with openings limited only to the smaller facades, oriented North and South. It features an inclined roof with an outer surface of clay tiles. The roof is constructed on a horizontal reinforced concrete slab, thermally insulated with a 10.0 cm expanded polystyrene (XPS) layer.

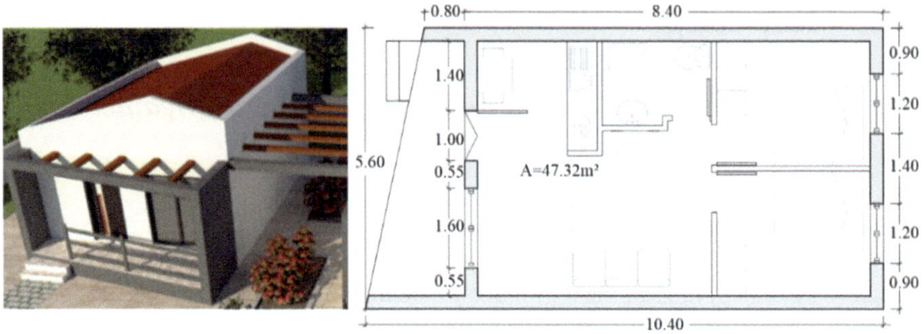

Fig. 12.1 External 3D view and plot of the examined house

Similarly, the floor of the house is made from reinforced concrete, and it is thermally insulated from the ground in contact, with a 10.0 cm layer of XPS. Regarding its vertical building envelope, the windows comprise of PVC frame with a double, low-e glazing. The U-value of the transparent elements is equal to 2.0 W/(m^2 K) which, according to the national thermal insulation requirements is considered as a well thermal protected one.

The energy performance of the building is evaluated for two different configurations of the vertical load-bearing elements involving both conventional and prefabricated construction methods (Fig. 12.2) (Tsikaloudaki et al. 2022). The former configuration involves conventional masonry made of fired-clay brick walls, while the latter consists of a prefabrication system that represents novel concepts on holistic and environmentally friendly building envelopes.

The prefabricated wall elements have been developed within the SU.PR.I.M. research project as a new approach of prefabricated wall elements with enhanced properties (Tsikaloudaki et al. 2022). It consists of two lightly reinforced, 5 cm concrete plates, which are positioned on either side of vertical (and occasionally diagonal) steel hollow elements. The distance between the vertical hollow steel elements varies between 0.70 m–1.00 m and the gap is filled in with expanded polystyrene. The prefabricated panels are connected to each other and to the metal columns/beams of

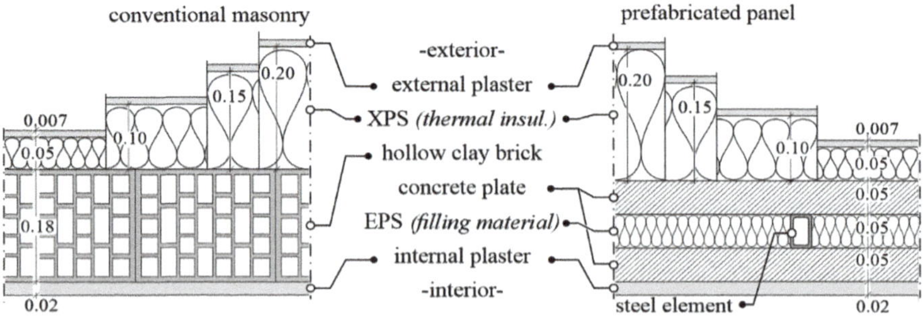

Fig. 12.2 Horizontal cross-section of the examined wall elements

the structural organization with anchors, developed especially to integrate the prefabricated elements into a robust and resilient building structure. In both alternative wall configurations, the thermal insulation levels are the same, in terms of the thermal insulation material and thickness, placed on the external side of the envelope.

12.2.3 Simulation Scenarios and Parameters

The energy performance of both examined building envelope alternatives is assessed with Energy Plus, a transient simulation software that is widely used for advanced energy analyses. The energy analysis concerns the annual heating and cooling needs of the buildings for different thermal insulation levels yet the same thickness in both wall configurations. The thermal insulation consists of XPS with a layer thickness ranging from 5 to 20 cm with a step of 5 cm, forming four thermal insulation scenarios. The climate conditions for the analysis are those of the city of Thessaloniki in northern Greece. The energy performance is evaluated both for current and future climatic conditions, using hourly data sets generated for the study according to the procedure that is presented in the following sections.

The operational characteristics of the house, involving occupancy, lighting, equipment, heating and cooling setpoints, ventilation and infiltration rates, conform with the respective values provided by the Technical Guides of the recast of the Hellenic Building Energy Performance Regulation. Since in local climatic conditions, natural ventilation during the warm period is considered as the most favorable passive cooling means, the simulation considers natural ventilation during the cooling period, provided that the outdoor air temperature is lower than the indoor one by at least 1.0 °C.

The simulations were conducted on an hour time step and an annual basis. The HVAC systems have been employed in the simulation model using the simplified model of the ideal load HVAC system. Domestic Hot Water (DHW) needs were also calculated, for all climatic zones of Greece, according to the Hellenic Regulation for Building Energy Performance, and the standards EN 15,316–3.2 and EN 15,316–3.3. Finally, for all the examined scenarios, the energy performance simulation has been performed using 20 °C and 26 °C thermostat ser for the heating and cooling setpoints, following the Provisions of the National Building Energy Performance Regulation.

12.3 Results and Discussion

12.3.1 *Evaluation of Climate Change and Future Air Temperature*

The simulated mean daily Tair for the reference period (1981–2000) and for the future period (2041–2060) is illustrated in Fig. 12.3. The comparison

of the parameters between the two datasets (reference and future) reveals a similar intra-annual variability, with the future projection presenting higher Tair values. More specifically, daily Tair during the period 1981–2000 ranges between 2.8 °C in winter (15/1) and 25.6 °C in summer (29–31/7). Similar are the results for the future period 2041–2060, with the lowest temperatures (2.8 °C) being reported during winter (15/1) and the highest (27.9 °C) during summer (16–17/7).

Figure 12.4 presents the monthly mean values of air temperature (Tair) for the two examined periods and the distribution of these values in terms of mean, median, interquartile range, minimum and maximum. Projected mean monthly temperature presents an increase of 1.1 °C–1.5 °C and 2.1 °C–2.8 °C during heating and cooling seasons, respectively, with an overall annual temperature rise of 1.7 °C. Peak maximum temperatures exceeding 2.5 °C are expected to occur in July and September, indicating significant risks for binding energy usage. The findings from the boxplots of the two datasets show the increase in temperature during the period 2041–2060.

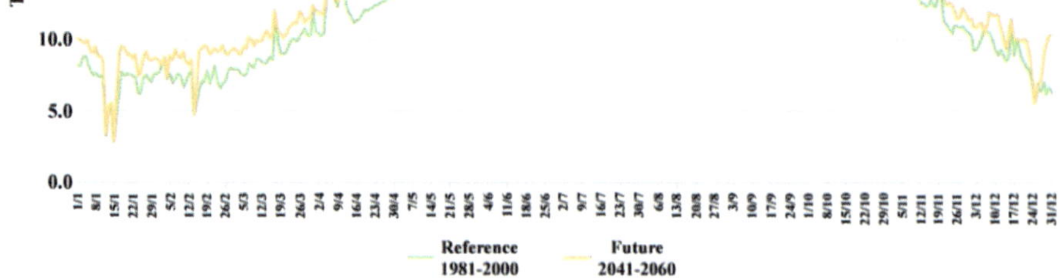

Fig. 12.3 Intra-annual variations of mean daily Tair values (°C) from RegCM4 simulations for the reference period (1981–2000) and the future period (2041–2060)

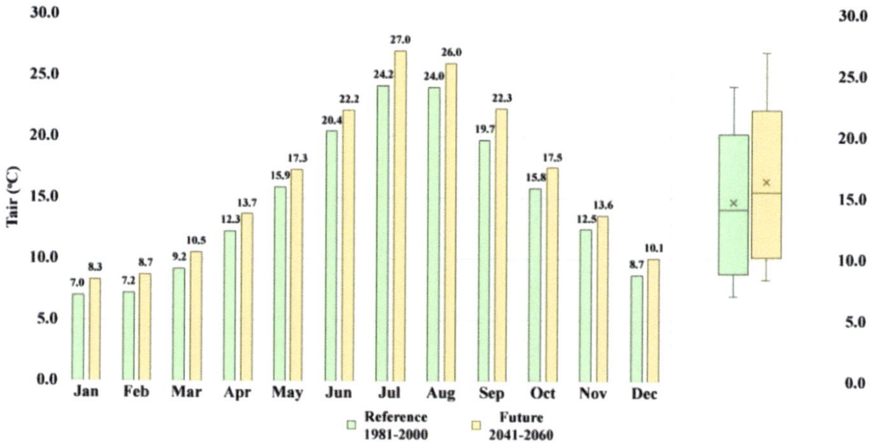

Fig. 12.4 Monthly mean Tair values (°C) from RegCM4 simulations for the reference period (1981–2000) and for the future period (2041–2060) and the corresponding boxplots

12.3.2 *Evaluation of the Energy Demand for the Conventional and Prefabricated Buildings*

The annual heating and cooling energy demand of the examined buildings of conventional (i.e., ETICS building) and prefabricated construction (i.e., PREFAB. building), for the reference period and future climatic conditions are shown in Figs. 12.5 and 12.6 respectively. Simulation results are presented in pairs with respect to the thermal insulation thickness and the wall configuration.

In terms of the simulation outcomes for the current climate conditions, the energy needs of the prefabricated building consistently prove to be lower compared to those of the conventional ETICS construction. The difference in total energy needs spans

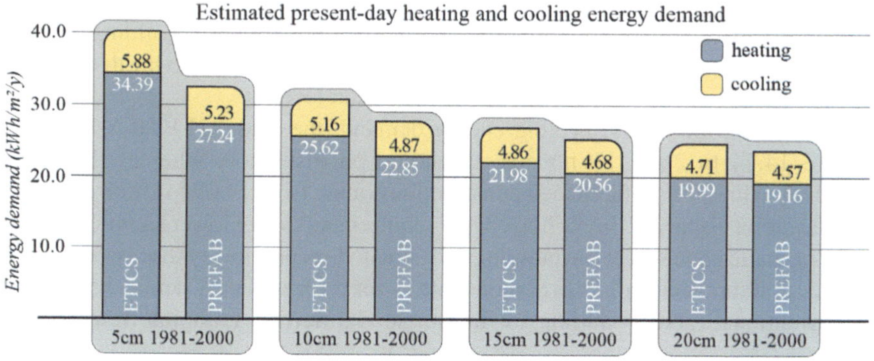

Fig. 12.5 Estimated, annual heating and cooling energy demand for the different, examined wall configurations, for the present-day climatic conditions

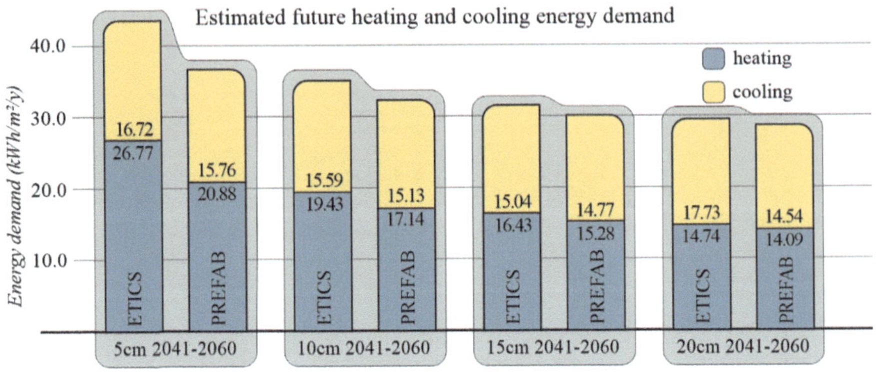

Fig. 12.6 Annual heating and cooling energy demand for the different, examined wall configurations, for the future climatic conditions

from 4 to 18%, with the heating energy demand being reduced by 3% to 17.5% depending on the thermal insulation thickness (Fig. 12.5).

Additionally, for both building construction methods, the heating energy demand exceeds cooling needs, as anticipated given the region's geographic location and climate. Maximizing thermal resistance in the walls reduces heat loss during winter but doesn't proportionally decrease cooling needs, as solar heat gains, primarily issued from transparent building elements, remain significant. Furthermore, the discrepancy in heating needs between the two construction types varies with insulation thickness; as insulation thickness increases, its impact on reducing heating load lessens.

Regarding the anticipated future energy needs, the simulation outcomes indicate a rising trend in cooling energy demands and a decrease in heating requirements for both types of construction, with the projected alterations varying based on the level of insulation thickness. The analysis suggested that, regardless of the insulation level, both conventional and prefabricated buildings witness a more than double increase in cooling energy requirements attributed to the higher future summer air temperature values. However, for the future period 2041–2060, prefabricated buildings exhibit lower total annual energy needs compared to conventional ones, with the discrepancy ranging from 4 to 16%, depending on the insulation level.

Concerning the annual heating energy demand, the analyzed ETICS buildings require 14.74 to 26.77 kWh/m^2/y, depending on insulation, whereas the corresponding values for prefabricated constructions range from 14.09 to 20.88 kWh/m^2/y. Once more, disparities in cooling energy requirements are of lower significance due to the climatic zone where the buildings are situated, as previously discussed. Nevertheless, prefabricated buildings demonstrate lower cooling energy demand compared to conventional ETICS buildings, with estimated differences ranging from 1.5% to 6%.

12.4 Conclusion

This chapter aimed to evaluate the impact of climate change on the heating and cooling energy requirements of prefabricated buildings in comparison to conventionally constructed buildings made of reinforced concrete and brick masonry. Given that the global prefabricate buildings market is expected to present a significant growth in the years to come, the present research aimed to assess their performance with respect to climate change and the anticipated future higher air temperatures.

Dynamic downscaling techniques, using the RegCM4 model on a spatial model resolution of 25 × 25 km, have been employed for the ftuture climatic projections; the analysis concerned the future period 2041–2060 and the intermediate RCP4.5 emission scenario has been considered. The generated future weather data were processed and organized to create suitable weather files for the dynamic simulations. The analysis suggested that the future mean monthly Tair is expected to rise by 1.1 °C–1.5 °C and 2.1 °C–2.8 °C during heating and cooling seasons respectively for the period 2041–2060 compared to the reference period of 1981–2000, which represents current climatic conditions (i.e. reference period). In addition, the average annual Tair value will be higher by 1.7 °C in the future, compared to the present-day conditions.

As future temperatures rise, there is a decrease in heating energy demand for both prefabricated and conventional buildings, while cooling energy demand is expected to be more than double. However, despite the projected increase in total annual energy demand, the examined prefabricated buildings were found to be more efficient compared to conventional ones, requiring 4%–16% less energy for heating and cooling purposes, depending on the insulation level. Especially for cooling, prefabricated buildings are expected to present lower future energy demand compared to conventional ETICS buildings, with estimated differences ranging from up to 7%.

Finally, it has to be mentioned that the current chapter only focused on one specific type of prefabrication (i.e. prefabricated concrete structures) without, however, examining other materials such as steel or wood. Moreover, only one building typology has been evaluated and one RCP emission scenario has been considered for future climate projections. The above-mentioned limitations undoubtedly suggest future research perspectives but still, the results of this study provided significant insight on the potential of prefabricated construction to present high energy performance in the context of ongoing climate change.

References

Aghasizadeh S, Tabadkani A, Hajirasouli A, Banihashemi S (2022) Environmental and economic performance of prefabricated construction: a review. Environ Impact Assess Rev 97:106897

Aye L, Ngo T, Crawford RH, Gammampila R, Mendis P (2012) Life cycle greenhouse gas emissions and energy analysis of prefabricated reusable building modules. Energy Build 47:159–168

Berardi U, Jafarpur P (2020) Assessing the impact of climate change on building heating and cooling energy demand in Canada. Renew Sustain Energy Rev 121:109681

Bhandari S, Riggio M, Jahedi S, Fischer EC, Muszynski L, Luo Z (2023) A review of modular cross laminated timber construction: implications for temporary housing in seismic areas. J Build Eng 63:105485

Chan A (2011) Developing future hourly weather files for studying the impact of climate change on building energy performance in Hong Kong. Energy Buildings 43(10):2860–2868

Chen Y, Zhu D, Tian Z, Guo Q (2023) Factors influencing construction time performance of prefabricated house building: a multi-case study. Habitat Int 131:102731

Cui P, Zhao H, Dong Z (2022) Modeling the occupational health risk of workers caused by environmental release during the production of PC components. Front Public Health 10:1076461

Dalalbashi A, Pinto J, Reis C, Pimenta F, Ferreira NO, Pereira NB (2024) A proposal for an alternative structural modular solution. J Build Eng 85:108709

Ekström M, Grose MR, Whetton PH (2015) An appraisal of downscaling methods used in climate change research. Wiley Interdiscip Rev: Clim Chang 6(3):301–319

Gibb AG (1999) Off-site fabrication: prefabrication, pre-assembly and modularisation. Wiley

Giorgi F, Mearns LO (1991) Approaches to the simulation of regional climate change: a review. Rev Geophys 29(2):191–216

Grotch SL, MacCracken MC (1991) The use of general circulation models to predict regional climatic change. J Clim 4(3):286–303

Hong J, Shen GQ, Li Z, Zhang B, Zhang W (2018) Barriers to promoting prefabricated construction in China: a cost–benefit analysis. J Clean Prod 172:649–660

Ismail FH, Shahrestani M, Vahdati M, Boyd P, Donyavi S (2021) Climate change and the energy performance of buildings in the future–a case study for prefabricated buildings in the UK. J Build Eng 39:102285

Jaillon L, Poon CS (2008) Sustainable construction aspects of using prefabrication in dense urban environment: a Hong Kong case study. Constr Manag Econ 26(9):953–966

Jeong G, Kim H, Lee H-S, Park M, Hyun H (2022) Analysis of safety risk factors of modular construction to identify accident trends. J Asian Arch Build Eng 21(3):1040–1052

Kamali M, Hewage K (2016) Life cycle performance of modular buildings: a critical review. Renew Sustain Energy Rev 62:1171–1183

Kontoleon K, Georgiadis-Filikas K, Tsikaloudaki K, Theodosiou T, Giarma C, Papanicolaou C, Triantafillou T, Asimakopoulou E (2022) Vulnerability assessment of an innovative precast concrete sandwich panel subjected to the ISO 834 fire. J Build Eng 52:104479

Kostopoulou E, Giannakopoulos C, Anagnostopoulou C, Tolika K, Maheras P, Vafiadis M, Founda D (2007) Simulating maximum and minimum temperature over Greece: a comparison of three downscaling techniques. Theoret Appl Climatol 90:65–82

Masui T, Matsumoto K, Hijioka Y, Kinoshita T, Nozawa T, Ishiwatari S, Kato E, Shukla PR, Yamagata Y, Kainuma M (2011) An emission pathway for stabilization at 6 Wm−2 radiative forcing. Clim Change 109:59–76

Moss RH, Babiker M, Brinkman S, Calvo E, Carter T, Edmonds JA, Elgizouli I, Emori S, Lin E, Hibbard K (2008) Towards new scenarios for analysis of emissions, climate change, impacts, and response strategies, Technical Summary. Intergovernmental Panel on Climate Change (IPCC), (Geneve), pp 25

Navaratnam S, Ngo T, Gunawardena T, Henderson D (2019) Performance review of prefabricated building systems and future research in Australia. Buildings 9(2):38

Nik VM, Kalagasidis AS (2013) Impact study of the climate change on the energy performance of the building stock in Stockholm considering four climate uncertainties. Build Environ 60:291–304

Pérez-Andreu V, Aparicio-Fernández C, Martínez-Ibernón A, Vivancos J-L (2018) Impact of climate change on heating and cooling energy demand in a residential building in a Mediterranean climate. Energy 165:63–74

Rangasamy V, Yang J-B (2024) The convergence of BIM, AI and IoT: reshaping the future of prefabricated construction. J Build Eng 108606

Riahi K, Rao S, Krey V, Cho C, Chirkov V, Fischer G, Kindermann G, Nakicenovic N, Rafaj P (2011) RCP 8.5—a scenario of comparatively high greenhouse gas emissions. Clim Change 109:33–57
Shahzad W, Mbachu J, Domingo N (2015) Marginal productivity gained through prefabrication: case studies of building projects in Auckland. Buildings 5(1):196–208
Shen P (2017) Impacts of climate change on US building energy use by using downscaled hourly future weather data. Energy Buildings 134:61–70
Shen Y, Xu M, Lin Y, Cui C, Shi X, Liu Y (2022) Safety risk management of prefabricated building construction based on ontology technology in the BIM environment. Buildings 12(6):765
Tam VW, Fung IW, Sing MC, Ogunlana SO (2015) Best practice of prefabrication implementation in the Hong Kong public and private sectors. J Clean Prod 109:216–231
Tavares V, Lacerda N, Freire F (2019) Embodied energy and greenhouse gas emissions analysis of a prefabricated modular house: the "Moby" case study. J Clean Prod 212:1044–1053
Thomson AM, Calvin KV, Smith SJ, Kyle GP, Volke A, Patel P, Delgado-Arias S, Bond-Lamberty B, Wise MA, Clarke LE (2011) RCP4. 5: a pathway for stabilization of radiative forcing by 2100. Clim Change 109:77–94
Tootkaboni MP, Ballarini I, Corrado V (2021) Analysing the future energy performance of residential buildings in the most populated Italian climatic zone: a study of climate change impacts. Energy Rep 7:8548–8560
Tsikaloudaki K, Theodosiou T, Tsoka S, Chastas P (2022) Sustainable prefabricated buildings: a holistic approach. Renew Energy Environ Sustain 7:11
Tsikaloudaki K, Tsoka S, Theodosiou T (2019) Development of a new building element for prefabricated constructions of advanced performance. In: Conference proceedings of interdisciplinary research on real estate CIRRE, Trondheim, Norway, pp 37
Tsoka S, Theodosiou T, Papadopoulou K, Tsikaloudaki K (2020) Assessing the energy performance of prefabricated buildings considering different wall configurations and the use of PCMs in Greece. Energies 13(19):5026
Tsoka S, Velikou K, Tolika K, Tsikaloudaki A (2021) Evaluating the combined effect of climate change and urban microclimate on buildings' heating and cooling energy demand in a Mediterranean City. Energies 14(18):5799
Tsz Wai C, Wai Yi P, Ibrahim Olanrewaju O, Abdelmageed S, Hussein M, Tariq S, Zayed T (2023) A critical analysis of benefits and challenges of implementing modular integrated construction. Int J Constr Manag 23(4):656–668
Van Vuuren DP, Stehfest E, den Elzen MG, Kram T, van Vliet J, Deetman S, Isaac M, Klein Goldewijk K, Hof A, Mendoza Beltran A (2011) RCP2. 6: exploring the possibility to keep global mean temperature increase below 2 C. Clim Change 109:95–116
Velikou K, Tolika K, Anagnostopoulou C, Zanis P (2019) Sensitivity analysis of RegCM4 model: present time simulations over the Mediterranean. Theoret Appl Climatol 136:1185–1208
Yuan Z, Wang H, Yang Y, Yi C, Huang D, Yu D (2023) Improving the construction accuracy of precast components in prefabricated buildings by analyzing relevant factors from the perspective of supply chain: a system dynamics model, Architectural Engineering And Design Management, pp 1–19
Yuan Z, Man Q, Guan Z, Yi C, Zheng M, Chang Y, Li HX (2024) Simulation and optimization of prefabricated building construction considering multiple objectives and uncertain factors. J Build Eng 108830
Zhang X, Du J, Sharples S (2023) A parametric analysis of future climate change effects on the energy performance and carbon emissions of a Chinese prefabricated timber house. Build Serv Eng Res Technol 44(2):167–185
Zhong C, Zhang W (2020) Quality management and safety evaluation for prefabricated buildings. Int J Perform Eng 16(5):811

Chapter 13
Green Urban Renewal Strategies in Relation to the Sustainable Urbanization Policies

Ivana Katurić and Ozana Palić

Abstract The concept of sustainable urbanisation is a wide term and very common in recent academic works and spatial planning policy documents, especially in relation to zero-net land discussion. One group of measures aimed at sustainable urbanisation is green urban renewal, which is addressed to varying degrees in different strategic documents at international and national levels. This paper aims to provide a more detailed overview of those strategic documents at the European Union level, providing the context of the new planning instrument for achieving sustainable urbanisation in the Republic of Croatia—green urban renewal strategies. The results of scientific research conducted within the ESPON SUPER project are presented, framing the sustainable urbanisation discourse and practices. SUPER project created a comprehensive framework for assessing the sustainability of different forms of urbanisation and produced a handbook for planners and decision-makers in spatial planning on sustainable urbanisation practices. A special focus of the article is placed on the relationship between green urban renewal strategies and the principles of New European Bauhaus. This relationship is examined through the lens of a handbook that has been developed to enhance the implementation potential of the strategies and promote its components and possible applications.

Keywords Sustainable urbanization · Green urban renewal · Spatial planning · Sustainable development policies · Circularity · Croatia

I. Katurić (✉)
University of Rijeka, Urbanex, Split, Croatia
e-mail: ivana.katuric@urbanex.hr

O. Palić
Urbanex, Zagreb, Croatia
e-mail: ozana.palic@urbanex.hr

M. Vukmirović et al. (eds.), *Digital Dreams, Urban Realities*, The Urban Book Series,
https://doi.org/10.1007/978-3-031-76353-3_13

13.1 Sustainable Urbanisation—The Term in Scientific Discourse

The topic of sustainable urbanisation is frequently discussed and promoted in various areas of the political discourse of the United Nations and the European Union, as well as in the scientific community. The contexts in which this topic is emphasised in both discourses are numerous and quite fragmented. The UN-HABITAT World Cities Report 2020 UN-Habitat (2020), for example, highlights current trends in urban development that combine the importance of urbanisation with the challenges of sustainable development. Based on these trends, cities are recognised as a central component of international development policy, continued urban expansion and escalating land consumption through construction projects are assumed, and the importance of migration dynamics and social inequalities that shape urban structures is highlighted. Taken together, these factors underline the complex interplay of various economic, ecological, social, technological and political-administrative elements that ultimately influence the sustainability of urbanisation. Due to this complexity, there are numerous ways of defining and approaching the concept of sustainable urbanisation in the scientific and policy literature.

In scientific research on sustainable urbanisation in the European context, particularly relevant is the project of the European Spatial Planning Observation Network (ESPON) entitled "Sustainable Urbanization and Land Use Practices in European Regions" (SUPER), in which the first author took part. The aim of the ESPON SUPER project was to provide evidence-based findings and recommendations to promote sustainable land use while mitigating and preventing unsustainable urbanisation in Europe, its cities, and regions ESPON (2020). Among other things, this project focused on creating a straightforward conceptual framework for understanding urbanisation and land use dynamics. It was, therefore, necessary to clearly define the concept of sustainable urbanisation.

The concept of sustainability itself is a multidimensional concept that is difficult to explain unambiguously. Sustainability is commonly defined by its temporal dimension—as a temporal balance, which means the ability to maintain the quality of life on earth. And not to use resources in a way that exceeds their recovery rate Neuman and Churchill (2015). Since land is a finite resource, this approach proves somewhat limited and promotes a narrow quantitative approach that focuses on slowing down or stopping urbanisation altogether (such as the European Union's goal of ending 'land take' by 2050). In this sense, a thematic definition of sustainability proves to be more plausible for planners. While sustainability originally emerged as a concept with an environmental dimension, its understanding has expanded to include various human activities Leitmann (1999). Therefore, three key dimensions are often emphasised: social, environmental and economic Purvis et al. (1999). In practice, sustainable development would mean progress in one or more of these dimensions without harming the others Campbell (2016). Sustainability can be measured using several models and approaches Neuman and Churchill (2015).

On the other hand, urbanisation is defined in the project as the spatial phenomenon of the transformation of land into urban land use ESPON (2020). It is defined as a spatial category that encompasses the entire spatial development of built-up land, in contrast to the traditional understanding of urbanisation in the geographical and planning literature, which focuses predominantly on the expansion of cities and the causes of this process. Urbanisation can take different forms depending on the areas in which it occurs, and it can even develop in the opposite direction, such as deurbanisation—the transformation of urban land into rural or natural areas.

The sustainability of urbanisation has been discussed in various ways in the academic literature in recent decades, but the number of papers specifically addressing the sustainability of different forms of urbanisation is quite limited. With the exception of the SUPER project, which systematically addressed this issue by developing a methodology to assess the sustainability of individual urban forms, such a methodology was first developed by Jabareen and presented in the paper from 2006 Jabareen (2006). Jabareen argued that the extent to which a city accommodates these human activities in an efficient and sustainable manner depends not only on its size but also on its form. In his work, Jabareen subjected four sustainable urban forms (neotraditional urban development, compact city, urban containment and eco-city) to a multi-criteria analysis using indicators from the relevant literature. In the SUPER project, a similar model was used to determine the sustainability of individual urban forms through a multi-criteria analysis. However, in contrast to Jabareen's approach, three urban forms were considered, none of which were classified a priori as sustainable or unsustainable. Instead, their sustainability was assessed on the basis of a range of economic, environmental and social criteria. The urban forms studied in the SUPER project are compact urbanisation, polycentric urbanisation and diffuse urbanisation.

Compact urbanisation emerges as a strategic response to urban containment Nelson et al. (2007). The implementation of urban containment policies involves establishing boundaries to restrict further urban expansion into surrounding non-urban areas. Within these boundaries, all urban development takes place on both developed and undeveloped urban land, characterised by medium to high building densities Nelson et al. (2004). In some cases, similar zoning extends to satellite settlements encircling cities, preserving green belts comprising agricultural land, meadows, and forests between them Millward (2006). Compact urbanisation is often associated with the concept of zero net land consumption, which was originally developed from an environmental point of view. This concept is about avoiding land take by reusing already used and/or sealed areas or compensating for land take through renaturation or unsealing projects.

Another form is polycentric (clustered) urbanisation, a clustered medium-density development around several cores. Its main characteristics are the economical use of land, the availability of transportation, and the fact that it is a mixed land use that allows the population to access basic services on foot Daniels and Lapping (2005), Handy (2005), Stewart et al. (2006). One of the most common strategies used to achieve polycentric urbanisation is transit-oriented development, which involves the development of a compact neighbourhood designed specifically for pedestrians and cyclists around a public transportation hub Cervero (1998), Van Lierop et al. (2017).

The third form of urbanisation is diffuse urbanisation, which is often referred to as urban sprawl. This is sporadic low-density development in rural areas, usually along road networks or near highway exits, especially with commercial activities. Diffuse urbanisation is primarily a consequence of infrastructure planning or tax policy, e.g. related to home ownership, and not a deliberate goal of planning policy Buitelaar and Leinfelder (2020). Its main characteristic is the lack of control over urban development. Nevertheless, diffuse urbanisation also has positive elements that can be used for sustainable development Stewart et al. (2006).

As part of the SUPER project, three groups of sustainability indicators were selected following a comprehensive analysis of the scientific literature on the effects of urbanisation on various spatial aspects. These indicators aim to capture the effects of urbanisation as comprehensively and differentiated as possible. They include economic sustainability, environmental sustainability and social sustainability. The sustainability of the three urban forms by the above groups of indicators, based on a comprehensive study of the literature, was assessed by the scale with mathematical operators. The values of these indicators come from the meta-evaluation, i.e. scientific evidence pointing to positive or negative effects on sustainability identified in scientific research on individual indicators Evers et al. (2020). The results are presented in Annex 4—Sustainability assessment and scenarios of the SUPER project [Evers et al. (2020), pp 14–23] and briefly outlined in the text below:

Compact urbanisation has relatively positive indicators for economic sustainability, as it benefits from economic agglomeration and the development of central business districts with highly skilled jobs, efficient mobility and lower energy consumption. However, challenges arise from the associated housing prices and a decline in available housing units. From an environmental perspective, compact cities are well suited to reducing car dependency and greenhouse gas emissions while protecting agricultural land, but dense development leads to a decrease in open space and biodiversity. Socially, compact urbanisation offers an active lifestyle with different building types and mixed-use areas, although social segregation exists alongside problems such as traffic noise, limited green space and high housing costs.

Polycentric urbanisation has slightly more positive sustainability indicators in all dimensions. It benefits from good accessibility to business areas through public transport, rational expansion of the housing stock, and lower transportation costs. Environmental sustainability includes less dependence on cars and improved access to green spaces, while social sustainability enables living in walkable neighbourhoods with ample recreational space. However, rising house prices due to poor transport links pose a challenge.

Conversely, diffuse urbanisation has the lowest sustainability indicators. While it can raise economic standards through new business parks, it also concentrates low-wage businesses that rely on car transport, leading to high transportation costs and negative environmental impacts. Socially, it offers affordable housing but is associated with social segregation, low social mobility and health problems due to air pollution.

Apart from the scientific value of the SUPER project, one of its most important practical results was the development of a guide for sustainable urbanisation and land use Cotella et al. (2020). This guide is based on a study of spatial planning practices in the European countries within the project and an assessment of their sustainability. It summarises the key elements needed to make land use practices sustainable and is therefore useful for all groups involved in spatial planning (planners, public administration, and stakeholders from the private and civil society sectors). The guide offers a wide range of ideas, perspectives and suggestions for balanced and sustainable urban development that could be derived from past and current spatial planning practices and lead to concrete interventions in future spatial development. Among other things, the handbook proposes measures for green regeneration and the possible application of various nature-based solutions (NBS) in compact urban development (containment), as the documents presented in the following sections show.

13.2 EU Policy Context Regarding Sustainable Urbanisation

In the political context of the application of sustainable principles in urbanisation within the European Union, the initiative and working method "Urban Agenda", which was launched by the ministers responsible for urban issues of the EU member states in Amsterdam in 2016, has provided an important impetus. The Urban Agenda comprises 14 different thematic areas, among which the thematic area of sustainable land use and NBS is particularly emphasised (https://ec.europa.eu/futurium/en/system/files/ged/leaflet_sulnbs_en.pdf). Against a backdrop of uncontrolled urban growth and expansion, air pollution and pressure on natural ecosystems, the Urban Agenda recognises the need to better integrate urban development and growth with the surrounding natural environment. This includes harnessing the potential of unused land, clean air and water, and green spaces in and around cities. Partnerships have been formed between the cities and states participating in this thematic area to promote and improve the sustainable use of natural resources, particularly in the urban environment. Particularly relevant partnerships for this topic are the Urban Agenda Partnership on Circular Economy and the Urban Agenda Partnership on Sustainable Use of Land and Nature-Based Solutions. Their aim is to develop legal frameworks, methodologies and recommendations on issues such as land take, development of unused land and brownfields, joint spatial development within functional urban areas and implementation of NBS. In the initial phase, the focus is on research, discussions with relevant interest groups, designing pilot projects and securing funding for them. The implementation of these goals is planned through nine actions: inclusion of land take and soil properties in impact assessment procedures, funding and financing guidelines for brownfield redevelopment, identification and management of underutilised land, indicators for land consumption, promotion of cooperation in functional urban areas (FUA) as a tool to curb urban sprawl, better regulation to promote NBS at European, national and local level, better financing of NBS, raising awareness in the areas of NBS and sustainable land use (urban sprawl) and agreement on common targets and indicators for NBS, urban green infrastructure, biodiversity and ecosystem services in cities.

One of the most significant outcomes of these partnerships so far is the development of the Handbook on Sustainable & Circular Re-use of Spaces & Buildings Urban Agenda Partnership on Circular Economy (2019). This handbook is intended to serve as the foundation for a comprehensive strategy for implementing new models of managing urban reuse while incorporating principles of circular economy. The models are accompanied by examples of good practice from various European cities. The handbook is also aligned with UN sustainable development goals 11 ("Make cities and human settlements inclusive, safe, resilient and sustainable") and 15 ("Life on land").

The European Union promotes sustainable urbanisation through various mechanisms, albeit to varying degrees. As part of the European Green Deal, which sets the direction for transforming Europe into a climate-neutral continent by 2050, the aim is to achieve integrated sustainable urban development. Such urban development requires that economic, environmental and social dimensions are interlinked in planning and that partnerships between stakeholders from different sectors are promoted (https://ec.europa.eu/regional_policy/policy/themes/urban-development_en). However, the specific measures to be taken in this context have not yet been precisely defined. Nevertheless, certain priority areas have been identified on the European Union's official website, outlining steps to be taken in the context of the European Green Deal. In the context of sustainable urbanisation, the New European Bauhaus stands out in particular. This is a creative, transdisciplinary initiative that aims to develop pleasant, sustainable and inclusive spaces for everyday life (https://new-european-bauhaus.europa.eu/index_en). In this context, the solutions offered by green urban regeneration are crucial. They are expected to contribute to the aesthetic value, sustainability and inclusiveness of spaces. In the context of the New European Bauhaus, the initiative "A Renovation Wave for Europe—greening our buildings, creating jobs, improving lives" has an important role to play (https://energy.ec.europa.eu/topics/energy-efficiency/energy-efficient-buildings/renovation-wave_en). This initiative, which is strategically planned until 2020, aims to promote the development of climate-friendly buildings, focusing on three main directions: tackling energy poverty and modernising the least energy-efficient buildings, renovating public buildings and decarbonising heating and cooling systems.

In February 2024, the European Parliament adopted the Nature Restoration Law European Parliament (2024). It sets a target for the EU to restore at least 20% of the Union's land and sea areas by 2030 and all degraded ecosystems by 2050. Urban ecosystems are particularly important in this context. The aim is to increase green spaces in urban areas so that they can provide ecosystem services. This includes regulating water drainage, reducing the urban heat island effect, and creating better conditions for the restoration of ecosystems in urban areas. Green infrastructure will be developed in cities, and the use of NBS, such as green roofs or walls, will be promoted. Stronger links between natural areas in cities and surrounding rural areas are also supported.

13.3 Sustainable Urbanisation in Croatian Planning Policies

The European Green Deal policy and the related documents and initiatives presented in the previous section are implemented in different ways in the strategic documents of the Member States. The Republic of Croatia has developed two strategic documents aimed at achieving sustainable urbanisation through the development of green infrastructure and a circular economy.

The Urban Green Infrastructure Development Program 2021–2030 (https://mpgi.gov.hr/UserDocsImages//dokumenti/EnergetskaUcinkovitost//Program_razvoja_zelene_infrastrukture_do_2030), presented by the Ministry of Physical Planning, Construction and State Assets in 2021, outlines three specific objectives related to green infrastructure: effective planning and management of green infrastructure development, improved, connected and easily accessible green infrastructure in urban areas, and increasing knowledge and social awareness of sustainable urban development through the promotion of green infrastructure. This program emphasises the importance of local-level initiatives and promotes mutual understanding between public, private and civil sector stakeholders on green infrastructure as a key public interest of local communities. Implementation measures include the formulation of specific strategic documents at a local level and the integration of green infrastructure considerations into the next generation of spatial plans, providing a roadmap for their development.

In 2021, the same ministry introduced the Circular Management of Spaces and Buildings Program 2021–2030 (https://mpgi.gov.hr/UserDocsImages//dokumenti/EnergetskaUcinkovitost//Program_razvoja_KG_prostorom_i_zgradama_2021-2030.pdf). It was developed to promote sustainable, inclusive, safe and resilient cities by promoting circular economy measures in the design of new buildings and establishing guidelines for the construction of buildings based on circular economy principles. Specific objectives include the development of circular economy systems for spaces and buildings, the circular refurbishment of unused spaces and buildings, and a high level of knowledge and social awareness about the circular economy of spaces and buildings. This document is in line with the EU initiative "A Renovation Wave for Europe—Greening our Buildings, Creating Jobs, Improving Lives" and provides for the improvement of the legal and technical conditions for the implementation of the goals of this initiative.

13.4 Green Urban renewal—a New Planning Instrument in Croatia

Emerging from the previously mentioned Urban Agenda Partnerships and the two national programmes, a new tool of importance for cities and municipalities has been developed to promote sustainability principles and ensure their implementation at the local level. This new instrument encourages the two most important components of compact city sustainability: using green infrastructure to combat soil sealing and repurposing land, particularly brownfield sites. The new viewpoint makes it possible to rethink urban development holistically by considering green systems and the circularity of land as a single organism. Green Urban Renewal Strategy refers to the achievement of goals for green infrastructure development, integration of NBS solutions, improvement of circular management of space and buildings, energy efficiency, adaptation to climate change, and strengthening resilience to risks.

In 2023, a Handbook for Implementation of Green Urban Renewal Strategies (https://mpgi.gov.hr/UserDocsImages/dokumenti/NPOO/zelena_buducnost/2023_12_12_Prirucnik_za_implementaciju_strategija_zelene_urbane_obnove.pdf) was developed through the project Support for Implementing Green Transition Measures Under Croatia's NRRP in Reconstructing Earthquake-Damaged Buildings. The project was funded by the European Union via the Technical Support Instrument (TSI) and implemented by the European Bank for Reconstruction and Development (EBRD) in cooperation with the European Commission's Directorate-General for Structural Reform Support (DG Reform). Through the project, technical support was provided to the Ministry of Physical Planning, Construction, and State Assets in the Republic of Croatia. The Handbook has been developed to facilitate processes of development and implementation and promote local-level capacity building and urban renewal and regeneration opportunities with specific emphasis on best practices and approaches.

The two relevant national programmes identified a lack of knowledge at the local level regarding important topics related to green urban renewal. As a result, the need for developing a handbook to provide guidance emerged. Moreover, in recent times, there has been an increased interest in green urban renewal, making the need for such a handbook even more pressing. The Handbook itself has been developed through participatory processes with the representatives of cities and municipalities aiming to identify key implementation challenges and bottlenecks of activities related to green urban renewal developed previously at the local level. The goal was to engage stakeholders who would be directly involved in implementing the strategies and gather their thoughts and needs for future processes.

The Handbook is organised in several thematically specific chapters providing insight into identified challenges, proposed general recommendations, and good practice examples from Croatia and abroad. The initial chapters of the Handbook (Green Infrastructure and Circular Spaces and Buildings Management at the National Level and Importance of GUR Strategies and Their Benefits at the Local Level) aim to provide users with a basic understanding of green urban renewal-related terms, such as green infrastructure and circularity. The chapter emphasises the significance of undertaking such projects in the context of sustainable development at the local level, improving the well-being of the population, boosting the economy, and addressing challenges unique to urban areas, such as environmental pollution, degradation of public open spaces, reduction in water permeability, and negative effects of climate change.

Establishment of a Working Team, Involvement of Experts and Fostering the Multi-sectoral Character of Green Urban Renewal highlights the importance of involving experts with specific knowledge bases (e.g., architects, landscape architects, urban planners, conservationists, energy engineers, construction engineers, forestry engineers) to work together and develop and implement integrated solutions related to green infrastructure, climate resilience, circular use of buildings, energy efficiency etc., as well as the significance of cooperation between public, private and civil sector. Beyond just the financial side, the private sector's involvement is crucial since it facilitates the acquisition of information about innovative solutions and processes, management, and entrepreneurship, while representatives of civil society organisations, including associations and foundations, can aid with citizens' mobilisation and encourage green urban renewal through bottom-up approaches.

Data collection, Monitoring, and Development of a Strategic Framework provide recommendations for the development of a strategic framework that will yield integrated projects. The Handbook advises guiding strategic framework towards solving issues specific to the spatial, ecological, social, and economic context of each local context and aligning it with the results of data analysis to aim for balance in measures and activities strategic the entire city and those related to specific locations and to combine infrastructural and soft measures. Furthermore, it stresses the importance of developing measures that take a holistic view of green infrastructure and circular management, as it is often possible to achieve the full potential of spatial, environmental, social, and economic benefits only through their integration.

Project Planning and Project Selection discuss issues like project prioritisation. It recommends prioritising projects that address the detrimental effects of climate change and natural hazards, as well as those concerning people's health and well-being and projects that are in high stages of preparation. The phasing of more complicated projects is mentioned as well. A specific section is dedicated to the importance of careful consideration while developing NBS so as to encourage fit-for-problem solutions and avoid applying solutions not suitable for specific contexts. This may include building a green roof as a substitute for building additional public green spaces on street level, planting trees that are not meant for a specific environment and climate, or utilising NBS without social benefits as a substitute for other viable options.

GUR Strategy in the Spatial Planning System highlights the importance of adjusting GUR strategies with relevant documents and plans within physical planning and strategic planning systems. Physical and strategic planning systems in the Republic of Croatia are separated and under the jurisdiction of two different ministries, while green urban renewal strategies have characteristics that could fit them under both. Yet primarily, GUR is considered as an expert basis that could inform physical planning processes, while at the same time should be aligned with strategic planning documents at the national level: Urban Green Infrastructure Development Programme and Circular Management of Spaces and Buildings Programme, therefore general recommendations are provided, navigating how to strengthen implementation potential of GURS through two planning systems.

The General Public and Encouraging Citizens' Participation chapter aims to promote how crucial citizens' engagement is for the development and implementation of targeted and well-accepted projects. During the process of handbook development, it was acknowledged that some individuals had little interest in participating in activities. This lack of participation in project planning and execution could possibly be caused by a loss of public confidence in local government resulting from the lack of capacity to carry out participatory procedures at the local level. The participatory process should not just be limited to gathering input and information in the first stages of GURS development. Instead, a continuous process should be developed that will allow different forms of participation even during the execution of less demanding and complex projects (e.g. tactical urbanism actions).

Enabling Mechanisms and Funding Opportunities from Different Sources offers an overview of possible financing and funding sources for green urban renewal, including EU funds, since during the participatory process included in the Handbook development, lack of financial resources and lack of knowledge in relation to financial sources has been identified as one of the most significant challenges. The Handbook provides an overview of relevant sources at the EU and national level, examples of previous funding opportunities related to topics of green urban renewal through European structural and investment funds, a brief explanation of EU programs and special instruments such as Horizon, Interreg, and LIFE, and a specific section dedicated to the New European Bauhaus, as an important enabling mechanism for green urban renewal. The chapter briefly describes possibilities for developing public–private partnerships and exploring possibilities for financial contributions from the private sector and interested public or cooperation with public institutions.

The Progress Monitoring chapter proposes several steps and activities for easier implementation. The first step is to develop a strategic framework and indicators, along with a tentative timeline and financial plan. To ensure success, all parties involved must communicate and coordinate effectively, and risk management measures should be established. As the project progresses, it's important to monitor and report on its implementation, make schedule adjustments if necessary, and ensure transparency throughout the process.

13.5 Green Urban Renewal in the Context of New European Bauhaus

Green Urban Renewal strategies, if developed and implemented appropriately, should in fact reflect the three core values of NEB i.e., sustainability, inclusiveness, and quality of experience. To make this link straightforward and understandable at the implementation level, NEB is presented in the Handbook as one of the enabling mechanisms for the implementation of GUR strategies.

Sustainability: Green Urban Renewal strategies are designed as a new policy instrument intended to promote and encourage sustainable practices in spatial development, i.e., to enable the development of green infrastructure in urban areas and reuse of brownfields and other inadequately used sites to prevent soil sealing and inefficient land use. All the activities and projects that should be developed within GURS reflect sustainability principles, as their content should reflect local ambitions to achieve green transition measures. The Handbook stresses the importance of considering the environmental, spatial, and social aspects of green urban renewal while developing activities and projects. Although urban renovation can be accomplished through a simple revitalisation of neglected buildings, it is critical to consider the environmental benefits that can be achieved by developing green areas along with the buildings. Moreover, it is equally important to reflect on the social benefits that can be attained by involving residents in the renovation process.

Inclusiveness: The development of GUR strategies is conditioned by the implementation of a participatory process, i.e. by surveying the public opinion of citizens and other key stakeholders, as its results are one of the inputs that should inform the drafting of the strategic framework. Two major topics of the Handbook directly address the principle of inclusiveness and guide the strategies, measures, and projects towards comprehensive and integrated development and implementation processes that rely on multiple connections and the coexistence of various stakeholders. "Establishment of a Working Team, Involvement of Experts and Fostering the Multisectoral Character of Green Urban Renewal " focuses on sustainable collaboration of different disciplines (such as architects, landscape architects, urban planners, energy engineers, forestry engineers, climatologists) and the importance of continuous participation of public, civil and private sector to enable just, well-informed and comprehensive processes. "General Public and Encouraging Citizens' Participation" highlights the benefits of deploying co-creation processes, such as determining the real needs and potential of public spaces by gathering input from its users, boosting openness and visibility of projects, leading to strengthen public confidence in local governance, enabling the implementation as a result of prompt information and improved collaboration between all parties and participants, providing citizens with opportunities to be an integral part of public space development and therefore achieve sense of belonging through different initiatives such as tactical urbanism.

Quality of experience: The Handbook advocates for enhancing the quality of experience in urban areas and developing activities and projects that take a holistic view of the themes of green urban renewal, not focusing only on the functionality and technical aspects of the solutions. Furthermore, it promotes initiatives and projects that have regenerated neglected and unappealing spaces in other European countries and contributed to creating a more pleasant urban environment for the local population, providing local authorities in Croatia with a guide towards creating balanced urban environments.

References

Buitelaar E, Leinfelder H (2020) Public design of urban sprawl: governments and the extension of the urban fabric in flanders and the Netherlands. Urban Planning 5(1):46–57

Campbell S (2016) The planner's triangle revisited: sustainability and the evolution of a planning ideal that can't stand still. J Am Plann Assoc 82(4):388–97

Cervero R (1998) The transit metropolis: a global inquiry. Island Press, Washington, D.C

European Commission. Sustainable use of land and nature-based solutions partnership, https://ec.europa.eu/futurium/en/system/files/ged/leaflet_sulnbs_en.pdf. Last accessed 2024/03/10

Cotella G, Evers D, Janin Rivolin U, Solly A, Berisha E (2020) ESPON SUPER—sustainable urbanisation and land-use practices in European regions: a guide to sustainable urbanisation and land-use. ESPON, Luxembourg

Daniels T, Lapping M (2005) Land preservation: an essential ingredient in smart growth. J Plan Lit 19(3):316–29

ESPON (2020) SUPER—Sustainable Urbanization and land-use Practices in European Regions https://www.espon.eu/super. Last accessed 2024/01/15

European Parliament. European Parliament legislative resolution of 27 February 2024 on the proposal for a regulation of the European Parliament and of the Council on Nature Restoration (COM(2022)0304–C9-0208/2022–2022/0195(COD)

Evers D, Katurić I, van der Wouden R, van Schie M, van Rijn F (2020) ESPON SUPER Final Report Annex 4—sustainability assessment and scenarios. ESPON, Luxembourg

Handy S (2005) Smart growth and the transportation-land use connection: what does the research tell us? Int Reg Sci Rev 28(2):146–67

Jabareen YR (2006) Sustainable urban forms: their typologies, models, and concepts. J Plan Educ Res 26(1):38–52

European Commission. Sustainable urban development, https://ec.europa.eu/regional_policy/policy/themes/urban-development_en. Last accessed: 2024/03/10

New European Bauhaus Homepage, https://new-european-bauhaus.europa.eu/index_en. Last accessed: 2024/03/10

European Commission. Renovation Wave, https://energy.ec.europa.eu/topics/energy-efficiency/energy-efficient-buildings/renovation-wave_en. Last accessed: 2024/03/10

Leitmann J (1999) Sustaining cities: environmental planning and management in urban design. McGraw-Hill, New York

Van Lierop D, Maat K, El-Geneidy A (2017) Talking TOD: learning about transit-oriented development in the United States, Canada, and the Netherlands. J Urban: Int Res Placemaking Urban Sustain 10(1):49–62

Millward H (2006) Urban containment strategies: a case-study appraisal of plans and policies in Japanese, British, and Canadian cities. Land Use Policy 23(4):473–85

Ministry of Physical Planning, Construction and State Assets of the Republic of Croatia, https://mpgi.gov.hr/UserDocsImages//dokumenti/EnergetskaUcinkovitost//Program_razvoja_zelene_infrastrukture_do_2030.pdf. Last accessed 2024/03/15

Ministry of Physical Planning, Construction and State Assets of the Republic of Croatia, https://mpgi.gov.hr/UserDocsImages//dokumenti/EnergetskaUcinkovitost//Program_razvoja_KG_prostorom_i_zgradama_2021-2030.pdf. Last accessed 2024/03/15

Ministry of Physical Planning, Construction and State Assets of the Republic of Croatia, https://mpgi.gov.hr/UserDocsImages/dokumenti/NPOO/zelena_buducnost/2023_12_12_Prirucnik_za_implementaciju_strategija_zelene_urbane_obnove.pdf. Last accessed 2024/03/15

Nelson AC, Dawkins CJ, Sanchez TW (2004) Urban containment and residential segregation: a preliminary investigation. Urban Studies 41(2):423–39

Nelson AC, Sanchez TW, Dawkins CJ (2007) The social impacts of urban containment. Ashgate, Aldershot, Burlington

Neuman M, Churchill SW (2015) Measuring sustainability. Town Plan Rev 86(4):457–82

Urban Agenda Partnership on Circular Economy, Urban Agenda Partnership on Sustainable Use of Land and Nature-Based Solutions (2019) Sustainable Circular Re-Use of Spaces and Buildings

Purvis B, Mao Y, Robinson D (2019) Three pillars of sustainability: in search of conceptual origins. Sustain Sci 14(3):681–95

Stewart D, Sirr L, Kelly R (2006) Smart growth: a buffer zone between decentrist and centrist theory? Int J Sustain Dev Plan 1(1):1–13

UN-Habitat (2020) The value of sustainable urbanisation. UN-Habitat, Nairobi

Chapter 14
Preserving the Authentic Appearance of Heritage Buildings Through Sustainable Restoration and Adaptive Reuse

Nenad Šekularac, Jelena Ivanović-Šekularac, Nikola Macut, and Tijana Žišić

Abstract A systematic and well-structured approach to preserving heritage buildings is essential for their restoration. This paper explores the integration of historic building restoration with sustainability, aiming to establish an effective methodology that addresses the restoration of historic buildings and their sustainable renovation, particularly regarding energy efficiency. The primary method employed in this study involves observing and analysing the restoration and utilisation of a historic building, with a specific emphasis on improving energy efficiency and conservation. The research concentrates on the Žica Monastery's Dining Room, exploring the outcomes of the restoration process to ensure energy refurbishment and the enhancement of cultural heritage. The paper concludes with recommendations for the restoration and adaptive reuse of historic buildings, aligning with contemporary standards for comfort and environmental conservation while maintaining the authentic appearance of the structure.

Keywords Heritage conservation · Energy refurbishment of historic buildings · Methodology of reconstruction · Conservation requirements

14.1 Introduction

The construction of historic and heritage buildings used to be carried out in accordance with bioclimatic design—specific resilience of architectural structures, durability of local construction materials, abundance of greenery and free land space around the buildings. After hundreds of thousands of years, the environmental endurance and durability of these construction materials testified to the sustainability and wisdom of ancient builders and architects (Šekularac et al. 2018; Günçe and Mısırlısoy 2019). Simultaneously, numerous studies confirm the energy-efficient

N. Šekularac · J. Ivanović-Šekularac · N. Macut · T. Žišić (✉)
University of Belgrade—Faculty of Architecture, Belgrade, Republic of Serbia
e-mail: tijana.zisic@arh.bg.ac.rs

M. Vukmirović et al. (eds.), *Digital Dreams, Urban Realities*, The Urban Book Series,
https://doi.org/10.1007/978-3-031-76353-3_14

characteristics of vernacular architecture, which are adapted to local environments and climates. Therefore, interventions aimed at improving energy efficiency in historic and vernacular buildings could be minimised, consequently reducing their potential impact on the appearance and function of heritage structures (Mileto et al. 2014). Over time, inevitable adaptive preservations require striking a delicate balance between conserving the authenticity and integrity of these structures while accommodating modern functional requirements (Franco and Magrini 2017). Moreover, preserving cultural heritage requires a diverse approach, such as protection from further deterioration, meticulous restoration, and seamless adaptation to contemporary needs. Crucially, interventions must be executed with utmost care to minimise disruption to the buildings and their broader context. The necessity to improve energy efficiency in historic buildings arises as one of the vital aspects of their conservation. Beyond the understandable reduction in energy consumption, energy retrofitting serves as a way to safeguard the material historical values and characteristics inherent in these structures, ensuring their preservation for the future (Webb 2017).

In addition to preserving and maintaining heritage buildings, all refurbishment efforts undertaken on this kind of architectural inheritance must strictly adhere to current directives and regulations. The European Union's Energy Efficiency Directives primarily target the reduction of energy consumption and the promotion of savings within existing building stock. Nevertheless, they lack explicit provisions addressing the complication of energy refurbishment in historical structures (Bionaz 2015). Given the diversity of buildings, it becomes evident that each country has the autonomy to determine whether to impose energy efficiency standards on existing and heritage structures. To bridge the disparity between heritage buildings and energy retrofitting, it is important for National Heritage Institutions to spearhead initiatives aimed at discovering effective approaches for retrofitting heritage buildings (Mazzarella 2015). However, certain historic buildings, particularly those falling under specific protection regimes, may be exempt from obligatory energy certification requirements if fulfilling such criteria would contravene the preservation of their state (appearance, form, function…).

The focus of the research is on restoring and enhancing comfort for users of the Žiča Monastery Dining room during their stay in the area. The aim is to improve energy efficiency, uphold environmental preservation, and maintain the authentic visual appearance and materialisation of the structure. While the stated law and recommendations contribute valuable insights into measuring and quantifying energy efficiency in existing buildings, there are limitations in guidance when applied to specific buildings with heritage significance, which also require compliance with conservation principles and standards (Hunt and Suhr 2008). Consequently, the notable disproportion between refurbishments and conservation interventions on heritage buildings emphasises the need to employ alternative tools and guidelines from the field of conservation.

To achieve a level of architectural restoration characterised by both preservation of the original appearance and function and optimal energy efficiency, a series of prescribed steps are recommended. These include:

- detailed on-site observations and analyses,
- application of locally sourced construction materials wherever possible,
- in accordance with the existing materialisation of the object
- improvement of thermal insulation of the structure,
- exploiting favourable solar exposure while reducing solar heat gains,
- incorporating energy-efficient heating systems in the refurbishment process and implementing natural ventilation strategies within the building structure.

When formulating refurbishment principles for a specific structure such as the Žiča Monastery Dining room, it's essential to consider its social and religious significance for its daily users and the broader spiritual community of the monastery. The preservation of authenticity, as presented in The Nara Document on Authenticity (ICOMOS 1994), assumes a critical role in guiding cultural heritage research and conservation planning processes. It states that the sustainability model in energy-efficient refurbishment should be established by preserving the building's authentic function and materialisation—appearance while introducing moderate improvements in comfort conditions. By retaining the Dining room's authenticity while enhancing comfort levels and reducing energy consumption, we are establishing a resilient model that respects its historical legacy and contemporary needs. Restoration of historic buildings and sustainability have to be meaningfully linked (Martins and Carlos 2014). Consequently, we must exercise caution when examining the typical initial step of energy-efficient refurbishment—applying thermal insulation to exterior walls, as preserving the appearance of historic buildings outweighs current requests for energy use. Inadequate insulation may lead to complications after the completion, including heightened energy consumption, elevated operational costs, and structural complications such as heat loss, condensation-related deterioration, and seasonal overheating (Šekularac et al. 2019). These issues not only jeopardise the structure of the building but also contribute to the increasing existence of mould and allergens, thus creating an unhealthy indoor environment for inhabitants and occupants. Therefore, a balanced and custom-tailored approach that prioritises authenticity while addressing modern energy consumption recommendations is crucial for the long-term preservation and viability of the Žiča Monastery Dining room and similar historic structures.

The research aims to:

- enhance understanding of the significance of preservation, restoration and revitalisation of historic buildings or their remains, and recognise the potential of these landmarks as valuable resources,
- conduct active research at the site to contribute to the protection and conservation of the monuments, alongside effective site management,
- identify construction strategies to enhance the energy efficiency of historic buildings,
- organise acquired insights into enhancing the energy efficiency of historic buildings, formulate recommendations for their conservation, restoration, and revitalisation concerning the postulates of the energy-efficient refurbishment of buildings, and disseminate the research findings to the academic community.

The hypotheses of this paper are:

Hypothesis 1 Through the application of appropriate conservation measures and the utilisation of authentic construction materials, it is feasible to restore deteriorated structures like the Žiča Monastery. This restoration can preserve the original aesthetic and architectural integrity while enhancing comfort conditions and maintaining harmony with the surrounding environment;

Hypothesis 2 The restoration of deteriorated building components, coupled with the implementation of energy-efficient refurbishment strategies, is anticipated to yield energy savings and enhance the overall energy efficiency of the structure. This approach aligns with environmental preservation objectives.

The goals of this paper are:

- utilising site conditions to safeguard and conserve the historic building while minimising environmental impact and effectively managing the site,
- enhancing understanding of the significance of preserving cultural heritage and historic structures, methods of restoration, their potentials, and use of authentic building materials in accordance with the conservation requirements,
- implementing construction measures in order to improve energy efficiency in the restoration of historic buildings and thereby contributing to the preservation of the environment of this site,
- formulating recommendations for the preservation, restoration, and adaptive reuse of historic buildings based on the results obtained from the application of specific materials in restoration and energy refurbishment efforts,
- disseminating the research findings to the academic community.

14.2 Methodology

During the research, the following assessments and tasks were carried out:

- on-site observations and evaluations of the structures and their remains,
- developing design plans and protective standards for construction based on previous property condition assessments and selected restoration methods,
- analysing the construction materials and techniques utilised in historical buildings,
- executing restoration efforts for historical buildings in compliance with conservation mandates, with a focus on preserving the authenticity of structural elements such as façade walls and roof structures,
- performing a case study to analyse restoration and energy retrofitting, specifically focusing on The Dining Room within the Žiča Monastery, with the aim of deriving scientifically substantiated conclusions,
- strategising and designing restoration plans for the building to adapt it to reuse, enhance energy efficiency, and reduce energy consumption,

- ensuring that all interventions for restoration, adaptive reuse, and energy efficiency enhancement are reversible and in line with conservation regulations,
- analysing and evaluating the results of restoration efforts in relation to energy retrofitting, adaptive reuse, and the integration of renewable energy sources to support energy provision and environmental protection.

14.3 The Dining Room of Žiča Monastery Property-Specific Characteristics

Established in the thirteenth century in the village of Žiča, near the present-day town of Kraljevo, Serbia—the Žiča Monastery is an architectural marvel steeped in historical and spiritual significance for Orthodox Serbs. Its origins trace back to the visionary leadership of Serbian King Stefan Nemanjić, renowned as Stefan Prvovenčani, who oversaw the construction of its most iconic edifice, the Church of the Holy Dormition, completed from 1206 to 1221 (Čanak-Medić et al. 2014).

Designated a cultural monument of exceptional importance in 1979, the Žiča Monastery is a testament to the enduring legacy of Serbian spirituality. Among its architectural treasures, the Dining Room occupies a pivotal position. It was constructed in 1935. adjacent to the outer narthex of the Church of the Holy Dormition, originally conceived as a sanctuary for weary travellers and devout pilgrims, the Dining Room continues to fulfil its hospitable role, offering refreshments meticulously crafted by the monastery's artisans (as presented in Fig. 14.1). Operating daily from 9 am to 5 pm, the Dining Room welcomes visitors with modern conveniences. Its functional layout encompasses a welcoming reception area, a dedicated kitchenette for beverage self-service and dishwashing, an administrative office, and well-appointed restroom facilities seamlessly integrated into the architectural function of the monastery.

Fig. 14.1 The Dining room of the Žiča Monastery complex, appearance from the monastery yard (provided photo taken by the author)

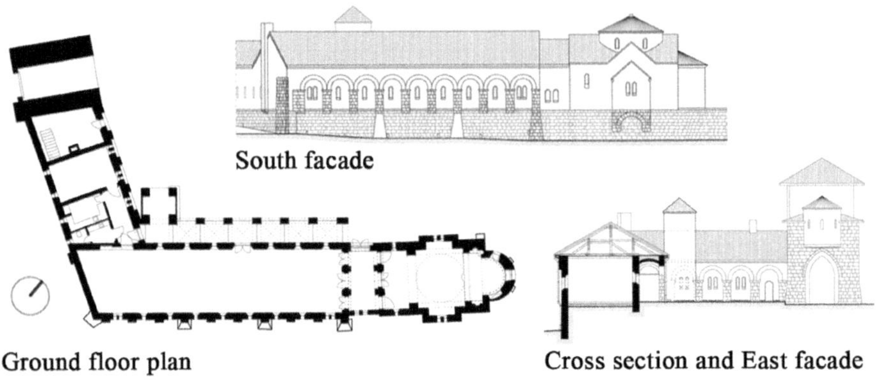

Fig. 14.2 Architectural drawings of the Dining Room building (provided by the author)

In 2019, Prof. Dr Nenad Šekularac and his team, authors of this paper, were commissioned by the representatives of the Žiča Monastery to undertake the structural rehabilitation project for both the Church of St. Sava and the Dining Room. This endeavour forms the cornerstone of the research presented in this paper, entailing an in-depth analysis of the current state of these historical structures and the development of a model for enhancing energy efficiency within the existing framework of the Dining Room. Detailed drawings of the building were meticulously crafted for this purpose, delineating the current state of the structure's construction elements. The drawings shown in Fig. 14.2 capture the authentic details of the preserved frescoes and interior inventory, providing invaluable insight into the historical and architectural integrity of the Dining Room.

14.4 Energy Efficient Restoration and Research Findings

Preserving the significant historical qualities of a building during restoration, keeping its original function, yet making it more comfortable while consuming less energy are all important factors for formulating a new sustainable model for restoration. Energy-saving efforts should not alter the authentic appearance of the historic building or the environment surrounding it, especially when addressing a culturally significant landmark like the Žiča Monastery. Sustainable construction and restoration are key parts of sustainable development. This includes (Šekularac et al. 2020):

- utilising materials and methods that were originally used when restoring a building and it is important to choose materials that are harmless for the surrounding environment and would not damage the site in any way,
- focusing on the energy efficiency of the existing building, as well as the future designed state, including the entire process and every used method and element,
- managing and reusing waste from construction and demolition.

This paper explores the energy efficiency and possibilities for improvement in the Dining Room situated within the Žiča Monastery property. The analysis is conducted by comparing the relevant parameters of the three carefully formed models:

- **M01: the first model** that represents the existing condition of the Dining Room, which underwent restoration in 2008 and 2009,
- **M02: the second model** outlines the proposed restoration plan for the Dining Room from 2019,
- **M03: the third model** illustrates the proposed improvements to the planned restoration of the Dining Room from 2019 by the authors.

With needed information from the first model, M01, the other two restoration models, M02 and M03, were meticulously developed, incorporating specific construction measures aligned with conservation requirements. These two models, their elements, and methods for improving energy efficiency on the existing structure were carefully examined and analysed throughout this paper to compare the results.

Simulation of the energy performance of these models was executed using OpenStudio, a dynamic energy simulation software based on EnergyPlus. Meteorological data sourced from the EnergyPlus website, authenticated by the World Meteorological Organization, was employed to facilitate the simulation process. Comprehensive scenarios were defined for Building Performance Simulation (BPS) utilising OpenStudio software, incorporating considerations for a typical meteorological year (TMY). This involved assessing the existing building model based on the restoration Model M01, as well as two proposed models (M02–M03). Leveraging technological advancements in BPS allowed for effective optimisation of energy efficiency and integration of renewable energy sources (RES). Verification and analysis of energy improvement were conducted through rigorous simulation and study of the three restoration models using BPS methods and energy efficiency optimisation techniques. BPS serves as a powerful predictive tool for assessing a building's dynamic behaviour, optimising energy efficiency, and integrating renewable energy sources. Parameters and schedules governing facility use were established based on heating system operation, facility occupancy, lighting, and utilisation of electrical appliances, grounded in observations of actual user behaviour obtained through fieldwork. Furthermore, the ventilation system of the building was carefully considered since the structure is naturally ventilated. An air change rate (ACR) value of 1 h^{-1} was uniformly assigned across all models (M01–M03), mirroring the ACR of well-refurbished older structures with good air tightness, thus minimising air infiltration into the object.

Following an in-depth on-site assessment, the state of both the Church of St. Sava and the Dining Room was thoroughly examined. The initial model in our research, labelled Model **M01**, represents the current condition of the Dining Room. The Church of St. Sava, constructed with bricks and mortar, features a southern wall integrated into the monastery's stone perimeter. During the spring of 2019, cracks were observed in the walls and vaults of the Church of St. Sava and the Dining Room, accompanied by structural dampness indicative of capillary moisture. The dining room's façade, constructed of brick façade, aligns with the monastery's perimeter and features window openings overlooking the surroundings, accompanied by a porch along the building's length facing the monastery yard. The structural assembly comprises longitudinal brick walls and sandstone column colonnades. Cracks are evident in the Dining Room's walls, some extending from the building's base to its ceiling, while others appear on the porch's ceiling. The ceiling is made of planks–panelling, so it is not possible to detect any deformations. The wooden roof structure, prone to snow-induced strain during winter, exhibits signs of distress, leading to wall cracks.

Despite the 2008–2009 restoration efforts, moisture persisted in the buildings, necessitating a new restoration and structural rehabilitation project in 2019, denoted as Model **M02**. Works to combat capillary rising damp were undertaken by injecting the perimeter brick walls of both structures, which were necessitated to safeguard their stability and preserve interior frescoes. The restoration involved replacing the complete floor surface area, as well as underfloor heating, and reconstructing the Dining Room's roof with shingles to prevent rain penetration and preserve frescoes. Per the restoration project specifications, a new floor was to be constructed, consisting of a reinforced concrete slab laid over a layer of lean concrete. Positioned above the slab was a waterproofing layer, safeguarded on its upper side by a cement screed. A thermal insulation layer, designed to accommodate the installation of heating hoses, was then applied over this protective screed. Subsequently, another layer of cement screed was added, serving as the base for laying granite ceramics and as the final floor covering in the Dining Room. A thermal insulation layer and a steam dam were strategically positioned between the bars over the ceiling boards from the upper side. This process facilitated the creation of a floor within the attic space; notably, thermal insulation was designed to be reversible. Following this, the roof was meticulously covered with shingles. As part of the restoration project, it was imperative to install fittings such as air vents when covering the roof planes to ensure adequate ventilation of the attic space. Preserving the architectural integrity of the building was one of the most important factors, and thus, all window openings on the façade were preserved in terms of shape and rhythm. The windows were upgraded to new wooden ones featuring a 9.0 cm frame and double thermal insulation glass filled with krypton gas. This resulted in a significant improvement in thermal transmittance, with U-values reduced to 1.5 W/m^2K. Additionally, the entrance door to the building, as well as the door between the church lobby and the Dining Room, were replaced with new solid wood doors. Despite the restoration efforts, the space retained its authenticity in terms of construction elements and function.

The third Model **M03**, designed for simulation, proposes enhancements to the 2019 restoration of the Dining Room. With a focus on maximising the potential offered by restoration while minimising construction interventions, this proposal aims to improve energy efficiency and overall comfort within the space. Retaining the authentic visual appearance of the building's facade walls, the brick exterior walls of the Dining Room were completely retained, with the addition of thermal insulation. Internally, these walls are plastered and frescoed, so it was important to maintain their authentic appearance without any additional changes. Additional thermal insulation is planned for installation in the attic space over the existing floor layers. This improvement aims to advance thermal efficiency within the attic area. Furthermore, the wall separating the office and storage room will undergo additional installation of thermal insulation and plastering on the side facing the storage room, thereby enhancing thermal efficiency towards the unheated space. The existing windows will be replaced with new wooden ones featuring an 11.0 cm frame and three-layer thermal insulation glass filled with xenon gas. This upgrade is expected to significantly enhance energy efficiency and elevate overall comfort and living quality within the space.

A comprehensive comparative analysis of all three examined models of the Dining Room within the Žiča Monastery provides insights into the impact of restoration and proposed energy refurbishment measures on the reduction of annual energy demands. The calculation of energy requirements involved analysing three models of the Dining Room facility (M01–M03), with simulations conducted to validate the effectiveness of the refurbishment approach. Parameters and usage patterns, including heating system performance, occupancy, lighting, and electrical appliance usage, were determined based on the actual needs of the building users. Consultations with the monastery's nuns helped define the visiting schedule, with the Dining Room open from 9 am to 5 pm daily, accommodating large tourist groups for short stays and refreshments. The frequency of light and equipment usage was established in accordance with the nuns' instructions and common experience. Heating energy demand simulations assumed a heat pump efficiency of 4.02 (CoP), delivering hot water at 32 °C to radiant floor heating, with a thermostat setting of 21 °C for winter. As cooling was not simulated due to the sole use of natural ventilation, the proposed energy efficiency measures are analysed only regarding heating demands. Model M02 was projected to reduce specific annual heating demands by 25.67% (Šekularac et al. 2018) compared to the current restored condition—Model M01. Further enhancements in Model M03 would result in a 51.94% reduction in specific annual heating demands compared to Model M01. If the proposed improvements in Model M03 were compared to those in Model M02, a 35.34% reduction in specific annual heating demands would be observed.

Considering consistent equipment usage, lighting, and occupancy rates, heating energy demand emerges as the most indicative aspect. Results from energy performance simulations indicate that Model M03 significantly surpasses Model M02 and represents an improvement over the baseline Model M01.

14.5 Discussion

Examining the Dining Room in the Žiča Monastery involved simulating three models and suggesting strategies for structural rehabilitation and energy refurbishment. The findings suggest that significant energy savings are achievable by implementing extensive thermal protection measures on the building's components while still adhering to conservation standards and preserving the building's authentic appearance. Analysis and comparison of total heat loss values across the three models yielded data indicating a reduction in specific annual heating demands. Field observations were utilised to test the hypotheses in conjunction with assessing the significance of the building within the Žiča Monastery premises and conducting energy efficiency tests for all models (M01–M03) using BPS (Building Performance Simulation). The results substantiated the hypothesis that it is feasible to enhance comfort conditions, improve energy efficiency, and achieve energy savings by implementing suitable rehabilitation and construction measures during the restoration of historic buildings. These measures are implemented with due consideration for conservation requirements and the preservation of the authentic appearance of the buildings. The primary objective of applying these measures is to uphold the original function of the Dining Room within the Žiča Monastery, which stands as a cultural monument of exceptional importance.

This research confirmed the conclusion made by Šekularac (Šekularac et al. 2018; Šekularac et al. 2020): In certain situations, it is necessary to avoid the introduction of thermal insulation as an envelope for the building over the external walls due to the retention of the authentic appearance of the building, which is imperative in reconstruction and conservation, and far more important than energy savings. Once more, the study underscores the significance of thoroughly analysing all aspects that could potentially detrimentally impact a heritage building during restoration. It emphasises that optimal energy efficiency of a historic building can only be achieved without compromising its architectural character and integrity (Franco and Magrini 2017).

Moreover, the results validate the conclusion drawn by Bionaz (Bionaz 2015), which suggests that ancient builders also considered the energy performance of buildings. Furthermore, it highlights that the knowledge and skills applied to historic buildings by these builders were unfairly undervalued, while modern efforts to enhance buildings to meet prescribed standards are sometimes overrated (Mazzarella 2015). The principles and measures of energy efficiency aimed at achieving energy refurbishment of a building within the Žiča Monastery property, a designated cultural monument of outstanding significance, are validated throughout the paper. This underscores the professional relevance and justification of the research both scientifically and in terms of practical application, with potential applicability to other locales possessing historic buildings of comparable structural composition and materialisation. The direct verification of the research through practical application at the site represents its most substantial contribution.

14.6 Recommendation for Increasing Re-usability Potential and Energy Efficiency of Historic Buildings in Preservation and Building Restoration Project

Energy efficiency measures for heritage buildings should include the following procedures following conservation requirements:

- thermal insulation of the building envelope,
- heating installation (favourably from reusable resources),
- heat loss reduction.

Considering the historical and architectural significance of heritage structures, in accordance with conservation guidelines, conventional methods of thermal insulation, such as exterior facade installation, are not recommended. Instead, the energy efficiency enhancements implemented for the refurbishment of the Dining Room within the Žiča Monastery property include (Šekularac et al. 2020):

- added thermal insulation of the perimeter walls, where possible and only from the inside, which was predicted for only one wall in the unheated storage room, while the rest of the walls in the main space of the Dining Room are facade walls that are fresco-painted on the inside and it is not possible to cover them with a layer of thermal insulation;
- added thermal insulation layer on floors on the ground,
- added thermal insulation on the mezzanine structure towards the unheated attic,
- replacement of facade joinery (windows and front doors),
- new heating system installation,
- predicted future energy management.

14.7 Conclusions

The restoration and adaptive reuse of heritage buildings, coupled with the integration of energy efficiency measures, demand a tailored approach to each unique historic structure. This study's primary contribution lies in the analysis and interpretation of research findings to ascertain the retrofitting potential through implementing energy efficiency measures in heritage buildings while adhering to conservation requirements and preserving the authentic character and integrity of the Dining Room in Žiča Monastery. The imperative to conserve the existing historical building, prevent further decay, and safeguard the fresco paintings adorning the interior walls and the authentic brick facade while maintaining the original purpose of the structure significantly influenced the restoration of the Dining Room within the Žiča Monastery property. By employing selected construction measures aligned with conservation standards, the enhancement of energy efficiency and comfort conditions within the building was achieved, facilitating a seamless connection between cultural heritage and modern necessities, all the while preserving the building's original purpose and authentic visual allure.

To elevate the energy efficiency of listed buildings during the restoration process, it is imperative to:

- conduct a comprehensive analysis of the building's current condition,
- determine the type and extent of restoration required,
- select appropriate construction measures to enhance energy performance while preserving the original function in line with conservation requirements,

- establish energy management protocols at the site and select the optimal method for determining total annual energy demands to achieve energy savings, efficiency, and environmental protection.

In conclusion, the design and execution of restoration works for historic and listed buildings, alongside the application of suitable energy efficiency measures, necessitate a bespoke approach tailored to each cultural monument. Various restoration models for this building were tested and analysed using the Building Performance Simulations (BPS) methodology. The primary significance of this study lies in the potential for practical validation of achieved energy refurbishment results through the application of energy efficiency principles and measures in the restoration of listed historic buildings.

References

Bionaz C (2015) Preservation and energy behaviour in Aosta Valley's traditional buildings. In: Vernacular architecture: towards a sustainable future, proceedings of the international conference of vernacular heritage, sustainability and earthen architecture, Valencia, Spain. CRC Press, Taylor, New York

Čanak-Medić M, Popović D, Vojvodić D (2014) Žiča Monastery—Manastir Žiča. Institute for the Protection of Cultural Monuments of Serbia—Belgrade

Franco G, Magrini A (2017) Historical buildings and energy. Springer International Publishing AG, Basel, Switzerland

Günçe K, Mısırlısoy D (2019) Assessment of adaptive reuse practices through user experiences: traditional houses in the walled city of Nicosia. Sustainability 11:540

Hunt R, Suhr M (2008) Old houses handbook. In: A practical guide to care and repair. SPAB, London, UK

ICOMOS (1994) The Nara document on authenticity. In: Proceedings of the world heritage convention, Nara, Japan

Martins AM, Carlos JS (2014) (2014) The retrofitting of the Bernardas' convent in Lisbon. Energy Build 68:396–402

Mazzarella L (2015) Energy retrofit of historic and existing buildings. The legislative and regulatory point of view. Energy Build Int J 95:23–31 Special Issue: Historic, historical and existing buildings: designing the retrofit

Mileto C, Vegas F, García-Soriano L, Cristini V (2014) Vernacular architecture: towards a sustainable future. CRC-Taylor & Francis Group, London

Šekularac N, Šumarac D, Čikić Tovarović J, Čokić M, Ivanović-Šekularac J (2018) Reuse of historic buildings and energy refurbishment analysis via building performance simulation (BPS)—a case study. Therm Sci Int Sci J 22(6):2335–2354

Šekularac N, Ivanović-Šekularac J, Petrovski A, Macut N, Radojević M (2020) Restoration of a historic building in order to improve energy efficiency and energy saving—case study—the dining room within the Žiča Monastery Property. Sustainability 12:6271

Šekularac N, Ristić ND, Mijović D, Cvetković V, Barišić S, Ivanović-Šekularac J (2019) The use of natural stone as an authentic building material for the restoration of historic buildings in order to test sustainable refurbishment: case study. Sustainability 11(15):4009 Special Issue Natural Stone and Architectural Heritage ISSN 2071-1050

Webb AL (2017) Energy retrofits in historic and traditional buildings: a review of problems and methods. Renew Sustain Energy Rev 77:748–759

Part VI
Micro Scale: Public Spaces that Encourage Healthy Lifestyles

Micro Scale: Designing the Details of Urban Life

The micro-scale represents the most intimate and human-centered level of intervention in the vast and complex world of urban design and planning. On this scale, the grand visions of sustainable, inclusive, and beautiful cities are translated into tangible spaces where people live, work, and interact. The microscale focuses on the design and organization of individual buildings, public spaces, and small urban elements that directly shape people's daily experiences within the built environment. The interaction between individuals and their immediate surroundings is palpable, and the quality of these interactions is crucial in determining the usability, comfort, and aesthetic appeal of urban spaces (Lynch 1960; Mehta 2013).

The micro-scale is where the ideals of the New European Bauhaus come to life in the most direct and personal way. This initiative emphasizes the creation of environments that are sustainable, inclusive, beautiful, and enriched. At the microscale, this means paying meticulous attention to the details that influence how people perceive and experience their surroundings, whether it is the layout of a room, the texture of a material, or the way light filters through a window. Through these details, the New European Bauhaus principles are realized, creating spaces that are not just functional, but also uplifting and inspiring (Montgomery 2013).

However, some critics argue that overemphasizing micro-scale design can lead to neglecting broader systemic issues that shape cities. While micro-scale design interventions are often celebrated for their capacity to enhance user experience and activate urban life, some scholars caution against an overreliance on these approaches, warning that it may obscure broader structural forces that shape cities. Cuthbert contends that focusing too narrowly on micro-scale interventions risks overlooking the more significant socio-economic and political factors critical in shaping urban environments. The mentioned can lead to superficial solutions that fail to address underlying structural issues within cities (Cuthbert 2006). Similarly, Harvey warned against technocratic approaches that prioritize aesthetics or functionality at the microscale without considering the broader social context, advocating for more inclusive and participatory urban design processes (Harvey 2000).

In the micro-scale section of the book, we explore how the micro-scale is essential for crafting urban environments that respond to the needs and desires of users. The design of individual buildings and their immediate surroundings is a primary focus, with considerations ranging from interior layouts that maximize functionality and comfort to façade designs that contribute to the identity and character of a neighborhood (Alexander et al. 1974). Attention to human scale and proportion ensures that spaces feel welcoming and are easy to navigate, thus enhancing the overall experience of the built environment (Gehl 2010; CABE 2001).

Public space design is another critical aspect of the microscale. Here, we delve into the creation of plazas, parks, and streetscapes that serve as the heart of urban life. Communities gather, interact, and form connections in these spaces, making their designs pivotal for fostering social cohesion and inclusivity. The microscale design in these areas considers everything from seating arrangements and landscaping to public art and accessibility, ensuring that public spaces are inviting, safe, and functional for all (Mehta 2013; Whyte 2001).

The microscale also emphasizes the importance of detail and materiality in design. The choice of materials, finishes, and construction techniques plays a significant role in the overall quality and feel of a space. Designers at this scale carefully select durable and functional materials that contribute to the aesthetic and sensory experience of the environment (Moughtin et al. 1999). Sustainability is crucial because it focuses on choosing eco-friendly materials, minimizing waste, and incorporating energy-efficient technologies.

Perhaps most importantly, the microscale is where the user experience is most directly influenced. This section explores how design can shape behavior, encourage social interaction, and create intuitive and engaging environments. By considering ergonomics, sensory experiences, and behavioral influences, designers can create spaces that are not only beautiful but also profoundly attuned to the needs and behaviors of their users (Dovey 2020).

The microscale section then highlights how microscale design can encourage healthy lifestyles through the visionary transformation of Belgrade's urban landscape, moving from car-centric to community-focused spaces. The chapters explore how sustainable behavior can be fostered in urban settings, emphasizing the importance of detail and materiality in creating functional, aesthetically pleasing environments that are responsive to human needs.

The discussion on affordable and sustainable social housing illustrates how microscale interventions can address social equity while enhancing the quality of life. By contrast, examining multi-criteria renovations in the post-war housing stock and modular renovations for energy saving in Bosnia-Herzegovina and Serbia shows how sustainability can be integrated into the minor details of urban living.

As we journey through the microscale, we will examine case studies and practices that illustrate how careful attention to detail can enhance the quality of urban life. From the design of intimate architectural spaces to the creation of vibrant public areas, the micro-scale is where the dreams of sustainable, inclusive, and beautiful cities are realized through the everyday experiences of the people who inhabit them.

References

Alexander C, Ishikawa S, Silverstein M (1974) A pattern language. Oxford University Press, New York

CABE DC (2001) The value of urban design. Thomas Telford. Accessed: Aug 28, 2024. Available: http://www.cabe.org.uk (Online)

Cuthbert AR (2006) The form of cities: Political economy and urban design. Blackwell Pub.

Dovey K (2020) Urban design thinking: a conceptual toolkit. Bloomsbury Publishing

Gehl J (2010) Cities for people, 1st edn. Island Press, Washington, Covelo, London

Harvey D (2000) Spaces of hope. University of California Press, Berkeley

Lynch K (1960) The image of the city. MIT Press, USA. Accessed: Jul 16, 2024

Mehta V (2013) The street. A quintessential social public space. Routledge, UK

Montgomery C (2013) Happy city : transforming our lives through urban design. Farrar, Straus and Giroux

Moughtin C, Oc T, Tiesdell S (1999) Urban design: ornament and decoration, 2nd edn. Elsevier Architectural Press, Oxford

Whyte WH (2001) The social life of small urban spaces. Project for Public Spaces, New York

Chapter 15
From Cars to Communities: A Visionary Shift in Belgrade's Urban Landscape

Milena Vukmirović, Senka Bekrić, Leontina Pezić, and Bojana Radovanović

Abstract Parklets have recently gained significant attention as a permanent solution for enhancing streets and public spaces. Originating as temporary installations, parklets are now recognised for their potential to transform urban environments into more pedestrian-friendly and socially vibrant areas. Considering the mentioned, this chapter delves into the evolution of parklets, their impact on urban spaces, and their potential for public space transformation in Belgrade. This chapter explores the potential for a visionary shift in Belgrade's urban landscape, moving from a car-centric design to a more community-oriented one. The paper discusses the current state of Belgrade's urban landscape, highlighting the negative impacts of a car-dominated city on the environment, public health, and social cohesion. It then presents a vision for a more sustainable and liveable city. It focuses on creating vibrant and inclusive communities by improving the sidewalks and introducing a street life zone developed by converting parking spaces into miniature urban areas for different user activities. This is presented following the catalogue structure to show other options instead of parking that will be more community-oriented. The paper also discusses the challenges and opportunities of such a transformation, including the need for political will, public participation, and innovative urban planning strategies. The authors propose a multi-faceted approach based on the parklet concept, including pedestrianisation, public transportation improvements, green spaces and public gathering areas. The potential benefits of this shift are also explored, such as reduced air pollution, improved public health, and increased social interaction. The paper concludes by emphasising the importance of collaboration and cooperation between various stakeholders to implement this visionary shift in Belgrade's urban landscape successfully.

Keywords Parklets · Pedestrian-friendly design · Urban transformation · Sustainable cities · Community-oriented design · Belgrade

M. Vukmirović (✉) · S. Bekrić · L. Pezić · B. Radovanović
Chair for Planning and Design in Landscape Architecture, Department of Landscape Architecture and Horticulture, University of Belgrade—Faculty of Forestry, Belgrade, Serbia
e-mail: milena.vukmirovic@sfb.bg.ac.rs

M. Vukmirović et al. (eds.), *Digital Dreams, Urban Realities*, The Urban Book Series,
https://doi.org/10.1007/978-3-031-76353-3_15

15.1 Introduction

Ambulatory Urbanism (Davis 2015) refers to a concept focusing on walking as a fundamental element in urban life and design. This approach views walking as a means of transportation and a way of experiencing and understanding the city. The approach emphasises the importance of pedestrian mobility, social interactions, and the human-scale experience of urban environments. It underscores the significance of designing walkable, pedestrian-friendly cities conducive to human activities and promoting a more sustainable, inclusive, and vibrant urban landscape. The author also highlights the transformative potential of prioritising walking in urban planning and design, aiming to create cities that are more liveable, accessible, and engaging for residents and visitors alike. Also, cities built around pedestrians rather than vehicles—define the most desirable cities of the world, such as European ones (Gehl 2010; Davis 2015).

Conversely, conventional planning previously entailed that every increase in demand should be followed by capacity building. The main task was identifying current and forecasting future transportation demands and providing them with appropriate capacity ("predict and provide"). However, this approach only resolves transport problems in the short term. Indeed, any increase in capacity generates new traffic demands due to a range of different environmental behaviour reactions, which further necessitate the construction (expansion) of capacities (Litman 2010).

The inability to solve long-term transport problems and expensive construction, development issues under limited resource conditions, and increasing concerns about environmental protection have led to a change in concept. The former idea of "adapting the city to traffic" has been replaced by the concept of "adapting transportation to the city" (Simićević 2013). This new approach recognises the negative impacts of car-dominated cities on the environment, public health, and social cohesion.

The new concept involves a balanced development of the transportation system, meaning using each mode of transport for what is best suited. This typically entails a greater reliance on non-motorized forms of transportation (pedestrian and bicycle traffic) and urban public transport while limiting (but not eliminating) car traffic (Milosavljevic et al. 2010).

The change in the management of the transportation system has also led to changes in parking management. The old concept involved meeting the general demand for parking by building a sufficient number of parking spaces. However, it became clear that the availability of parking spaces (and their unrestricted use) directly influences the generation of parking demands (travel) and that this policy contributes to traffic congestion and hurts air quality, economic development, pedestrian environment, etc. (Weinberger et al. 2010).

Discussing the future of streets and the appearance of new transport options, such as different types of autonomous vehicles (AV), (Sevtsuk and Davis 2019) conclude that "without active efforts to prioritise shared, public, and active mobility, the introduction of automated modes for transporting both humans and goods in an already contested public realm could mean a reversion to Modernist practices of privileging speed, efficiency, and function over human-scale interactions and serendipity." Instead of embracing the latest transport innovations in street design, city planners should carefully assess how these changes align with existing priorities of fostering people-friendly, sustainable, and community-centric streets. Streets encompass intricate economic, social, and political dynamics facilitating communal activities and democratic engagement (Sevtsuk and Davis 2019). While historically rooted in religious, cultural, and social functions rather than pure mobility needs, contemporary streets remain vital as spaces for face-to-face interactions among pedestrians, as platforms for public gatherings or performances, as avenues for economic activity such as shops or vendors, as playgrounds for children; and even as areas providing livelihoods for marginalised communities like the homeless (Rykwert 1978; Loukaitou-Sideris and Ehrenfeucht 2009).

As we understand them today, parklets first emerged in San Francisco with temporary installations to extend sidewalks' social life and pedestrian space into parking spaces (Brozen and Loukaitou-Sideris 2013). These small public spaces, created by repurposing parking spots or sidewalk areas, aim to enhance community engagement and support local businesses. San Francisco's Pavement to Parks Program exemplifies efforts to transform parking spaces into new, reversible public spaces such as parklets and plazas (Panganiban and Ocubillo 2014).

Since then, the concept of parklets has been adopted in major cities worldwide, including Los Angeles, New York, Vancouver, and Melbourne. Parklets transform street-side parking spaces into miniature public parks (Birdsall 2013). This global adoption underscores the growing recognition of parklets as innovative solutions for reclaiming excess road space for pedestrian use. Research has examined how cities implement parklet programs, including considerations for the departments involved, permitting issues, and design guidelines (Brozen and Loukaitou-Sideris 2013).

Parklets also foster increased social connections, public vibrancy, and support for local businesses. They respond to community desires to repurpose parking stalls for communal spaces, as seen in cities like San Francisco and Vancouver (Corey 2014). Moreover, parklets are part of a broader movement towards creating pedestrian-friendly urban environments, including pedestrian zones, encounter zones with pedestrian priority, and various designated areas for outdoor dining, wellness, and cultural activities. The effects of this kind of transformation could also be seen from the perspective of living streets.

Living Streets represents a paradigm shift in urban design, emphasising the need for streets prioritising pedestrian use and social interaction over vehicular traffic. This approach is characterised by blurring boundaries between sidewalks and roadways, creating spaces where pedestrians and cyclists have priority. Implementing encounter zones with speed limits and pedestrian priority, as studied in Basel, Switzerland, showcases the positive impact of such designs on social relations in urban neighbourhoods (Sauter and Huettenmoser 2008).

The literature on parklets and liveable streets (Appleyard et al. 1981) explores the concept of parklets to enhance urban liveability through community participation and the repurposing of public spaces. Parklets are small public seating platforms created by converting curbside parking into vibrant community spaces, promoting safety, active transportation, and pedestrian-friendly environments (Owens et al. 2018). These initiatives aim to address the need for wider sidewalks, increase pedestrian traffic, and provide spaces for social interaction and community identity. Parklets are a cost-effective way for cities to improve urban spaces, enhance public life, and create places for active living (Owens et al. 2018). The research emphasises the importance of community involvement in the design process to tailor urban spaces to local needs and foster a sense of belonging and community cohesion. Overall, parklets represent a creative solution to urban design challenges, offering simple yet impactful ways to transform streets into more liveable and engaging spaces for residents (Owens et al. 2018; Ahmed et al. 2019; Shokry 2020). Parklets have emerged as a promising strategy for enhancing urban liveability and creating people-centric cities (Koue 2013).

Parklets also foster increased social connections, public vibrancy, and support for local businesses. They respond to community desires to repurpose parking stalls for communal spaces, as seen in cities like San Francisco and Vancouver (Corey 2014). Moreover, parklets are part of a broader movement towards creating pedestrian-friendly urban environments, including pedestrian zones, encounter zones with pedestrian priority, and various designated areas for outdoor dining, wellness, and cultural activities. The emergence of parklets has sparked a visionary shift in the urban landscape, transforming car-dominated spaces into vibrant and inclusive communities.

Parklets are part of a larger framework to enhance urban infrastructure to promote pedestrian activity, community engagement, and environmental sustainability. This includes developing urban pedestrian zones, pathways, and connectors that facilitate walking, cycling, and connecting various city points of interest. For instance, the "City Greenways" concept proposed in Portland's 2035 Comprehensive Plan calls for a citywide network of park-like pedestrian and bicycle-friendly streets (Liu et al. 2018).

On the other hand, community play streets and other community-led initiatives demonstrate the potential of temporary street closures for recreational activities, highlighting the benefits of such interventions for children's play and family recreation (Mitra and Abbasi 2020). These initiatives reflect a growing recognition of the importance of community involvement in creating and maintaining public spaces that cater to local needs and preferences.

This chapter explores the progression of parklets, their influence on city environments, and their potential to reshape public spaces in Belgrade. The chapter also discusses the current state of Belgrade's urban landscape, highlighting the negative impacts of a car-dominated city on the environment, public health, and social cohesion. It then presents a vision for a more sustainable and liveable city, focusing on creating vibrant and inclusive communities. The authors propose a multi-faceted approach that centres around the concept of parklets and includes elements such as pedestrianisation and creating green spaces and public gathering areas. This chapter also proposes a Catalogue of small urban places that could replace parking to make more liveable streets in densely populated areas such as Vračar. This catalogue aims to provide a practical framework for implementing the visionary shift in Belgrade's urban landscape by identifying locations where parking spaces can be transformed into vibrant community spaces. The Catalogue aims to outline design options, potential amenities, and the overall impact of transformed spaces on the surrounding urban environment and community. The proposed shift in Belgrade's urban landscape also aligns with the New European Bauhaus initiative's objectives, which focus on sustainability, aesthetics, and inclusivity in the built environment. This research strives to promote sustainable urban development and foster a sense of community and inclusivity in urban spaces by reimagining parking spaces as vibrant community areas.

15.2 Impact of Parklets on Urban Spaces: Enhancing Pedestrian-Friendliness and Social Vibrancy

Parklets have become recognised for their potential to transform urban environments into more pedestrian-friendly and socially vibrant spaces. The evolution of parklets from temporary installations to permanent features of urban landscapes signifies a shift towards more inclusive, pedestrian-oriented urban planning. By repurposing underutilised spaces for public use, parklets contribute to creating vibrant, socially connected communities. Alongside living streets and other pedestrian-priority zones, parklets play an essential role in reimagining urban spaces as places of interaction, recreation, and cultural expression. Integrating parklets and similar initiatives into urban design strategies will be crucial for fostering a sustainable, liveable urban environment as cities evolve.

Informed by the works of Jane Jacobs, William H. Whyte, and Jan Gehl, the quality of open public urban spaces is intricately linked to their usage level. Interventions that aim to boost the vibrancy of these spaces by providing diverse opportunities for activities, like converting parking spaces into seating areas or leisure spots, are crucial for enhancing the vitality and functionality of public spaces. This approach aligns with the idea that the dynamism and attractiveness of public spaces are directly tied to the range of activities they offer, fostering a sense of community and engagement among individuals within urban environments.

This research focuses on how people use public spaces, especially sidewalks, instead of passive vehicle occupation. This aligns with Jacobs's diversity in uses and people, the look at the sidewalks as a crucial public space, and the spontaneous order. Within the first, by providing flexible public spaces, parklets can support a variety of uses—from casual seating and dining to public performances and community gatherings—thereby contributing to the diversity and dynamism of the urban environment. For Jacobs, sidewalks were the most important public spaces in the city, serving as "the stage for the daily ballet of urban life" (Jacobs 1961). Parklets extend the functionality and capacity of sidewalks, enhancing their role as places for social interaction, observation, and participation in public life. Finally, from the spontaneous order that emerges from the complex interactions of individuals and businesses in a vibrant urban environment, parklets can be seen as platforms that facilitate spontaneous encounters and activities, enriching the social fabric of neighbourhoods.

Jan Gehl's principles focus on creating vibrant, liveable urban areas that encourage people to spend time outside, engage with their surroundings, and interact with each other (Gehl 1971). These principles are highly relevant to parklets, small public seating areas, or green spaces created by converting parking spots or other underutilised road areas. Parklets are designed to enhance the pedestrian experience, provide additional public space, and improve the overall quality of urban environments. They also aim to reclaim space for people in urban areas dominated by vehicles. They serve as extensions of the sidewalk, offering places for rest, leisure, and social interactions, thus contributing to more lively and engaging streetscapes. By prioritising human needs and experiences in urban design, parklets align with Gehl's advocacy for cities that support healthy, active, and community-oriented lifestyles.

By observing urban life, Gehl (Gehl and Svarre 2013) demonstrates how street activities occur in outdoor spaces and how places' physical characteristics (e.g., buildings, sidewalks, and street furniture) influence people's behaviour. Gehl introduces a list of 12 key quality criteria that will ensure safe, comfortable, and enjoyable city spaces based on research about how people use public spaces or city activities. Within the comfort requirements, the city must offer places where its users can walk, stand, or stay, sit, see, talk, listen, play, and exercise. He also states that "many things do not need to happen at once, but rather for the "stage" to be set so that it can be used for many different things sequentially: from quiet square to marketplace to open-air cinema, and so on" (Gehl 2006, p. 106).

Whyte's study of the use and layout of small urban spaces offers valuable insights into how these spaces can improve social interactions and community connections (Whyte 2001). Through observing people's behaviour in city settings and the design factors that influence public space usage, Whyte outlines several principles, including "settable" spaces, sunlight, trees, and water features, as well as accessibility and visibility. Additionally, he emphasises the importance of flexibility and adaptability, the role of food in attracting individuals to these spaces, the significance of experimentation and observation, and attention to detail.

Observations indicate that people are more likely to utilise public spaces when they can sit, linger, and watch the world go by. Parklets meet this demand by repurposing parking spots into welcoming areas with seating like benches and chairs, making city environments more appealing and functional. By integrating elements such as plants, trees, and occasionally water features, parklets cater to the public's attraction to spaces that provide sunlight, greenery, and water, enhancing the urban setting.

The transformation of parking spaces into parklets allows cities to convert parts of the street for communal use, enriching sidewalks' social and recreational roles and contributing to a livelier street scene. Positioned directly on the street and accessible to everyone, parklets invite pedestrians to pause, relax, and interact within the space, fostering community engagement. Whyte championed the idea of versatile urban areas capable of hosting diverse activities. Parklets inherently embody this versatility, offering spaces for socialising, eating, working, or simply unwinding.

The author also noted food's significant role in attracting people to public spaces. Consequently, many parklets are strategically placed near eateries and cafes, facilitating outdoor dining experiences that draw individuals in and promote extended visits. This interaction between parklets and nearby businesses revitalises the public domain and bolsters the local economy. The presence of others is a crucial factor in the appeal of public spaces.

Parklets serve as arenas for public life and become venues where individuals can observe and be observed, generating a cycle of attraction. This highlights the need to design open, welcoming parklets encouraging collective gatherings and interactions. When thoughtfully designed, Whyte showed that even small urban areas can significantly influence a city's social dynamics. Parklets illustrate this concept by demonstrating how minor changes to the urban landscape can forge significant communal spaces that boost community engagement and city vibrancy.

The iterative development of parklets, involving temporary or pilot projects evaluated and refined based on their use and impact, enables cities to evolve and enhance public spaces grounded in real-world observations and community input. The meticulous design of parklets is crucial to creating comfortable, attractive, and functional spaces catering to the community's needs. By balancing the requirements of pedestrians, vehicles, and natural elements, a well-planned street can support a variety of uses, including cafes, social hubs, playgrounds, and public congregating spots. Moreover, the design of an effective street should prioritise pedestrian access and convenience above all. This approach ensures that streets are not just thoroughfares for vehicles but are vibrant, multifunctional spaces that enhance urban life.

The thoughtful integration of parklets into urban design reflects a commitment to creating more human-centred cities. These small yet impactful interventions offer a blueprint for reimagining urban spaces in ways that prioritise social interaction, community well-being, and environmental aesthetics. By fostering places where people can gather, relax, and enjoy the outdoors, parklets play a crucial role in enhancing the quality of urban life. Their flexibility and adaptability make them an invaluable tool for urban planners and community members seeking to inject vitality into city streets and create more inclusive, engaging, and liveable urban environments.

The collaborative nature of parklet projects—often involving local businesses, residents, and city officials—exemplifies a community-driven approach to urban development. This participatory process ensures that parklets meet the specific needs and desires of the community but also strengthens the sense of ownership and pride among local stakeholders. As cities continue to evolve, the principles underpinning the design and implementation of parklets offer valuable lessons in creating spaces that celebrate and accommodate the diverse ways people use and experience public spaces.

Ultimately, parklets' success in activating public spaces underscores the importance of designing cities with people in mind. By reimagining the potential of underutilised urban areas, parklets contribute to a broader vision of more sustainable, resilient, vibrant, and inclusive cities. Through careful planning, design, and community engagement, parklets can continue to serve as vital components of the urban landscape, enriching cities' social and cultural fabric worldwide.

Bringing together and studying 28 instances of transformed streetscapes from 14 European cities, the authors found that "the potential for increased cycling and walking, along with more greenery and permeable surfaces, can only be realised by reducing space for cars" (Koch et al. 2024). This underscores the significance of developing high-quality urban areas, prioritising people and community life over car-centric environments. This aligns with transforming streets into vibrant, inclusive spaces that foster social interaction, cultural diversity, and a sense of belonging within urban settings. Unlike traditional approaches to street design, which prioritise vehicular traffic, this perspective emphasises creating urban spaces focused on people and community life. It involves designing streets and public areas to shift attention from cars toward highlighting human interactions and activities (Signer 2020), contrasting with conventional practices prioritising vehicle movement over pedestrian well-being.

The statements by Signer (Signer 2020) regarding the importance of creating appealing and high-quality spaces and atmospheres have motivated other authors who also recognise that street life is central to society. Furthermore, streets are essential for daily activities, serving as a spatial platform for various economic and cultural endeavours. This perspective has led to further research and discussions on street design principles (Casas-Valle et al. 2023). As part of this development, the authors identified 56 activities that could take place within the streets while establishing ten future street design principles.

Consequently, interventions in public spaces can enhance their usage by focusing on crucial aspects such as community activity, visual elements, accessibility, safety, well-being, and maintenance. It is essential to design public spaces that cater to the needs of residents and users of all ages, ensuring maximum access and usability, especially for individuals with special needs. Public areas can attract more people and encourage diverse activities by creating safe, welcoming, and visually appealing spaces. Additionally, interventions should aim to increase the functionality and vibrancy of public spaces, fostering a sense of community engagement and long-lasting positive outcomes (Lleshi et al. 2022; Zhang et al. 2022).

One approach to understanding the impact of converting parking spaces into public areas is through the lens of tactical urbanism, as discussed in the development of the Pattern Painter tool. This tool empowers communities to design revitalisation projects, including converting parking spaces into more vibrant community spaces. The experiment with Pattern Painter showed that users could effectively communicate and create scenarios that enhance public space vitality, suggesting that community-led interventions can significantly improve urban revitalisation (Cooney and Raghavan 2020).

The analysis of public park usage using mobile GPS data, as seen in the study of Central Park, offers insights into how specific interventions in public spaces can promote social interaction and community engagement. The study found that amenities within the park acted as beacons, attracting users and fostering social interactions. This suggests that converting parking spaces into areas with attractive amenities enhances social interaction and the overall vibrancy of urban public spaces (Xu 2018).

Also, the space-centric approach to studying public space activeness in urban residential areas underscores the importance of population density and proximity to commercial facilities in determining the utilisation and vibrancy of public spaces. This supports the idea that converting parking spaces into public seating or leisure areas in strategic locations—especially those with high population density and near commercial facilities—can significantly enhance the activeness and attractiveness of urban public spaces (Lau et al. 2021). The study's findings suggest that such conversions can lead to increased public engagement and vitality, aligning with the principles of creating vibrant, human-centred urban environments.

These references contribute to a deeper understanding of how urban design can be reimagined to create more liveable, equitable, and beautiful cities. They offer a blend of theoretical insights, empirical evidence, and practical examples that illuminate the potential of parklets to transform urban environments. By focusing on European experiences and incorporating broader global perspectives, this curated list of references not only enriches the discourse on urban design but also aligns with the transformative vision of the New European Bauhaus.

15.3 Parklets and the New European Bauhaus: Aligning Concepts for Urban Revitalization

The New European Bauhaus is an initiative by the European Union designed to support the transformation of societies in alignment with the European Green Deal. It focuses on three fundamental values:

Sustainability, from climate goals to circularity, zero pollution, and biodiversity.

Inclusion, valuing diversity, and ensuring accessibility and affordability; and Aesthetics is a quality of experience beyond mere functionality (European Commission 2021).

The initiative aims to create a movement by bringing together citizens, experts, businesses, and institutions to reimagine sustainable living in Europe and beyond while supporting projects that are beautiful, sustainable, and inclusive, providing access to EU funding to facilitate this transformation (European Commission 2022).

While there may not be a direct and explicit connection between the concepts of parklets and the New European Bauhaus initiative, underlying principles and objectives align with both concepts. The New European Bauhaus initiative aims to reimagine sustainable living in Europe by leveraging green and digital challenges to transform lives for the better. It seeks to address complex societal problems through co-creation and bridging different backgrounds and disciplines (Di Liscia 2020). As small urban spaces prioritising human needs and fostering community, Parklets align with the New European Bauhaus's objective of creating sustainable and inclusive living environments. Their focus on enhancing social interactions and community bonds in urban spaces can contribute to the vision of the New European Bauhaus initiative. Both the Initiative and the approach emphasise the importance of creating pedestrian-friendly environments that prioritise people over cars, and they aim to transform cities into more liveable and sustainable spaces. Parklets can serve as tangible examples of how urban design principles promoted by the New European Bauhaus, such as sustainability, inclusivity, and aesthetics, can be implemented at a smaller scale to create positive changes in the urban landscape.

Sustainability is a vital principle of parklets and the New European Bauhaus (Koue 2013). They aim to create more sustainable urban environments by prioritising climate goals, circularity, zero pollution, and biodiversity (Di Liscia 2020). By integrating nature and greenery into urban spaces, parklets contribute to biodiversity and help combat climate change by reducing heat island effects. Furthermore, parklets often incorporate sustainable features such as rainwater harvesting, recycled materials, and energy-efficient lighting. These sustainable elements align with the goals of the New European Bauhaus, which seeks to promote sustainable living practices and integrate sustainability into all aspects of society. Both parklets and the New European Bauhaus Initiative emphasise sustainability in the built environment. Parklets contribute to sustainability by repurposing underutilised spaces, promoting active transportation, integrating greenery, and fostering social interaction within communities. Similarly, the New European Bauhaus Initiative seeks to advance sustainability by promoting the integration of environmental, social, and economic considerations into the design and implementation of built projects and incorporating sustainable design elements, such as rainwater harvesting, permeable pavement, or renewable materials, to minimise environmental impact and promote long-term resilience.

Inclusivity is another principle shared by parklets and the New European Bauhaus. Both initiatives aim to create inclusive spaces that are accessible and affordable for all community members. Parklets provide accessible and comfortable seating areas, ramps for wheelchair access, and spaces for people of all ages and abilities to gather. Similarly, the New European Bauhaus Initiative seeks to ensure everyone has equal access to well-designed and sustainable spaces regardless of socioeconomic background or skills. Parklets are often developed through community engagement processes that involve collaboration between local stakeholders, including residents, businesses, and government agencies. Similarly, the New European Bauhaus Initiative emphasises the importance of participatory approaches and co-creation processes that engage communities in project design, planning, and implementation. Both concepts recognise the value of involving diverse voices and perspectives in shaping the built environment to better meet the needs and aspirations of communities.

Aesthetics plays a significant role in the concept of parklets and the New European Bauhaus Initiative by prioritising the integration of aesthetics into urban design to create visually appealing and aesthetically pleasing spaces. They aim to enhance the quality of experience and style of the built environment beyond mere functionality.

The New European Bauhaus Initiative seeks to integrate aesthetics into sustainable design solutions beyond functional requirements. At the same time, parklets aim to create visually appealing and inviting public spaces that contribute to the overall aesthetic of the urban landscape. This emphasis on aesthetics helps to develop a sense of place and identity, fostering a deeper connection between individuals and their surroundings. In both contexts, aesthetics is seen not as superficial considerations but as integral aspects of design that contribute to urban spaces' overall quality, experience, and identity. By prioritising aesthetics alongside other factors such as sustainability, inclusivity, and community engagement, parklets and the New European Bauhaus Initiative seek to create more beautiful, harmonious, and meaningful environments that enrich inhabitants' lives and contribute to the cultural vitality of cities and communities.

Both concepts encourage innovation and creativity in urban design by providing opportunities for experimentation with alternative uses of space, materials, and amenities. In line with that, the New European Bauhaus Initiative fosters innovation and creativity through its support for interdisciplinary collaboration, research, and experimentation in sustainable design practices. Both concepts promote a culture of innovation that seeks to address contemporary challenges and inspire new ways of thinking about the built environment.

While parklets and the New European Bauhaus Initiative may not be directly linked, they share common goals. By aligning efforts and leveraging synergies between these concepts, cities and communities can create more liveable, resilient, and equitable urban spaces that enhance the quality of life for all. Furthermore, the emphasis on pedestrianisation and creating green spaces and public gathering areas align with the initiative's goal of integrating nature and sustainable design into the urban fabric. This multi-faceted approach demonstrates a holistic understanding of urban planning that addresses environmental, social, and cultural aspects, reflecting the principles of the New European Bauhaus initiative.

15.4 European Experiences in Permanent Parklets and Urban Space Enhancement

Several European cities have implemented the concept of permanent parklets or plan to do so. While each city's approach may vary based on its unique urban landscape, regulations, and community needs, some common characteristics and principles are often associated with implementing permanent parklets or removing parking spaces for street life enhancements in European cities. One common characteristic is the prioritisation of people over cars (Berggren 2017).

The Porto-Saragozza neighbourhood in Bologna saw the inauguration of the first parklet, a project by City Space Architecture (Bravo et al. 2020). It aimed to foster community involvement and establish versatile learning areas across disciplines. Situated outside the Public Space Museum as a venue for developing public space culture through art and technology, this intervention represented an interdisciplinary approach to investigating and engaging with public spaces.

Furthermore, European cities such as Brussels, London, Bristol, Stockholm, Rotterdam, Berlin, Oslo, and Vienna are embracing the parklet concept as a novel addition to urban environments, indicative of growing interest and adoption in these locations. These examples showcase how parklets have been successfully implemented in European cities, transforming urban spaces, promoting community engagement, and enhancing the quality of public areas.

Brussels has embraced the parklet concept by investing 140,000 euros in converting unused parking spaces into green spaces or other functional areas to support sustainable urban mobility strategies (Park4SUMP 2018a). The city has also allocated funds to develop various types of parklets at different locations, transforming unused parking spaces into vibrant community areas where people can socialise, children can play, and individuals can enjoy additional green spaces.

The United Kingdom's first officially approved parklet, backed by a local council, was established in 2015. Modeled after the parklets in San Francisco, this urban space was created in Hackney, London, through a collaboration between Hackney Council, Sustrans, and Cyclehoop (We Are Possible 2021). The parklet included bike racks, seating areas, greenery installations, and solar-powered lighting by repurposing an unused loading bay. Embraced by pedestrians and cyclists in Hackney, it became a favoured location for socialising and leisure. Other examples of UK parklets showed up when Bristol was named European Green Capital when Sustrans collaborated with Stonebridge Park residents to create and test them on their street. A survey showed that traffic patterns changed after the parklet's installation, with regular rat-run users now avoiding the street. As a result, the parklet became permanent and remains in place today. The COVID-19 pandemic has led to the emergence of parklets in high streets across the UK as shops and restaurants adapt to using outdoor space. Parklets provide a solution for increasing seating capacity while maintaining social distancing measures, increasing their presence in town centres across the UK.

Sweden implemented a hyperlocal approach to urban planning known as the "one-minute city," focusing on redesigning streets at a neighbourhood level. The initiative, called Street Moves, allowed local communities to co-design their streets, prioritising public spaces over parking (O'Sullivan 2021). Tested in Stockholm and set to expand, the goal was to transform every street in Sweden into healthy, sustainable, and vibrant spaces by 2030. This micro-focus aimed to foster more comprehensive urban transformation by reimagining streets as critical connecting spaces for communities. While its approach may have seemed romantic, it drew inspiration from American street furniture models like the parklet.

Rotterdam's initiatives to promote off-street parking primarily involved implementing pricing policies. In addition to this measure, the city also decreased the number of on-street parking spots to enhance the use of public spaces for activities benefiting people rather than solely accommodating parked vehicles. Under the Park4SUMP project, the goal was to reduce 3000 parking spaces in the inner city by 2020. These spaces were substituted with bike lanes, bicycle parking garages, or terraces.

This idea came during the COVID-19 pandemic, when Rotterdam introduced circular decks, enabling entrepreneurs to repurpose parking spaces into venues for business endeavours. Since June, numerous parking areas across the city were transformed into restaurant extensions or terraces until November 1st, constituting a supportive measure for local businesses while ensuring public safety. This initiative, surpassing previous endeavours like reallocating 125 parking spots across 75 locations for summer terraces, extended to winter terraces until April 1st, 2021 (Park4SUMP 2022). Within specified parameters, bars and restaurants were permitted to enclose these spaces with walls up to 1.20 m high and utilise heaters, albeit with a mandate to maintain environmental sustainability.

The metamorphosis of Bergmannstraße in Berlin ensued from an extensive discourse between the Friedrichshain-Kreuzberg district and its constituents regarding the envisioned future of the thoroughfare. During a one-year assessment phase, 15 parklets provided by Vestre were strategically deployed along the street to mitigate traffic flow by substituting parking spaces with vibrant urban activities. As the initial twelve-month period concluded in October 2019, deliberations are underway to relocate the parklets to a different site within Berlin (Vestre 2021).

Studio Oslo Landscape Architects and Vestre Furniture AS have developed Norway's first parklet system, fulfilling urban functions such as outdoor dining, display, transport, planting, gathering and commerce (NLA 2020). Two pilot projects in Rosenkrantzgate and Nedre Slottsgate in Oslo significantly contribute to the Car-Free City Life initiative in Oslo municipality, where the area within Ring 1 is set to be closed to private cars by 2020. The Parklets 2.0 is an adjustable frame that can adapt to various curb heights for universally accessible access. Different floor modules expand the sidewalk area and provide space for more urban life. Standardisation of elements means that the parklet system can be delivered, installed, moved, and reused quickly with minimal time and effort—significantly transforming Oslo into one of the world's most forward-thinking cities.

Some examples were part of the Park4SUMP project funded by the European Union's Horizon 2020 research and innovation program. The project focuses on leveraging parking management as a strategic tool to support broader urban development and mobility goals, including reducing congestion, freeing up public space, supporting local businesses, and promoting sustainable transport modes (Park4SUMP 2018b; CIVITAS 2019; CORDIS European Commission 2019). One of the critical objectives of Park4SUMP is to use parking management not just as a tool for controlling vehicle storage but as a means to achieve broader sustainable urban mobility and quality of life improvements. This includes freeing up public space currently used for parking, investing parking revenues into sustainable transport options, developing more human-centred neighbourhoods, and making cities more attractive and safer (Park4SUMP 2018b).

The project involved a collaboration of 16 cities, categorised into Leading Cities and Follower Cities. Leading Cities, which include Vitoria Gasteiz, Rotterdam, Freiburg, Krakow, and Trondheim, share their knowledge and experiences in innovative parking management with the Follower Cities, facilitating knowledge transfer and capacity building across Europe. The partnership aims to demonstrate the benefits of strategically managed parking and how it can be effectively integrated into SUMPs (CORDIS European Commission 2019).

They were banning cars from cities, and while a significant decision, it yielded numerous benefits. Cities worldwide encouraged residents to combat climate change by altering their daily habits, with cars being a notable option due to their pollution and noise. Removing cars also provided more space for cyclists and pedestrians, fostering a more bike-friendly environment and promoting public transportation. Additionally, it created room for new green areas and parks while avoiding the maintenance costs associated with car-centric infrastructure.

Many European cities have shifted their focus from private cars to more environmentally friendly and citizen-centred mobility options. Cities like Hamburg, Helsinki, Madrid, and Oslo have announced plans to become partially car-free, while Brussels, Copenhagen, Dublin, Milan, and Paris are also working hard to reduce motorised traffic by implementing car-free days, opting for cycling infrastructure and pedestrian areas, reducing parking space availability, and improving public transport services (Cathcart-Keays 2015; Nieuwenhuijsen and Khreis 2016).

Initially, Oslo proposed to become the first major European city to ban cars from its centre but faced criticism over the suddenness and potential impact on local businesses. Instead, Oslo opted to eliminate parking spots in the centre, part of its broader strategy to discourage car travel in the area. This shift resulted in large portions of Oslo being car-free, particularly around schools and parks, with over 700 parking spaces replaced by bike lanes, greenery, and seating areas. Concurrently, the city significantly enhanced its public transportation system, resulting in no pedestrian or cyclist deaths reported in Oslo and Helsinki in 2020.

The City Council of Vitoria-Gasteiz implemented Barcelona's Superblocks as part of their SUMP to mitigate negative impacts from passenger vehicles, including air and noise pollution while improving road safety and pedestrian mobility. Superblocks involve restructuring city blocks to prioritise pedestrian and cyclist movement, reducing vehicular traffic. Between August 2019 and May 2020, 'Médico Tornay Street in the 'Judimendi' neighbourhood underwent a redesign. Former parking spaces were repurposed to create pedestrian areas totalling 5500 square meters. Furthermore, 68 parking spaces were eliminated, significantly reducing private car traffic. Subsequent analysis revealed a 14% increase in pedestrian traffic, a 50% decrease in motor vehicle traffic, and a 13.6% reduction in overall speed. Noise levels decreased by 16%, and surveys indicated positive community reception to the interventions.

While the specific public spaces mentioned earlier may have yet to transform parking spaces directly, Barcelona implemented initiatives that repurposed parking spaces or limited vehicular access to create pedestrian-friendly areas. Examples include Superblocks, extended sidewalks and plazas, and outdoor dining areas. The concept of Superblocks involved reconfiguring city blocks to prioritise pedestrian and cyclist mobility while limiting vehicular traffic (Postaria 2021). Some streets within superblocks had their parking spaces removed or repurposed to create space for walking, cycling lanes, greenery, and public amenities. Barcelona also widened sidewalks and created new pedestrian plazas by reallocating previously used parking spaces. This enhanced the pedestrian experience, supported outdoor dining, and created vibrant gathering spaces. Additionally, the city converted parking spaces into outdoor dining areas for restaurants and cafes, particularly in high-foot-traffic areas, revitalising previously underutilised spaces and contributing to the city's culinary scene and street life.

In Paris, various initiatives have been launched to revitalise public spaces and reduce the dominance of cars in the city centre. These include programs like Paris Respire, Reinventing Paris, placemaking efforts, low-traffic neighbourhoods, and pedestrianisation projects. For example, the Paris Respire program temporarily turns selected streets into pedestrian zones on Sundays and holidays to encourage outdoor activities and cultural events. The Reinventing Paris initiative focuses on sustainable urban development that creates vibrant public spaces by repurposing parking lots or underutilised areas for mixed-use developments and community hubs. Moreover, placemaking strategies involve residents creating temporary parks or seating areas from parking spaces for community engagement and cultural expression. Additionally, low-traffic neighbourhoods (Quartiers apaisés) restrict vehicle access to prioritise pedestrians and cyclists while improving air quality and reducing noise pollution by removing on-street parking.

Central to the vision of Paris is implementing a comprehensive traffic plan to incentivise walking, cycling, and public transportation. A key component involves the removal of a significant portion of on-street parking spaces, with approximately 72% slated for elimination. Specifically, out of the 83,500 existing on-street parking spaces, a total of 60,000 will be removed, signalling a substantial commitment to reallocating urban space to accommodate cyclists and promote sustainable modes of transportation (Park4SUMP 2021).

Mutually, these cases and cities exemplify a European shift towards sustainable mobility and urban spaces that celebrate the human scale. They have shown commitment to environmental principles, innovation in public space management, and a forward-thinking embrace of sustainable transportation policies. These transformations are about reducing traffic and fostering vibrant urban life where public spaces serve as platforms for social interaction, economic activity, and cultural expression. They demonstrate that cities can be adaptive and resilient, offering valuable lessons for a future where urban spaces are increasingly designed with people and the planet in mind.

15.5 Exploring the Transformation of Belgrade's Urban Landscape Through Parklets

Parking is a significant issue throughout Belgrade's urban area. The problem has emerged due to a notable mismatch between the demand for parking and the available parking spaces, resulting from historical city structures, planning deficiencies, and inadequate management of parking capacities. A key feature of parking in Belgrade, including its central zone, is that most of the capacity is in street parking lots.

Finding parking spaces in a city with approximately 1.6 million inhabitants, such as Belgrade, is a formidable challenge. This is underscored by 664,092 registered vehicles bearing the BG license plate on the capital's streets in 2022 (Republicki zavod za statistiku 2023). The exact number of all parking spaces in Belgrade has yet to be discovered, but the provision of zoned parking spaces for a city of one million is not at a commendable level. According to data from the public utility company "Parking Service," there are approximately 25,000 parking spaces in the zoned area of Belgrade, with around 9000 located in garages and parking lots managed by this enterprise. Belgrade has 23 cars per zoned parking space (Tomic 2019). Compared to the situation from 2009, the total static capacity of street parking lots in the zone was 11,056 parking spaces distributed across different zones: 1,052, 5,049, and 4955, respectively (Milosavljevic et al. 2010).

Street parking lots are always occupied, and illegal parking is even being realised. In contrast, off-street parking lots are insufficiently filled (on average 37%), and at any given moment, a free parking spot can be found (maximum utilisation is 74%). The reason for this structurally specific imbalance lies in the tariff system, as the cost of parking in off-street parking lots is several times higher than on-street (Simićević 2013).

Excessive on-street parking can negatively affect the quality of life and environment due to pollution, noise, and occupying spaces that could otherwise be used for public amenities. Professor Lipovac, from the Faculty of Traffic Engineering, Department of Traffic Safety, highlights that the problem of parking space in Belgrade also involves improper marking. Lipovac (Tomic 2019) emphasises that the city has made efforts to mark parking spaces, but the markings often need to be done more correctly to follow regulations. Lipovac further argues that improperly marked parking spaces detract from the quality of life, as they obstruct movement and contribute to congestion in the city.

According to SUMP (City of Belgrade—Department of Transport 2020), the absence of organised parking is a characteristic feature of present-day Belgrade. This inherited condition has worsened due to the continual increase in mobility and motorisation rates. The demand for vehicle parking far exceeds current capacities. The distribution of parked vehicles according to parking location in the central zone reveals that only 8% of cars are parked in garages, 2% in organised open parking lots, and 90% on street fronts. The lack of parking capacity is one of the causes behind the emergence of so-called "irregular parking," manifested by vehicles parking on sidewalks, pedestrian areas, and even certain less congested parts of residential streets. Irregular parking can also be observed in some available green spaces.

In the context of the documented scarcity of parking spaces, the Belgrade Sustainable Urban Mobility Plan (SUMP) outlines a strategic intervention known as the relocation of vehicular parking from public surfaces (8.4) (City of Belgrade—Department of Transport 2020). This measure, integrated within the broader objective of enhancing the appeal and quality of the urban environment, entails the transfer of parking facilities positioned along street fronts in the central urban core. The primary aim of this initiative is to alleviate pedestrian zones currently obstructed by vehicles due to inadequacies in parking provisions.

15.5.1 Park(Ing) Vračar

In line with the stated measure, students in the Master course "Open Public Space Design" of the Module Landscape Architecture at the University of Belgrade—Faculty of Forestry have been testing the scenario of removing the on-street parking at the Vračar Municipality in Belgrade. Despite the unfavourable parking situation for many car owners, the research aimed to change the perspective on this situation altogether, shifting the focus to other modes of transportation, namely improving the environment that would support them. In particular, inspired by the Park(ing) Day initiative, the research by design aims to explore how the unfunctionally inherent in parking spaces can be effectively transformed into a multifunctional paradigm, thereby maximising its utility and ensuring a broader spectrum of societal benefits.

For research purposes, the municipality of Vračar was chosen due to its status as the smallest and most densely populated district. With greater precision, as per the statistics provided by the Republic Institute of Statistics for the year 2023, it is reported that within an area measuring 3 square kilometres, there resides a population of 55,477 individuals, yielding a density of 18,492 inhabitants per square kilometre (Republicki zavod za statistiku 2023).

The area is characterised by a predominantly traditional block layout, with parking arrangements primarily established along the streets. Off-street parking is observed within the interior block spaces of residential structures, private intra-block garage facilities, and at select sites under the management of the public enterprise Parking Servis. The parking facilities known as Slavija 1, comprising 87 parking spaces, and Slavija 2, encompassing 108 parking spaces, along with the parking garage located at Baba Visnjina, accommodating 351 parking spaces, and the parking facility situated at the premises of the Hilton Hotel, referred to as Garage Kralja Milutina, which offers 45 parking spaces. Furthermore, the entire expanse of Vračar is encompassed by a structured system of paid parking zones allocated to on-street parking spaces, whereby the classification of each parking spot pertains to either the yellow or green zone (Fig. 15.1).

In response to the challenge of navigating through traffic to locate parking spaces, a strategic intervention has been deployed within the central thoroughfares of Belgrade. Specifically, an infrastructure enhancement initiative has been undertaken whereby more than 5200 parking spaces have been equipped with advanced parking sensors. These sophisticated sensors facilitate motorists' parking endeavours by furnishing real-time information about the availability of parking spots. Complementing this technological advancement, informational boards have been strategically placed at intersections to relay the current count of available spaces. Moreover, to ensure accessibility and convenience, the data regarding parking availability is also made accessible through a dedicated mobile phone application. In addition, the area in

Fig. 15.1 Zone system of parking in Belgrade's central area. *Source* P. E. Parking service

question accommodates 24,750 registered automobiles (Republicki zavod za statistiku 2023). Moreover, the intricacies surrounding parking are exemplified by the noteworthy observation that this particular locale boasts the highest recorded sales price for parking spots.

This scenario has engendered multifaceted challenges characterised by the abundant presence of vehicles on public surfaces, whether in motion or stationary. Consequently, public spaces have been configured to allocate scant room for pedestrians and cyclists, precipitating conflicts on sidewalks. Furthermore, the situation has necessitated the removal of green spaces and tree-lined avenues to accommodate parking demands, amplifying pollution levels. In such a situation, the proliferation of impervious surfaces for parking spaces has led to extensive areas that exacerbate the flooding of roadways during heavy rainfall, thereby impeding traffic flow. Additionally, it has contributed to the aesthetic deterioration of open urban public areas.

In line with those mentioned above, an analysis of green areas was conducted for the territory of the Vračar municipality for the research, juxtaposed with an analysis of parking surfaces in open public and private spaces (see Fig. 15.2). This investigation aimed to assess the permeability levels of surfaces within the Municipality by categorising parking areas as surfaces with impermeable coverings, in contrast to green areas treated as permeable surfaces. The review conducted in this manner has determined that the indicative ratio between porous and non-porous surfaces stands at 35% to 65%, with non-porous surfaces prevailing.

Currently, precise information regarding the porosity ratio is unavailable for the Vračar municipality and the entirety of the urban area within the City of Belgrade. However, the predominant situation appears to be further exacerbated, as evidenced by graphical data obtained through methodologies employed by the Copernicus platform (see Fig. 15.3).

A survey (Urbel 2021) conducted to inform the formulation of the Draft General Urban Plan for 2041 revealed that three paramount values for the future development of Belgrade—namely, environmental cleanliness, the integration of natural

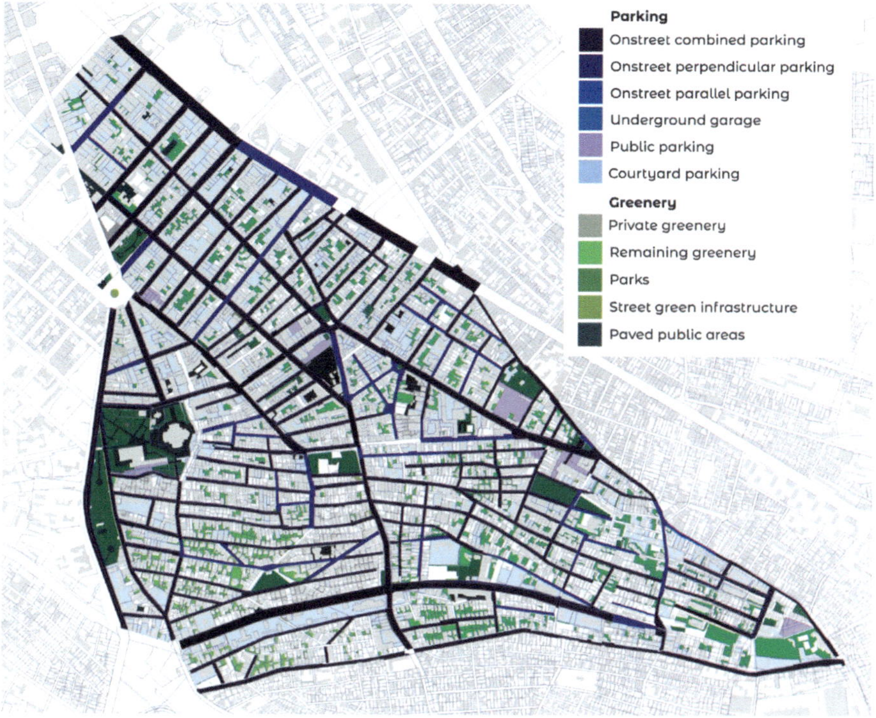

Fig. 15.2 Actual state of surface permeability according to the type of area

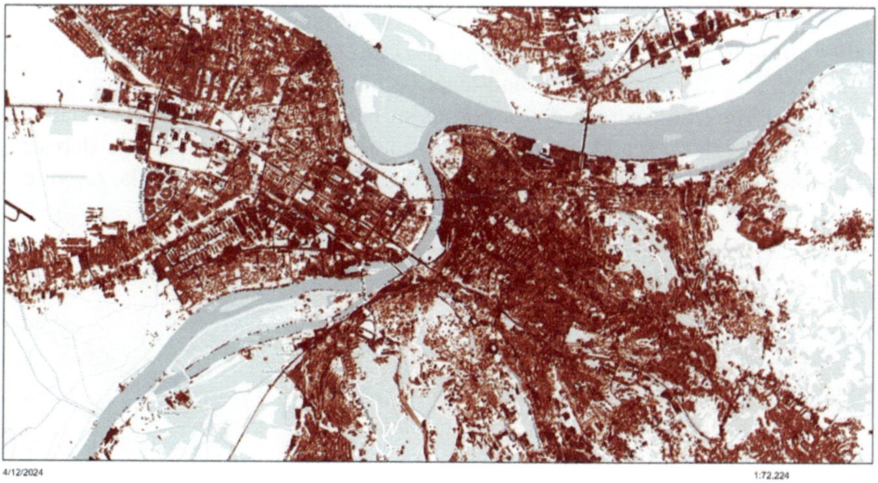

Fig. 15.3 Imperviousness density of the part of the City of Belgrade territory in 2018. *Source* Copernicus

elements within the urban fabric, and the promotion of holistic health and well-being—were deemed highly significant by over 80% of respondents. Furthermore, nearly a quarter of participants suggested the addition of "expanded green spaces, recreational zones, and environmental purity" as an essential value to be included in the list of considerations for public feedback.

Furthermore, within the realm of enhancing the transportation system of Belgrade, key aspects identified as paramount include the enhancement of public urban passenger transport (significant 73.5%, necessary 20.3%) and the amelioration of pedestrian mobility conditions, along with an emphasis on augmenting pedestrian safety as road users (significant 60.1%, important 33.1%). Additionally, there is a highlighted necessity for developing vehicle parking infrastructure and the systematic resolution of this challenge by implementing diverse transportation policy measures (significant 66.7%, important 27.4%). When considered comprehensively, the pivotal priorities delineated for the next two decades in urban planning include transportation (ranking first), green spaces and natural elements (ranking second), with the network and accessibility of public urban areas positioned at sixth place (Urbel, 2021).

The primary objective of the future state is to address parking issues within the municipality of Vračar. It involves redeveloping parking spaces to create open urban areas to increase soil permeability and enhance ecosystem services in the urban environment. The parking challenge is viewed through its spatial occupation. A typology of parking spaces has been formed (see Fig. 15.4), dividing them into point (a) individual parking spaces), linear (b and c) and surface parking. Each parking area occupies part of the public space, and the future state envisages various purposes to enhance public space, introduce new amenities, and improve the urban environment's quality.

The pursuit of an intelligent redesign for parking spaces within Vračar Municipality prompts a critical examination of the current urban fabric. Recognising the adverse effects of expanding ground-level parking on open space quality and the detrimental impact of increased non-porous surfaces on the urban microclimate, the latter phase of the research proposes a paradigm shift. This involves a departure from conventional approaches towards parking utilisation to promote the re-wilding

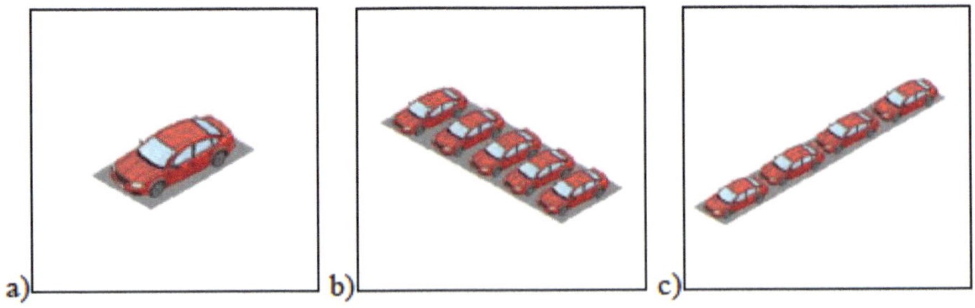

Fig. 15.4 Typology of parking spaces: **a** point, **b** and **c** linear

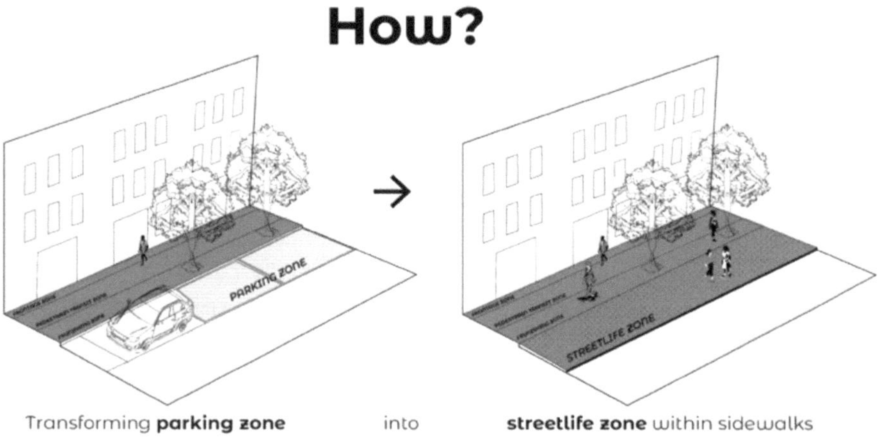

Fig. 15.5 The approach of transforming the parking zone into the street life zone within sidewalks

of urban streets and the creation of loveable, human-scaled micro places, reinvigorating densely built environments while turning back the street life to the sidewalks (Fig. 15.5).

Instead, the study advocates for targeted design interventions and installations to repurpose oversized parking areas within densely constructed, historically significant inner-city neighbourhoods. The overarching objective is to foster the resurgence of natural elements within urban thoroughfares and establish a network of intimate, human-scale microenvironments. Such initiatives aspire to rejuvenate the densely built environment while enhancing its liveability and ecological sustainability.

Aligned with this foundational principle, the initial segment of the research endeavour encompasses the establishment of a Catalogue of small urban spaces poised to supplant parking facilities, thereby fostering enhanced liveability within Vračar's thoroughfares. Although still in its nascent stage of development, this Catalogue holds potential as an illustrative tool showcasing the multifaceted utilisation of parking places.

The preliminary stage of the study involved identifying the potential range of functions that could be accommodated within an area equivalent to a single parking space. Consequently, an analysis was conducted within the parameters delineated by six distinct urban purposes, encompassing green spaces, recreational amenities, sporting facilities, communal areas, commercial and service-oriented establishments, cultural and educational venues, and pedestrian pathways (Fig. 15.6).

In this manner, 23 potential applications for such space were identified (see Fig. 15.7), ranging from simple lawns and permeable surface pathways to installing green canopies providing shaded seating areas, spaces for playing table tennis, or exhibition areas. The ensuing phase of the study encompassed an examination of diverse activities feasible for users within the envisioned microenvironments of open public urban spaces. Twelve distinct activities were identified: walking, standing,

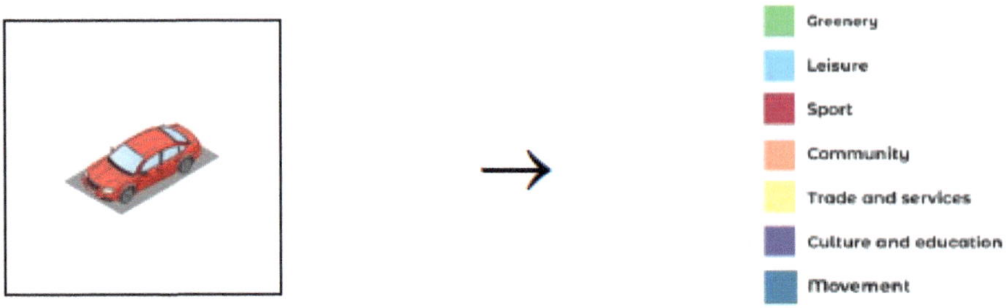

Fig. 15.6 Functions that could be accommodated within the area of the single parking space

sitting, observing pedestrians, conversing, exercising, relaxing, entertainment, shopping, working, dining, and engaging in recreational play (Fig. 15.8).

Through an analysis of the various activities that can take place in a single parking space instead of parking, efforts have been made to determine the capacity for multipurpose use of these micro areas. Such activities do not necessarily have to co-occur. However, they offer more options for users of open spaces, both in terms of timing throughout the day and about different user groups and their needs.

A similar analysis of various uses of parking areas was also conducted for linear spaces created by multiplying parallel or perpendicular parking spaces. These spaces can be treated as a patchwork of individual transformed parking areas, but they can also be viewed as a whole, defining an ambient entity or a zone with some dominant content. That ambient could be a tree avenue, exhibition place, or linear seating area instead of an on-street parallel parking space (Fig. 15.9).

On the other hand, due to its surface area and organisation, the linear space obtained from perpendicular parking offers slightly more options. These possibilities include an open-air cinema, a café terrace, an urban beekeeping space, a community gathering area, a green oasis, green walls, or an outdoor market (Fig. 15.10).

The proposed solutions have been devised following the sustainable development goals (SDGs), focusing on providing amenities for movement and relaxation for all user groups. Special attention has been directed towards introducing new locations where the provision of ecosystem service offerings will be expanded.

Consistent with this, as well as with data from other documents such as the General Regulation Plan for Green Areas of the City of Belgrade, the research-through-design approach has encompassed a scenario for the development of the Vračar Municipality based precisely on the principles of increasing the presence of green spaces. This entails creating an urban environment supporting sustainable transportation modes, such as pedestrian movement and cycling.

Such a scenario encompassed a thorough evaluation of identified challenges and primarily focused on amalgamating extant green areas by instituting green corridors and repurposing parking zones (Fig. 15.11). Through a meticulous examination of parking facilities, they were categorised into six specific types: surface courtyard parking, surface public parking, parallel parking, perpendicular parking, and combined parking. Linear parking lots are earmarked for the establishment of green

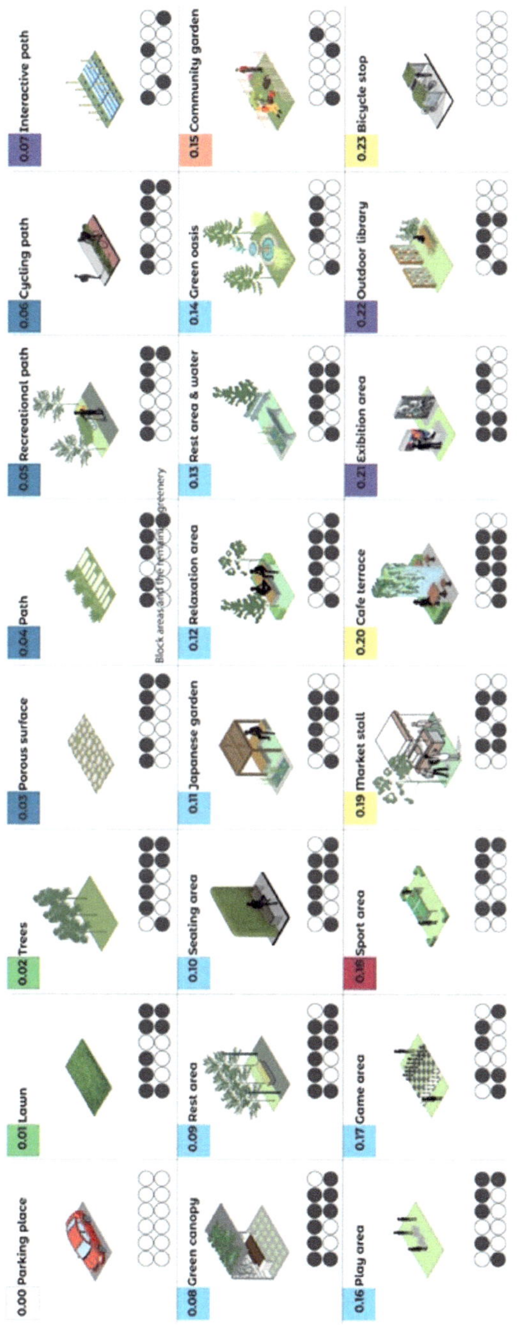

Fig. 15.7 Catalogue of small-scale public space proposals for single parking space

Fig. 15.8 Various activities that can take place within the open public space

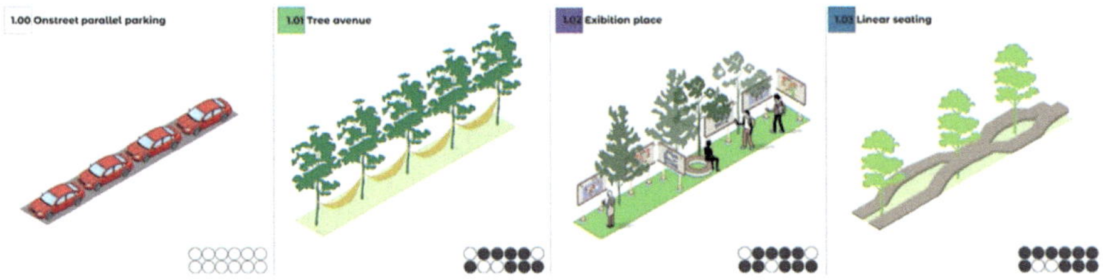

Fig. 15.9 Various purposes of linear spaces instead of parallel parking on the street

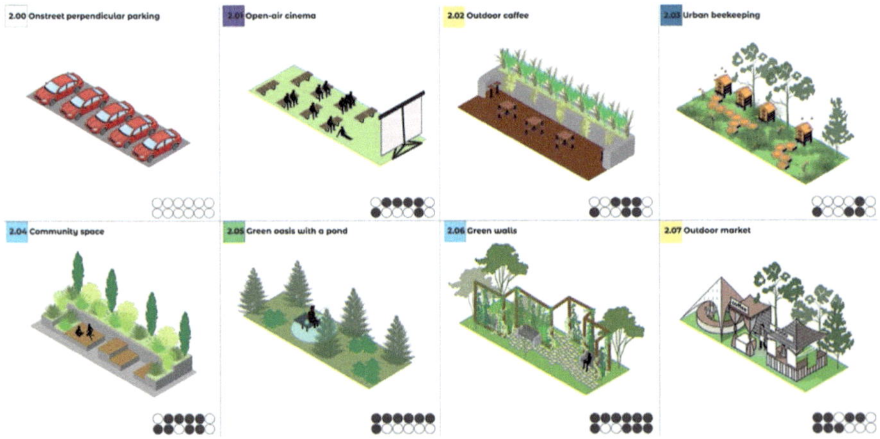

Fig. 15.10 Various purposes of linear spaces instead of on-street perpendicular parking

corridors. In contrast, surface parking areas are allocated to form urban pockets, squares, and plazas and improve block interiors.

The prospective scenario of transforming parking areas within the territory of the Vračar Municipality has also been analysed regarding the provision of specific ecosystem services. These services are depicted in Fig. 15.12 and correlate with the

Fig. 15.11 Future scenario of the Municipality of Vračar development

typology of open public spaces the solution envisages. Based on this, it is concluded that the newly proposed interventions, especially in the form of various greening initiatives, would contribute to activating multiple ecosystem services, particularly in the regulatory and cultural domains.

In addition, contributions can be made towards various sustainable development goals (SDGs) such as SDG3 (Good Health and Well-being), SDG11 (Sustainable et al.), SDG12 (Responsible Consumption and Production), and SDG13 (Climate Action).

15.6 Discussion

A key element in this transformative approach is the concept of parklets. Parklets are innovative urban elements that reclaim and convert parking spaces into small, vibrant community spaces. These micro-urban pockets typically extend out from the sidewalk at the level of parking spaces and offer an array of amenities, such as seating, greenery, art installations, or spaces for outdoor dining. This re-purposing

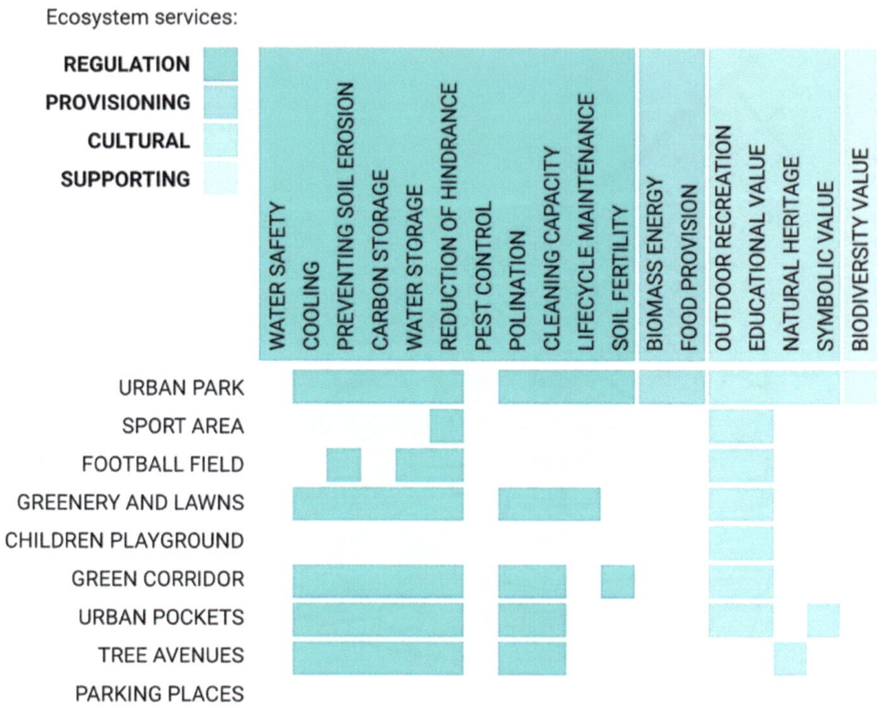

Fig. 15.12 Typologies of open public space interventions and ecosystem services

endeavours to solve the dual challenge of lack of public space and the dominance of cars in urban centres.

Parklets serve as a cost-effective strategy for enhancing the quality and aesthetics of urban environments and are instrumental in fostering a sense of community and place. They encourage a shift from viewing streets merely as thoroughfares for cars towards seeing them as multi-use public spaces that facilitate safety, interaction, and community liveliness.

The paper specifically discusses the transformative potential of parklets on Belgrade's urban landscape, suggesting a move from the car-dependent city design towards one that fosters community and pedestrian engagement. This transition recognises a vehicle-focused city's adverse effects on the environment, public health, and overall social cohesion.

15.6.1 Community-Oriented Design

Through parklets and other community-oriented designs, Belgrade can aim to create more sustainable and liveable communities. The envisioned urban landscape is comprised of vibrant and inclusive communities, achieved by transforming previously car-dominated spaces into dynamic areas thriving with social activity and greenery.

This approach in Belgrade and other cities represents a holistic shift towards a more equitable and environmentally conscious urban planning paradigm. It reflects a growing awareness of the necessity for cities to be adaptable, resilient, and centred around the well-being of their inhabitants. Discussing such urban transformations can illuminate the potential benefits for other cities looking to revitalise their urban spaces similarly.

Parklets represent a notable shift in urban planning, transitioning from a car-centric model to one emphasising pedestrian-friendly spaces. Originating as temporary installations, they have evolved into permanent urban features that significantly enhance streets and public spaces. By reclaiming parking spots and transforming them into vibrant communal areas, parklets provide a novel public realm experience that fosters social interaction, supports local businesses, and contributes to the aesthetic improvement of urban environments.

15.6.2 Relationship with New European Bauhaus

This approach aligns with the principles of the New European Bauhaus initiative, an interdisciplinary movement aimed at combining sustainability, aesthetics, and inclusiveness in the built environment. It calls for creative strategies to reimagine urban spaces, making them more liveable and attractive while addressing critical environmental and societal challenges.

In the specific context of Belgrade, like many other urban areas, a high vehicle-to-parking-space ratio leads to a scarcity of parking, contributing to congestion, pollution, and reduced urban quality of life. At the same time, the city's streets and public spaces are overly car-dominated, resulting in a less accommodating environment for pedestrians and a general lack of accessible and interactive public spaces.

By introducing parklets and similar interventions focused on people rather than cars, Belgrade can work towards alleviating its parking problems. Converting parking spots into pedestrian zones can reduce the attraction of city centres for car use, thereby lowering the demand for parking. This reshaping of urban spaces can contribute to a more balanced and sustainable transport ecosystem, prioritising public transport, cycling, and walking, which aligns with the New European Bauhaus ethos.

Implementing such transformations in Belgrade poses challenges, including balancing the interests of various stakeholders, ensuring public acceptance, securing funding, and navigating political complexities. Nonetheless, by moving towards a community-oriented urban design, the city can enhance its public spaces, improve environmental conditions, and create a more resilient and liveable urban landscape.

This research also highlights the significant social and environmental benefits that can result from a visionary shift in urban planning, particularly in Belgrade's context. From a social perspective, transforming parking spaces into pedestrian areas can increase pedestrian traffic, as specific European examples show. This can have a positive effect on community interaction and enhance the vibrancy of the urban space. Regarding community reception, the restructuring of city blocks to prioritise pedestrian and cyclist movement can be well-received by the community, as evidenced by surveys indicating positive reactions to the interventions made in European cities. With reduced motor vehicle traffic and lower overall speeds, the safety of pedestrian zones can be improved, making streets more liveable and inviting for residents and visitors.

Environmental benefits could be seen with the reduction in motor vehicle traffic and speed, a decrease in noise levels, and a potential impact on the urban heat island effect. Minimising the number of cars on the road can lead to substantial environmental improvements, such as decreased emission levels, directly affecting air quality. The restructuring efforts in urban areas have also reduced noise levels. This benefit contributes to the well-being of city dwellers and creates a more serene urban environment. While not explicitly dealing with within this research, integrating more green spaces and pedestrian-oriented areas could be widely recognised for combating the urban heat island effect, which can be inferred as a potential environmental benefit of adapting Belgrade's public spaces.

15.6.3 Presented Case Studies

The research also provides examples from other cities that support these benefits and suggests that Belgrade could embrace similar initiatives to improve its urban landscape. The push toward pedestrianised zones and repurposed parking spaces reduces traffic and creates vibrant urban life where public spaces can thrive as platforms for social interaction, economic activity, and cultural expression. These changes reflect an intent to design urban spaces focusing on people and the planet, aiming for a future where cities are increasingly adaptive and resilient.

The research covered case studies and examples from cities like Barcelona and Paris to understand and demonstrate the transformative impact of reallocating urban space from cars to pedestrian and cyclist infrastructure. Those cities have displayed a dedication to transforming urban public spaces into areas that prioritise pedestrian and cyclist mobility. Their vision acknowledges that cities should not merely function but also promote their inhabitants' health, well-being, and happiness.

In line with that, the paper suggests that Belgrade could adopt strategies similar to those successfully implemented in Barcelona and Paris. By restructuring city blocks and repurposing parking spaces, Belgrade can enhance its urban environment—increasing pedestrian space and activities, reducing noise and environmental pollution, and invigorating its community spaces. The discussion highlights how these examples inspire Belgrade as they consider creating a more liveable, sustainable, and community-focused urban landscape.

Successful urban transformation requires overcoming several potential roadblocks. Each city has unique challenges, but common hurdles include political will, effective parking management, public participation, and innovative planning strategies. The paper discusses how other European cities have navigated these challenges and provides insights into how Belgrade might do the same.

Political Will: Political commitment is essential for advancing urban transformations. Projects can continue with decisive action from policymakers and local authorities. Observing other cities like Barcelona and Paris, it becomes apparent that solid leadership and a clear vision are vital for translational action.

Effective Parking Management: A primary challenge in urban planning is managing the demand for parking. As the paper suggests, effective parking management is becoming critical with the increasing number of registered vehicles in Belgrade. Barcelona's and Paris' initiatives to repurpose parking spaces into valuable community assets demonstrate that comprehensive parking policies, which balance the needs for accessibility with broader urban sustainability and liveability goals, are crucial.

Public Participation: Transformative urban projects succeed when they incorporate input and engagement from the broader community. An open dialogue with residents, business owners, and other stakeholders ensures that projects are well-received and supported. Lessons from Paris' participatory approaches to development and Barcelona's commitment to engaging residents in the superblock concept indicate the importance of public buy-in and its positive impact on project outcomes.

Innovative Planning Strategies: A transition toward pedestrian-friendly spaces often requires creative solutions and a break from traditional urban planning methods. For instance, Paris and Barcelona have integrated eco-friendliness and aesthetics in rethinking urban spaces, aligning with the New European Bauhaus initiative's objective of creating sustainable, inclusive, and beautiful spaces.

Belgrade can leverage these insights from other European cities by:

Building Political Consensus: Engaging political stakeholders and fostering multi-party support for pedestrian-friendly initiatives.

Introducing Progressive Parking Policies: Reducing dependence on personal vehicles by offering alternative transportation options, adjusting parking fees, and setting parking restrictions could help manage the existing parking demand.

Engaging the Public: Utilizing participatory planning processes, Belgrade can ensure that public space transformations reflect the needs and desires of its citizens, potentially through public forums, surveys, and workshops.

Fostering Innovation: Embracing new technologies and urban planning concepts can help Belgrade address its specific issues creatively and effectively, looking beyond traditional approaches to finding unique solutions for enhancing the urban landscape. This might include deploying innovative city technologies to monitor traffic, manage public spaces, and facilitate better urban mobility.

Belgrade has the potential to learn from the successes and experiences of other European cities. By studying case studies from Barcelona, Paris, and others, it can adopt bespoke solutions that reflect its unique character while still aspiring to the sustainable and inclusive urban design principles that have proven effective elsewhere.

Moreover, collaboration with urban planners, architects, environmentalists, and the private sector can bring fresh perspectives and resources. Drawing on a diverse pool of expertise helps to craft informed, well-rounded strategies that account for the complex interplay of factors inherent in urban development.

Ultimately, while the challenges may appear daunting, cities worldwide have demonstrated that with a concerted effort, creativity, and a willingness to embrace change, it is possible to transform urban landscapes in ways that benefit all residents and the environment. Belgrade's journey towards this goal may well be informed by these global experiences, adapting them into a unique vision that meets the city's specific needs and aspirations.

15.6.4 Belgrade's Vision

The vision for Belgrade as a sustainable and liveable city, inspired by the transformational principles outlined in the paper, places community-oriented design at its core. This vision imagines a city landscape where the priority shifts from vehicular traffic to pedestrian zones, public spaces, and green areas that foster community interaction and environmental well-being. In such a vision, Belgrade would be a tapestry woven with vibrant communal spaces, facilitating interactions, economic vitality, and cultural expression. Crucially, introducing new transport technologies, such as electric and autonomous vehicles, would support this urban fabric, serving as an enhancement rather than a centrepiece. The strategy for integrating these technologies with an emphasis on shared and public mobility could involve the following aspects:

Complementary Role of AVs: Autonomous vehicles could be integrated into a mixed-modal transport network, complementing walkability and cycling infrastructure rather than competing with them. They could be used for public transportation or as shared vehicles to solve the "last mile" problem, efficiently connecting residents to transit hubs without increasing personal vehicle use.

Shared Mobility Platforms: Encouraging car-sharing and ride-sharing platforms incorporating new transport technologies could reduce the number of vehicles needed and lower the demand for parking spaces.

Pedestrian and Cyclist Prioritization: Consistent with the principles seen in Barcelona and Paris, Belgrade could prioritise infrastructure development for pedestrians and cyclists over personal car use, even when those cars are autonomous.

Innovative Traffic Management: Utilizing AI and IoT technologies for intelligent traffic systems that optimise routes for AVs, reduce congestion, and prioritise public transportation, bicycles, and pedestrian flow.

Policy and Regulation Frameworks: Implementing robust policy regimes and regulatory frameworks to ensure that integrating new transport technologies aligns with broader urban design goals of sustainability and liveability while also being adaptable to future advancements.

Public Engagement: Ensuring ongoing public participation and engagement in the planning and implementation phases to align everyone's interests and garner public support.

Considering these aspects, Belgrade can become a more sustainable and attractive metropolitan centre for its residents and visitors. The focus on community-oriented urban design, supported by innovative transport solutions that enhance shared and public mobility, is at the heart of creating this future-facing, resilient cityscape responsive to changing needs and technologies.

The paper emphasises the need for broad collaboration among diverse stakeholders to shift towards a pedestrian-focused urban landscape successfully. This collaborative approach is essential to addressing urban transformation's complex dynamism and ensuring the shift reflects a shared vision for Belgrade's future.

To achieve a sustainable and pedestrian-friendly urban environment, the involvement of the following stakeholders is crucial: government agencies, urban planners and architects, business and investors, residents and community groups and civil society and NGOs. Government Agencies because they play a crucial role in setting policies, providing funding, and facilitating the legislative environment necessary for the successful implementation of urban transformation projects. Additionally, they can coordinate across different departments to ensure a fully integrated approach. Professionals in urban design and architecture are instrumental in drafting the blueprints of transformation, ensuring that these plans are conducive to pedestrian mobility, considering the human scale, and fostering community spaces. The active participation of the private sector can bring in vital investment and innovation. Businesses, particularly local ones, stand to benefit from increased pedestrian traffic and can support the vision by investing in the community and adapting to new urban designs. Public participation is essential for gaining insights into the needs and desires of the people using these spaces regularly. Residents have a vested interest in the development of their communities and can offer valuable perspectives and feedback. Organisations focused on environmental sustainability, social inclusion, or economic development can advocate for community interests and help to identify best practices from other cities and projects.

Collaboration among these groups can facilitate the exchange of ideas, ensure that plans are resilient and adaptable, and incorporate a wide range of expertise and perspectives. The success of transitioning to a pedestrian-focused city requires more than just the redesign of physical spaces—it entails a collective change in mindset and a cooperative effort to reimagine Belgrade's urban ecosystem. Cooperation and partnership are, therefore, not just beneficial but necessary to realise the vision of transforming Belgrade into a more sustainable, liveable, and community-friendly city.

15.7 Conclusion

The study argues for converting parking spaces into multifunctional green areas in the Vračar Municipality to promote sustainable urban development and tackle parking space scarcity. This transformation enhances the area's aesthetics and contributes to sustainability, health, well-being, and climate action goals. It envisions a more environmentally friendly and community-focused urban landscape by creating green corridors, urban pockets, and plazas from parking zones. These initiatives optimise urban space utilisation while providing ecosystem services for a healthier environment.

The research demonstrates innovative methods for tailored urban renewal by addressing identified challenges through greening efforts. The proposed interventions set an example for potential transformations of other urban spaces with careful planning consideration that maximises utility for society and the environment.

While parklets offer benefits like creating vibrant community spaces and encouraging pedestrian activity, it is essential to acknowledge opposing arguments against their implementation. Critics raise concerns about exacerbating parking scarcity in cities like Belgrade by converting limited parking infrastructure into parklets, which could lead to increased congestion, affect residents' ability to find parking quickly and cause financial strain on local governments due to potential revenue loss.

Some argue that parklets may not effectively address underlying issues, such as inadequate pedestrian infrastructure or lack of accessibility options, without comprehensive public transportation improvements in broader urban planning strategies. Additionally, opponents highlight possible risks associated with gentrification resulting from revitalisation efforts, which can lead to displacement of existing communities and loss of diversity in neighbourhoods where property values rise significantly due to these projects.

The research emphasises the importance of repurposing parking areas into green spaces to create a more sustainable and community-centred environment.

The case study of Vračar Municipality illustrates how urban spaces can be reimagined in alignment with sustainable development goals. Parklets have the potential to address parking scarcity and enhance urban spaces, but their implementation should mitigate challenges such as congestion, limited accessibility, gentrification, and displacement of communities. Parklets could experiment with new solutions and facilitate meaningful dialogue before making permanent decisions.

References

Ahmed N, Elshater A, Afifi S (2019) The community participation in the design process of livable streets. Lecture notes of the institute for computer sciences, social-informatics and telecommunications engineering, LNICST, vol 296, pp 144–157. https://doi.org/10.1007/978-3-030-34863-2_13/COVER

Berggren CC (2017) Parklets: urban sustainability one parking space at a time | smart cities dive, smart cities dive. Available at: https://www.smartcitiesdive.com/ex/sustainablecitiescollective/parklets-urban-sustainability-one-parking-space-time/207861/. Accessed 19 Apr 2024

Birdsall M (2013) Parklets: providing space for people to park... themselves. Institute of Transportation Engineers. ITE J 85(5):36–39. Available at: https://www.proquest.com/openview/a4e017c08687ad2a39221a963034f961/1?pq-origsite=gscholar&cbl=42116. Accessed 19 Apr 2024

Bravo L, McCormick M, Hillary F (2020) Reviewing and speculating on public space futures through a new lens. J Public Space 5(4):1–6. https://doi.org/10.32891/JPS.V5I4.1445

Brozen M, Loukaitou-Sideris A (2013) Reclaiming the right of way a toolkit for creating and implementing Parklets. Los Angeles

Casas-Valle D, Oliveira I, Dias CB (2023) The future design of streets. In: The future design of streets association. Available at: https://thefuturedesignofstreets.eu/book/. Accessed 26 Mar 2024

Cathcart-Keays A (2015) Will we ever get a truly car-free city?, The Guardian. Available at: https://www.theguardian.com/cities/2015/dec/09/car-free-city-oslo-helsinki-copenhagen. Accessed 2 Apr 2024

City of Belgrade—department of transport (2020) SUMP (Sustainable Urban Mobility Plan) Belgrade. Belgrade

Civitas (2019) PARK4SUMP. Available at: https://civitas.eu/projects/park4sump. Accessed 2 Apr 2024

European Commission (2021) New European Bauhaus: beautiful, sustainable, together. Available at: https://new-european-bauhaus.europa.eu/index_en?prefLang=it. Accessed 28 Mar 2024

European Commission (2022) European bauhaus: major catalyst of the European Green Deal. Available at: https://ec.europa.eu/commission/presscorner/detail/en/ip_23_203. Accessed 28 Mar 2024

Cooney S, Raghavan B (2020) Tactical Patterns for Grassroots Urban Repair. arXiv:2010.02401 [Preprint]

CORDIS European Commission (2019) Focus on parking management boosts urban planning, CORDIS. Available at: https://cordis.europa.eu/article/id/442744-focus-on-parking-management-boosts-urban-planning. Accessed 2 Apr 2024

Corey K (2014) Making space: an exploration of parklets in North America and Vancouver. University of Guelph. Available at: http://hdl.handle.net/10214/8081. Accessed 19 Apr 2024

Davis DE (2015) 'Ambulatory Urbanism', in Europe City. In: Barcelona and Zurich: CCCB and Lars Mueller Publishers

Donald A, Gerson SM, Lintell M (1981) Livable Streets. University of California Press

Gehl J (1971) Life between buildings: using public space. Danish Architectural Press, Copenhagen

Gehl J (2006) New city life. Danish Architectural Press, Copenhagen

Gehl J (2010) Cities for people, 1st edn. Island Press, Washington, Covelo, London

Gehl J, Svarre B (2013) How to study public life, how to study public life. Island Press-Center for Resource Economics. Available at: https://doi.org/10.5822/978-1-61091-525-0/COVER

Jacobs J (1961) The death and life of great American cities. Random House, New York

Koch P et al (2024) Die neue Öffentlichkeit | New Public Spaces Europäische Straßenräume des 21. Jahrhunderts | European Urban Streetscapes in the 21st Century. M BOOKS

Koue C (2013) The unrealized potential of parklets, TheCityFix. Available at: https://thecityfix.com/blog/unrealized-potential-parklets-world-parking-day-new-urbanism-camille-koue/. Accessed 19 Apr 2024

Lau BPL et al (2021) The study of urban residential's public space activeness using space-centric approach. IEEE Internet Things J 8(14):11503–11513. https://doi.org/10.1109/JIOT.2021.3051343

Di Liscia V (2020) As Part of EU's $872 Billion Recovery Plan, a Call For a "New Bauhaus", Hyperallergic. Available at: https://hyperallergic.com/590455/european-union-new-bauhaus/. Accessed 19 Apr 2024

Litman T (2010) Parking pricing implementation guidelines: how more efficient parking pricing can help solve parking and traffic problems, increase revenue and achieve other planning objectives

Liu JHI et al (2018) Understanding the accessibility, economic and social equity impacts of urban greenway infrastructure. USA: National Institute for Transportation and Communities (NITC). Available at: https://doi.org/10.21949/1503647
Lleshi B, Jashari R, Fáczányi Z (2022) Interventions in spontaneous developed public spaces. Pollack Periodica 18(1):19–24. https://doi.org/10.1556/606.2022.00559
Loukaitou-Sideris A, Ehrenfeucht R (2009) Sidewalks: conflict and negotiation over public space. Cambridge, Massachusets: the MIT Press. Available at: https://books.google.rs/books?hl=sr&lr=&id=XEAAAQBAJ&oi=fnd&pg=PR9&dq=Loukaitou-Sideris,+A+(2011)+Sidewalks:+Conflict+and+Negotiation+over+Public+Space,+MIT+Press&ots=Y4dgKI1heB&sig=XFufqVNx-PlvLr7bGrUP6os2zSk&redir_esc=y#v=onepage&q=Loukaitou-Sideris%2C%20A%20(2011)%20Sidewalks%3A%20Conflict%20and%20Negotiation%20over%20Public%20Space%2C%20MIT%20Press&f=false. Accessed 27 Mar 2024
Milosavljevic N, Simicevic J, Maletic G (2010) Vehicle parking standards as a support to sustainable transport system: Belgrade case study. Technol Econ Dev Econ 16(3):380–396. https://doi.org/10.3846/TEDE.2010.24
Mitra R, Abbasi Z (2020) Bringing back play to urban streets. In: Transport and children's wellbeing, pp 237–252. Available at: https://doi.org/10.1016/B978-0-12-814694-1.00011-7
Nieuwenhuijsen MJ, Khreis H (2016) Car free cities: pathway to healthy urban living. Environ Int 94:251–262. https://doi.org/10.1016/J.ENVINT.2016.05.032
NLA (2020) Parklets 2.0, Norske landskapsarkitekters forening. Available at: https://landskapsarkitektur.no/prosjekter/parklets-2-0. Accessed 2 Apr 2024
O'Sullivan F (2021) A tiny twist on street design: the one-minute city, Bloomberg City Lab. Available at: https://www.bloomberg.com/news/features/2021-01-05/a-tiny-twist-on-street-design-the-one-minute-city. Accessed 2 Apr 2024
Owens LB, Cha M, Owens B (2018) A tale of two cities: the Parklet implementation processes in Los Angeles and San Francisco. Occidental College
Panganiban J, Ocubillo RA (2014) Citywide assessment of parklets & plazas: summary of data collected for summer 2014 public life study
Park4SUMP (2018a) Convert unused parking spots into green spaces. Available at: https://park4sump.eu/news-events/news/convert-unused-parking-spots-green-spaces. Accessed 28 Mar 2024
Park4SUMP (2018b) Objectives, Park4SUMP. Available at: https://park4sump.eu/index.php/about/objectives. Accessed 2 Apr 2024
Park4SUMP (2021) Why Paris is eliminating 72% of its on-street parking spaces, Park4SUMP. Available at: https://park4sump.eu/news-events/news/why-paris-eliminating-72-its-street-parking-spaces. Accessed 2 Apr 2024
Park4SUMP (2022) Parking spots turned into shop extensions or terraces in Rotterdam, Park4SUMP. Available at: https://park4sump.eu/news-events/news/parking-spots-turned-shop-extensions-or-terraces-rotterdam. Accessed 2 Apr 2024
Postaria R (2021) Superblock (Superilla) Barcelona—a city redefined. Public Realm, Cities Forum. Available at: https://www.citiesforum.org/news/superblock-superilla-barcelona-a-city-redefined/. Accessed 2 Apr 2024
Republicki zavod za statistiku (2023) Opstine i regioni u Republici Srbiji
Rykwert J (1978) The street: the use of its history. In: Anderson S (ed) Institute for architecture and urban studies on streets: streets as elements of urban structure, 1st edn. The MIT Press
Sauter D, Huettenmoser M (2008) Liveable streets and social inclusion. Urban Des Int 13(2):67–79. https://doi.org/10.1057/UDI.2008.15/METRICS
Sevtsuk A, Davis DE (2019) Future of streets. In: Modeling and simulation in science, engineering and technology, pp 537–546 https://doi.org/10.1007/978-3-030-12381-9_29
Shokry H (2020) Are street urban spaces can Invite City life? Parklets; small places with big impacts'. In: ICWSAUD2019 A joint conference with 4th international conference on engineering, innovation and technology. Penang: University Sains Malaysia. Available at: https://www.researchgate.net/publication/342145782_Are_Street_Urban_Spaces_Can_Invite_City_Life_Parklets_Small_Places_with_Big_Impacts. Accessed 19 Apr 2024

Signer S (2020) Methods to design squares teoksessa Die neue Öffentlichkeit | New Public Spaces–Europäische Stadtplätze des 21. Jahrhunderts–European Urban Squares in the 21st century. In: Barz-Malfatti H, Signer S (eds) M BOOKS. Available at: https://www.m-books.eu/store/die-neue-oeffentlichkeit-europaeische-stadtplaetze-des-21-jahrhunderts/. Accessed 26 Mar 2024

Simićević JR (2013) Prilog definisanju politike parkiranja u zavisnosti od nivoa usluge na saobraćajnoj mreži, Универзитет у Београду. Available at: https://nardus.mpn.gov.rs/handle/123456789/2671. Accessed 26 Mar 2024

Tomic J (2019) Na jedno zonirano parking mesto 23 registrovanih automobila u Beogradu, Danas. Available at: https://www.danas.rs/vesti/beograd/na-jedno-zonirano-parking-mesto-23-registrovanih-automobila-u-beogradu-video/. Accessed 5 Apr 2024

Urbel (2021) Ka Beogradu—Inicijalna anketa, Urbel. Available at: https://forms.office.com/Pages/AnalysisPage.aspx?id=tkqp7dknvEW-OkUdHiVIN_qcwOHF0_5KhILF097oIL5UMUU3OFFaUEJTUDVLTkU0NkNLOVBOSFFEOS4u&AnalyzerToken=U647XiwQUDHJgTNuWzXMC8zUEkTZeFus. Accessed 8 Apr 2024

Vestre (2021) Bergmannstraße, Berlin, Vestre. Available at: https://vestre.com/case-studies/bergmannstrasse-berlin. Accessed 2 Apr 2024

We Are Possible (2021) A history of parklets—possible, possible. Inspiring Climate Action. Available at: https://www.wearepossible.org/latest-news/a-history-of-parklets. Accessed 2 Apr 2024

Weinberger R, Kaehny J, Rufo M (2010) U.S. parking policies: an overview of management strategies. Institute for Transportation & Development Policy, New York. Available at: www.itdp.org. Accessed: 19 Apr 2024

Whyte WH (2001) The social life of small urban spaces. Project for Public Spaces, New York

Xu W (2018) Urban explorations: analysis of public park usage using mobile GPS Data. arXiv: 1801.01921 [Preprint]

Zhang Y et al (2022) The impact of interventions in the built environment on physical activity levels: a systematic umbrella review. Int J Behav Nutr Phys Act 19(1):156. https://doi.org/10.1186/S12966-022-01399-6

Chapter 16
Sustainable Dimensions of Healthy Behaviour in Urban Settings

Darja Kobal Grum and Bojan Grum

Abstract This chapter delves into the nuanced relationship between the built environment and the health-oriented behaviour of its users, presenting a novel perspective by extending beyond the conventional focus on physical health to include mental health as well. Unlike previous studies that predominantly explored the physical health implications of the built environment, our research uncovers a broader phenomenon that integrates both psychological and social aspects of behaviour within such environments. By employing a scoping review method, we analysed documented research from the past decade across various digital databases. Our major contribution to the field is the development and validation of a hypothetical model aimed at promoting healthy lifestyles within the built environment. This model is underpinned by our findings, which highlight the complexity of health-seeking behaviour as it intersects with the psychological and social dimensions of individuals' interactions with their surroundings. In the concluding section of the chapter, we articulate various domains of health-oriented behaviour in relation to the built environment and propose recommendations for qualitative transformations and design interventions. These suggestions are geared towards fostering environments that not only support physical well-being but also enhance mental health, thereby contributing significantly to the discourse on creating more holistic, health-promoting built environments.

Keywords Built environment · Health-oriented behaviour · Physical health · Mental health · Scoping review

D. K. Grum (✉)
University of Ljubljana, Ljubljana, Slovenia
e-mail: darja.kobal@ff.uni-lj.si

B. Grum
New European University, Nova Gorica, Slovenia

M. Vukmirović et al. (eds.), *Digital Dreams, Urban Realities*, The Urban Book Series,
https://doi.org/10.1007/978-3-031-76353-3_16

16.1 Introduction

The exploration of how the environment impacts human health has long intrigued scholars from various fields. Recent advancements in environmental psychology have spotlighted a crucial element: the built environment, which plays a significant role in influencing human health, both positively and negatively (Bower et al. 2023; Khaddar et al. 2023). Notably, the emphasis has been on the physical aspects of the built environment—such as its impact on physical health—while the psychological dimensions, particularly mental health, have received comparatively less attention (James et al. 2017; Plikas et al. 2017; Thornton et al. 2017). However, this study diverges by also incorporating mental health, acknowledging its importance alongside physical well-being in the context of the built environment (James et al. 2017; Knight et al. 2014). The primary research question of this study focuses on identifying health-oriented behaviours within the built environment. It aims to uncover the characteristics of mental health behaviours related to the built environment. Furthermore, it seeks to determine the key factors that drive health-oriented behaviours in the built environment, which most effectively bolsters the mental health of its occupants. This approach marks a significant contribution to the field by broadening the understanding of health in the built environment beyond its physical implications, offering insights into how the design and structure of our surroundings can foster better mental health outcomes. The main hypothesis of our research is that characteristics and living conditions are significantly related to healthy lifestyle behaviour in both directions: the environment influences mental health, but individuals' behaviour in the urban environment also influences their own mental health. This means that we need to think in both directions: how can we create an environment in which individuals feel comfortable, and how can we encourage the behaviour of users of the built environment so that they experience it as a pleasant and high-quality environment in which to live?

The built environment in urban centres, characterised by its potential to inhibit human well-being and health, necessitates a nuanced understanding of its role as a stressor. Psychological research (Tao et al. 2024) underscores a pronounced disparity in stress levels associated with lifestyle between urban and rural settings. To counteract the adverse effects of urban stress, health advocates strongly recommend the adoption of health-oriented behaviours, notably maximising time spent in green spaces, such as city parks, and minimising engagement with smart devices (Bilotta and Evans 2013). Such practices are shown to have a beneficial impact on an individual's mental and physical well-being, serving as a preventive measure against stress responses. This acknowledgement that the built environment inherently acts as a stressor, influencing residents' physical and mental health in diverse ways, introduces the concept of environmental or urban stress (Bilotta and Evans 2013).

To circumvent environmental stress, health-oriented behaviours that reduce environmental stress should be promoted. Integrating green spaces within urban planning, ensuring accessible recreational areas, and enhancing the overall quality of both living and neighbourhood conditions are essential strategies (Hobbs et al. 2021). These interventions not only address the direct and indirect stressors inherent in urban environments but also underscore the importance of a holistic approach to urban planning that prioritises health-oriented behaviours. Through such measures, the built environment can transform from a source of stress to a facilitator of well-being, enabling residents to lead healthier, more fulfilling lives.

Our research builds on Cradock and Duncan's (2014) model, which identifies the following relationships between built environment characteristics and health-oriented behaviour: physical activity, healthy diet, no alcohol and tobacco use, psychological well-being, injury prevention behaviour, and road safety. The authors (Cradock and Duncan 2014) note that the built environment can promote physical activity among residents, particularly when it includes facilities that provide recreation and other physical activities. They argue that the provision of cycle paths, safe walking, and walking routes to key buildings such as schools, nurseries, health centres, post offices, shops, etc., as well as bus stops in the urban environment, is of paramount importance. To promote healthy eating, they recommend locating restaurants, shops and other food outlets that offer healthy—not just fast food—and non-sugary drinks. The most important thing to reduce or even eliminate tobacco and alcohol consumption is to discourage schoolchildren from using such substances, which is, of course, best supported by the example of parents. In Slovenia, it is not legal to sell alcohol and tobacco to minors, but young people can easily obtain these substances through a variety of tricks. If the establishments where these substances are sold are far enough away from children's schools and homes, this is already one of the first preventive measures against their abuse. In fact, a large body of research shows that the availability of tobacco and alcoholic beverages in proximity encourages young people to use them (Koh et al. 2011; West et al. 2010).

Research (Cradock and Duncan 2014) also shows that the built environment plays a crucial role in this regard, especially for residents with lower socio-economic status. Among these groups, children and adolescents experience the highest number of injuries and traffic accidents, which researchers attribute to the disorganised conditions of the built environment where residents live in deprived socio-economic circumstances. In any case, the installation of ramps at intersections on busy roads and clear marking of all traffic hazard locations are recommended.

Psychological well-being is intricately linked to the urban living environment. The association between negative mood and the adverse factors of the built environment, as identified by James et al. (2017), highlights the profound impact of urban design on psychological states. Conversely, the strategic organisation of urban spaces, including accessible recreational facilities, walking paths, and children's play areas, plays a crucial role in fostering socialisation. This, in turn, significantly mitigates social isolation and loneliness, particularly among the older demographic of urban populations (Oliver et al. 2015). Research by Rollings et al. (2017), along with Ottoni et al. (2016), underscores that an environment conducive to social interaction and physical activity is essential for the mental health and overall well-being of its inhabitants.

The critical importance of psychological well-being in the built environment is rooted in the understanding that urban design influences not only physical health but also the social and emotional life of communities. As urban populations continue to grow, the need to create spaces that support mental health becomes increasingly urgent. Theories focusing on environmental psychology emphasise the role of sensory experiences, social interaction, and physical activity in urban settings in shaping mental health outcomes. Such perspectives advocate for urban planning and policies that prioritise human-centred design, integrating natural elements, promoting walkability, and facilitating community engagement.

Based on the problem of the study and the hypothetical model presented above (Cradock and Duncan 2014), we systematically reviewed the literature dealing with health-oriented behaviour in the built environment. The aims of our study were to identify which types of health-promoting behaviour have been most researched in the last ten years and to identify how the built environment can contribute to the healthy behaviour of its occupants. By reviewing the literature, we aimed to provide a general overview of the relationship between the built environment and health behaviour and to provide theoretical evidence for the further development of environmental research.

16.2 Materials and Methods

16.2.1 Search Protocol

We conducted a scoping review of the literature following the PRISMA protocol (Moher et al. 2009; Page et al. 2021). A scoping review is designed to thoroughly explore the literature on a broad topic area, aiming to identify key concepts, theories, sources of evidence, and gaps in the research. We employed the methodological framework proposed by Arksey and O'Malley (2005). We searched for scholarly articles in the databases EBSCOhost, specifically in the databases Academic Search Complete, Business Source Premier, APA PsycINFO, SocINDEX with Full Text, PubMed and GreenFILE using the search thread (without restrictions as to where the words were found) "healthy behaviour" AND "built environment*". The searches were undertaken between December 2023 and February 2024. Before determining the final search term, we familiarised ourselves with the results of various other search terms that proved to be too general or too narrow to gain insight into the concept of healthy behaviour in the context of the built environment. A review of hits with the final search thread "healthy behaviour AND built environment*" was confirmed as stable.

16.2.2 Inclusion and Exclusion Criteria

The initial search (entirely without search criteria) using the search thread "healthy behaviour" AND "built environment*" yielded 2097 hits.

In the next step, using automated tools, we narrowed the search using the following criteria:

- Time: We covered the period of the last ten years, from January 2013 to February 2024,
- Type of source: Works published in academic journals or in scientific books;
- Accessibility: Fully accessible works;
- Quality: Peer-reviewed works;
- Language: Work published in English.

16.2.3 Data Extraction

Under the automatically determined criteria, 636 works remained. Of these, 34 works were removed from the system as exact duplicates. This left 245 articles, which, in methodological terms, we can classify as review and original scholarly articles reporting on research conducted using quantitative and qualitative methods and theoretical articles. In the second step, we reviewed the 245 articles by title, abstract and subject indicators. In the first stage of the screening process, we excluded 165 works. The reasons for exclusion were as follows: Type of scholarly article (Editorials); Theoretical works that are too general or insufficiently focused on specific objectives; Duplicates that the system did not automatically exclude. Results are presented in Fig. 16.1.

We tried to obtain the remaining 114 articles in their entirety, but it turned out that 20 articles were not accessible, while among the other 94 articles, 27 had content that did not match the aims and research questions of the review. We therefore included in the review of the literature 67 articles, all of which we presented in the chapter Results and Discussion (Table 16.1). The entire process of searching the literature is presented in the PRISMA diagram in Fig. 16.1, which shows how, from 2097 identified articles, we gradually selected 67.

16.3 Results and Discussion

A scoping review of the literature shows that the built environment can be both a facilitator and a barrier to active leisure for its inhabitants (Benton et al. 2016; Serrano et al. 2023). In selected articles, the built environment includes both buildings such as blocks of flats, condominiums, shops, and other infrastructure, as well as other facilities in a settlement or city such as parks, playgrounds, walking trails, bike paths,

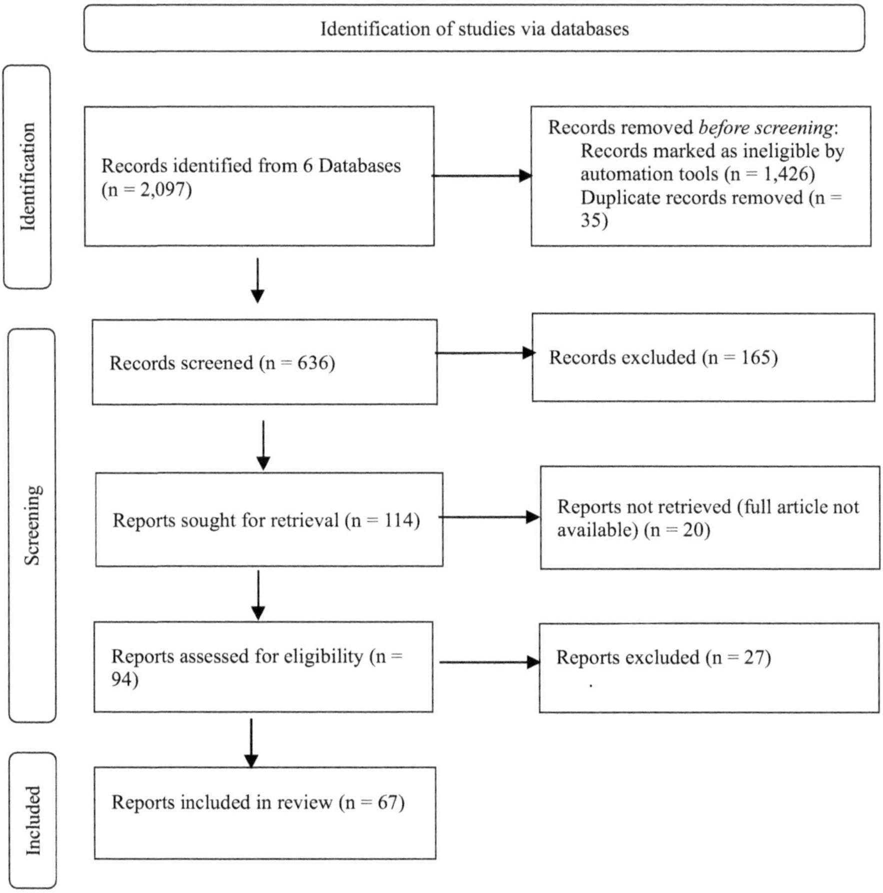

Fig. 16.1 PRISMA diagram for the search protocol, and the inclusion and exclusion of reviewed articles

etc. (Sallis et al. 2012). The extracted articles offer insights into diverse aspects of health in the built environment, from the impact of urban design on health outcomes to psychological effects. Each of these studies contributes to a better understanding of the complex interactions between the built environment and health, offering potential pathways for improving public health through planning and policies. However, our analysis shows two factors related to the built environment and health-oriented behaviour: (1) physical activity and a sense of security, and (2) mental health and psychological well-being. The results are presented in Table 16.1.

We present some of the latest findings, showcasing how the built environment can act both as a facilitator and a barrier to active leisure time for its inhabitants. The selected articles cover the built environment, including residential buildings, shops, and other infrastructure, as well as community or city objects like parks, playgrounds, walking paths, cycling paths, etc. (Armstrong et al. 2015; Benton et al. 2016; Serrano

Table 16.1 A priori focus levels of the review and identified categories (research topics)

Focus level of research and practice	Identified categories (research topics)	Sources
Physical activity and a sense of security	Walking and recreation Dimensions of the built environment Safety Body mass index and obesity prevention Age groups Gender Indoor environment Mobility/pedestrian needs Social support	Araya León et al. (2023); Armstrong et al. (2015); Atella and Kopinska (2014); Azim et al. (2023); Badland et al. (2017); Benton et al. (2016); Berge et al. (2023); Bilotta and Evans (2013); Bower et al. (2023); Bracy et al. (2014); Buszkiewicz et al. (2021); Carroll et al. (2016); Cerin et al. (2013); Cradoic and Duncan (2014); Ding et al. (2023); Daniels et al. (2021); Ferguson and Evans (2023); Figueiredo et al. (2023); Finlay et al. (2015); Foster et al. (2016); Frehlich et al. (2022); Glonti et al. (2016); Gutiérrez et al. (2024); Han et al. (2024); Haines et al (2016); Hobbs et al. (2021); Hohl and Lotfata (2022); Hughes et al. (2022); Khaddar et al. (2023); Kerr et al. (2016); Lisboa et al. (2021); Martin et al. (2014); OECD 2023a; 2023b; Oliver et al. (2015); Ottoni et al. (2016); Oyeyemi et al. (2016); Petersen et al. (2020); Pliakas et al. (2017); Putra et al. (2022); Rollings et al. (2017); Roy et al. (2023); Sadali et al. (2023); Serrano et al. (2014; 2023); Suminski et al. (2023); Sun et al. (2021); Topothai et al. (2023); Zenk et al. (2019); Zhou et al. (2020);
Mental health and psychological well-being	Age groups Loneliness Stress Dimensions of the built environment Safety Social relations	Araya León et al. (2023); Astell-Burt et al. (2014); Badland et al. (2017); Bilotta and Evans (2013); Bower et al. (2023); Christian et al. (2015); Cradoic and Duncan (2014); Dutton, (2014); Engel et al. (2016); Ferguson and Evans (2023); Figueiredo et al. (2023); Finlay et al. (2015); Hobbs et al. (2021); Hogan et al. (2016); Hooper et al. (2015); James et al. (2017); Knight et al. (2014); Liu et al. (2021); Ottoni et al. (2016); Plikas et al. (2017); Rollings et al. (2017); Tao et al. (2024); Thornton et al. (2017); Xu and Wang (2015)

et al. 2023). Transitioning from the overarching insights gleaned from our scoping review, which highlights the dual role of the built environment in promoting or hindering active leisure among its inhabitants, we delve deeper into the specific aspects of how this environment impacts individual behaviours and perceptions. We have extended the model introduced by Cradoic and Duncan (2014) by including other areas that have received more attention from researchers in the last ten years. We launch a renewed model in Fig. 16.2.

The model delineated in Fig. 16.2 encapsulates four core concepts critical to understanding the multifaceted relationship between health-oriented behaviours and the

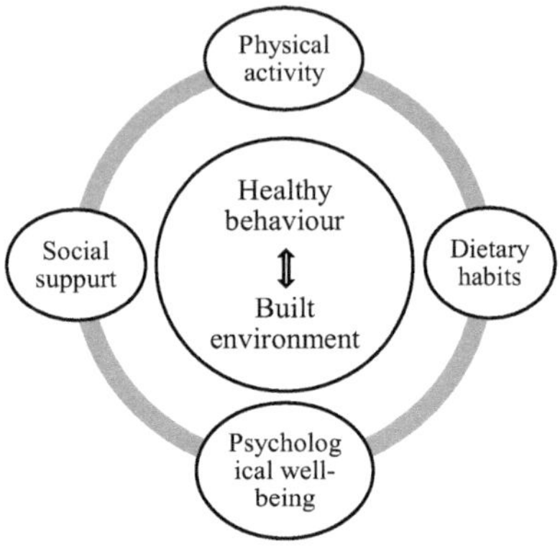

Fig. 16.2 Hypothetical model of the relationship between health-oriented behaviour and the built environment, derived from our scoping review based on the model by Cradoic and Duncan (2014)

built environment. The first of these concepts, "physical activity," is explored through the lens of urban design and its impact on residents' participation in physical activities and their perceptions of safety. The connection between the physical layout of urban areas and the personal experiences of those who navigate through them becomes particularly pertinent when considering the interplay of physical activity and security. As we examine the intricate ways in which the built environment influences residents' engagement in activities such as cycling, running, or walking, the focus shifts towards understanding the nuanced perceptions of safety and security that accompany these activities. This nuanced exploration not only builds upon the foundational knowledge established by the scoping review but also sets the stage for a detailed analysis of how urban design and policy interventions can enhance the quality of life for city dwellers. Notably, the latest findings from the Organisation for Economic Co-operation and Development (OECD 2023a) delineate Slovenia as a paragon among member countries, showcasing a populace that not only actively participates in physical activities but also perceives a heightened sense of security, particularly against crime. However, this perception is somewhat attenuated when considering traffic and environmental safety, with specific concerns directed towards air pollution.

An instrumental development in this arena is the work of Kerr et al. (2016) and Cerin et al. (2013), who introduced the Neighbourhood Environment Walkability Scale-Abbreviated (NEWS-A). This tool, an abridged version of its predecessor, evaluates seven dimensions of the built environment that bolster physical activities, chiefly walking. Employed in a myriad of studies, the NEWS-A scale has been pivotal in assessing and tailoring the built environment to enhance pedestrian mobility. The application of this scale across 17 cities has illuminated the potential for adjusting urban environments to meet residents' mobility needs, thus emphasising the importance of incorporating these needs into urban planning and development strategies.

Furthering this discourse, three seminal studies (Azim et al. 2023; Khaddar et al. 2023; Suminski et al. 2023) merit attention for their examination of transport, walkability, and indoor environmental factors affecting health across diverse population groups. Khaddar et al. (2023) delve into the interplay between travel activity, physical activity, and the built environment's attributes on health outcomes. Through the development of a latent segmentation-based logit model, this study discerns the heterogeneous impacts of travel and physical activity on health between younger suburban and older urban dwellers. The findings elucidate the nuanced relationship between vehicle reliance and health detriments in younger cohorts versus the health-affirmative correlation with the built environment experienced by older urban residents. This research offers critical insights for crafting transport policies aimed at fostering healthier urban habitats.

Simultaneously, Suminski et al. (2023) evaluate the predictive efficacy of the National Walkability Index (NWI) and Walk Score regarding physical activity in urban locales. Despite the significant correlation between Walk Score, employment/household occupancy mix, and pedestrian activity, the NWI's predictive capability on physical activity outcomes was found wanting. This discrepancy underscores the preference for a Walk Score and a modified NWI in urban planning and research endeavours. Lastly, Azim et al. (2023) address the paucity of synthesised knowledge concerning indoor environments and their impact on the daily routines of older adults. Highlighting the significance of campus, buildings, and furnishings, this systematic review accentuates the role of indoor features in promoting physical activity among the elderly. Such insights are invaluable for the conceptualisation of future housing models and research aimed at augmenting the living quality of older populations through enhanced environmental design.

Collectively, these studies unravel the complex dynamics between the built environment and health outcomes (Araya León et al. 2023; Badland et al. 2017; Oyeyemi et al. 2016; Pliakas et al. 2017; Rollings et al. 2017; Thornton et al. 2017). They advocate for integrated approaches in urban planning, transport policy, and housing design to nurture health and well-being across life stages and communities. This body of work underscores the imperative for multidisciplinary strategies that reconcile the intricate relationships between physical activity, sense of security, and health within the built environment.

Transitioning to the second concept, "a balanced diet," our discussion extends beyond the physical structure of urban spaces to consider the nutritional aspects of health and well-being. This segment underscores the critical role of dietary habits in maintaining a healthy lifestyle and its interplay with the built environment. A balanced diet also contributes to a healthy body weight and, thus, to better psychophysical well-being (Berge et al. 2023; Martin et al. 2014; Roy et al. 2023). The results of the OECD country survey (OECD 2023b) show that Slovenians are relatively well positioned compared to OECD countries in terms of a balanced diet, especially when it comes to the daily consumption of fruit and vegetables.

In addition to physical activity, maintaining an optimal body weight also contributes to a healthy lifestyle. It is already common knowledge that more exercise leads to a more balanced body weight. However, this also requires an appropriate diet and a lifestyle that, in addition to physical activity, also includes other stimulating factors for maintaining physical and mental health (Bower et al. 2023). In the built environment, it is very important that residents proactively pursue such behaviour.

Obesity is a major problem in the developed world. It is increasing not only in adults but also in children. The scoping review we conducted in this study shows that research on the relationship between the built environment and obesity has increased greatly over the past decade, demonstrating that partial changes in the built environment can provide sustainable care for a balanced population body weight (Carroll et al. 2016; Gutiérrez et al. 2024; Han et al. 2024; Sadali et al. 2023; Topothai et al. 2023). Our analysis shows that the selected studies generally emphasise three factors that have the strongest influence on body mass index, namely structural (Buszkiewicz et al. 2021; Zenk et al. 2019; Zhou et al 2020) and demographic (Atella and Kopinska 2014; Hughes et al. 2022) environment and the social (Glonti et al. 2016; Serrano et al. 2014; 2023) environment in which a person lives.

Moving on to the third concept, "social support," we explore the indispensable role of interpersonal relationships and community networks in promoting healthful behaviours. This section delves into the significance of social ties in encouraging physical activity and supporting dietary choices, highlighting the complex interplay between social environments and individual health outcomes. In exploring the significant social factors influencing physical activity and balanced diet, particularly the lack of support for movement and balanced nutrition from family members and cohabiting with a partner with obesity, several studies have made notable contributions. Some studies (Haines et al. 2016; Petersen et al. 2020) focused on the association between parent and child physical activity, emphasising the role of family structure and measurement methods of parent and child physical activity. These studies highlighted the complexity of these relationships and employed a rigorous methodological approach to assess the quality and findings of included studies, indicating that family dynamics significantly impact children's physical activity levels.

Another study by Lisboa et al. (2021) analysed the social support from family and friends for adolescents to practice physical activity. Using structural equation modelling on data from 2710 Brazilian adolescents from urban settings, it found that greater social support from both friends and family correlates with increased physical activity among adolescents. This study provides evidence that social support is crucial for encouraging physical activity, underscoring the importance of supportive relationships in promoting healthier lifestyles. Research on the built environment's impact on obesity has highlighted several critical factors contributing to weight-related health outcomes (Putra et al. 2021). The availability of fast-food outlets, the scarcity of stores offering nutritionally balanced foods, and the lack of recreational spaces such as parks, playgrounds, and walking and cycling paths play significant roles in shaping dietary and physical activity behaviours that can lead to obesity. Some studies (Ding et al. 2023; Sun et al. 2021) show that certain aspects like land use mix and urban sprawl showed some association with overweight or obesity, but the overall evidence remains weak. This indicates a need for further research to fully understand the complexities of how the built environment influences weight status. Some studies (Daniels et al. 2021; Hobbs et al. 2021; Hohl and Lotfata 2022; Sun et al. 2021) also emphasise the methodological diversity and geographical coverage of studies. They highlight the role of environmental factors related to physical activity, such as walkability and green spaces, and the food environment, illustrating a multifaceted approach to tackling obesity through urban planning and policy interventions .

These studies (Daniels et al. 2021; Ding et al. 2023; Haines et al. 2016; Hobbs et al. 2021; Hohl and Lotfata 2022; Lisboa et al. 2021; Petersen et al. 2020; Putra et al. 2021; Sun et al. 2021) collectively underscore the multifaceted nature of health-oriented behaviours in the built environment, particularly highlighting the pivotal role of social support from family and friends in fostering physical activity. They suggest that interventions aimed at improving physical activity and dietary habits should consider enhancing social support mechanisms within families and social networks to effectively promote healthier lifestyles. These findings underline the importance of designing and implementing urban environments that promote more active lifestyles and healthier food choices as part of a comprehensive strategy to combat obesity. By focusing on creating spaces that encourage physical activity and provide easier access to nutritious food options, urban planners and public health officials can work together to address the obesity epidemic from an environmental perspective.

Lastly, we address the fourth concept, "psychological well-being," focusing on the mental and emotional aspects of health within the context of the built environment. This part of the discussion brings to light the vital connections between psychological health, the physical surroundings, and the broader social context, emphasising how the design and infrastructure of urban areas can influence overall well-being. The relationship between psychological well-being and the built environment has increased over the last ten years (Araya León et al. 2023; Astell-Burt et al. 2014; Ferguson and Evans, 2023 Rollings et al. 2017). Research has mainly focussed on the life stage of inhabitants of the built environment. Our scoping review highlighted three aspects of the studies: the first focuses on late adulthood, the second on childhood, and the third on the whole life course. Specifically, we found that research (Christian et al. 2015; Dutton 2014; Engel et al. 2016; Figueiredo et al. 2023; Hogan et al. 2016; Hooper et al. 2015; Knight et al. 2014) most frequently examines the psychological well-being of children and people in late adulthood, but often covers the entire life course of specific groups. Finlay et al. (2015) find that the built environment has a significant impact on health and psychological well-being, specifically in the older population. Ottoni et al. (2016) conducted a study in which they interviewed 28 participants aged 61–89 years to determine the relationship between the built environment, their mobility within it and their psychological well-being. They found that the participants who walked more were those who had the opportunity to rest on the pavements. Through this research, the authors found that it is not only enough to have green spaces and well-maintained playgrounds and walkways but that older walkers should also be provided with park benches where they can rest. In this context, particular reference was made to the need to consider micro-factors in the built environment, such as benches in the present study, which should be considered by the planners of such environments. Such interventions are also among the most affordable.

In a systematic review study, Christian et al. (2015) analysed 32 articles on the effects of the built environment on children. They found that proximity and accessibility to various facilities for children and their parents' activities, such as library, kindergarten, playgrounds, theatre, etc., are positively associated with children's physical development, psychological well-being, and social competence. In addition, parents' sense of living in an environment that is safe for their children also contributes significantly to children's well-being and socio-emotional development. A review of research (Hogan et al. 2016) across the lifespan highlights four factors in the built environment that are most important to the psychological well-being of residents, regardless of their age. These are safety, opportunities for physical activity, especially pedestrian routes, good neighbourliness, and other social relationships, such as the opportunity to socialise outside the home and peace and quiet from noise and traffic. Other studies also report related factors but add to the attractiveness and appeal of the built environment (Bracy et al. 2014; Kerr et al. 2016).

This comprehensive exploration of the built environment through the lenses of physical activity and security, balanced diet, social support, and psychological well-being offers a holistic view of the multifarious ways in which urban design and policy can influence health outcomes. By integrating these four concepts, our model advocates for a multi-dimensional approach to urban planning that prioritises the physical, nutritional, social, and emotional dimensions of well-being.

16.4 Conclusion

This scoping review has illuminated the intricate relationship between the built environment and its impact on the health and well-being of its inhabitants. Through a comprehensive analysis of selected literature spanning various aspects of urban design, community infrastructure, and residential environments, we have identified key factors that either promote or hinder active lifestyles, psychological well-being, and optimal body weight management. The evidence suggests that the built environment plays a pivotal role in facilitating physical activity and a sense of security among its residents. Well-designed urban spaces that prioritise traffic safety and protection against crime not only encourage more people to engage in physical activities like walking, cycling, and running but also contribute significantly to the inhabitants' overall sense of well-being (Foster et al. 2016). The introduction and application of tools such as the NEWS-A (Kerr et al. 2016) scale highlight the importance of assessing and adapting urban environments to meet the mobility needs of residents, thereby promoting healthier lifestyles through enhanced walkability and cycling opportunities. Moreover, the built environment's influence extends beyond physical activity to encompass factors critical to managing body weight and fostering psychological well-being. Our review underscores the significance of structural and demographic environments, alongside the social contexts in which individuals live, in shaping health-oriented behaviours. The role of social support from family and friends in promoting physical activity and balanced dietary habits cannot be overstated, emphasising the need for interventions that enhance social support mechanisms within communities.

Our findings also shed light on the critical challenges posed by obesity, highlighting the necessity for urban planning and policy interventions that address the availability of nutritious food options and recreational spaces. Addressing these challenges requires a multifaceted approach that combines urban planning, transport policy, and housing design to create environments that support healthy living across different stages of life and communities. The relationship between psychological well-being and the built environment further emphasises the need for thoughtful urban design that considers the life stages and needs of its inhabitants. From providing resting spaces for older adults to ensuring accessibility to facilities that support children's physical and socio-emotional development, the studies reviewed highlight the importance of designing inclusive spaces that cater to the diverse needs of the community.

In conclusion, this review reinforces the necessity of adopting multi-layered strategies in the planning and development of urban environments. By prioritising the creation of spaces that promote physical activity, ensure safety, and offer social support, urban planners and policymakers can significantly enhance the health and well-being of urban populations. Future research should continue to explore the complex interactions between the built environment and health outcomes, with a focus on developing innovative solutions that address the evolving needs of diverse communities.

16.5 Data Availability Statement

The raw data supporting the conclusions of this article will be made available by the authors without undue reservation.

16.6 Conflicts of Interest

The authors declare that the research was conducted in the absence of any commercial or financial relationships that could be construed as a potential conflict of interest.

Funding Research was supported by the Slovenian research agency (grant numbers J5-3112, J7-4599, and P5-0110).

Institutional Review Board Statement Not applicable.

References

Araya León MJ, Guasch R, Estévez AT, Peña J (2023) Interaction between the interior built environment and the human being is an integrative review in relation to perception, health, and well-being. Theor Issues Ergon Sci 24(6):698–728. https://doi.org/10.1080/1463922X.2022.2134940

Arksey H, O'Malley L (2005) Scoping studies: towards a methodological framework. Int J Soc Res Methodol 8(1):19–32. https://doi.org/10.1080/1364557032000119616

Armstrong B, Lim CS, Janicke DM (2015) Park density impacts weight change in behavioural intervention for overweight rural youth. Behav Med 41(3):123–130. https://doi.org/10.1080/08964289.2015.1029428

Astell-Burt T, Mitchell R, Hartig T (2014) The association between green space and mental health varies across the lifecourse: a longitudinal study. J Epidemiol Community Health 68(6):578–583. https://doi.org/10.1136/jech-2013-203767

Atella V, Kopinska J (2014) Body weight, eating patterns, and physical activity: the role of education. Demography 51(4):1225–1249. https://doi.org/10.1007/s13524-014-0311-z

Azim FT, Ariza-Vega P, Gardiner PA, Ashe MC (2023) Indoor built environment and older adults' activity: a systematic review. Can J Aging 42(2):241–258. https://doi.org/10.1017/S0714980822000241

Badland H, Foster S, Bentley R, Higgs C, Roberts R, Pettit C, Giles-Corti B (2017) Examining associations between area-level spatial measures of housing with selected health and wellbeing behaviours and outcomes in an urban context. Health Place 43:17–24

Benton JS, Anderson J, Hunter RF, French DP (2016) The effect of changing the built environment on physical activity: a quantitative review of the risk of bias in natural experiments. Int J Behav Nutr Phys Act 13:1–18

Berge JM, Hazzard VM, Trofholz A, Noser AE, Hochgraf A, Neumark-Sztainer D (2023) Longitudinal associations between family meal quality and quantity: does one matter more for child, parent, and family health and well-being or are they synergistic? Appetite 191:1–10. https://doi.org/10.1016/j.appet.2023.107080

Bilotta E, Evans GW (2013) Environmental stress. In: Steg L, Ven den Berg AE, de Groot JIM (eds) Environmental psychology: an introduction. BPS Blackwell, pp 27–35

Bower M, Kent J, Patulny R, Green O, McGrath L, Teesson L, Jamalishahni T, Sandison H, Rugel E (2023) The impact of the built environment on loneliness: a systematic review and narrative synthesis. Health Place 79:1–15. https://doi.org/10.1016/j.healthplace.2022.102962

Bracy NL, Millstein RA, Carlson JA, Conway TL, Sallis JF, Saelens BE, King AC (2014) Is the relationship between the built environment and physical activity moderated by perceptions of crime and safety? Int J Behav Nutr Phys Act 1:11–13. https://doi.org/10.1186/1479-5868-11-24

Buszkiewicz JH, Bobb JF, Hurvitz PM, Arterburn D, Moudon AV, Cook A, Mooney SJ, Cruz M, Gupta S, Lozano P, Rosenberg DE, Theis MK, Anau J, Drewnowski A (2021) Does the built environment have independent obesogenic power? Urban form and trajectories of weight gain. Int J Obes 45(9):1914–1924. https://doi.org/10.1038/s41366-021-00836-z

Carroll SJ, Paquet C, Howard NJ, Coffee NT, Taylor AW, Niyonsenga T, Daniel M (2016) Local descriptive norms for overweight/obesity and physical inactivity, features of the built environment, and 10-year change in glycosylated haemoglobin in an Australian population-based biomedical cohort. Soc Sci Med 166:233–243. https://doi.org/10.1016/j.socscimed.2016.08.031

Cerin E, Conway TL, Cain KL, Kerr J, De Bourdeaudhuij I, Owen N, Reis RS, Sarmiento OL, Hinckson EA, Salvo D, Christiansen LB, Macfarlane DJ, Davey R, Mitáš J, Aguinaga-Ontoso I, Sallis JF (2013) Sharing good NEWS across the world: developing comparable scores across 12 countries for the Neighborhood Environment Walkability Scale (NEWS). BMC Public Health 13:309. https://doi.org/10.1186/1471-2458-13-309

Christian H, Zubrick SR, Foster S, Giles-Corti B, Bull F, Wood L, Boruff B (2015) The influence of the neighborhood physical environment on early child health and development: a review and call for research. Health Place 33:25–36. https://doi.org/10.1016/j.healthplace.2015.01.005

Cradock AL, Duncan DT (2014) The Role of the built environment in supporting health behavior change. In: Riekert KA, Ockene JK, Pbert L (eds) The handbook of health behavior change. Springer Publishing Company, pp 437–462

Daniels KM, Schinasi LH, Auchincloss AH, Forrest CB, Diez Roux AV (2021) The built and social neighborhood environment and child obesity: a systematic review of longitudinal studies. Prev Med: Int J Devoted Pract Theory 153:106790. https://doi.org/10.1016/j.ypmed.2021.106790

Ding Y, Ma L, Jia P, Xue H, Xu F, Zhao L, Xue B, Wang Y, Wang Y (2023) Longitudinal effects of school policies on children's eating behaviors and weight status: findings from the childhood obesity study in China megacities. Int J Obes 47(1):17–23. https://doi.org/10.1038/s41366-022-01206-z

Dutton R (2014) The built housing environment, wellbeing, and older people. In: Cooper R, Burton E, Cooper CL, Cooper R, Burton E, Cooper CL (eds) Wellbeing and the environment, vol II. Wiley-Blackwell, pp 335–372. https://doi.org/10.1002/9781118539415.wbwell067

Engel L, Chudyk AM, Ashe MC, McKay HA, Whitehurst DT, Bryan S (2016) Older adults' quality of life—exploring the role of the built environment and social cohesion in community-dwelling seniors on low income. Soc Sci Med 164:1–11. https://doi.org/10.1016/j.socscimed.2016.07.008

Ferguson KT, Evans GW (2023) Built environment, family processes, and child and adolescent health and well-being. In Ortiz SE, McHale SM, King V, Glick JE (eds) Environmental impacts on families: change, challenge, and adaptation, vol. 12. Springer Nature Switzerland AG, pp 87–127 https://doi.org/10.1007/978-3-031-22649-6_5

Figueiredo M, Eloy S, Marques S, Dias L (2023) Older people perceptions on the built environment: a scoping review. Appl Ergon 108:1–10. https://doi.org/10.1016/j.apergo.2022.103951

Finlay J, Franke T, McKay H, Sims-Gould J (2015) Therapeutic landscapes and wellbeing in later life: impacts of blue and green spaces for older adults. Health Place 34:97–106. https://doi.org/10.1016/j.healthplace.2015.05.001

Foster S, Hooper P, Knuiman M, Christian H, Bull F, Giles-Corti B (2016) Safe RESIDential Environments? A longitudinal analysis of the influence of crime-related safety on walking. Int J Behav Nutr Phys Act 13(1):22. https://doi.org/10.1186/s12966-016-0343-4

Frehlich L, Christie CD, Ronksley PE, Turin TC, Doyle-Baker P, McCormack GR (2022) The neighbourhood built environment and health-related fitness: a narrative systematic review. Int J Behav Nutr Phys Act 19:124. https://doi.org/10.1186/s12966-022-01359-0

Glonti K, Mackenbach JD, Ng J, Lakerveld J, Oppert JM, Bárdos H, McKee M, Rutter H (2016) Psychosocial environment: definitions, measures and associations with weight status–a systematic review. Obesity Rev 17(Suppl 1):81–95. https://doi.org/10.1111/obr.12383

Gutiérrez GE, García SM, Pastor BR, Fernández de Larrea BN, Rollán GA, Peñalver AB, Peña RI, Pollán M, Pérez GB (2024) A nation-wide analysis of socioeconomic and geographical disparities in the prevalence of obesity and excess weight in children and adolescents in Spain: results from the ENE-COVID study. Pediatr Obes 19(1):e13085. https://doi.org/10.1111/ijpo.13085

Haines J, Rifas-Shiman SL, Horton NJ, Kleinman K, Bauer KW, Davison KK, Walton K, Austin SB, Field AE, Gillman MW (2016) Family functioning and quality of parent-adolescent relationship: cross-sectional associations with adolescent weight-related behaviors and weight status. Int J Behav Nutr Phys Act 13:1–12

Han Y, Sung H, Choi Y, Kim YS (2024) Trends in obesity, leisure-time physical activity, and sedentary behavior in Korean adults: Korea national health and nutritional examinations survey from 2014 to 2021. PLoS ONE 19(1). https://doi.org/10.1371/journal.pone.0296042

Hohl A, Lotfata A (2022) Modeling spatiotemporal associations of obesity prevalence with biking, housing cost and green spaces in Chicago, IL, USA, 2015–2017. J Transp Health 26:1–10. https://doi.org/10.1016/j.jth.2022.101412

Hobbs M, Kingham S, Wiki J, Marek L, Campbell M (2021) Unhealthy environments are associated with adverse mental health and psychological distress: cross-sectional evidence from nationally representative data in New Zealand. Prev Med: Int J Devoted Pract Theory 145:106416. https://doi.org/10.1016/j.ypmed.2020.106416

Hogan MJ, Leyden KM, Conway R, Goldberg A, Walsh D, McKenna-Plumley PE (2016) Happiness and health across the lifespan in five major cities: the impact of place and government performance. Soc Sci Med 162:168–176. https://doi.org/10.1016/j.socscimed.2016.06.030

Hooper CM, Ivory VC, Fougere G (2015) Childhood neighbourhoods as third places: developing durable skills and preferences that enhance wellbeing in adulthood. Health Place 34:34–45. https://doi.org/10.1016/j.healthplace.2015.03.017

Hughes AM, Sanderson E, Morris T, Ayorech Z, Tesli M, Ask H, Reichborn-Kjennerud T, Andreassen OA, Magnus P, Helgeland Ø, Johansson S, Njølstad P, Davey Smith G, Havdahl A, Howe LD, Davies NM (2022) Body mass index and childhood symptoms of depression, anxiety, and attention-deficit hyperactivity disorder: a within-family Mendelian randomization study. eLife, 11:e74320 https://doi.org/10.7554/eLife.74320

James P, Hart JE, Banay RF, Laden F, Signorello LB (2017) Built Environment and depression in low-income African Americans and Whites. Am J Prev Med 52(1):74–84. https://doi.org/10.1016/j.amepre.2016.08.022

Kerr J, Emond JA, Badland H, Reis R, Sarmiento O, Carlson J, Sallis JF, Cerin E, Cain K, Conway T, Schofield G, Macfarlane DJ, Christiansen LB, Van Dyck D, Davey R, Aguinaga-Ontoso I, Salvo D, Sugiyama T, Owen N, Mitáš J, Natarajan L (2016) Perceived neighborhood environmental attributes associated with walking and cycling for transport among adult residents of 17 cities in 12 countries: the IPEN study. Environ Health Perspect 124(3):290–298. https://doi.org/10.1289/ehp.1409466

Khaddar S, Rahman Fatmi M, Winters M (2023) How daily activities and built environment affect health? A latent segmentation-based random parameter logit modeling approach. Travel Behav Soc 33:100624. https://doi.org/10.1016/j.tbs.2023.100624

Knight KR, Lopez AM, Comfort M, Shumway M, Cohen J, Riley ED (2014) Single room occupancy (SRO) hotels as mental health risk environments among impoverished women: the intersection of policy, drug use, trauma, and urban space. Int J Drug Policy 25(3):556–561. https://doi.org/10.1016/j.drugpo.2013.10.011

Koh HK, Alpert HR, Judge CM, Caughey RW, Elqura LJ, Connolly GN, Warren CW (2011) Understanding worldwide youth attitudes towards smoke-free policies: an analysis of the Global Youth Tobacco Survey. Tob Control 20(3):219–225. https://doi.org/10.1136/tc.2010.038885

Lisboa T, Silva WRD, Silva DAS, Felden ÉPG, Pelegrini A, Lopes JJD, Beltrame TS (2021) Social support from family and friends for physical activity in adolescence: analysis with structural equation modeling. Cad Saude Publica 37(1):e00196819. https://doi.org/10.1590/0102-311X00196819

Liu C, Zhang L, Ahmed F, Shahid M, Wang X, Wang Y, Wang J, Guo J (2021) Family care, economic stress, and depressive symptoms among Chinese adults during the COVID-19 outbreak: difference by urban and rural areas. Front Psych 12:700493. https://doi.org/10.3389/fpsyt.2021.700493

Martin A, Ogilvie D, Suhrcke M (2014) Evaluating causal relationships between urban built environment characteristics and obesity: a methodological review of observational studies. Int J Behav Nutr Phys Act 11(1):142. https://doi.org/10.1186/s12966-014-0142-8

Moher D, Liberati A, Tetzlaff J, Altman DG, PRISMA Group (2009) Preferred reporting items for systematic reviews and meta-analyses: the PRISMA statement. PLoS Medicine 6(7):e1000097. https://doi.org/10.1371/journal.pmed.1000097

OECD (2023a) Health at a Glance 2023: OECD Indicators, OECD Publishing, Paris. https://www.oecd.org/health/health-at-a-glance/

OECD (2023) How's life? 2023: measuring well-being. OECD Publishing, Paris. https://doi.org/10.1787/how_life-2017-en

Oliver M, Mavoa S, Badland H, Parker K, Donovan P, Kearns RA, Lin E-Y, Witten K (2015) Associations between the neighbourhood built environment and out of school physical activity and active travel: an examination from the Kids in the City study. Health Place 36:57–64. https://doi.org/10.1016/j.healthplace.2015.09.005

Ottoni CA, Sims-Gould J, Winters M, Heijnen M, McKay HA (2016) "Benches become like porches": Built and social environment influences on older adults' experiences of mobility and well-being. Soc Sci Med 169:33–41. https://doi.org/10.1016/j.socscimed.2016.08.044

Oyeyemi AL, Kasoma SS, Onywera VO, Assah F, Adedoyin RA, Conway TL, Moss SJ, Ocansey R, Kolbe-Alexander TL, Akinroye KK, Prista A, Larouche R, Gavand KA, Cain KL, Lambert EV, Aryeetey R, Bartels C, Tremblay MS, Sallis JF (2016) NEWS for Africa: adaptation and reliability of a built environment questionnaire for physical activity in seven African countries. Int J Behav Nutr Phys Act 13(1):33. https://doi.org/10.1186/s12966-016-0357-y

Page MJ, McKenzie JE, Bossuyt PM, Boutron I, Hoffmann TC, Mulrow CD, Shamseer L, Tetzlaff JM, Akl EA, Brennan SE, Chou R, Glanville J, Grimshaw JM, Hróbjartsson A, Lalu MM, Li T, Loder EW, Mayo-Wilson E, McDonald S, Moher D (2021) The PRISMA 2020 statement: an updated guideline for reporting systematic reviews. BMJ 372.https://doi.org/10.1136/bmj.n71

Petersen TL, Møller LB, Brønd JC, Jepsen R, Grøntved A (2020) Association between parent and child physical activity: a systematic review. Int J Behav Nutr Phys Act 17(1):67. https://doi.org/10.1186/s12966-020-00966-z

Pliakas T, Hawkesworth S, Silverwood RJ, Nanchahal K, Grundy C, Armstrong B, Casas JP, Morris RW, Wilkinson P, Lock K (2017) Optimising measurement of health-related characteristics of the built environment: comparing data collected by foot-based street audits, virtual street audits and routine secondary data sources. Health Place 43:75–84. https://doi.org/10.1016/j.healthplace.2016.10.001

Putra IGNE, Astell-Burt T, Feng X (2022) Association between built environments and weight status: evidence from longitudinal data of 9589 Australian children. Int J Obes 46(8):1534–1543. https://doi.org/10.1038/s41366-022-01148-6

Rollings KA, Wells NM, Evans GW, Bednarz A, Yang Y (2017) Housing and neighborhood physical quality: children's mental health and motivation. J Environ Psychol 50:17–23. https://doi.org/10.1016/j.jenvp.2017.01.004

Roy A-S, Mazaniello-Chézol M, Rueda-Martinez M, Shafique S, Adams AM (2023) Food systems determinants of nutritional health and wellbeing in urban informal settlements: a scoping review in LMICs. Soc Sci Med 322:1–14. https://doi.org/10.1016/j.socscimed.2023.115804

Sadali UB, Kamal KKBN, Park J, Chew HSJ, Devi MK (2023) The global prevalence of overweight and obesity among nurses: a systematic review and meta-analyses. J Clin Nurs 32(23–24):7934–7955. https://doi.org/10.1111/jocn.16861

Sallis JF, Floyd MF, Rodríguez DA, Saelens BE (2012) Role of built environments in physical activity, obesity, and cardiovascular disease. Circulation 125(5):729–737. https://doi.org/10.1161/CIRCULATIONAHA.110.969022

Serrano M, Torres R, Pérez CM, Palacios C (2014) Social environment factors, diet quality, and body weight in 12-year-old children from four public schools in Puerto Rico. P R Health Sci J 33(2):80–87

Serrano N, Leider J, Chriqui JF (2023) Pedestrian-oriented zoning moderates the relationship between racialized economic segregation and active travel to work, United States. Prev Med: Int J Devoted Pract Theory 177:1–4. https://doi.org/10.1016/j.ypmed.2023.107788

Suminski RR, Kelly K, Plautz E (2023) The associations between place-based measures of walkability and physical activity across a range of diverse streetscapes. Prev Med: Int J Devoted Pract Theory 169:1–8. https://doi.org/10.1016/j.ypmed.2023.107454

Sun X, Zhao B, Liu J, Wang Y, Xu F, Wang Y, Xue H (2021) A 3-year longitudinal study of the association of physical activity and sedentary behaviours with childhood obesity in China: the childhood obesity study in China mega-cities. Pediatr Obes 16(6):e12753. https://doi.org/10.1111/ijpo.12753

Tao Y, van Ham M, Petrović A (2024) Changes in commuting mode and the relationship with psychological stress: a quasi-longitudinal analysis in urbanizing China. Travel Behav Soc 34:100667. https://doi.org/10.1016/j.tbs.2023.100667

Thornton CM, Kerr J, Conway TL, Saelens BE, Sallis JF, Ahn DK, Frank LD, Cain KL, King AC (2017) Physical activity in older adults: an ecological approach. Ann Behav Med 51(2):159–169. https://doi.org/10.1007/s12160-016-9837-1

Topothai T, Tangcharoensathien V, Edney SM, Suphanchaimat R, Lekagul A, Waleewong O, Topothai C, Kulthanmanusorn A, Müller-Riemenschneider F (2023) Patterns and correlates of physical activity and sedentary behavior among Bangkok residents: a cross-sectional study. PLoS ONE 18(10):e0292222. https://doi.org/10.1371/journal.pone.0292222

West JH, Blumberg EJ, Kelley NJ, Hill L, Sipan CL, Schmitz KE, Ryan S, Clapp JD, Hovell MF (2010) Does proximity to retailers influence alcohol and tobacco use among Latino adolescents? J Immigr Minor Health 12(5):626–633. https://doi.org/10.1007/s10903-009-9303-2

Xu Y, Wang F (2015) Built environment and obesity by urbanicity in the U.S. Health Place 34:19–29. https://doi.org/10.1016/j.healthplace.2015.03.010

Zenk SN, Tarlov E, Wing C, Slater S, Jones KK, Fitzgibbon M, Powell LM (2019) Does the built environment influence the effectiveness of behavioral weight management interventions? Prev Med 126:105776. https://doi.org/10.1016/j.ypmed.2019.105776

Zhou Y, Buck C, Maier W, von Lengerke T, Walter U, Dreier M (2020) Built environment and childhood weight status: a multi-level study using population-based data in the city of Hannover, Germany. Int J Environ Res Public Health 17(8):2694. https://doi.org/10.3390/ijerph17082694

Part VII
Micro Scale: Affordable and Sustainable Social Housing

Chapter 17
Extending the Lifespan of the Post-War Housing Stock—Multi-criteria Renovation

Ljiljana Đukanović and Bojana Lević

Abstract This research presents the possibilities of extending the lifespan of multi-family residential buildings constructed during the mass construction period through the application of comprehensive renovation. Buildings constructed between 1946 and 1970 constitute one-quarter of Serbia's total multi-family housing stock; however, their lifespan is questioned due to their energy, spatial, and functional deficiencies. The study analyses the application of the comprehensive renovation methodology through volumetric additions, aiming to achieve energy, spatial, and functional improvements to the building to enhance residential comfort and extend its lifespan. The research includes an analysis of a selected freestanding multi-family residential building and two variants of comprehensive building renovation using different volumetric additions: continuous and partial additions. The study examines the possibilities of achieving the highest possible energy class by improving the building's thermal envelope, increasing the usable heated area of residential units, spatial reorganisation of apartments, forming private open spaces for apartments, as well as enhancing vertical building communication by adding an elevator. The research aims to determine the advantages of implementing multi-criteria building renovations to extend the building's lifespan.

Keywords Multi-family housing stock · Renovation · Energy efficiency · Spatial comfort · Volumetric additions

17.1 Introduction

This research explores the possibilities of multi-criteria renovation of multi-family residential buildings constructed during the mass construction period in Serbia. Multi-criteria renovation encompasses energy, spatial, and functional renovation of the building.

L. Đukanović (✉) · B. Lević
Faculty of Architecture, University of Belgrade, Belgrade, Serbia
e-mail: djuli@arh.bg.ac.rs

M. Vukmirović et al. (eds.), *Digital Dreams, Urban Realities*, The Urban Book Series,
https://doi.org/10.1007/978-3-031-76353-3_17

The energy renovation of the existing building stock is a current topic in Europe. The renovation of the existing housing stock is a key factor in the task set by the European Commission for the period up to 2050 because the share of existing buildings is much higher than that of newly designed ones in the total building stock of the European Union. Many renovation strategies are based on energy-saving measures, such as retrofitting existing buildings to improve the building's thermal envelope (European Commission 2020). Serbia, although not a member of the EU, has committed to the "Green Agenda for the Western Balkans" as a concrete plan to expand the European Green Deal to Southeast Europe. The objectives of the Green Agenda in Serbia are to enhance the strategic and legislative framework and co-finance the implementation of innovative pilot projects in practice that contribute to Serbia's green transformation (European Commission 2020). One of the most important aspects of this strategy is the utilisation of existing buildings, as this sector has a high potential for primary energy consumption savings. The renovation of the existing housing stock has significant potential for implementing measures to increase energy efficiency and save energy (Acuner and Onaygil 2015).

Increasing the energy efficiency of existing buildings is one of the most common reasons for building renovation. However, in addition to improving the energy aspect of the building, it is also crucial to enhance other aspects in order to enable a longer lifespan of the building. The lifespan of a building can be observed through physical parameters, such as the durability of embedded materials and components, the durability of the structure and its stability, and functional parameters related to the functionality of the building. It is closely linked to current building standards and user needs (Thomsen and Straub 2018).

Multi-family residential buildings constructed between 1946 and 1970 constitute one-quarter of the total multi-family housing stock in Serbia (Jovanović Popović et al. 2013). Based on analyses of the energy characteristics of existing multi-family buildings conducted in the Atlas of multi-family buildings in Serbia, the majority of buildings from this period are characterised by high energy consumption for heating due to poor thermal insulation (Jovanović Popović et al. 2013). During their construction, there were no thermal regulations for residential buildings. From today's perspective, these buildings do not meet the requirements for energy efficiency, and it is recommended that energy retrofitting be improved by at least one energy class (Ministry of Environmental Protection and Government of the Republic of Serbia 2011). In addition to significant energy deficiencies, these buildings also exhibit significant spatial and organisational shortcomings in their residential units. Due to the need to build a large number of apartments during the mass construction period after World War II, these buildings are characterised by cost-cutting measures in construction, resulting in small and cramped apartments, rigid spatial organisation, small windows, lack of balconies, and elevators (Lj 2021). From today's perspective of spatial comfort in residential buildings, these buildings have significant shortcomings (Lević et al. 2023) that affect the lifespan of the building in terms of functionality.

The paper analyses the application of the comprehensive renovation methodology through volumetric addition, i.e., increasing the usable heated space, to achieve energy, spatial, and functional improvement of the building. The research includes an analysis of a selected freestanding residential multi-family building and two variants of complex building renovation using different volumetric additions to the building: partial and continuous additions to the building facades and adding an additional floor to the building roof in both variants. The focus of the research is the analysis of possibilities for achieving the highest possible energy class by improving the building's thermal envelope by increasing the usable heated area of residential units. The paper also analyses and compares the advantages and disadvantages of the two-volume edition variants and spatial reorganisation of apartments, aiming to enhance spatial comfort and extend the functional lifespan of the building. In addition to improving residential units by increasing usable space, spatial reorganisation, and adding private open spaces, the functional lifespan of the building is extended by improving vertical building communications through the addition of an elevator. The aim of the research is to determine the advantages of implementing multi-criteria building renovation to extend the building's lifespan.

17.1.1 Strategy of Volumetric Addition to an Existing Building

The strategy of adding volume to an existing building represents integrated renovation aimed at improving the building from multiple aspects. The term "volumetric addition" can be explained as adding a new structure to an existing building (Lević et al. 2023). This strategy can be applied if there is a possibility of expanding the facade or upgrading the existing roof and if urban parameters allow for increasing the building's volume. By adding missing functional elements to the existing building, such as elevators, fire escapes, or balconies, its spatial-functional quality can be significantly improved, thus enhancing residential comfort (Guidolin 2018).

17.2 Research Methodology

The research examines the possibilities of applying the volumetric addition strategy to a typical freestanding multi-family building constructed during the mass construction period. Firstly, an analysis of the spatial organisation, functionality, and construction of the existing condition of the selected building is conducted. Using the Knaufterm software to calculate energy efficiency and determine the energy class of buildings in Serbia, the energy performance of the building is assessed. Subsequently, two variants of building improvement solutions are analysed by applying different volumetric additions to the facades of the building. One variant demonstrates the application of continuous volumetric additions to all floors of the building, while the other variant shows the application of partial volumetric additions. Both variants also apply continuous volumetric addition to the roof of the building. The energy performance is calculated for both improvement variants, and the obtained results are compared. Then, a comparative analysis of the spatial comfort of the existing state of the building and both improvements is conducted. Finally, the advantages and disadvantages of applying different variants of the volumetric addition strategy to the existing building are determined.

17.2.1 Strategy of Volumetric Addition to an Existing Building

Volumetric additions can be divided into volumetric additions applied to the building facades and on the building roof. By applying volumetric additions to the building facades, the existing usable living space can be increased, allowing for the improvement of spatial comfort in minimal-sized apartments, typical for buildings from the mass construction period. By applying volumetric additions to the building's roof, new housing units can be formed. Increasing the number of building floors is possible if static calculations show that the building is suitable for such intervention. When it comes to volumetric additions to building facades, the division can be made into continuous and partial volumetric additions. Continuous volumetric additions directly rely on the ground and represent independent structures as they pass through all building floors. However, partial volumetric additions represent structures added only to certain segments of the facade on individual floors, thus depending on the structure and stability of the building. In addition to volumetric additions, which aim to increase the usable heated space, users often demand the expansion of open private spaces (balconies and loggias). Adding surfaces to form open private spaces can also be continuous through all floors, relying on the ground and columns, or partially cantilevered on individual floors (Fig. 17.1). Due to the complexity of static calculations of building stability and the general feasibility of executing partial volumetric and surface additions, this research considers the implementation of these addition variants through the gradual placement of added structures on floors, allowing for direct support of the structure on the ground and walls of the volumetric additions.

17.2.2 The Selected Freestanding Multi-family Residential Building

The selected building was constructed in 1965 on the outskirts of Belgrade, where the construction of a large residential area with multi-family residential buildings was planned. This settlement was intended for the working class and consisted of typical residential blocks and freestanding buildings surrounded by large green areas. There are 25 identical freestanding multi-family buildings within an open city block (Fig. 17.2a).

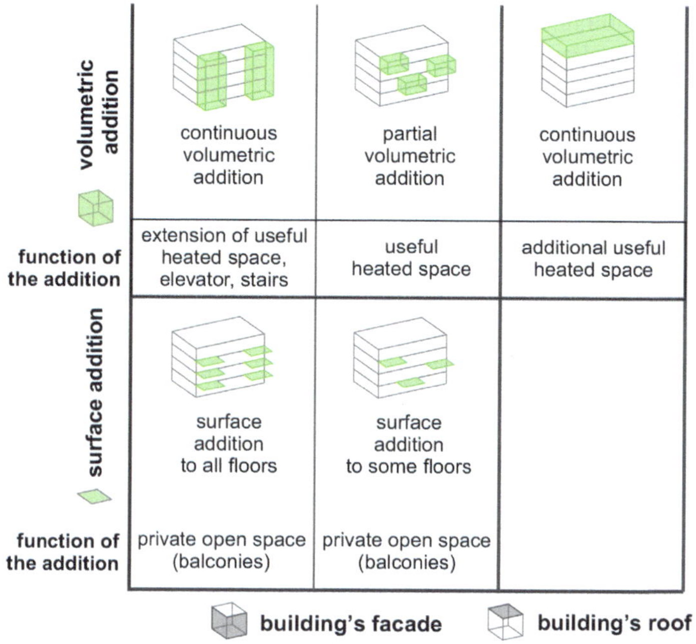

Fig. 17.1 Variants of positions of volumetric additions on an existing building

a b

Fig. 17.2 **a** Grouping of 25 identical freestanding multi-family buildings in Belgrade; **b** the existing state of the selected freestanding multi-family building

The selected building is freestanding, with five floors for residential purposes and an unheated basement. The building contains a centrally located staircase providing access to four one-bedroom apartments and four studios on each floor. The studios are around 21.5 m^2 and consist of an entrance area, a small bathroom, and one combined living room. One-bedroom apartments are around 38.5 m^2 and comprise an entrance area, a small bathroom, a kitchen, and a dining area leading to a combined living and sleeping room. Studios have a terrace, while one-bedroom apartments do not. The apartments are cramped, non-functional, and built to a minimum spatial standard. The building does not have an elevator.

The building was constructed using a massive structural system with load-bearing walls made of hollow bricks, 29 cm thick on the facades and 19 cm thick in the interior. The system was further reinforced for seismic effects by installing vertical reinforced concrete ties measuring 20/50 cm and horizontal ties at ceiling level. The inter-floor structure consists of reinforced concrete slabs 12 cm thick, supported in two directions. The partition walls are 7 cm thick. The roof structure is wooden, with a slight pitch, covered with flat sheet metal. Originally, thermal insulation was installed below the basement ceiling and above the ceiling slab, with a thickness of 5 cm. The facade walls are not thermally insulated. The windows are originally double wooden frames, and the doors are wooden.

Over time, the need for increased living space led to the enclosure of balconies and, in some cases, the addition of attic floors, but this did not alleviate the housing shortage faced by residents. As a result of individual interventions by residents, minimal energy renovations were carried out on most buildings, with the partial addition of thermal insulation to facade walls but of insufficient thickness, failing to meet modern energy efficiency standards. Additionally, some residents replaced wooden windows with PVC windows (Fig. 17.2b). Aesthetically, this has compromised the appearance of the buildings because the facade renovation was not uniform but partial, resulting in variations in facade colour and different types of windows. Consequently, the buildings have become eyesores, and the fundamental problem has not been solved, neither in terms of spatial comfort nor thermal comfort.

A proposal for comprehensive renovation of these buildings would involve expanding the living space to enhance residential comfort, as well as energy retrofitting aimed at improving thermal comfort. Given that the buildings are free-standing and located within open green areas, it is possible to expand the building footprint without disrupting the surroundings.

17.2.3 Improvement Solutions for Extending the Lifespan of the Selected Building

The first variant demonstrates the application of continuous volumetric addition. In larger apartments, two spatial structures are added to create apartments with two bedrooms, while the studio apartment is enlarged to the width of the existing balconies. Large open private spaces are planned between the added structures, serving as green gardens. All apartments have access to a private green garden (Fig. 17.3).

The second variant illustrates the application of partial volumetric addition. Spatial structures are placed in every other structural bay. This creates apartments with

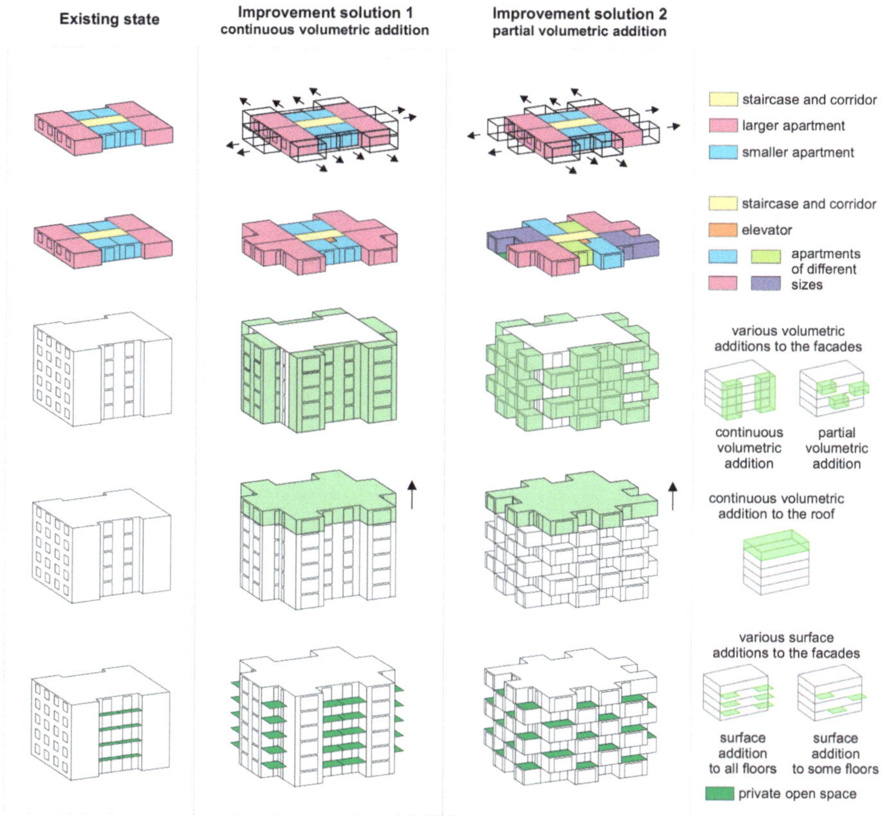

Fig. 17.3 Diagram of the application of different variants of volumetric addition to improvement solutions

different spatial structures and sizes. From the original two types of apartments, four types of apartments with different spatial structures are obtained. On the next floor, the spatial structures are shifted by one structural bay so that the added structure rests on the walls of the spatial structures below it. As in the previous variant, large open private spaces—green gardens—are planned between the added spatial structures (Fig. 17.3).

In both variants, the spatial organisation of residential units has been redesigned. In improvement solution 1, the existing apartment type A, sized 38.5 m^2, was increased to 68.2 m^2. The apartment type A now features a large living area with a kitchen, dining area, and living room, as well as a separate sleeping area with two bedrooms. The living room is connected to a large open balcony space. The existing apartment type B in this variant was expanded to the width of the original balcony, increasing its area from 21.5 m^2 to 27.3 m^2. The spatial organisation remains the same, but the living room has been enlarged. Additionally, the apartment now has access to a large balcony (Fig. 17.4).

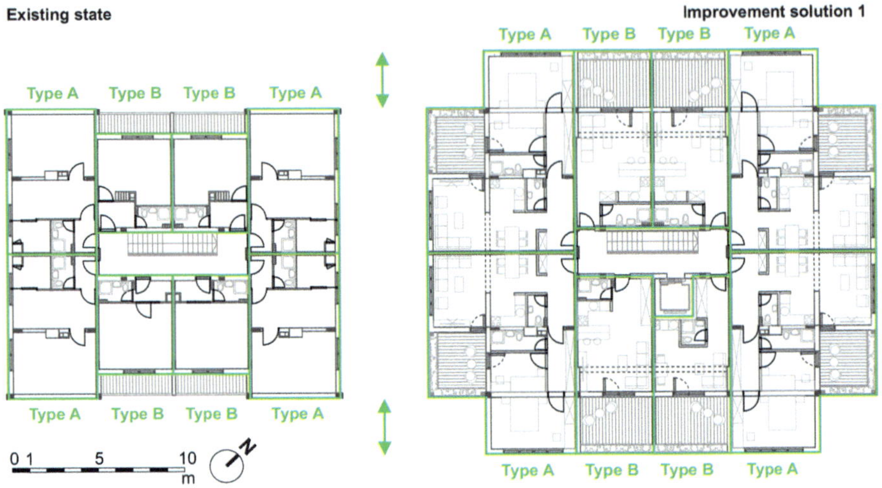

Fig. 17.4 Typical floor plan of the existing state of the building and the improvement solution 1

In improvement solution 2, four types of apartments were formed. Apartment types A and D were created from the existing apartment type A. The difference between these apartments in improvement solution 2 is that apartment type D has two bedrooms and is 68.2 m^2, while apartment type A has one bedroom and is 50.4 m^2 but also has two large balconies. This difference arose due to the rhythmic alternation of volumetric and surface additions. Apartment types B and C were formed from the existing apartment types B and C. Apartment type B is now 40.3 m^2, and apartment type C is 27.5 m^2. All apartments feature modern standards of spatial comfort with an open living area connected to the kitchen, dining area, and living room (Fig. 17.5).

The addition of an elevator in the central part of the building hallway is planned in both improvement solution variants.

17.3 Results

The application of the volumetric addition strategy to the renovation of existing buildings has multi-criteria benefits. In addition to improving the spatial organisation of residential units, an important aspect of renovation is enhancing the energy efficiency of the building by improving the thermal envelope. Various parameters influence the calculation of heating energy consumption: the building envelope surface area, the ratio of the building volume to the building envelope surface area, the total usable heated area, and the building's compactness (Ministry of Environmental Protection and Government of the Republic of Serbia 2011).

In both improvement variants, the overall thermal envelope of the building has been enhanced so that the heat transfer coefficients for all positions of the building

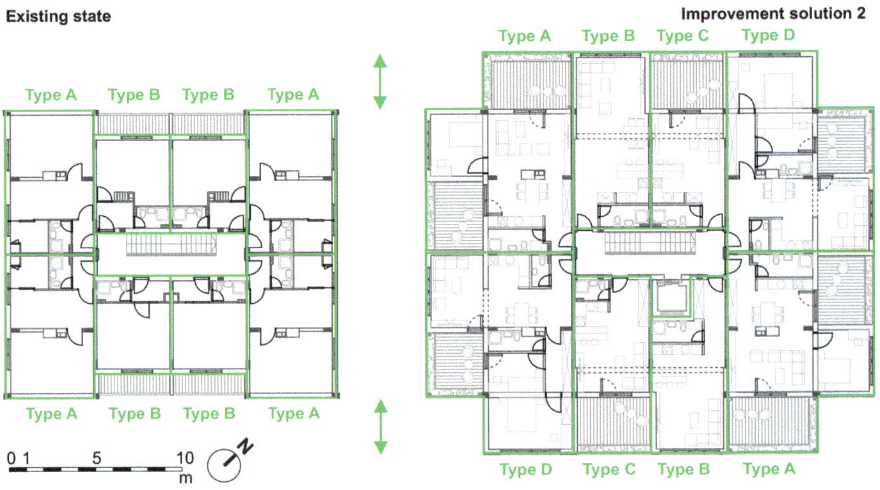

Fig. 17.5 Typical floor plan of the existing state of the building and the improvement solution 2

envelope meet the maximum requirements specified in the Rulebook on Energy Efficiency in Serbia. The same improvements to the thermal envelope of the building were applied in both variants. The existing facade wall has been improved by adding 15 cm of stone mineral wool insulation. The inter-floor ceiling towards the unheated basement has been enhanced with an additional 10 cm of stone mineral wool. The wall towards the unheated staircase has also been improved by adding 5 cm of stone mineral wool. Aluminium-frame windows with enhanced thermal breaks and triple glazing are planned for the windows. Metal-insulated doors are planned for doors leading to unheated spaces. In addition to the mentioned positions of the thermal envelope of the building, which are present in the existing building's thermal envelope and in both improvements, additional thermal envelope elements are introduced in the improvements. The new thermal envelope elements are the facade walls, the inter-floor construction towards the exterior, and the flat roof, which are elements of volumetric additions applied in both improvement variants. High-energy-efficient assemblies with low heat transfer coefficients are also applied for these positions. Significant energy savings for heating were achieved in both improvement variants compared to the existing state's energy performance due to the improved thermal envelope of the building. The existing building belongs to energy performance class F, which is one of the lowest. In improvement solution 1, the energy required for heating was reduced by over 76% compared to the existing state, and the energy performance class increased by four levels, reaching level B. A similar situation is observed in improvement solution 2, where savings of 68% compared to the existing state were achieved, and the energy performance class increased by three levels, reaching level C (Fig. 17.6).

By applying volumetric additions, the total usable heated area of the building increases while simultaneously increasing the volume of the building. Since the

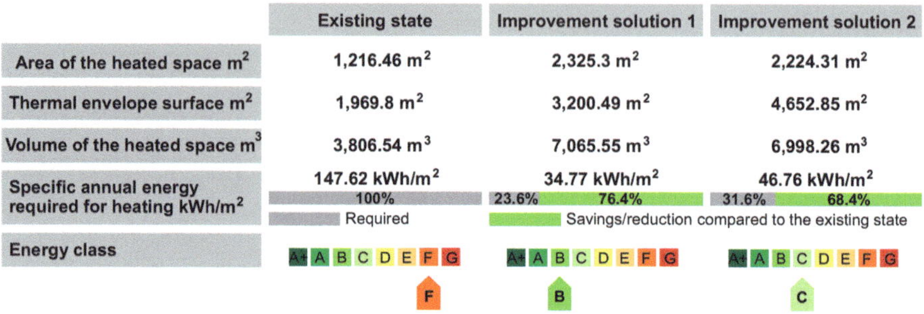

Fig. 17.6 A comparative analysis of the energy required for heating and energy performance class of the existing state and two improvement solutions

building's compactness is higher and the surface area of the thermal envelope is smaller in variant one compared to variant 2, better results are obtained in the calculation of heating energy requirements for variant one compared to variant 2.

17.4 Discussion

The advantages and disadvantages of the improvement solutions can be identified by conducting a comparative analysis of the spatial-organizational and functional parameters of the existing building state and the proposed improvement variants using volumetric additions (Table 17.1).

By adding volume to the facades and roof of the building, the heated area is nearly doubled in both variants compared to the existing condition. This has led to significant changes and improvements in the size and spatial organisation of residential units. When comparing the number of apartment types in the improvement variants, it can be noticed that in improvement solution 1, there are only two types of apartments, while in improvement solution 2, there are four types of apartments. The reason for this difference lies in the application of different variants of volumetric addition. With continuous addition, the spatial structure is added on each floor at the same location, resulting in the formation of apartments with the same surface area. With the partial addition of volume, there are differences between floors, resulting in different types of apartments. Since the total number of apartments in the building is 48 in both improvement variants, a better solution is to create more types of apartments with different sizes, as the market demands a greater choice.

When comparing the formation of open private spaces in the solution variants, it is observed that in improvement solution 1, each apartment has access to a large balcony. However, in improvement solution 2, apartment type B does not have access to a balcony, while apartment type A has two balconies. The reason for this disadvantage lies in the rhythmic repetition of added spatial structures in improvement solution 2. Since the window of apartment type A is directed towards the balcony between

Table 17.1 A comparative analysis of spatial-organizational and functional parameters of the existing state and two improvement solutions

	Existing state	Improvement solution 1	Improvement solution 2
Area of the heated space (m^2)	1216.46	2325.3	2224.31
Number of floors	5	6	6
Number of apartments	40	48	48
Apartments' size	Type A—38.5 m^2 Type B—21.5 m^2	Type A—68.2 m^2 Type B—27.3 m^2	Type A—50.4 m^2 Type B—40.3 m^2 Type C—27.4 m^2 Type D—68.2 m^2
Open private space	Type A—no balconies Type B—5 m^2	Type A—12.5 m^2 Type B—13.5 m^2	Type A—13.2 m^2 + 16.8 m^2 Type B—no balconies Type C—13.5 m^2 Type D—12.5 m^2
Elevator	No	Yes	Yes

apartments A and B (Fig. 17.5), it was not possible to provide apartment type B with access to the balcony.

It can be concluded that both improvement variants have some disadvantages, but there are far more positive characteristics for both, as such complex renovations can enhance spatial comfort, thermal comfort, functionality, and accessibility of the building. By adding spatial structures, the usable space would increase, allowing for the formation of open private spaces. Improving the thermal envelope would result in significant energy savings and better thermal comfort. Adding an elevator would greatly improve the functionality and accessibility of the building. As a synthesis of all the mentioned aspects, the lifespan of the building can be extended.

17.5 Conclusions

The research shows that complex renovation using volumetric additions can hypothetically enable the extension of the building's lifespan, both physically and functionally. This renovation strategy requires prior careful considerations: analysis of the existing building's stability, determination of site conditions, as well as the consent of all apartment owners in the building, which poses significant challenges in the prephase for further design of improvement solutions. In addition to the above, there is a need to raise awareness of the importance of the built environment and to empower governmental mechanisms to implement the renovation of residential buildings that are on the verge of obsolescence. Considering the scope achievable through such renovation, where residential stock is improved on multiple fronts, aligning tenants' comfort with modern standards of living, these complex interventions represent a way to preserve old buildings and extend their lifespan while ensuring all necessary living qualities within them.

References

European Commission (2020) A renovation wave for Europe—greening our buildings, creating jobs, improving lives. https://ec.europa.eu/commission/presscorner/detail/en/IP_20_1835. Last accessed 05 Mar 2024

European Commission (2020) Guidelines for the implementation of the green agenda for the western Balkans. An economic and investment plan for the western Balkans. https://neighbourhood-enlargement.ec.europa.eu/system/files/2020-10/green_agenda_for_the_western_balkans_en.pdf. Last accessed 05 Mar 2024

Acuner E, Onaygil S (2015) Comparative analysis on building energy efficiency for the Balkan world: is it really developing? European council for an energy-efficient economy. https://www.eceee.org/library/conference_proceedings/eceee_Summer_Studies/2015/5-energy-use-in-buildings-projects-technologies-and-innovation/comparativeanalysis-on-building-energy-efficiency-for-the-balkan-world-is-it-really-developing/. Last accessed 05 Mar 2024

Thomsen A, Straub A (2018) Lifespan assessment of dwellings. In: Conference papers of the European network for housing research (ENHR 2018): more together, more apart: migration, densification, segregation ENHR. TU Delft

Jovanović Popović M, Ignjatović D, Radivojević A, Rajčić A, Đukanović Lj, Ćuković Ignjatović N, Nedić M (2013) National typology of residential buildings in Serbia. Faculty of Architecture, University of Belgrade, Belgrade

Jovanović Popović M, Ignjatović D, Radivojević A, Rajčić A, Đukanović Lj, Ćuković Ignjatović N, Nedić M (2013) Atlas of multi-family buildings in Serbia. Faculty of Architecture, University of Belgrade, Belgrade

Ministry of Environmental Protection, Government of the Republic of Serbia (2011) Rulebook on energy efficiency of buildings. Official Gazette of RS. 61/2011. Last accessed 05 Mar 2024. https://www.paragraf.rs/propisi/pravilnik_o_energetskoj_efikasnosti_zgrada.html

Lj Đ (2021) Komfor u beogradskim stambenim zgradama. Univerzitet u Beogradu Arhitektonski fakultet, Belgrade

Lević B, Đukanović L, Ignjatović D (2023) Approaches for complex and integrated refurbishment to improve energy efficiency and spatial comfort of the existing post-war mass housing stock in Serbia. Sustainability 15:13884. https://doi.org/10.3390/su151813884

Lević B, Đukanović L, Radivojević A, Dimitrijević B (2023) Potentials for a complex and integrated refurbishment of the post-war housing stock of Serbia. Energ Effi 16:93. https://doi.org/10.1007/s12053-023-10171-z

Guidolin F (2018) Adaptive socio-technical devices - social inclusion as a rehabilitation tool. In: Sustainable and resilient building design, approaches, methods and tools, pp 229–247. TU Delft Open

Chapter 18
Sustainable Modular Renovations of Buildings Envelope from the Prosperous Years for the Purpose of Energy Saving in Bosnia-Herzegovina and Serbia

Darija Gajić, Budimir Sudimac, Aleksandar Rajčić, Slobodan Peulić, and Jelena Kljakić

Abstract Following the concept of sustainability in the development of new building systems and envelopes based on renewable and natural materials, which is recently obligatory for the EU member countries through the new Renovation Wave Strategy, the authors present research on the topic of energy renovation of buildings using renewable materials in modular systems and its possibilities in two developing countries. This research presents energy savings through a case study of deep renovation of buildings using timber modular systems for envelopes on the characteristic example of residential buildings. Research focuses on the Prosperous Years of construction period between 1961 and 1980 due to a considerable number of modular buildings designed during the peak of the construction progress and promotion of prefabricated buildings. The total potential of energy savings for such building types was considered according to national typologies of residential buildings in Bosnia and Herzegovina (B&H) and Serbia. Through a comparative analysis of these countries, the research presents the parameters of buildings built in Prosperous Years, which indicate the energy condition of buildings before and after the renovation of the envelope using standard measures from national typologies and the application of wooden modular envelope systems, which have the potential for in-depth renovation of buildings.

Keywords Renovation strategies · Deep energy refurbishment · Residential buildings · Timber systems · Modular envelopes

D. Gajić (✉) · S. Peulić · J. Kljakić
Architecture Technologies Department, Faculty of Architecture, Civil Engineering and Geodesy, University of Banja Luka, Banja Luka, Bosnia and Herzegovina
e-mail: darija.gajic@aggf.unibl.org

B. Sudimac · A. Rajčić
Architecture Technologies Department, University of Belgrade—Faculty of Architecture, Belgrade, Serbia

M. Vukmirović et al. (eds.), *Digital Dreams, Urban Realities*, The Urban Book Series,
https://doi.org/10.1007/978-3-031-76353-3_18

18.1 Introduction

In countries with higher energy intensity (units of primary energy consumption per unit of GDP of the country), energy consumption in buildings is even higher than 50%, and that is why the greatest energy savings are expected when renovating buildings. The energy intensity was estimated at 0.42 for B&H (https://www.iea.org/countries/bosnia-and-herzegovina) and at 0.35 for Serbia (https://www.iea.org/countries/serbia) (tone of oil equivalent (toe)/1000 USD of GDP) according to data from the International Energy Agency from 2021. B&H and Serbia are even the countries with the highest energy intensity of all the countries of the former Yugoslavia, to which they also belonged together with North Macedonia [0.24], Montenegro [0.23], Croatia [0.15] and Slovenia [0.14].

In order to achieve the planned total reductions of greenhouse gas emissions in the EU of at least 55% by 2030, it is necessary to reduce emissions from buildings by 60% and their energy consumption by 14%, while the use of energy for heating and cooling must be reduced by 18%. The European Union Renovation Wave Strategy, published by the European Commission (2020), aims to reduce greenhouse gas emissions, increase material reuse and recycling, stimulate economic recovery after the COVID-19 pandemic, reduce energy poverty and support achieving the EU's goal of becoming climate neutral by 2050. The renovation of the building envelope, which is guided by organic materials, such as wood, was presented in this research through a case study of energy renovation of the existing buildings, which were built in the Prosperous Years from 1961 to 1980. The Prosperous Years refer to industrialisation and accelerated construction of residential buildings, which occurred in both countries after 1960.

In those years in B&H and Serbia, as prefabrication was already current in Europe, buildings were built from 1961 to 1970 with imitation of facade wall prefabrication because they have many of the same repeating elements on the facade of the building, and from 1971 to 1980 pure reinforced concrete prefabricated buildings are being built.

The Strategy for the Renovation of Buildings in B&H (by entities, the Strategy for the Renovation of Buildings in The Republic of Srpska until 2050 and Directive 2012/27/EU) has not yet been officially adopted. Strategies of B&H and Serbia are guided by cost-optimal analyses prescribed by Regulations No 244/2012 (European Union Parliament and Council 2012). However, they do not emphasise the refurbishment of building envelopes with renewable and organic products and materials, but only the use of renewable energy sources (Government of the Republic of Serbia, Ministry of Civil Engineering, traffic and infrastructure (MGSIRS) 2022; Sudimac et al. 2022). The starting point for the formation of the Strategy in B&H is the cost-optimal analyses for residential and non-residential buildings done during 2016–2017. The B&H Strategy states that the greatest energy savings, about 60%, will be when we apply all valid new U-values on the envelope (Rulebook on energy efficiency; Rulebook on minimal requirements on Energy Efficiency in Buildings). By applying all measures to the building envelope, the price of the investment is twice as high, while energy consumption has been reduced by five times. In Serbia, the formation was based on residential and non-residential buildings done during 2019–2020, developing five potential renovation scenarios according to the level of energy-saving potential—Scenario 4 in Serbia is considered a deep renovation because Scenario 5 is intended for nZEB, and as B&H requires deep renovation of the entire building envelope resulting from cost-optimal analysis and following the requirements of the New Renovation Wave Strategy, it is necessary to point out possible deep renovations envelope of buildings with organic materials that will reduce the value of the energy indicator, the energy need for heating, by 60%.

18.2 Modular Envelope Renovations for Residential Buildings

Over the years, renewable materials have not found a path towards serious use in analysed countries. The reason for this lies in three main factors: higher price of renewable materials, poor industrial capacity and lack of knowledge, and lack of public awareness on the importance of natural, bio-based materials in energy renovation (Gajić et al. 2021). Traditional systems with contact façades are most accepted because they are the cheapest and most often used; however, over time, the question of the durability of such solutions, maintenance and certainly extremely bad impact on the environment arose. On the surging side, ecologically acceptable solutions are not represented in the analysed countries (B&H and Serbia), mainly due to the lack of interest, but also legislation that defines the carbon footprint of materials used in construction or missing understanding and rules on life cycle analysis. Building stock in both countries has become old and in a condition that is already questionable as to whether it is suitable for renovations; however, national programs should enhance processes and funding for energy renovations. These programs can be based on four main renovation models adjusted on a specific case-by-case basis: external composite insulation system, ventilated system, partially prefabricated systems and modular systems (Peulić et al. 2023). When it comes to the selection of the most adequate solution, there are several influential factors—the state of the building envelope, the time frame for the execution of works, the complexity of the details, i.e. the existence of modules in the building envelope, the development of the local construction industry and financial structures, etc. (Peulić and Gajić 2022).

Developed EU countries succeeded in implementing these solutions in AEC practice. The topic of modular prefabrication has gained serious attention through several projects developed in the EU. One of the pioneering projects was TES—Timber Energy Systems, where a systematic typological classification of facilities suitable and in need of deep renovation, design methodology, factory production and installation of modules was presented. This project advanced to additional study on the extension of buildings or their segments with the help of prefabricated spatial structures. Further research developed in recent years promoted different approaches to methodological solutions with emphasis on several key points presented in Table 18.1.

Considering the pedigree of the projects and participant countries (Finland, Germany, Norway), it is possible to note similar approaches with different outcomes.

Smart-TES EXTENSIONS is the continuation of the previous project (TES Energy) and is a step forward to building extensions rather than classic renovation

Table 18.1 Overview of the key outputs from projects developed in the European Union

Project	Type of improvement	ECBCS	BERTIM	More connect	TES energy façade	Smart-TES
Thermal comfort improvements	Opaque envelope	+	+	+	+	+
	Transparent envelope	+	+	+	+	+
Layout improvements	Elevator	–	–	–	–	+
	Balconies and/ or terraces	–	–	+	–	+
	Extending window area	–	–	+	–	+
Structure survey	3D laser scanning	+	+	NA	+	NA
	Total station	–	+	NA	–	NA
Developed system	Prefabricated vertical panels	+	+	+	+	+
	Prefabricated horizontal panels	+	–	–	–	+
	Prefabricated volumes	–	–	–	–	+
Building services	Integrated ventilation systems with heat recovery	+	+	+	+	+
	Panels pre-equipped with HVAC systems	–	+	+	+	+
Software	Automated panel distribution	–	+	–	–	–
	Off-site fabrication	+	–	+	+	+
	Robotic manufacturing and new algorithms	–	+	–	–	–

(usually narrowed to the envelope level), Table 18.1. The project results booklet showed several types of these extensions—horizontal, annexe, vertical annexe, terrace/balconies annexe, etc. Since the name is about the extensions, often several stories high, this project explained load transfer—via TES facade and extensions, via extensions, via extensions and existing building, via existing building loadbearing structure (Heikkinen et al. 2009). In addition, functional differentiation is considered since horizontal annexes typically affect existing flats and rooms (layout, ventilation, insolation, etc.), compared with vertical annexes that can perform independently. Prefabricated space modules require special attention to the transport requirements of different countries.

ECBCS (Prefabricated Systems for Low Energy Renovation of Residential Buildings) is a project of 7 countries (Austria, Czech Republic, France, Netherlands, Portugal, Sweden, Switzerland) where combines best approach in both on-site and software approaches, going with the case-by-case methodology rather than unification of overall process, Table 18.1 (Mark 2012). Reduce energy consumption needs to the range of 30–50 kWh/m^2 per year for heating and domestic hot water. Thus, panels can be slightly optimised accordingly, which can result in better overall building performance and economic benefit of the investment. The project resulted in 6 successful demonstration sites (*until the 2012 report) in Austria, the Netherlands and Switzerland. Energy consumption goals were fully achieved; furthermore, with solar installations, these demands were reduced to almost zero.

In the desire to bring to life the deep renovation of buildings, again using wood as a substructure, six countries (France, Spain, Sweden, Germany, Norway, and Poland) in the period from 2015 to 2019 created the project BERTIM (Building Energy Renovation through Timber Prefabricated Module), Table 18.1. This project tries to automatise the surveying process, as well as modular coordination and communication. An algorithmic approach is a step forward in thinking about modular coordination. However, there are certain gaps between different processes, such as tolerances ranging from 5 to 10 mm in surveying, compared with 3 mm tolerance in prefabricated panel production, which can mislead the results to be adjusted manually and on-site. Robotic manufacturing and RENO BIM software were developed to quickly design the optimal panel shape and distribution. During this project, an algorithm was developed to reduce the time needed to manually design and arrange modules. As in the previous project, prefabricated modules are designed to be pre-equipped with heating, cooling, ventilation and domestic hot water systems (BERTIM 2019).

MORE CONNECT (MOdular REtrofitting and CONNECTions) sets a number of user-friendly criteria to influence users' everyday life during renovation as little as possible and tries to solve the aforementioned problems by developing prefabricated, multifunctional renovation elements for the total building envelope (façade and roof) and installation/building services, from period 2015–2018 participated seven countries (Latvia, Estonia, Czech, Netherlands, Denmark, Portugal, Switzerland). The project has been launched to develop energy efficiency, hygrothermal performance and aesthetics of buildings and demonstrate technologies of prefabricated modular renovation elements, including the prefab integration of multifunctional components, e.g. for climate control, Table 18.1 (Volf et al. 2018). A set of basic modular facade and roof elements including renewable energy production and integration of HP insulation.

The presented projects show a variety of methodologies in research on the renovation of prefabricated modules. It is interesting to see different approaches in setting typologies, how to classify them in categories, ways to survey buildings, do building construction and envelope analysis, and see how to set exact aspects that renovation should accomplish. Some of the projects were answered affirmatively. However, some remained unclear on the methodology of surveying or building typology classification. Each of the projects gave some interesting conclusions, and this research will try to present a similar methodology that could possibly be implemented in the case studies in Bosnia-Herzegovina and Serbia.

18.3 Energy Savings Potential in Modular Envelope Renovations of Prefabricated Residential Buildings in B&H and Serbia

Data on the comparison of residential building stock in B&H and Serbia were collected through the methodological framework for researching the typology of residential buildings, based on the European international research project "TABULA" (EPISCOPE.EU 2023; EPISCOPE.EU 2023) Both project Typologies of Residential Buildings in B&H (Arnautović-Aksić et al. 2016) and Serbia (Popović et al. 2013) showed absolute and specific energy need for heating was calculated for the total of 29 representative residential buildings in B&H and 39 representative residential buildings in Serbia, which represent six categories of buildings classified into six and eight periods of construction. For the purpose of comparing countries according to the TABULA methodology (Loga et al. 2016), it was reduced to 22 buildings for B&H divided into four categories and six periods, and for Serbia, 31 buildings divided into four categories and eight periods. Adequate buildings for the application of prefabricated timber panels are selected according to three criteria: layout, which allows modular division of the facade envelope; the period of construction and amount of such buildings. Multi-family residential buildings (MFH and AB), in contrast to individual residential buildings (SFH and TH), have larger envelope area which needs to be renovated and can be heat-upgraded with the same pre-fabricated elements of organic materials, which are also the subject of this analysis.

18.3.1 Existing Residential Building Stock for Energy Renovations Modular Envelope

The potential of construction period between 1961 and 1980 was investigated. The construction of prefabricated reinforced concrete systems in the EU began in 1960 (Giebeler et al. 2009), while in B&H and Serbia after 1970, but the influence of EU construction leads to the imitation of prefabrication on buildings made of cast concrete and masonry construction in Bosnia and Herzegovina, (see Fig. 18.1).

Today, such buildings are adequate for renovation with modular panels, which would be made in industrial conditions and installed on site in its entirety on the existing casing. The characteristics of such residential buildings are that they were built as free-standing (MFH) or lamellas (AB), (see Fig. 18.2), and that their number of floors is 4 or more with at least 20 apartments within such structures.

In addition, they are usually connected to inefficient district heating. Building renovation strategies usually refer to buildings whose renovation can generate energy savings of 60% or more, but priority must be given to buildings where renovation would solve health and energy poverty problems for users (https://energy.ec.europa.eu/system/files/2015-07/INSIGHT_E_Energy%2520Poverty_Appendices_0.pdf). From the Typology, we can see how many buildings there are free-standing (MFH) and slats (AB) and what their ratio is compared to other types of buildings. The potential building stock over which the building envelope could be upgraded has been examined through a comparative analysis of data on the building stock of both countries. Periods 1961–1970 and 1971–1980 for the

Fig. 18.1 Residential buildings of Bosnia and Herzegovina (left—cast concrete building period 1961–1970, right—prefabricate concrete building period 1971–1980)

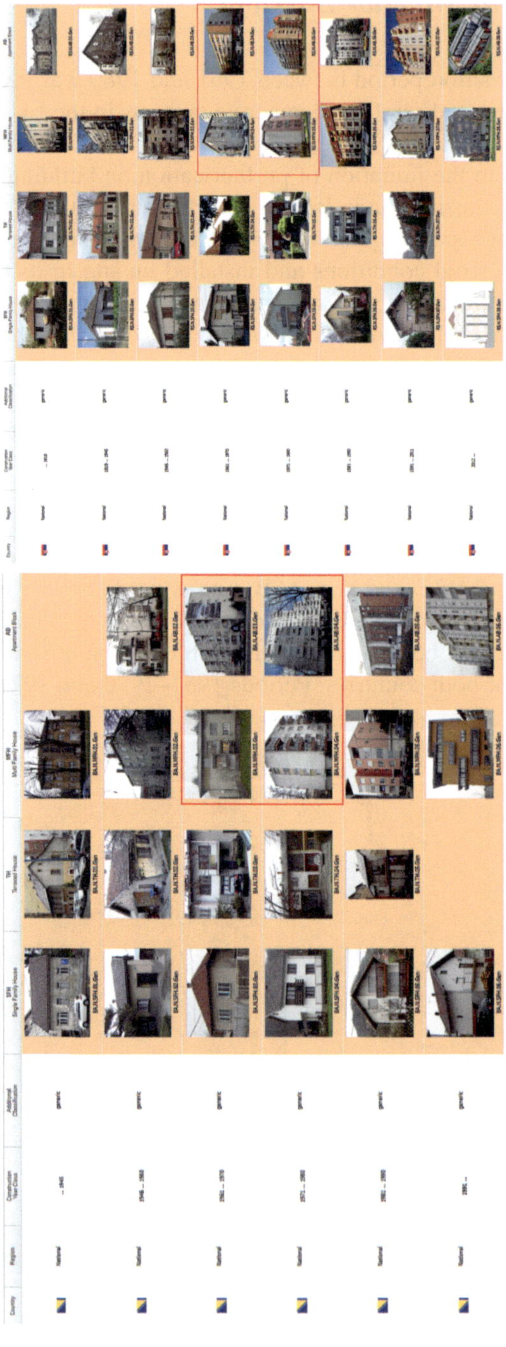

Fig. 18.2 Typology of residential buildings of Bosnia and Herzegovina (left) residential building typology of Bosnia and Herzegovina (EPISCOPE.EU 2023)

residential buildings (MFH and AB) have mutually in B&H 9355 buildings, while in Serbia, this number, 2.5 times higher than the number of buildings in B&H, is 24,372 buildings. However, looking at the number of dwelling units within these buildings, the ratio decreases because the number of dwellings in individual buildings is 66.47% in B&H and 73% in SRB, respectively, which is shown in Table 18.2.

Comparative analysis indicates that although Serbia has a larger number of buildings of all types. In B&H, the number of dwellings in types MFH and AB of the period from 1961 to 1980 in relation to all periods of collective housing buildings is 74.2%, while in relation to all types and periods of buildings, it is 18.38%. In Serbia, the number of dwellings in the types of MFH and AB periods from 1961 to 1980 in relation to all periods of collective housing buildings is 46.8%, while in relation to all types and periods of buildings, it is 12.63%. Although both types of buildings are suitable for modular envelope renovation, this research presents the possibility of energy renovation for the MFH type from the period 1961–1970 and the AB type from the period 1971–1980. For calculation requirements, it was assumed that the entire building surface used for residential purposes was heated. In regional countries, it was estimated that only 50% of households heated over 50% of the conditioned area (Csoknyai et al. 2016), whereas indicators for the EU countries are somewhat better (Atanasiu et al. 2014). Multi-family housing buildings are important for this research, and the energy required for heating is estimated. Expert calculations estimate the energy needed for heating individual buildings of all types, as shown in Table 18.3. It presents a comparison of B&H and Serbia by types MFH and AB for the construction period of 1961–1970 and 1971–1980.

Table 18.2 Number of residential buildings, apartments and gross surface in comparative countries

	Bosnia and Herzegovina			Serbia		
	No. of buildings	No. of apartment	Gross area of buildings (m^2)	No. of buildings	No. of apartment	Gross area of buildings (m^2)
SFH and TH	841,543	1,076,240	124,683,708	2,186,246	2,327,707	176,048,838
MFH and AB	20,422	542,945	38,244,921	60,074	860,707	113,638,882
Total	861,965	1,619,185	162,928,629	2,246,320	3,188,414	289,687,720
MFH and AB (1961–1980)	9535	297,644	21,923,226	24,372	402,891	49,160,278
MFH and AB (1961–1970)	5176	119,474	7,985,112	9640	138,955	15,930,173
MFH and AB (1971–1980)	4359	178,170	13,938,114	14,732	263,936	33,230,105
MFH (1961–1970)	3012	50,541	3,184,346	7285	91,109	8,690,967
AB (1971–1980)	2156	125,298	10,350,518	4752	131,135	19,899,758

Table 18.3 Comparison value of energy needed for heating on representative residential buildings in Bosnia and Herzegovina and Serbia (kWh/m^2 a)

Country	Bosnia and Herzegovina—$Q_{h,nd}$		Serbia—$Q_{h,nd}$	
Building type	Multi-family house/MFH	Apartment block/ AB	Multi-family house/MFH	Apartment block/ AB
1961–1970	188.44	153.06	172.00	159.00
1971–1980	146.80	130.90	191.00	158.00

An assessment of such indicators of the energy need for heating shows that although the same name and period of construction, especially since the two countries were under the same legal regulations and requirements for the building envelope until the 1980s, can only indicate higher shape factors of the selected representative examples of MFH and AB buildings for the period 1971–1980 in Serbia. Comparison value of energy needed for heating, for the period 1961–1970, Serbia has a 9% lower value for MFH and a 4% higher value for AB, and for the period 1971–1980, Serbia has a 23% higher value for MFH and a 17% higher value for AB.

18.3.2 Case Study of Energy Savings Potential with Modular Envelope Renovations of Residential Building

The case study shows a sample of buildings in Banja Luka (Bosnia and Herzegovina). Representative samples of the existing residential building are determined by a detailed energy audit—determining the specific energy consumption for heating using a standard which is still in force in B&H, EN ISO 13790:2008—Energy performance of buildings—Calculation of energy use for space heating and cooling. Sample 1 of the characteristic period (1961–1970) led to the conclusion that the real Sample from 1964 has a lower specific energy need for heating—QH, and than the representative building of AB from the same period of construction from Typology, Table 18.4.

Representative Sample 1, which was built in the period after the Second World War rather than in 1964, is a masonry construction of five-story brickwork, only with horizontal reinforced concrete ring beams. After the earthquake in Banja Luka in 1969, we were given vertical reinforced concrete columns and additional reinforced concrete walls for refurbishment. This building does not possess thermal insulation in the walls, floor, and roof, and it has a couple of wooden windows—windows glazed with two single glasses. Sample 1 is a compact building, rectangular in shape with no sunshade element like an overhang, balcony or loggias, with the form factor of the building (A/V ratio) of 0.40. A detailed energy audit was conducted for the building. The calculated value of the energy needed to heat the existing selected sample corresponds to the average value stated for typical buildings in the Typologies of B&H and Serbia. Table 18.4. By applying cost-optimal measures on the sample

Table 18.4 Comparative review of the energy needs for heating of representative sample 1 of the existing residential building before and after renovation of the building envelope

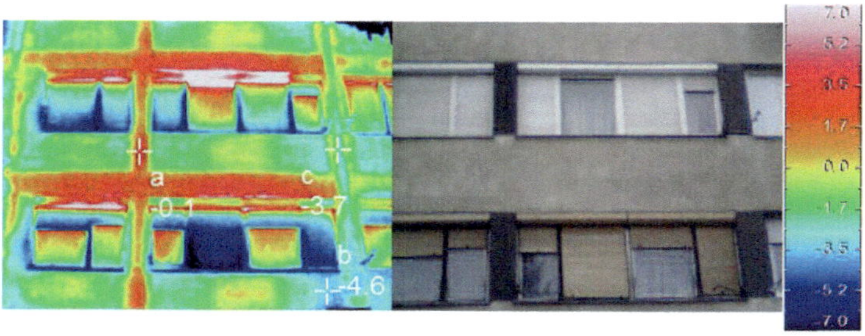

Thermal-vision image before renovation

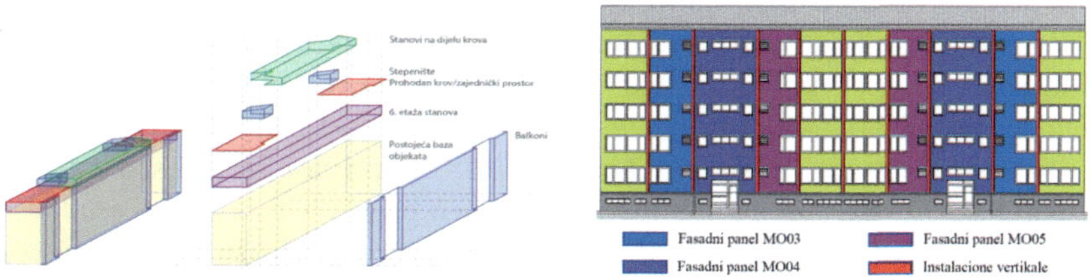

Design for the renovation of the building envelope—3D view and segment of the façade (Huseinbašić et al.)

Period		1964	
Dimension		10 × 42 m	
Heated space area		2025 m^2	
No. of floors		P + 4	
Heated space volume		5670 m^3	
Heat capacity	Wh/m^2 a	72	
Metabolic heat from person	W/m^2	3.8	
Orientation		NW—SE	
		Before	After
U-value walls	W/m^2 K	2.03	0.30
U-value windows	W/m^2 K	3.12	1.60
U-value roof	W/m^2 K	1.64	0.20
U-value floor	W/m^2 K	1.02	0.30
G-value	–	0.49	
A/V ratio	–	0.40	
Percentage of window area	%	23.70	
Infiltration	1/h	0.60	0.50

(continued)

Table 18.4 (continued)

		Before	After
Internal temperature	°C	20.0	20.0
Setback temperature	°C	16.7	16.7
Internal heat gains			
Ventilation	kWh/m^2 a	0.0	0.0
Lighting	kWh/m^2 a	2.6	2.6
Various equipment	kWh/m^2 a	13.5	13.5
Energy need for heating	kWh/m^2 a	164.4	31.2

envelope, it is possible to lower the value of the energy needed for heating below 40 kWh/m^2 (31.2 kWh/m^2). Sample 2 of the characteristic period (1971–1980) led to the conclusion that real Sample 2 from 1974 has a lower specific energy need for heating—QH, and than the representative building of AB from the same period of construction from Typology, Table 18.5. Representative Sample 2, an existing residential five-story building, which was built in the period 1971–1980 rather than 1974, is a skeleton structural system Zezelj—"IMS" of precast prestressed columns and floors. Facade walls are prefabricated sandwich panels with 3 cm of insulation pressed cane and wooden couple windows—windows glazed with two single glasses. Sample 2 is rectangular in shape, but the building has an overhang and lots of loggias, with the form factor of the building (A/V ratio) of 0.46. A detailed energy audit was conducted for the building. The calculation was guided by the design parameters of the building envelope characteristic for the specified period, with data characteristic of the real environment of the building (climate data and built environment) and the use of the building (building users and devices). The calculated value of the energy needed for heating the existing selected sample corresponds to the average value stated for typical buildings in the Typologies of B&H and Serbia. Table 18.5. By applying cost-optimal measures on the sample envelope, it is possible to lower the value of the $Q_{H,nd} = 40$ kWh/m^2 (39.3 kWh/m^2). Sample 1 shows five modules were defined as total façade envelope, and in the second case, for Sample 2, three modules and five submodules were defined as total façade envelope (façade panels made with wooden substructure filled with thermal insulation with wooden frame windows) with energy characteristics defined by cost-optimal analysis in B&H. The renovation of the building envelope was done from the conceptual design to the details, with energy analysis bill of quantities and recalculation of works, in order to determine energy savings and economic viability of the investment. The solution is guided by prices from 2021. It was concluded that in addition to the renovation of the envelope, it is necessary to upgrade the building with new dwellings and the addition of new technical systems for energy production to enable their initial investment in such renovation.

Table 18.5 Comparative review of the energy needs for heating of representative sample 2 of the existing residential building before and after renovation of the building envelope

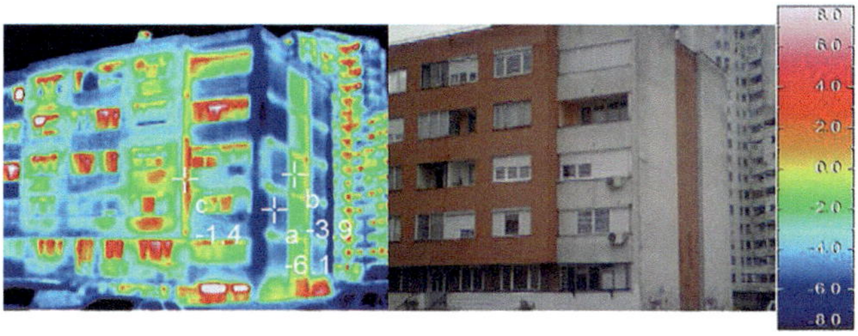

Thermal-vision image before renovation

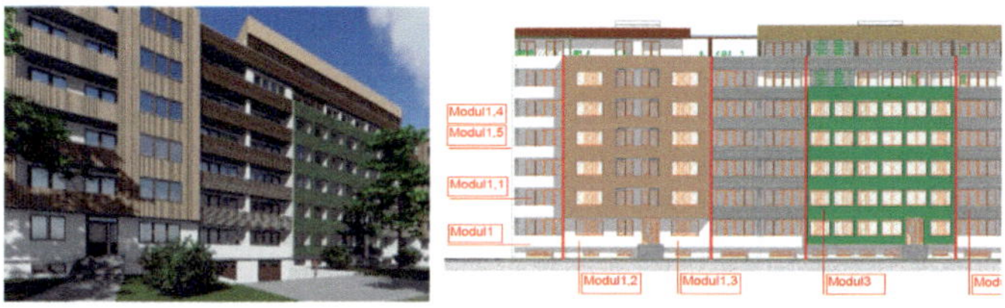

Design for the renovation of the building envelope—3D view and segment of the facade (Murguz et al.)

Period		1974	
Dimension		13.65 × 106 m	
Heated space area		6315 m^2	
No. of floors		P + 4	
heated space volume		17,682 m^3	
Heat capacity	Wh/m^2 a	46	
Metabolic heat from person	W/m^2	4.9	
Orientation		E—W	
		Before	After
U-value walls	W/m^2 K	1.35	0.30
U-value windows	W/m^2 K	3.08	1.60
U-value roof	W/m^2 K	0.70	0.20
U-value floor	W/m^2 K	1.04	0.30
G-value	–	0.34	
A/V ratio	–	0.46	
Percentage of window area	%	19.90	
Infiltration	1/h	0.50	0.50

(continued)

Table 18.5 (continued)

		Before	After
Internal temperature	°C	20.0	20.0
Setback temperature	°C	16.6	16.6
Internal heat gains			
Ventilation	kWh/m^2 a	0.0	0.0
Lighting	kWh/m^2 a	2.5	2.5
Various equipment	kWh/m^2 a	12.0	12.0
Energy need for heating	kWh/m^2 a	157.7	39.3

18.4 Comparative Analysis of Energy Savings

Following the same standard EN ISO 13790 standard, energy requirements of buildings are calculated and expressed in Typologies of buildings in Bosnia-Herzegovina and Serbia. In typologies of building are stated energy savings after applying the measures on the building envelope, which are governed by the valid country regulations and called the standard improvement of building energy performance.

Standard improvement measures in typology of residential building defined in accordance with usual measures applied during building reconstruction in the territory of B&H (improvement of thermal characteristics of walls and ceilings by technically common procedures—added thermal insulation 10 cm thickness with $\lambda =$ 0.041 W/mK) as well as a possible replacement of existing windows with new ones (defined minimal U-value 1.60 W/m^2 K). In Serbia, the measures applied from valid regulation for existing buildings (external walls 0.4 W/m^2 K, roof 0.2 W/m^2 K, windows 1.5 W/m^2 K and floor 0.4 W/m^2 K). Tables 18.4 and 18.5 analyse the current condition of buildings and their values of QH and representative types; examples of the case study from B&H are closer to the values of QH and types of Serbia.

Applying these measures to restore the envelope in typologies would create a 60% lower QH in buildings in B&H, while in Serbia, it is possible to have a 48% lower QH in buildings. The reason for this can be found in the fact that in B&H type AB, the existing condition has a lower value of QH. In case studies, Sample 1 and Sample 2 were treated with cost-optimal measures in the B&H area, which is listed in Tables 18.4 and 18.5, and which leads to savings of 81% for MFH type and 75% for AB type. For these characteristic types of buildings, which could be renovated in a modular system, we can analyse the possibilities of energy savings in MWh/a by country, applying standard measures from Typology and cost-optimal measures applied in the case study.

Tables 18.6 and 18.7 shows that the amount of energy can be further saved by applying cost-optimal measures compared to standard measures listed in the Typologies—in B&H for type MFH an additional 63,632 MWh/a, which represents an additional about 23.3%, while in Serbia for type MFH it can be saved another 424,148 MWh/a which is an additional 43.4% on savings from standard measures.

Table 18.6 Comparative representation of $Q_{H,and}$ in kWh/m^2 of representative examples of MFH buildings, before and after applying standard measures in typologies of B&H and Serbia and from case study

Type	MFH	MFH
	Before measure	After
1961–1970 B&H	188.44	67.86
1961–1970 Serbia	172.00	55.00
1964 (case study)	164.40	31.20

Table 18.7 Energy need for heating of MFH in B&H and Serbia before and after standard measures and case study measures (MWh/a)

MFH	MFH	MFH	MFH	MFH
Before measure	After measure from typology	Saving energy	After measure from case study	Saving energy
327,081	117,787	209,294	54,155	272,926
1,532,704	981,213	551,941	556,615	976,089

Tables 18.8 and 18.9 shows that the amount of energy can be further saved by applying cost-optimal measures compared to standard measures listed in the Typologies - in B&H for type AB an additional 129,381 MWh/a, which represents an additional 13.6%, while in Serbia for type AB it can be saved another 961,343 MWh/a energy, which is an additional 35% on savings from standard measures.

Table 18.8 Comparative representation of $Q_{H,nd}$ in kWh/m^2 of representative examples of AB buildings, before and after applying standard measures in typologies of B&H and Serbia and from case study

TYPE	AB	AB
	Before measure	After measure
1971–1980 B&H	130.90	51.80
1971–1980 Serbia	158.00	81.00
1974 (case study)	157.70	39.30

Table 18.9 Energy need for heating of AB in B&H and Serbia before and after standard measures and case study measures (MWh/a)

AB	AB	AB	AB	AB
Before measure	After measure from typology	Saving energy	After measure from case study	Saving energy
1,354,882	536,156	818,726	406,775	948,107
3,642,500	1,867,357	1,775,143	906,014	2,736,486

18.5 Conclusion

Currently, the new EU strategy for sustainable renovation of buildings favours deep renovation, with energy savings of about 60%. The analysis shows that cost-optimal measures, with slightly more demanding U-values for the non-transparent part of the envelope (external walls 0.30 W/m^2 K, roof 0.20 W/m^2 K and floor 0.30 W/m^2 K) than standard measures in Serbia and Bosnia and Herzegovina, and even less demanding U-value for windows (1.60 W/m^2 K), for MFH buildings can save 272,926 MWh/a in B&H and 976,089 MWh/a in Serbia and for AB buildings can save 948,107 MWh/a in B&H and 2,736,486 MWh/a in Serbia. This case study showed that it is possible to lower the value of energy needed for heating below 40 kWh/m^2 or, according to the requirements of the Strategy, to create savings of 60% (from case studies, 81% for MFH and 75% for AB). In addition, as the New Renovation Wave Strategy in one of the key principles of building renovation towards 2030 and 2050 expands the use of organic building materials, case studies have shown the application of cost-optimal measures through modular systems in organic materials, systems that could accommodate new technical systems, which should be considered in future research in the case of renovation of buildings according to nZEB standards.

References

International Energy Agency, Overview for Bosnia and Herzegovina. (On-line). Available: https://www.iea.org/countries/bosnia-and-herzegovina

International Energy Agency, Overview for Serbia. (On-line). Available: https://www.iea.org/countries/serbia

European Commission (2020) Renovation wave strategy. Available online: https://ec.europa.eu/energy/sites/ener/files/eu_renovation_wave_strategy.pdf

European Union Parliament and Council; Regulatory Commission; Regulation No 244/2012 of 16 January 2012 supplementing Directive 2010/31/EU of the European Parliament and of the Council on the energy performance of buildings by establishing a comparative methodology framework for calculating cost-optimal levels of minimum energy performance requirements for buildings and building elements

Government of the Republic of Serbia, Ministry of Civil Engineering, traffic and infrastructure (MGSIRS) (2022). https://www.mgsi.gov.rs/sites/default/files/Predlog%20Dugorocne%20strategije%20PUONFZRS.pdf

Sudimac B, Gajić D, Peulić S (2022) Energy renovation of residential building envelope using organic materials for the level of cost-optimal improvement/upgrade in Bosnia-Herzegovina and Serbia. In: STEPGRAD2022 conference, Banja Luka, Bosnia and Herzegovina

Republic of Srpska Goverment, Energy Efficiency Law, https://ekofondrs.org/sadrzaj/dokumenti/ee/zakon_o_energetskoj_efikasnosti.pdf

Republic of Srpska Goverment, Regulation on minimum requirements for energy performance of buildings, https://ekofondrs.org/sadrzaj/dokumenti/ee/Pravilnik_o_minimalnim_zahtjevima_za_energetske_karakteristike_zgrada.pdf

Gajić D, Peulić S, Mavrič T, Sandak A, Tavzes Č, Malešević M, Slijepčević M (2021) Energy retrofitting opportunities using renewable materials—comparative analysis of the current frameworks in Bosnia-Herzegovina and Slovenia. Sustainability 13(2):603

Peulić S, Gajić D, Sandak A, Tavzes Č, Mavrič T, Rašović J, Slijepčević M, Antunović B, Malešević M, Okilj U (2023) Towards deep energy retrofitting: an overview and possibilities for Slovenia and Bosnia-Herzegovina. In: ENEF 2023 conference, Banja Luka, Bosnia and Herzegovina

Peulić S, Gajić D (2022) Influential parameters on the adaptability of taller timber buildings. COST ACTION CA 20139, holistic design of taller timber buildings (HELEN), STATE OF THE ART REPORT. COST Association, Brussels, Belgium

Heikkinen P, Kaufmann H, Winter S, Larsen KE (2009) Energy Façade—prefabricated timber-based building system for improving the energy efficiency of the building envelope, research project from 2008–2009. Wood wisdom-Net, Helsinki

Zimmermann Mark (2012) ECBCS Annex 50 prefabricated systems for low energy renovation of residential buildings: project summary report

BERTIM (2019) Innovative wood prefabrication for energy efficient renovations. Project Results Booklet

Volf M, Hejtmánek P, Lupíšek A, Kalamees T, Christensen F, Olesen S, van Oorschot JAWH, Mørck OC, Borodinecs A, Šenfeldr P (2018) Development and advanced prefabrication of innovative, multifunctional building envelope elements for modular retrofitting and connections. A set of basic modular facade and roof elements, including renewable energy production and integration of HP insulation

EPISCOPE.EU. BA Bosnia and Herzegovina—country page, National Building Typologies. Accessed 3 May 2023. https://episcope.eu/building-typology/country/ba/

EPISCOPE.EU. RS Serbia—country page country page, National Building Typologies. Accessed 3 May 2023. https://episcope.eu/building-typology/country/cy/

Arnautović-Aksić D, Mladen B, Nijaz D, Darija G, Petar G, Džana K, Milovan K, Erdin S, Darko T, Nermina Z (2016) Typology of residential buildings of Bosnia and Herzegovina. Faculty of Architecture, University of Sarajevo, Sarajevo

Popović J, Milica ID, Ana R, Aleksandra R, Ljiljana Đ, Nataša ĐI, Miloš N (2013) National typology of residential buildings in Serbia. Faculty of Architecture University of Belgrade, Beograd

Loga T, Stein B, Diefenbach N (2016) TABULA building typologies in 20 European countries—making energy-related features of residential building stocks comparable. Energy Build 4–12

Giebeler G, Fisch R, Krause H, Musso F, Petzinka K-H, Rudolphi A (2009) Refurbishment manual. Birkhäuser, Basel, Boston, Berlin

Pye S, Dobbins A. Energy poverty and vulnerable consumers in the energy sector across the EU: analysis of policies and measures. https://energy.ec.europa.eu/system/files/2015-07/INSIGHT_E_Energy%2520Poverty_Appendices_0.pdf

Csoknyai T, Hrabovszky-Horváth S, Georgiev Z, Jovanovic Popovic M, Zeković B, Villatoro O, Szendrő G (2016) Building stock characteristics and energy performance of residential buildings in eastern-European countries. Energy Build 39–52

Atanasiu B, Kontonasiou E, Mariottini F (2014) Alleviating fuel poverty in the EU. Investing in home renovation, a sustainable and inclusive solution. In: Buildings performance institute Europe (BPIE)

Huseinbašić E, Jojić A, Petrović S, Jović D, Radaković M. Design for the renovation of building envelope_Distribution of characteristic TES modules on the facade envelope (course "Team project" on Master study "Energy Efficiency of Buildings" at University of Banjaluka)

Murguz A, Malešević N, Šućur A. Design for the renovation of building envelope_Distribution of characteristic TES modules on the facade envelope (course "Team project" on Master study "Energy Efficiency of Buildings" at University of Banjaluka)

Chapter 19
Bridging a Gap Between Dreams and Reality: The Adaptability of Urban Interface Design to 5G Millimetre-Wave Mobile Communication

Tianyu Hu and Tie Wang

Abstract There is a gap between urban reality and digital dreams. As the grandest vision of the digital dream, the Internet of Things is premised on the ubiquitous coverage of 5G mobile communication base stations. 5G millimetre wave signals have high frequency, poor penetration, and will be blocked when encountering physical obstacles in urban space, so millimetre wave relies on line-of-sight transmission. In the case of unchanged urban space, mobile communication professionals cannot fundamentally solve the problem of millimetre wave occlusion in space and can only obtain the continuity of signal coverage by increasing BSs. The energy consumption and number of superposition 5G base stations are several times that of 4G base stations, and the deployment cost of 5G is high. Based on the above problems and research status, this study aims to clarify the relationship between urban space and mobile communication channels represented by millimetre waves and explore how to integrate the demand for urban space coverage of millimetre wave channels into the existing urban design methodology, that is, to explore the urban design methodology with millimetre wave channel as a new reference factor.

Keywords Urban Interface · 5G millimetre wave · Adaptability · New factor · Methodology

---

T. Hu (✉) · T. Wang
Central Academy of Fine Arts, Beijing, China

Breuer Marcell Doctoral School, University of Pécs, Pécs, Hungary

M. Vukmirović et al. (eds.), *Digital Dreams, Urban Realities*, The Urban Book Series,
https://doi.org/10.1007/978-3-031-76353-3_19

19.1 Introduction

19.1.1 The Contradiction

As early as the 1960s, when mobile communication was not widely used by the general public, Lewis Mumford keenly observed the networked nature of cities, referring to the invisible world filled with cables, wires, connections, encoding, protocols, and capital as the "invisible city" (Mumford 1961). "Smart Cities" in 2009. IBM states that Smart Cities employ advanced information and communication technologies (ICT) to integrate the various core systems of urban operations, including people, businesses, transportation, communication, water, and energy, transforming them into a "system of systems" that operates in a more intelligent manner, thereby promoting sustainable urban development (IBM Institute for Business Value 2009).

The ultimate goal of human communication is to enable anyone to exchange any type of information with anyone, anytime, and anywhere, known as "Internet of Things." Mobile communication serves as the essential pathway to achieve this goal.

As a new infrastructure, 5G serves as the foundation for the Internet of Things and smart cities. The vision of the Internet of Things relies on the ubiquitous coverage of 5G mobile communication base stations. However, the current urban space poses obstacles to the efficient coverage of 5G mobile communication signals.

The use of high-frequency millimetre waves (mmWave) in 5G communication technology enables the transmission of large amounts of data over short distances (Bai and Heath 2015). However, the atmospheric absorption and high penetration loss of mmWave pose challenges to the coverage radius of 5G base stations (BSs), which is significantly smaller compared to previous technology layers such as 2G, 3G, and 4G (Maccartney et al. 2013; Sulyman et al. 2014).

Furthermore, the dense distribution of buildings in urban areas restricts the propagation and coverage of 5G signals. According to Pi and Khan (2011), even common building materials like brick and concrete exhibit high penetration loss for mmWaves in urban areas, resulting in significant indoor-outdoor isolation (Palizban et al. 2017). This phenomenon of penetration loss in 5G propagation is known as the Line-of-Sight (LOS) effect (Maccartney et al. 2013; Al-Falahy and Alani 2019; Qamar et al. 2019). Therefore, mmWave experiences high penetration and path loss (Maccartney et al. 2013; Lee et al. 2018), susceptibility to interference (Bai and Heath 2015), and sensitivity to environmental blockage (Sulyman et al. 2014). It is essential to consider LOS coverage in the target areas to ensure satisfactory Quality of Service (QoS) for 5G (Rappaport et al. 2013).

Clearly, the propagation characteristics of mmWave present significant investment cost challenges for the deployment of 5G. Telecom operators must deploy ultra-dense base stations in urban areas to achieve satisfactory service coverage. In recent years, energy-efficient technologies (Buzzi et al. 2016) and the upgrade of 4G BSs to 5G BSs (Sharma et al. 2018) have gained considerable attention in reducing the construction and operation costs of 5G cellular networks. The initial 5G rollout often involves Non-Standalone (NSA) deployment using existing 4G base stations. However, to

Fig. 19.1 Barcelona telecom tower. (*Source* © 2024 Foster + Partners)

minimise construction costs, it is crucial to cover the target area with the minimum number of base stations. The total investment for deploying a new 5G base station in China is approximately $53,000. Upgrading 4G base stations to 5G can reduce costs, but additional base station locations are still required to meet the coverage demands of 5G services (Figs. 19.1 and 19.2).

19.1.2 Current Countermeasures

Research on the obstruction of millimetre-wave mobile communication in urban spaces can be divided into two categories: studies within the field of mobile communication and studies outside the field of mobile communication.

Mobile Communications Within the field of mobile communication, researchers investigate the physical properties of millimeter waves and develop methodologies to adapt the millimeter wave channel to urban physical spaces.

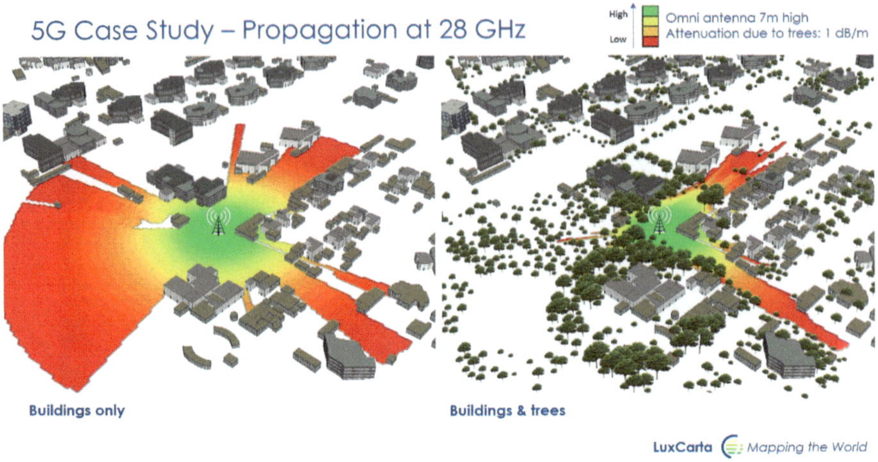

Fig. 19.2 Propagation diagram of 28 GHz millimetre waves in an urban environment. (*Source* © LuxCarta)

The research on the obstruction of millimetre-wave mobile communication in urban spaces within the field of mobile communication aims to develop methodologies for millimetre-wave base station deployment by studying the physical properties of millimetre waves and the morphology and materials of urban spaces. Mobile communication professionals are not the direct shapers of urban spaces, nor are they researchers in the field of materials. Urban spaces exist in mobile communication research as unalterable limiting conditions. Therefore, the propagation characteristics of millimetre waves present significant investment cost challenges for the introduction of 5G. To achieve satisfactory service coverage, 5G telecom operators must deploy ultra-dense base stations in urban areas. This limitation is inherent in the fifth generation of mobile communication development, as it cannot fundamentally solve the problem of signal coverage through technological advancements alone. Hence, the limitations imposed by space and materials require supplementation from disciplines outside the field of mobile communication. It is imperative to conduct research in architecture and landscape architecture to explore scene design from the perspective of 5G.

Other Majors Research on spatial optimisation models for cellular network planning has a long history, focusing on maximising service coverage. However, existing heuristic models face challenges in simulating line-of-sight (LOS) propagation and coverage of 5G signals in urban areas (Fig. 19.3).

Most models are designed for 2G/3G/4G networks and do not consider the high penetration loss of 5G signals. Recent studies have utilised GIS to optimise 5G network planning but lack global optimisation capabilities. Spatially-implicit heuristic algorithms have been developed to minimise penetration loss, but they overlook the three-dimensional morphology of urban buildings. Current modelling

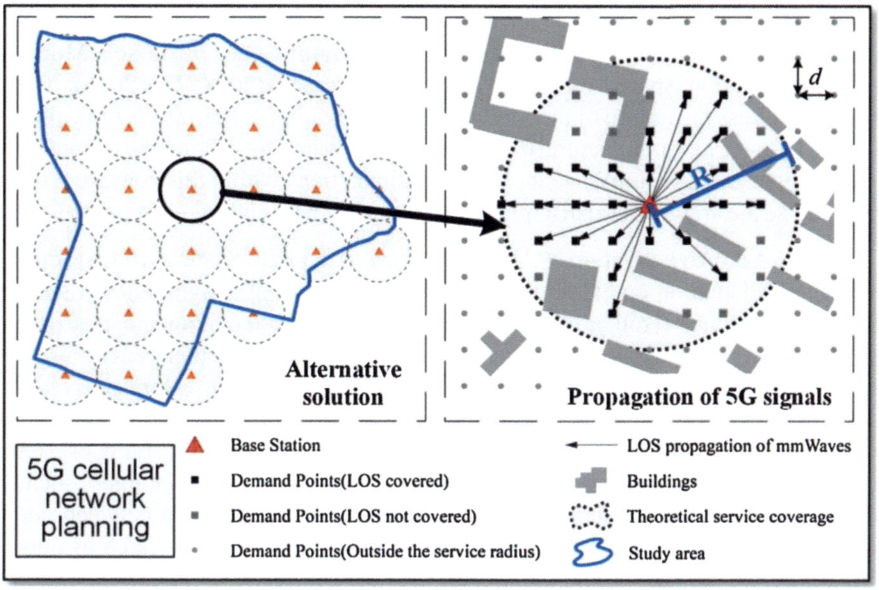

Fig. 19.3 Barcelona telecom tower. (*Source* © Qi Wang)

algorithms simplify buildings into two-dimensional planes, leading to inaccurate results (Wang et al. 2020). The lack of active modification of the millimetre wave coverage environment hinders achieving both low cost and high efficiency. Overcoming these challenges is crucial for informed decision-making and improving millimetre wave coverage efficiency.

19.2 Theoretical Construction

19.2.1 Theoretical Foundation

Infrastructure Architecture The concepts of "Architecture as Infrastructure" and "Infra-architecture" have similarities but differ in emphasis. "Architecture as Infrastructure" involves designing architecture as part of urban infrastructure for sustainable and smart cities, focusing on the connection between architecture and urban infrastructure. In contrast, "Infra-architecture" integrates infrastructure and architecture, creating a new form that provides various functions and aims for a more adaptable urban environment. This study examines the interface of urban public spaces, particularly in relation to mmWave mobile communication. Mobile communication, as grey infrastructure, possesses inherent infrastructure attributes. Infrastructure architecture research and theories serve as the foundation for this study. Scholars have developed frameworks for infrastructure, including concepts like "Thick 2D" and "Infrastructural Urbanism" by Stan Allen, "Terrain Vague" by de Sola Morales, and "Organization Space" by Keller Easterling (Varnelis 2008; Tan 2016). These theories address the failure of modernism and the autonomy of infrastructure, aiming to reconstruct urban architectural discourse. This research focuses on the adaptability of urban interfaces to mmWave mobile communications within this context, aiming to revitalise the architectural discipline's engagement with cutting-edge urban issues.

Complex Adaptive System (CAS) Complex Adaptive System (CAS) theory combines Adaptation and Complexity theory, representing a significant branch of complexity science (Holland 2000). Adaptation refers to the adjustment of organisms' genetic composition and characteristics to align with their environment's development. It spans various fields, ranging from individuals to entire ecosystems. Organisms are considered adaptive agents, while the environment serves as the object of adaptation. Ecological adaptation initially applied to natural ecosystems has been analogously extended to human systems, including economics, society, and culture. Economic adaptation studies analyse the relationship between adaptive agents and economic development, while social adaptation research examines changes in psychological and behavioural patterns in response to the social environment. Cultural adaptation research explores how regions adjust their behaviour according to the natural environment.

In "The New Science of Cities," Michael Batty argues that cities should be viewed as complex systems composed of networks and flows (Batty 2013). Within the urban CAS, mmWave mobile communication is a crucial infrastructure for enhancing network strength and supporting necessary flows to meet human needs.

19.2.2 Visual Analysis of Millimeter Wave Coverage

Information Collection The coverage of millimetre-wave mobile communications depends on Line-of-Sight (LOS) and Non-Line-of-Sight (NLOS) scenarios (Fig. 19.4). Controlled variable experiments, similar to the feather and lead ball falling freely in a vacuum, are used to demonstrate scientific principles. This study focuses on the physical properties of millimeter-wave, such as signal attenuation, base station height, coverage distance, range, and signal reflection from different surfaces.

The attenuation of millimetre-wave mobile communication signals can be described using a path loss model, where the attenuation value is a function of distance and other relevant factors. One commonly used path loss model is the free space propagation model, with the following formula:

$$\mathrm{Attenuation(dB)} = 20 * \log 10(\mathrm{d}) + 20 * \log 10(\mathrm{f}) + \mathrm{C}$$

where

Attenuation	The loss of signal during propagation, measured in dB.
D	Transmission distance, i.e., the distance over which the signal propagates, measured in meters.
f	Signal frequency is measured in Hertz.
C	Constant that incorporates other factors related to the propagation environment and system, such as antenna gain, environmental attenuation, etc.

According to the attenuation formula of the free space propagation model, it can be observed that as the distance increases, the attenuation value also increases. This is because the signal encounters free space path loss during propagation, and with increasing distance, the signal energy weakens inversely proportional to the square of the distance, as Fig. 19.5 presented. Therefore, this thesis focuses on urban interface and public space design within the coverage of a single base station.

Parametric Analysis After collecting the aforementioned data, we constructed a Grasshopper analysis model to simulate urban space signal coverage. By constructing a parametric analysis model, we calculated the actual coverage of a single-millimetre wave base station.

Specifically, the current situation analysis is divided into three parts: the first is to determine the location of the millimetre-wave base station, the second is to obtain the LOS line-of-sight area and NLOS non-line-of-sight area of the target space scene,

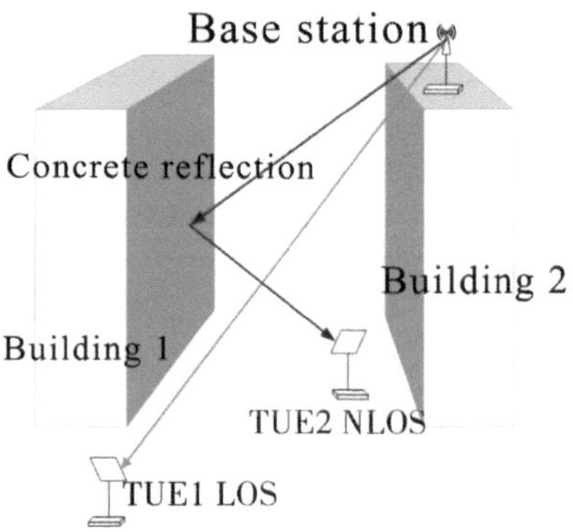

Fig. 19.4 Comparison of LOS and NLOS channels. (*Source* © Gao Shuai)

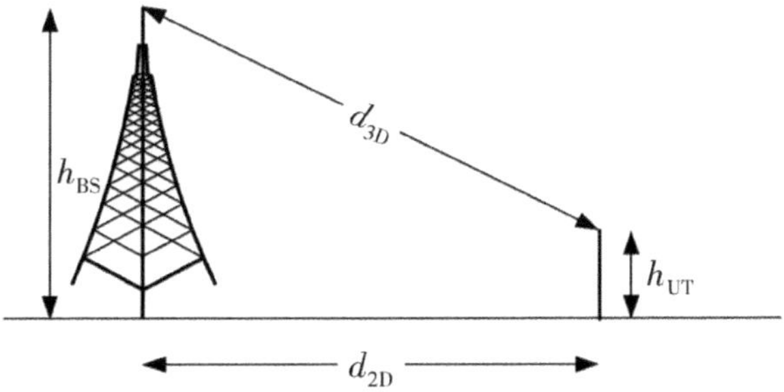

Fig. 19.5 Outdoor line-of-sight (LOS) propagation diagram in an urban environment. (*Source* © Gao Shuai)

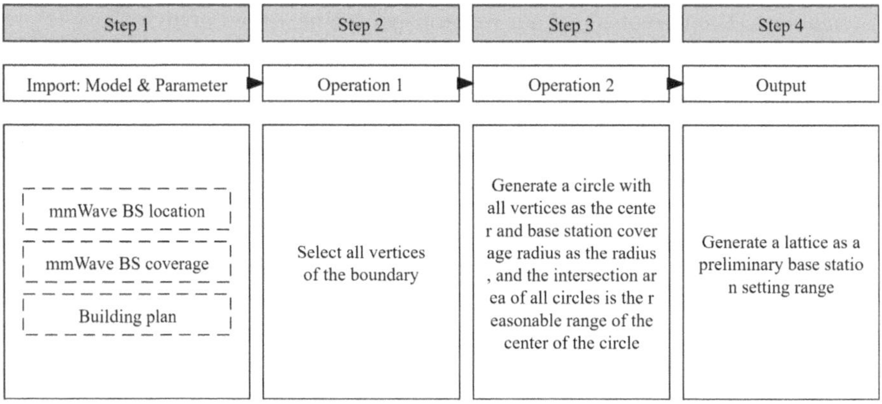

Fig. 19.6 The technical methodology for conducting Line-of-Sight (LOS) analysis of the scene. (*Source* © Authors)

and the last is to calculate the direct signal coverage area in the line-of-sight area and calculate the effective reflected signal area under ideal reflection conditions (from Figs. 19.6, 19.7 and 19.8).

In the analysis combined with actual cases, there will be problems that the analysis process is not clear because the model is too complicated. In order to display the above analysis methods more intuitively, the simplified version of the model is taken as an example for visual display. The analysis process will be omitted, and only the results will be presented when the actual case studies are conducted in subsequent chapters.

The analysis results are presented in the quantified form of effective coverage of millimetre waves. As shown in Fig. 19.9, processes 1–3 show how to determine the location of the millimetre-wave base station based on the site's outline. Meanwhile, processes 4–6 present the LOS analysis and 7–8 are the specular reflection analyses.

19.2.3 Case Analysis

The analysis model imports a simplified version of Sanlitun Taikooli into Grasshopper for Rhinoceros. It includes three-dimensional models of buildings and urban landscape infrastructure. The analysis evaluates line-of-sight (LOS) and first-order mirror-reflection transmission of millimeter-wave signals. LOS transmission assesses obstructions between arbitrary locations and the base station, while mirror reflection transmission calculates signals reflected by building facades, reaching outdoor terraces and rooftops without secondary obstructions. The analysis simplifies by disregarding the horizontal dimensions of architectural elements, focusing on the signal obstacles of buildings.

Fig. 19.7 Top, the technical methodology for conducting line-of-sight (LOS) analysis of the site. (*Source* © Author)

Fig. 19.8 Bot, the technical methodology for conducting line-of-sight (NLOS) analysis of the site. (*Source* © Authors)

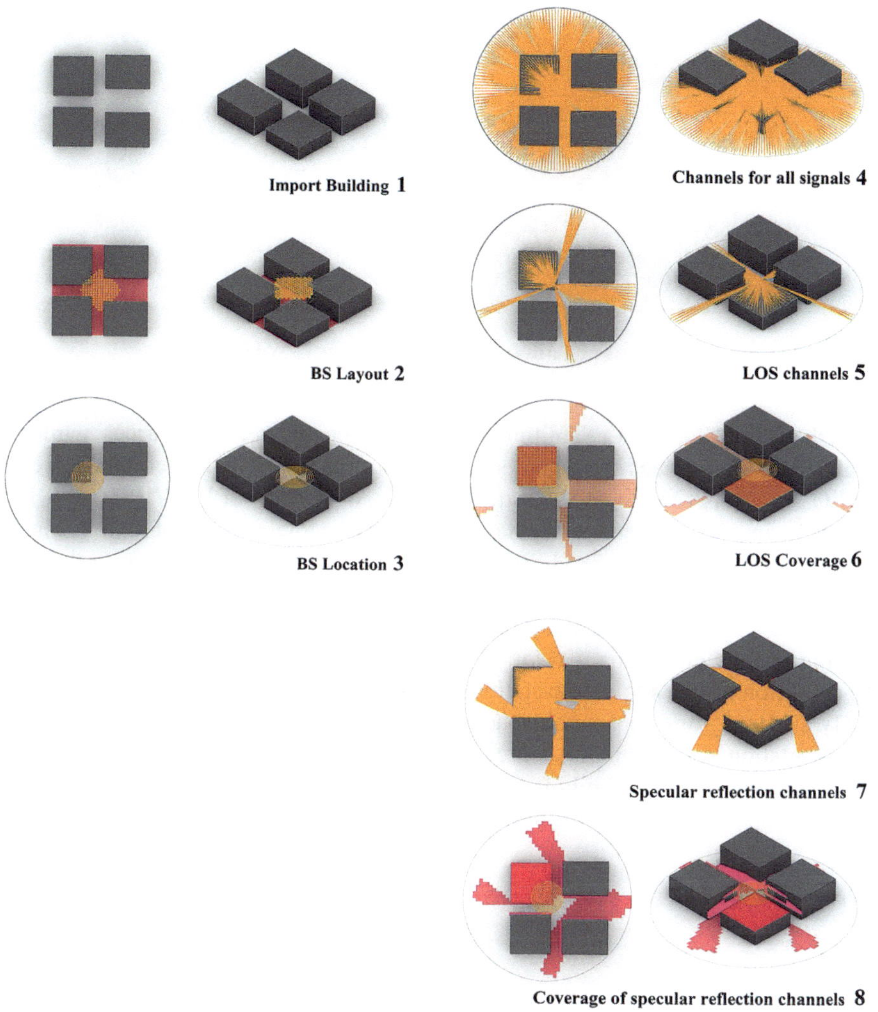

Fig. 19.9 The technical methodology for conducting line-of-sight (LOS) analysis of the scene. (*Source* © Authors)

The base station deployment locations are imported into the analysis model as spatial coordinate points. It should be noted that the determination of base station deployment locations also needs to satisfy certain criteria, requiring the construction of an independent parameterised analysis model, as presented in Fig. 19.6.

The analysis model includes the base station's horizontal coverage range, which is determined by its height. Outdoor rooftops are accessible to pedestrians. The parametric model in Grasshopper outputs the number of matrices occupied by LOS and NLOS signal coverage, allowing data such as signal coverage area and ratio to be obtained (Fig. 19.10).

Fig. 19.10 Parameter analysis of TaiKoo Li Sanlitun. (*Source* © Wikipedia)

In the case of TaiKoo Li Sanlitun, the LOS area occupies 1257 matrices, while the total number of matrices in the outdoor reachable area is 10904. Therefore, the area of the LOS region is $1257 * 4 = 5028$ square meters, with a ratio of $1257/10904 \approx 0.115$. Minus the invalid area outside the Sanlitun block, the total number of matrices in the outdoor reachable area is 7854. Subtracting the overlapping portion from the LOS coverage area, the total signal coverage area is 1556 matrices, corresponding to an area of 6224 square meters, with a ratio of approximately 0.198. Thus, the maximum coverage rate of millimetre wave in the Sanlitun South Region is about 19.8% under the condition of the single base station.

By analysing different urban spaces with varying plot ratios and morphologies, the correlation and correlation degree between millimetre-wave signal coverage and factors such as plot ratio, building spacing, and building height can be obtained. Due to limited space, this section is not presented in this paper (Figs. 19.11 and 19.12).

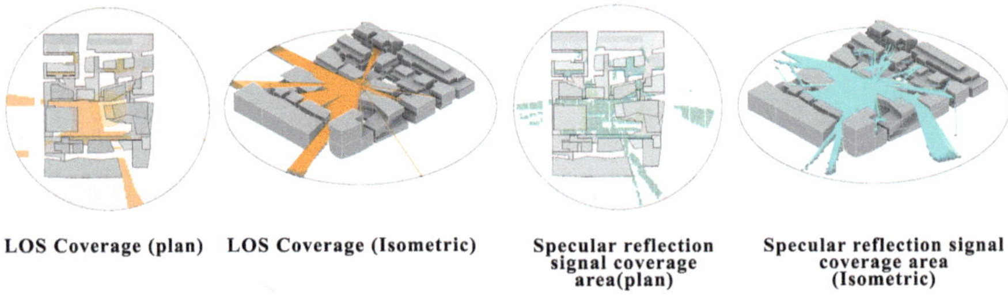

Fig. 19.11 Parameter analysis of TaiKoo Li Sanlitun. (*Source* © Authors)

Fig. 19.12 Parameter analysis of TaiKoo Li Sanlitun. (*Source* © Authors)

19.2.4 Signal Envelope

Regarding the incremental design methods for urban public space interfaces from the perspective of millimeter-wave adaptability, the concept of the "solar envelope" can be referenced in the context of solar-responsive urban design.

In 1974, Nourse published his first book titled "Energy and Form: The Ecological Approach to Urban Growth," in which he introduced the design method of the "solar envelope" for controlling the interface form of buildings based on solar adaptability to regulate the overall form of the city (Knowles 1974), as Fig. 19.13 shown. Forty-two years later, in September 2016, the Harvard Graduate School of Design (GSD) organised a two-day conference called "Heliomorphism," the inaugural event of the Urbanization Lab created by architect and urbanist Charles Waldheim. The conference served as a platform to explore Nourse's research on the "solar envelope," which is a legal provision that, if implemented as policy, would ensure the utilisation of solar energy in urban areas.

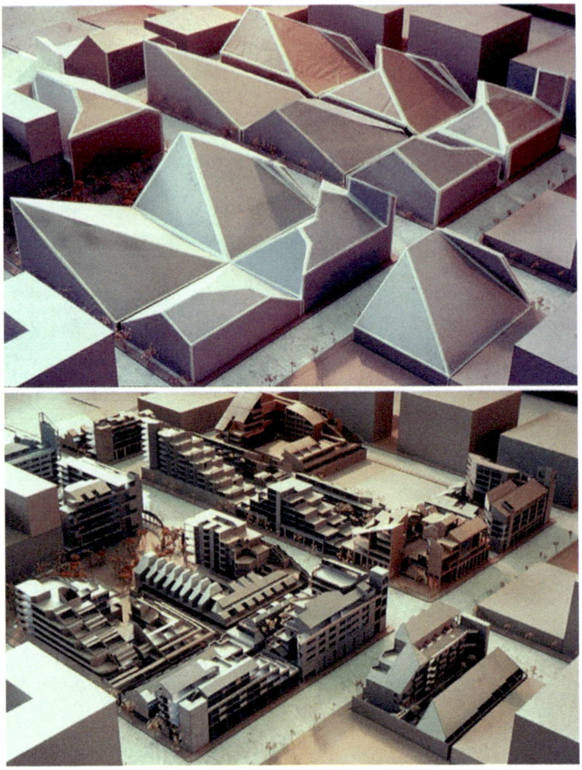

Fig. 19.13 A "Solar envelope". (*Source* © Knowles R L)

This paper defines the adaptation design of urban public space interfaces for millimetre-wave (mmWave) communications as the "signal envelope".

The underlying logic of the "signal envelope" stems from the essence of the study, which involves striking a balance between the requirements of the mmWave mobile communication channel and the spatial demands of mmWave applications in urban environments. This necessitates the allocation of mmWave coverage based on specific spatial needs, whereby the initial design incorporates the spatial requirements of mmWave application scenarios along with other elements of spatial design (such as traffic flow, green spaces, recreational areas, and other infrastructure needs). The mmWave coverage range is then delineated around the base station locations, forming a three-dimensional space known as the Line-of-Sight (LOS) mmWave channel. The generation logic of the signal envelope is shown in Fig. 19.14.

The negative space within this three-dimensional envelope represents the "signal envelope." In the design of buildings and landscape structures, as long as they do not exceed the boundaries of the "signal envelope," the alignment between the spatial requirements of mmWave applica- tion scenarios and the needs of the mmWave channel can be ensured. The architectural inter- face design of signal envelope based on Taikoo Li Sanlitun is shown in Fig. 19.15.

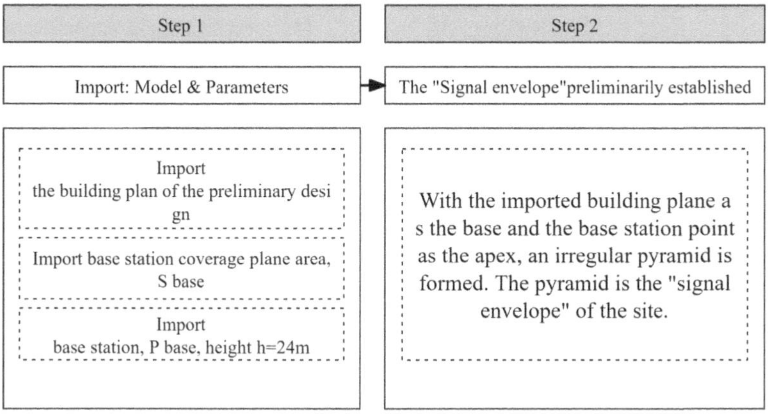

Fig. 19.14 Logic structure of "Signal envelope". (*Source* © Authors)

After obtaining the signal envelope of Sanlitun, it is refined by means of coordination with the floor area ratio, as shown in Fig. 19.16. The refined signal cover was imported into the parametric analysis operator and its LOS coverage area was calculated and compared with the analysis results of the current model, as shown in Table 2.1. It can be seen that when the channel cover is used as the exterior contour of the building, the LOS coverage rate of a single millimetre wave base station is 100 per cent, while the outer contour of the current Sanlitun block, this value is 11.5%. The effect diagram of the signal envelope is shown in the Table 2.1.

At the same time, the channel cover can also be coordinated with other elements, such as the intersection of the channel cover and the sunshine cover, to achieve comprehensive constraints on urban space design, as shown in Figs. 19.17, and 19.18.

19.3 Conclusions

Millimeter Waves have limited coverage due to propagation and penetration characteristics, posing challenges for building penetration in urban areas. Blocking 5G millimetre wave signals by urban buildings, plants, and other interfaces has resulted in more dense base station deployments, resulting in large investments and increased energy consumption. The existing MMwave base station deployment optimisation model oversimplifies urban buildings into two-dimensional planes, resulting in inaccurate 3D coverage calculation in line-of-sight (LOS) scenarios. The lack of analysis of height variations and complex structures further hampers accuracy.

This paper proposes an innovative approach based on infrastructure architecture and complex adaptive systems to enhance mmWave coverage efficiency. It introduces a simulation system and analysis tools to quantify the relationship between spatial

	Millimeter-wave LOS signal coverage plan	Millimeter wave LOS signal coverage axonometry
Current LOS area coverage: 11.5%	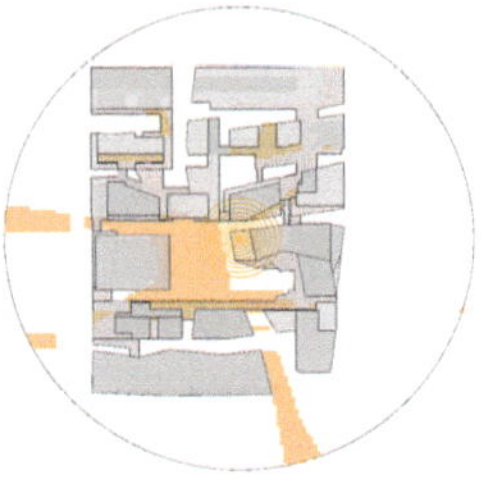	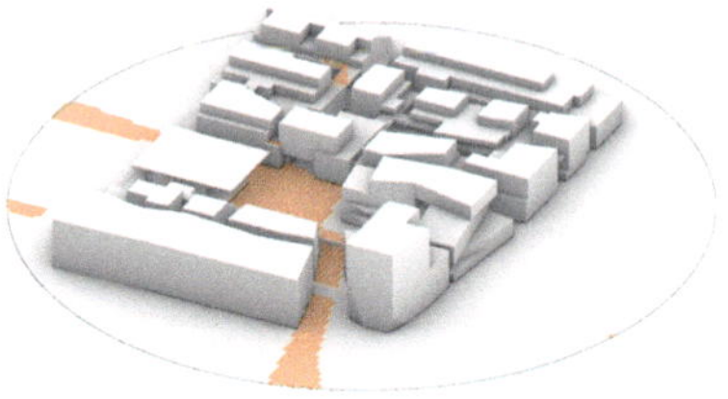
LOS area coverage under signal envelope constraints:100%	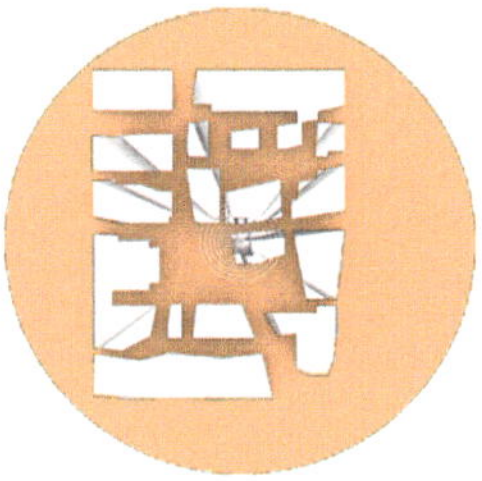	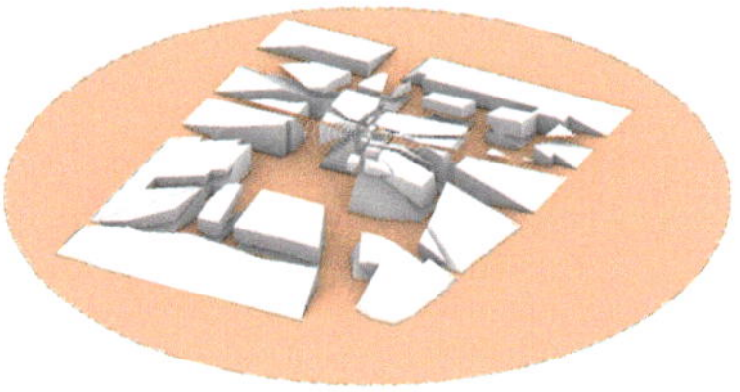

Fig. 19.15 A LOS coverage ratio comparison between the signal envelope and the current urban space. (*Source* © Authors)

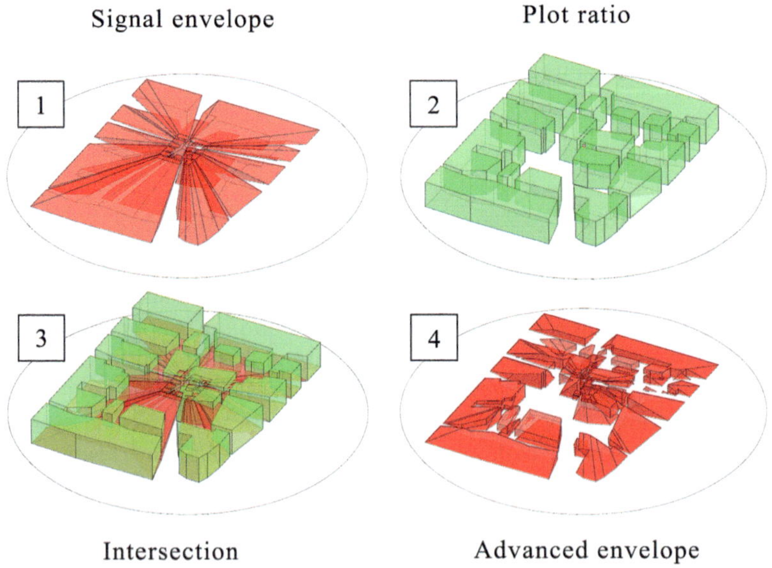

Fig. 19.16 Integration of channel cover and standard plot ratio. (*Source* © Authors)

Fig. 19.17 Signal envelope design concept figure: the above is the channel cover model, and the following is the concept effect. (*Source* © Authors)

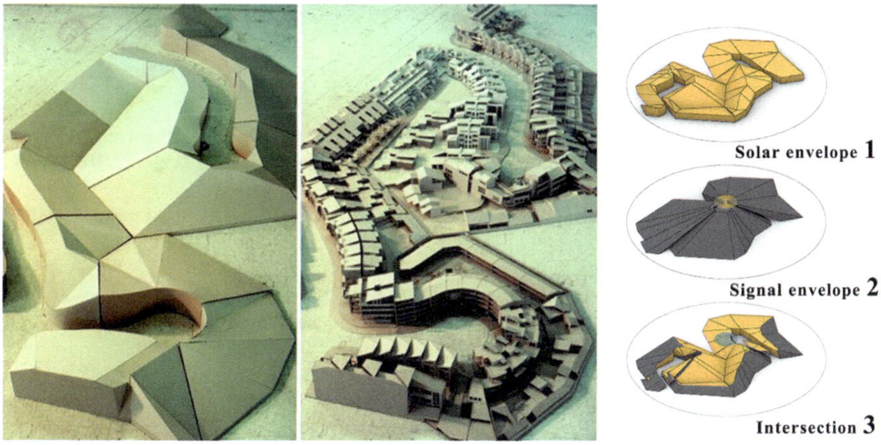

Fig. 19.18 Common constraints of Solar envelope and signal envelope. (*Source* © Knowles R L, *Source* © Authors)

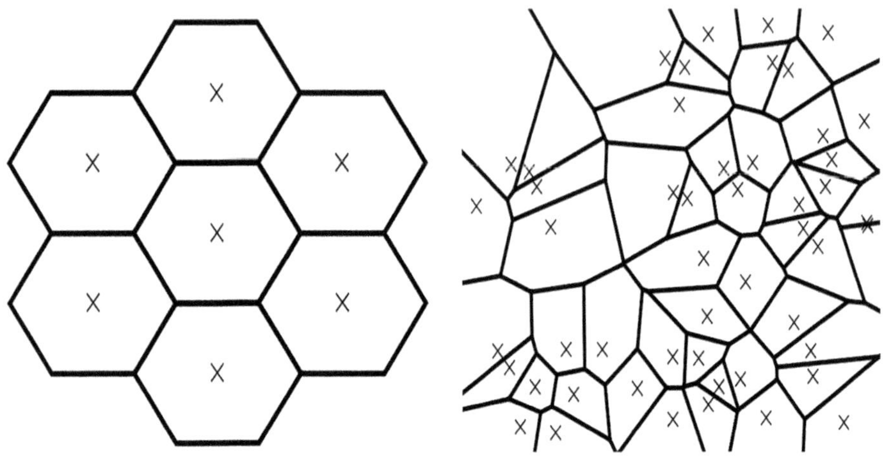

Fig. 19.19 Evolution of cellular network layout. On the left is the traditional layout, and on the right is the 5 g layout. (*Source* © Buzzi, S)

morphology and LOS scenarios, enabling further research on adaptive design. The "signal envelope" theory is proposed, constructing an ideal model for incremental design adaptive to millimetre waves. This paper proposes that the layout of 5G mobile communication can be returned from an irregular network structure to a regular cellular network structure through the constraint of channel cover, as presented in Fig. 19.19.

This paper advances a cutting-edge view of urban space design in the context of new infrastructure construction and urges architecture and urban design to assume more social responsibilities in promoting urban reality as a digital dream.

References

Al-Falahy N, Alani O (2019) Coverage and capacity improvement of millimetre wave 5G network using distributed base station architecture. IET Netw 8(4):246–255. https://doi.org/10.1049/iet-net.2018.5059

Bai T, Heath RW (2015) Coverage and rate analysis for millimeter-wave cellular networks. IEEE Trans Wirel Commun 14(2):1100–1114. https://doi.org/10.1109/TWC.2014.2364267

Batty M (2013) The new science of cities. MIT Press

Buzzi S., I C-L, Klein TE, Poor HV, Yang C, Zappone A (2016) A survey of energy-efficient techniques for 5G networks and challenges ahead. IEEE J Sel Areas Commun 34(4):697–709. https://doi.org/10.1109/JSAC.2016.2550338

Holland JH (2000) Emergence: from chaos to order. OUP Oxford

IBM Institute for Business Value (2009) Smart earth. Oriental Publishing House, Shanghai

Knowles RL (1974) Energy and form: an ecological approach to urban growth

Lee J, Tejedor E, Ranta-Aho K, Wang H, Lee K-T, Semaan E, Mohyeldin E, Song J, Bergljung C, Jung S (2018) Spectrum for 5G: Global status, challenges, and enabling technologies. IEEE Commun Mag 56(3):12–18. https://doi.org/10.1109/MCOM.2018.1700818. Retrieved from

Maccartney GR, Zhang J, Nie S, Rappaport TS (2013) Path loss models for 5G millimetre wave propagation channels in urban microcells. In: GLOBECOM—IEEE global telecommunications conference, pp 3948–3953. https://doi.org/10.1109/GLOCOM.2013.6831690. Retrieved from

Mumford L (1961) The city in history. Harcourt Brace, New York, pp 563–567

Palizban N, Szyszkowicz S, Yanikomeroglu H (2017) Automation of millimetre wave network planning for outdoor coverage in dense urban areas using wall-mounted base stations. IEEE Wirel Commun Lett 6(2):206–209.https://doi.org/10.1109/LWC.2017.2659732

Pi Z, Khan F (2011) An introduction to millimeter-wave mobile broadband systems. IEEE Commun Mag 49(6):101–107. https://doi.org/10.1109/MCOM.2011.5783993

Qamar F, Hindia N, Dimyati K, Noordin KA, Majed MB, Rahman TA, Amiri IS (2019) Investigation of future 5G-Iot millimetre-wave network performance at 38 GHz for urban microcell outdoor environment. Electronics (Switzerland) 8(5). https://doi.org/10.3390/electronics8050495

Rappaport TS, Gutierrez F Jr, Ben-Dor E, Murdock JN, Qiao Y, Tamir JI (2013) Broadband millimeter-wave propagation measurements and models using adaptive-beam antennas for outdoor urban cellular communications. IEEE Tran Antennas Propag 61(4, 1, SI):1850–1859. https://doi.org/10.1109/TAP.2012.2235056

Sharma SK, Bogale TE, Le LB, Chatzinotas S, Wang X, Ottersten B (2018) Dynamic spectrum sharing in 5G wireless networks with full-duplex technology: recent advances and research challenges. IEEE Commun Surv Tutorials 20(1):674–707. https://doi.org/10.1109/COMST.2017.2773628

Sulyman A, Nassar A, Samimi M, Maccartney G, Rappaport T, Alsanie A (2014) Radio propagation path loss models for 5G cellular networks in the 28 GHZ and 38 GHZ millimeter-wave bands. IEEE Commun Mag 52(9):78–86. https://doi.org/10.1109/MCOM.2014.6894456. Retrieved from

Tan Z (2016) Seeking parameters of modernity: infrastructure architecture. Times Archit 2016(02):6–13. https://doi.org/10.13717/j.cnki.ta.2016.02.003

Varnelis K (2008) The infrastructural city: networked ecologies in Los Angeles. Actar (03). ISBN-13:978-8496954250

Wang Q, Zhao X, Lv Z et al (2020) Optimising the ultra-dense 5G base stations in urban outdoor areas: coupling GIS and heuristic optimisation. Sustain Cities Soc 63:102445

Part VIII
Micro Scale: Sustainable Digital and Technological Solutions

Chapter 20
The Science of Play: An STS Analysis of Serious Games and Co-production

Coline Senior, Bradley Loewen, and Alenka Temeljotov Salaj

Abstract This chapter explores the concept of co-production in the context of science and technology studies (STS) and its application to serious games in smart cities. Co-production, which highlights the interdependence of science, technology, and society, challenges traditional notions of expertise and emphasises the active involvement of diverse actors in the design and implementation of technologies. We examine the role of serious games in public participation in smart cities, focusing on their outcomes and limitations from an STS perspective. By focusing on a case study from the Norwegian University of Science and Technology, we delve into how a serious game has been designed for children's participation in urban planning, underscored by STS principles. The concept of co-production in STS emphasises the importance of recognising the complex interactions between scientific knowledge, societal practices, and technological innovation. STS scholars argue that scientific knowledge is shaped by social and cultural contexts, and they highlight the importance of understanding the social and political dimensions of scientific knowledge production in policy-making processes. The benefits of serious games are documented in the literature for increasing engagement, understanding complex urban issues, and promoting collaborative decision-making. However, it has also been challenged by critical perspectives from STS scholars who question the transformative potential of serious games, citing evidence that some games reinforce existing socio-technical systems rather than fostering radical change. Results of our study show the importance of a critical approach to serious games, recognising them not only as tools but also as objects of study that reflect broader social and cultural values. Overall, this chapter contributes to the field of STS and Urban Facilities Management by examining the intersection of co-production, serious games, and smart cities,

C. Senior (✉) · B. Loewen · A. Temeljotov Salaj
Norwegian University of Science and Technology, Trondheim, Norway
e-mail: coline.senior@ntnu.no

B. Loewen
e-mail: bradley.loewen@ntnu.no

A. Temeljotov Salaj
e-mail: alenka.temeljotov-salaj@ntnu.no

M. Vukmirović et al. (eds.), *Digital Dreams, Urban Realities*, The Urban Book Series,
https://doi.org/10.1007/978-3-031-76353-3_20

providing a valuable perspective on the potential and limitations of serious games in fostering children's participation and shaping sustainable urban development.

Keywords Serious games · Science and technology studies · Public participation · Sustainable cities

20.1 Introduction

The intersection of science, technology, and society has become increasingly relevant in the context of urban planning and smart cities (Hommels 2020). Science and technology studies (STS) explore how society, culture, and politics interact with scientific and technological development. The concept of co-production is central to STS, which considers science, technology, and society as intimately linked and co-constructed. Smart cities, with their integrated and interconnected technological systems, present an ideal context for exploring co-production through serious games, such as Field (Cavada and Rogers 2020). This chapter explores the intersection of STS and serious games within the domain of smart cities, focusing on their potential and limitations for fostering children's participation.

The starting point for this study is an ongoing research project at the Norwegian University of Science and Technology's Department of Civil and Environmental Engineering, focused on children's participation in smart city development. During the first year of the project, the researchers developed a serious game prototype to teach children about sustainable urban transformation and enable their participation in decision-making processes. There is a significant body of scientific literature on the use of serious games for public participation that documents their manifold benefits. However, the concept of co-production seems to be overlooked in the engineering field, and there is a lack of understanding of how these serious games are shaped by diverse knowledge systems.

After defining the terms of smart cities as well as public and children's participation, the first section will present the research landscape of co-production in STS, examining how scientific knowledge, societal practices, and technological innovations are intertwined, particularly within the context of sustainable urban transformation. In our theoretical framework, we present the evolution of co-production within the STS field, highlighting the dynamic interplay between technology, society, and science. This serves as a foundation for a critical analysis of serious games, emphasising their role as educational tools but also as sociotechnical artefacts embedded within complex urban ecosystems.

Employing a design research methodology, we examine a specific case study from the Norwegian University of Science and Technology. This investigation not only showcases the practical implementation of serious games in smart city contexts but also offers insights into their design, objectives, and impact on children's engagement. The third section broadens the discourse to critically reflect on the role of serious games in public participation processes, challenging the notion of technological determinism and exploring the sociotechnical systems that underpin urban development.

It is essential to delimit the context of this study, which concerns only "serious" games, which refer to games developed for purposes other than pure entertainment (Susi et al. 2007). These emerged from the field of education and training as a way to improve knowledge transfer and increase the engagement of participants (Laamarti et al. 2014). This chapter does not address the body of literature dedicated to games designed primarily for entertainment, such as traditional video and computer games.

20.2 Theoretical Framework

20.2.1 Smart Cities and Public Participation

A substantial amount of literature has been produced on the concept of the smart city since the term was first introduced in 1994, with a significant increase in attention following its adoption as a cornerstone of the European Union's development strategy in 2010 (Mangnus et al. 2022; Collins et al. 2021). The European Union continues to intensify its efforts and funding capacity for smart city projects. In total, the funding program Horizon Europe aimed to invest 360 million euros in research and innovation actions related to the mission of climate-neutral and smart cities (European Commission 2021).

The development of smart cities is a complex process that requires the involvement and engagement of citizens to ensure sustainability and success (Castelnovo 2016). The United Nations Sustainable Development Goal (SDG) 11 emphasises the importance of sustainable urbanisation, which includes making cities inclusive, safe, resilient, and sustainable. To achieve this goal, cities must adopt participatory approaches that engage citizens in decision-making processes, such as those related to urban planning and transportation infrastructure (Plüschke-Altof et al. 2024; Prabowo et al. 2024). Similarly, SDG 13 focuses on climate action and the need to reduce greenhouse gas emissions. Citizen engagement and upskilling are essential in achieving this goal as well, as individuals need to be aware of their impact on the environment and understand how they can reduce their carbon footprint (Preston et al. 2020; Schleicher and Schmidt 2020). Citizens' lack of motivation to engage in participatory processes makes it challenging to recruit representative samples (Jowkar et al. 2022; Lim and Yigitcanlar 2022; Salaj et al. 2020). In addition, the choice of participation method determines which demographics are capable of engaging or not. At the end of the process, the results of public participation can be difficult to implement in the plans, especially if essential knowledge about the issue at stake is not sufficiently transferred to participants. A change towards more sustainable behaviours is also necessary in order to achieve the goals of sustainable urban development (Khansari et al. 2014).

Public participation is a broad and multi-dimensional concept that has been discussed extensively in various fields of research, including urban planning, public policy, and environmental studies (Bovaird 2007; Innes and Booher 2007; Reed 2008; Rowe and Frewer 2000). According to Arnstein (1969), public participation is a process that allows citizens to actively engage in decision-making processes and to have a genuine impact on the outcomes of those processes. This definition emphasises the importance of power dynamics between the government and citizens, with public participation serving as a means to manipulate, collaborate with or empower citizens, and to ensure that their interests are taken into account. Several scholars have proposed typologies or models to conceptualise the different levels and types of public participation. For instance, Pretty (1995) proposed a ladder of citizen participation, which ranges from "manipulative participation" to "interactive participation and self-mobilization," with each rung representing a different level of citizen involvement and influence in decision-making processes. Rowe and Frewer (2000) identified three levels of public participation: passive, where citizens receive information but do not provide feedback; consultative, where citizens provide feedback but do not have decision-making power; and collaborative, where citizens and decision-makers work together to create a mutually beneficial solution. Regarding the participation of children, the ladder of children's participation is built upon the recognition that children cannot effectively participate in processes designed and led by adults unless they receive sufficient knowledge to understand what is at stake, have the opportunity to make a meaningful contribution, and receive feedback on it (Hart 2008, 1992). As shown in Fig. 20.1, in order to reach the basic degrees of genuine participation, namely "Assigned but informed" and "Consulted but informed", it is a pre-requisite that children fully understand the project and the process.

20.2.2 The Concept of Co-production in Science and Technology Studies

Co-production is a concept that has gained significant attention in Science and Technology Studies (STS) over the past few decades. Co-production is based on the premise that knowledge and expertise are distributed among different actors and that successful innovation requires their active involvement. In the context of STS, co-production is particularly relevant because it challenges the traditional notion of expertise and the role of scientists and engineers in the innovation process. Instead, it emphasises the importance of public participation and the incorporation of diverse knowledge systems in the design and implementation of technologies. According to STS scholars, scientific knowledge production and societal practices are interdependent. This refers to the idea that scientific knowledge is not simply a product of objective observation and experimentation, but rather, it is shaped and constructed through interactions between scientists, the technologies they use, and the broader social and cultural contexts in which they are embedded. Jasanoff and Kim (2009)

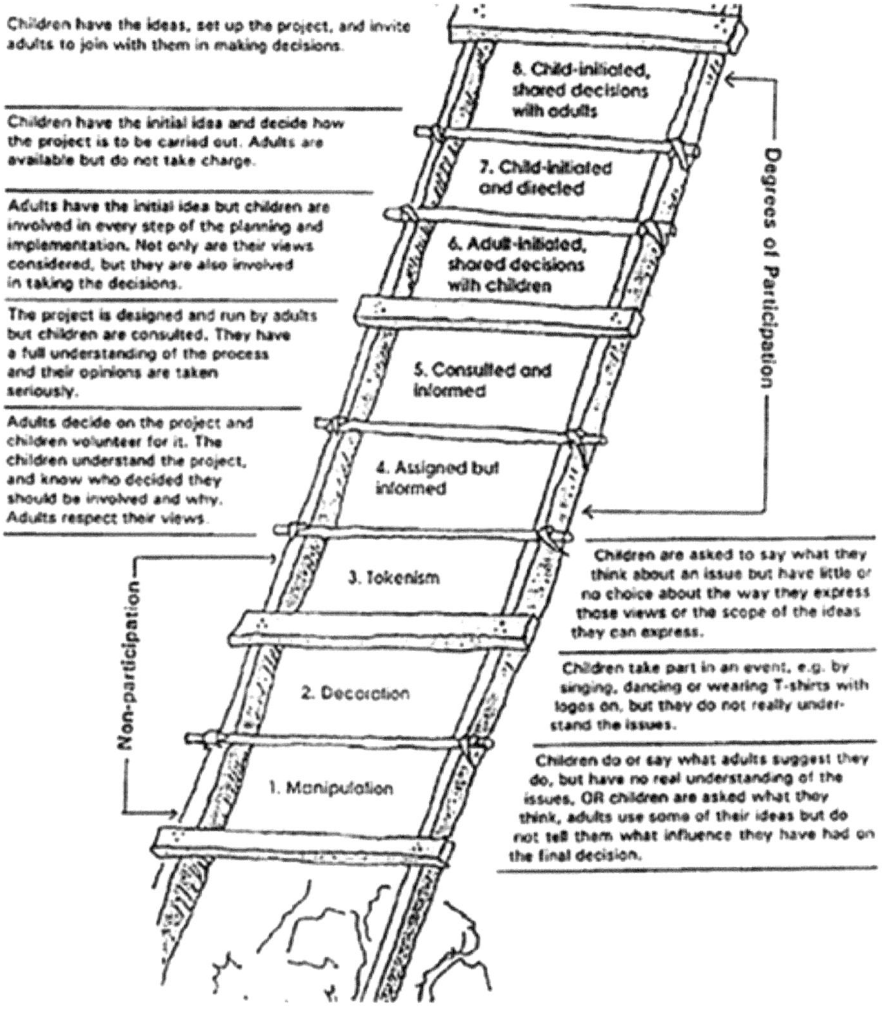

Fig. 20.1 Roger Hart's (1992) ladder of young people's participation in projects

introduced the concept of "sociotechnical imaginaries" to show how knowledge and social order are co-produced, that is, how social arrangements and institutions influence the production of knowledge. This understanding of co-production has been influential in the study of science and technology, as it emphasises the importance of recognising the social and political dimensions of scientific knowledge production in policy-making processes.

STS is a field of study that explores the interrelations between society, technology and science, thereby providing a useful framework for analysing the socio-technical dimensions of co-production in smart cities. According to STS scholars, technological innovations are not solely the result of technical expertise but rather emerge through a complex process of negotiation and co-creation between experts and users (Layton 2007). This process involves the recognition of the social and cultural context in which technologies are conceived and implemented, as well as the co-construction of knowledge and values. Hess and Sovacool (2020) argue that in the literature on public participation, STS perspectives depart from the broader literature by analysing how participatory and consultative processes construct publics rather than simply represent them. Similarly, STS perspectives challenge underlying assumptions about definitions of participation, public, and democracy. By examining how different types of publics are constructed through various participation processes, they have developed alternative designs for participation processes. This approach shows that concepts such as "the public" or "public participation" are actively constructed through social processes. The insights gained from this approach can inform the practice of public engagement by leading to alternative approaches to the design of engagement and participation.

Participation has been an important part of design for many years; however, it is a concept that is vaguely defined and polyvocal. Latour's Actor-Network Theory (ANT) provides a way to frame participation as a "matter of concern" and to challenge the concept and practice of participation (Andersen et al. 2015). In their study, Andersen et al. (2015) investigate how ANT can be used to better understand participation in participatory design by looking at an example project, Teledialogue, which used technology to strengthen the dialogue between children living in foster care and their social workers. ANT emphasises the mediated and processual aspect of reality and how facts are produced and constituted through hybrid actor networks. It also shows how participation is not limited to design events or premised by physical presence or intentional interaction, and that there is no gold standard for a priori evaluation of the quality of participation. According to Andersen et al. (2015), participation is a relational and heterogeneous network achievement that goes beyond specific design processes and projects. Through the Teledialogue project, it is shown that children always participate with others and that their participation is manifested in action. ANT challenges the notion of participation as a means to an end and enables novel ways to discuss and practice design as a network achievement.

The longstanding problems in social theory about agency and object have been extensively discussed in the STS literature (Woolgar 1990). One of the traditional views in social theory has tended to see agency and object as separate entities, with agency being attributed to human actors and objects being seen as passive and devoid of agency. This has been criticised in STS, which argues for the co-production of agency and object in shaping social reality. As Jasanoff (2004) noted, the co-production perspective challenges the notion that humans are the only active agents and instead emphasises the active role played by things in shaping the social and political world. For example, in the context of smart cities and public participation, technologies such as sensors and data analytics are not simply passive objects that enable human action but rather are active agents that shape and influence human behaviour and decision-making (Foth et al. 2015). Foth et al. (2015) investigated the changing role of people from passive members of the plebs to active instigators of change and prompted Human–Computer Interaction (HCI) designers to consider social, situational, cultural, and contextual factors when designing for community and civic engagement.

The concept of co-production in STS, therefore, challenges traditional views of agency and object and calls for a more nuanced understanding of the complex interactions between human actors and non-human objects in shaping social reality. This view is not only more accurate, but it also has important implications for policy and practice, as it highlights the need to take into account the active role played by technologies and infrastructures in shaping social outcomes (Jasanoff 2004). By recognising the co-production of agency and object, policymakers and practitioners can design more effective interventions that take into account the complex interplay between human and non-human actors in shaping social outcomes.

20.2.3 Serious Games as a Tool for Co-production

One way to facilitate co-production in smart cities is through the use of serious games, which can be used to simulate real-world situations and enable users to experiment with different scenarios and solutions. In the context of smart cities, co-production is essential in ensuring that citizens have a voice in the development and implementation of tech-driven solutions that impact their communities. Serious games are games that have a purpose beyond entertainment, such as education or training (Scurati et al. 2020). For smart cities, serious games can be used to engage citizens in urban planning and decision-making processes (Angelidou and Psaltoglou 2019; West et al. 2019; Wolff et al. 2017). This can help to increase transparency and accountability and to improve the quality of life in cities.

When it comes to their effectiveness, serious games have proved to lead to higher levels of engagement and learning compared to traditional methods, especially by leveraging the benefits of experiential learning, also known as "learning by doing" (Bakhanova et al. 2020). They can help citizens to better understand complex urban issues and provide a means for them to express their preferences and concerns (Angelidou and Psaltoglou 2019; Horgan and Dimitrijević 2019). Additionally, serious games have the potential to reduce the power imbalance between citizens and government and to promote more collaborative decision-making processes (Koplin et al. 2017). Several case studies demonstrate the effectiveness of serious games in fostering public participation in smart cities. Gamification of processes is one way the public authorities have explored to engage the younger public in the co-creation of urban spaces; the most institutionalised example is the UN habitat program "Block by Block," using Minecraft for participatory design (Delaney 2022).

The benefits of games and serious games are that they allow learners to experience situations and engage in learning activities that are otherwise impossible or difficult to implement and experience in the real world (Corti 2006). There is evidence that the use of games supports the development of skills such as strategic thinking, planning, communication, collaboration, group decision-making, and negotiation skills (Camilleri 2022). Games in education can also bring difficult-to-understand abstract concepts closer to students and could facilitate the understanding of several scientific concepts (Gabriel and Schmölzer 2021). In their study, Gugerell et al. (2018) demonstrate that serious games that incorporate environmental storylines can serve as effective entry points for promoting environmental awareness, interest, and knowledge, even among individuals who may initially have lower levels of intrinsic motivation or environmental attitudes.

The review of the scientific literature in the field of serious games and public participation in smart cities reported mostly positive outcomes and benefits. However, when exploring the field of serious games in energy transition, we retrieved a study that largely nuances this almost unanimous consensus. Wagner and Gałuszka (2020) conducted an extensive study on the role of serious games in energy transition, in which they explored the concept of sociotechnical imaginaries and their relationship to "serious gaming". They investigated whether these imaginaries could potentially be used to influence players' attitudes towards green solutions, in which case serious games could introduce new ideas of energy governance. However, based on their analysis of 51 energy-themed serious games, they did not find evidence of such claims. On the contrary, they found that energy imaginaries presented in the games comforted the established "socio-technical regimes and its gradual evolution, rather than supporting a radical change in the energy socio-technical system" (Wagner and Gałuszka 2020). The findings from this study are important to consider in the development of a serious game for smart and sustainable urban development. They are even more relevant when developing a game that targets children, if we want to avoid fostering the same "old socio-technical systems" but rather encourage them to explore new possibilities and shape alternative futures.

Through the lens of STS, serious games are not only tools but also objects of study. They reflect broader social and cultural values and can shape the way people think about and interact with technology. STS scholars draw our attention towards a critical approach to serious games that takes into account their social and political implications.

20.3 Case Study

The ByMaker project was initiated in 2021 at the Norwegian University of Science and Technology in Trondheim, Norway. ByMaker is a serious game, designed by researchers from the fields of civil engineering, urban planning, architecture, psychology and computer science to teach children from eight to twelve years old about sustainability issues in urban planning. This section describes the most important aspects of the design, development, and testing of the first prototype.

20.3.1 Context and Need

With climate change consequences increasingly affecting cities, there is a growing need to accelerate the implementation of the United Nations Sustainable Development Goals. Projections indicate that by 2050, 70% of the global population will reside in urban areas, presenting both challenges and opportunities for sustainable development (United Nations 2019). Engaging youth in shaping the cities is critical, as they will inherit the consequences of today's decisions. In parallel, there has been a shift in young people's concerns, as reported in the 2023 Norwegian Youth report, which revealed that the top concerns for young people are "increasing prices," "conflict, instability, and challenges to democracies," and "climate and environment" in that order (Vågsland et al. 2022). This marks a change from previous years where "climate and environment" consistently ranked highest. It indicates a greater focus on issues related to social and economic aspects. In the Norwegian context, public participation in urban planning is required by law, and the Planning and Building Act further specifies that municipalities must give particular consideration to involving underrepresented groups such as children and youth (Norwegian Ministry of Environment 2008).

20.3.2 Design Science Research Methodology

Design Science Research (DSR) is a methodological approach that focuses on developing and evaluating innovative artefacts, such as digital games, to address practical problems and create new knowledge fields (Peffers et al. 2007). The development of a serious game for children's understanding and participation in sustainable urban development involves several key steps. The use of the DSR methodology is particularly relevant in this case because it emphasises the creation of technological solutions that can have a real impact on society. By following the principles of DSR, researchers can design and develop a digital gaming solution that specifically targets the challenges of children's participation in sustainable urban development and evaluate its effectiveness in improving their understanding and engagement. The iterative nature of the DSR process also allows for continuous improvement and refinement of the game based on user feedback, ensuring its relevance and effectiveness in addressing the identified challenges. Figure 20.2 illustrates the six steps of the DSR methodology undertaken in this research.

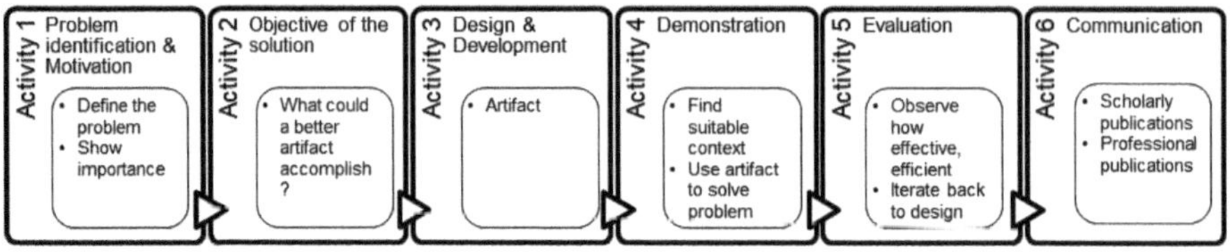

Fig. 20.2 Activities of the design science research methodology (adapted from Peffers et al. 2007)

20.4 The Artifact: The ByMaker Game

The ByMaker game leverages the engaging nature of serious gaming to cultivate a more profound comprehension of sustainability among younger individuals with the ultimate goal to provide them with sufficient knowledge and competences to make a meaningful contribution to sustainable urban projects. This section explores the details concerning the game's objectives, characteristics, user interface, and co-production process.

20.4.1 Objectives

ByMaker's primary objective is to improve eight to twelve years old pupils' understanding of sustainability by showcasing complex urban planning issues in an interactive and captivating gaming environment. The game seeks to educate them about environmental sustainability but also shed light on the more abstract social and economic aspects. By prompting players to make decisions in diverse urban development situations, ByMaker aims to cultivate a cohort of well-informed young citizens with a comprehensive view of sustainability.

20.4.2 Features and Gameplay

ByMaker's core design is based on a comprehensive 3D model of the pupils' city, in our case, the city of Trondheim in Norway. The geolocation of the game in the children's own city is crucial to provide them with a familiar environment and a tangible reality that they can easily relate to. This digital representation of the city serves as the playground where young minds can experiment with different scenarios affecting urban development, such as transportation systems, green spaces, residential zones, and public spaces. The game includes interactive features and familiar

Fig. 20.3 Screenshot of the ByMaker interface displaying the different clickable scenario options and the visual feedback on the sustainability pillars, social, environmental and economic on the bottom-left of the screen

landmarks, nurturing the players' sense of responsibility towards their local environment. Players engage with these features to make choices that reflect possible development scenarios within the urban area. Each choice influences the city's sustainability gauge, displayed visually to offer instant feedback on the outcomes of their decisions, as shown in Fig. 20.3. The sustainability gauge evolves according to the player's choices, and the characters symbolising each sustainability pillar will display different emotions according to how good the score is (i.e. a sad character means a poor score; a slightly smiling one a middle score; and an overly happy one is a perfect score). This visual representation is essential in helping players grasp how their decisions affect the overall sustainability gauge of the city.

20.4.3 User Interface

ByMaker's user interface is specifically designed to meet the requirements and preferences of the young players, focusing on simplicity, ease of use, and interactivity. The design encourages children to freely explore and engage with different elements within the game. Furthermore, it takes into account potential physical restrictions that children may have, including slower typing capabilities and less precise control over a mouse, in order to ensure accessibility and enjoyment for all players. ByMaker is built on the educational principle of experiential learning, which involves learning through hands-on experience. This learning approach is especially relevant for teaching sustainability concepts because it goes beyond traditional memorisation techniques and enables children to interact directly with the content.

20.4.4 Testing and Collecting Pupils' Feedback

The game prototype was tested in three classes from two distinct primary schools in Trondheim in a controlled setting, during which two researchers involved in the design of the game came to the classroom and briefly introduced the context of their work (i.e. research on children's participation in smart cities). After giving a basic introduction to sustainable development, the pupils were asked to answer the question, "Which words come to your mind when you hear about sustainable development?". The pupils were answering orally, and environmental considerations such as energy, wind power, water power, electric vehicles, public transportation, sorting trash, and preserving forests dominated the discussion in all three classrooms. This was explained later by the teachers, as the focus had been on ecological topics following the curriculum.

As a next step, the pupils got access to the game prototype on their own laptops and tablets via a link. They could then play freely through the game while the researchers and the class teachers went around to answer any questions and provide guidance. At this stage, the researchers documented the following observations:

- The game was best suited for pupils in the age of eight to ten as the older ones complained about the childish design and the simplistic gameplay.
- Some technical bugs resulted in frustration as some of the pupils were unable to pick-up where they left of if the game froze.
- The gauge's visual level, from nearly perfect to perfect score, was not clear enough, which resulted in a misunderstanding as to why they could not unlock the next part of the game.
- The difference of pace at which the pupils completed the game and thus the end of the assigned task created some difficulties to keep them engaged in the session as they waited for their peers to finish.

During the last part of the session, a collective summary was organised, during which the researchers went through the gameplay and showed the solution. The pupils were encouraged to interact and provide insights as to why each scenario could have positive or negative consequences on the sustainability gauge. This showed that the different scores for each scenario were made too obvious and displayed in a way that did not challenge them to think about the reasons behind it. Indeed, some pupils just made their choices to obtain the highest score without reflecting on them.

The test sessions were discussed and analysed by the research team, and the observations and feedback collected revealed positive outcomes in enhancing children's understanding of the balanced approach needed between environmental, social, and economic factors in urban planning. However, the weaknesses listed above, such as a too-obvious display of the scores for each scenario, make it challenging to assess whether the game actually fulfilled its purpose on its own. Input from the pupils resulted in additional improvements to the game by enhancing visualisation, interactive elements, and the clarity of scenario explanations. This step-by-step approach highlighted the significance of involving users in designing educational games.

20.4.5 Further Improvements

Based on the observations during the three sessions, the researchers created a roadmap for future testing sessions to include a live survey of the audience, pre- and post-gameplaying to measure learning outcomes. As shown in Fig. 20.4, the roadmap for future sessions was divided into seven sequences to keep the pupils' engaged throughout the whole 45 min, corresponding to the standard duration of a class period in Norway.

In addition, to address the shortcomings of serious games for participation identified in the literature, two feedback forms were created to let pupils influence the future development of the game. The first form, the "questionnaire", was designed with Likert scale questions to evaluate the user experience and user interface (Fig. 20.5).

The second form aims to engage the pupils in co-production and let them influence the future development of the game. As shown in Fig. 20.6, the form to be distributed during the "Design" sequence asks pupils to give input into the content of the game and suggest scenarios to be implemented in the future prototype. The medium for expressing ideas is free, and pupils can either draw or write on it. This form will also

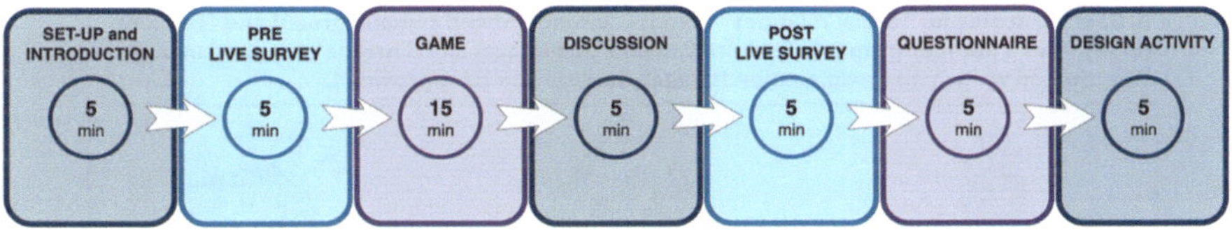

Fig. 20.4 Roadmap for future testing sessions

Fig. 20.5 Questionnaire for UX and UI testing with pupils

serve to measure learning outcomes as the pupils should assign sustainability scores to their own ideas.

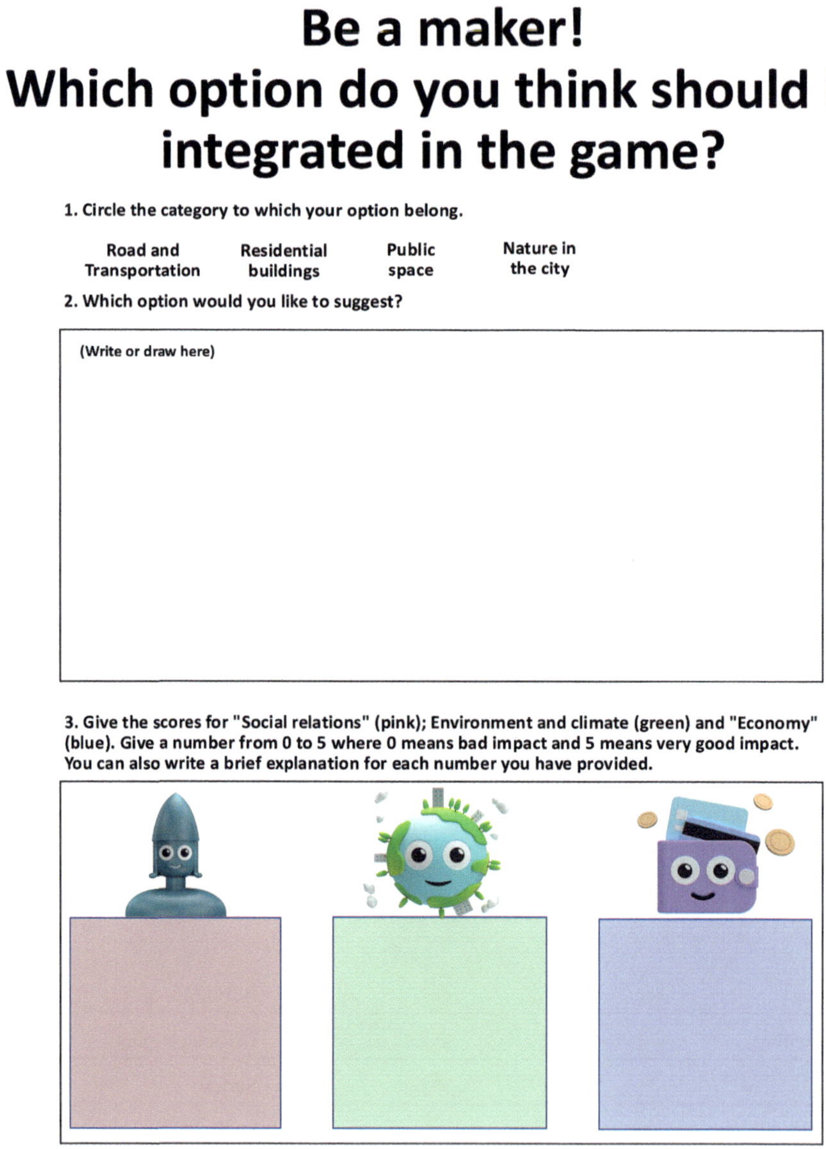

Be a maker!
Which option do you think should be integrated in the game?

1. Circle the category to which your option belong.

Road and Transportation | Residential buildings | Public space | Nature in the city

2. Which option would you like to suggest?

(Write or draw here)

3. Give the scores for "Social relations" (pink); Environment and climate (green) and "Economy" (blue). Give a number from 0 to 5 where 0 means bad impact and 5 means very good impact. You can also write a brief explanation for each number you have provided.

Fig. 20.6 "Design" form for co-production with pupils

20.5 Reflections and Conclusion

One of the challenges of serious games is that they tend to be designed by experts and professionals without much input from the users and stakeholders. This can lead to a narrow and limited understanding of the complex socio-technical issues at stake and can result in technologies that do not meet the needs and expectations of the users and stakeholders. STS scholars have also pointed out that serious games can create a false sense of participation and democracy, as they tend to be designed in a top-down manner, with little room for negotiation and co-creation among the participants. Serious games can also reinforce existing power structures and inequalities, as they tend to favour certain groups of stakeholders over others and can perpetuate the marginalisation of minority voices and perspectives.

To address these limitations and challenges, STS scholars call for a more critical and reflexive approach to the design and implementation of serious games. This approach involves recognising the diversity and complexity of the stakeholders and their knowledge systems and incorporating their perspectives and values in the design and implementation of technologies. In practice, for us, prototyping serious games for participation means creating a more equal and democratic platform for negotiation and co-creation among the participants early on and ensuring that the outcomes of the process reflect the needs and expectations of all stakeholders, including end-users.

In conclusion, serious games present an ideal context for exploring co-production in the context of public participation in smart cities. However, their implementation requires a more critical and nuanced approach, which takes into account the diversity and complexity of the stakeholders and their knowledge systems and ensures that the outcomes of the process reflect the needs and expectations of all stakeholders (Ertiö 2015). STS provides a useful framework for analysing the socio-technical dimensions of co-production in smart cities and for developing alternative approaches to the design and implementation of serious games that are more inclusive, democratic, and reflexive.

As suggested in the title, the purpose of this chapter is to introduce an analysis of the use of serious games for public participation in smart cities by examining the design and implementation phases through the lens of Science and Technology Studies. It is an attempt to open the minds of engineers working on such topics and encourage them to look beyond their approach to prototyping and user testing. Further work is needed to examine the power relations at play in participatory processes using serious games and to discuss the literature on the different levels of participation from a co-production perspective.

Acknowledgements This study is part of the research project "CaPs-Citizens as Pilots of Smart Cities", funded by Nordforsk (project number 95576) and the pilot project ByMaker, funded by "NTNU Discovery."

References

Andersen LB, Danholt P, Halskov K, Hansen NB, Lauritsen P (2015) Participation as a matter of concern in participatory design. CoDesign 11(3–4):250–261. https://doi.org/10.1080/15710882.2015.1081246

Angelidou M, Psaltoglou A (2019) Social innovation, games and urban planning: an analysis of current approaches. Int J Electron Gov 11(1):5–22. https://doi.org/10.1504/IJEG.2019.098807

Arnstein SR (1969) A ladder of citizen participation. J Am Inst Plann 35(4):216–224. https://doi.org/10.1080/01944366908977225

Bakhanova E, Garcia JA, Raffe WL, Voinov A (2020) Targeting social learning and engagement: what serious games and gamification can offer to participatory modeling. Environ Model Softw 134:104846. https://doi.org/10.1016/J.ENVSOFT.2020.104846

Bovaird T (2007) Beyond engagement and participation: user and community coproduction of public services. Public Adm Rev 67(5):846–60. https://doi.org/10.1111/J.1540-6210.2007.00773.X

Camilleri P (2022) Esports: a new frontier for inclusion through competitive game engagement. In: Inclusive digital education. Educational communications and technology: issues and innovations, pp 293–309. https://doi.org/10.1007/978-3-031-14775-3_19

Castelnovo W (2016) Co-production makes cities smarter : citizens' participation in smart city initiatives. In: Fugini M, Bracci E, Sicilia M (eds) Co-production in the public sector. Springer International Publishing, pp 97–117. https://doi.org/10.1007/978-3-319-30558-5

Cavada M, Rogers CDF (2020) Serious gaming as a means of facilitating truly smart cities : a narrative review. Behav Inf Technol 39(6):695–710. https://doi.org/10.1080/0144929X.2019.1677775

Collins D, Johansen A, Kalsaas BT, Temeljotov-Salaj A, Hamdy M (2021) Brought by degrees: a focus on the current indicators of lean 'smartness' in smart cities. In: IGLC 2021—29th annual conference of the international group for lean construction—lean construction in crisis times: responding to the post-pandemic AEC industry challenges, pp 167–76. https://doi.org/10.24928/2021/0156

Corti K (2006) Games-based learning: a serious business application. Informe De PixelLearning 34(6):1–20

Delaney J (2022) Minecraft and playful public participation in urban design. Urban Plann 7(2):330–342. https://doi.org/10.17645/up.v7i2.5229

Ertiö TP (2015) Participatory apps for urban planning—space for improvement. Plan Pract Res. https://doi.org/10.1080/02697459.2015.1052942

European Commission (2021) Climate-neutral and smart cities. https://research-and-innovation.ec.europa.eu/funding/funding-opportunities/funding-programmes-and-open-calls/horizon-europe/eu-missions-horizon-europe/climate-neutral-and-smart-cities_en

Foth M, Tomitsch M, Satchell C, Hank Haeusler M (2015) From users to citizens: some thoughts on designing for polity and civics. In: OzCHI 2015: being human—conference proceedings, pp 623–33. https://doi.org/10.1145/2838739.2838769

Gabriel S, Schmölzer B (2021). Climate4Kids : a gamified app teaching about climate change. Eur Conf Game Based Learn 236–243

Gugerell K, Funovits P, Ampatzidou C (2018) Daredevil or socialiser? Exploring the relations between intrinsic motivation, game experience and player types in serious games with environmental narratives. In: Participatory design theory, pp 157–178. https://doi.org/10.4324/9781315110332-12

Hart R (1992) Children's participation: from tokenism to citizenship. Unicef: innocenti essays. Vol 4 Florence: International child development centre. http://www.unicef-irc.org/publications/pdf/childrens_participation.pdf

Hart R (2008) Stepping back from 'the ladder': reflections on a model of participatory work with children. In: Reid A, Jensen BB, Nikel J, Simonovska V (eds) Participation and learning. Springer, Dordrecht, pp 19–31. https://doi.org/10.1007/978-1-4020-6416-6

Hess DJ, Sovacool BK (2020) Sociotechnical matters: reviewing and integrating science and technology studies with energy social science. Energy Res Soc Sci 65:101462. https://doi.org/10.1016/J.ERSS.2020.101462
Hommels A (2020) STS and the city: techno-politics, obduracy and globalisation. Sci Culture 29(3):410–416. https://doi.org/10.1080/09505431.2019.1710740
Horgan D, Dimitrijević B (2019) Frameworks for citizens participation in planning: from conversational to smart tools. Sustain Cities Soc 48. https://doi.org/10.1016/j.scs.2019.101550
Innes JE, Booher DE (2007) Reframing public participation: strategies for the 21st century. Plann Theory Pract 5(4):419–436. https://doi.org/10.1080/1464935042000293170
Jasanoff S (2004) States of knowledge: the co-production of science and the social order. In: States of knowledge: the co-production of science and the social order. https://doi.org/10.4324/9780203413845
Jasanoff S, Kim SH (2009) Containing the atom: sociotechnical imaginaries and nuclear power in the United States and South Korea. Minerva 47(2). https://doi.org/10.1007/s11024-009-9124-4
Jowkar M, Temeljotov-Salaj A, Lindkvist CM, Støre-Valen M (2022) Sustainable building renovation in residential buildings: barriers and potential motivations in Norwegian culture. Constr Manag Econ 40(3):161–172. https://doi.org/10.1080/01446193.2022.2027485
Khansari N, Mostashari A, Mansouri M (2014) Impacting sustainable behavior and planning in smart city. Int J Sustain Land Use Urban Plann 1(2). https://doi.org/10.24102/ijslup.v1i2.365
Koplin M, Siegert S, Neuvonen A, Salo K, Kerney K, Skelton C, Culén A, Nadelkovski I (2017) The smart city as shared design space. In: Smart cities in the mediterranean, pp 153–74. https://doi.org/10.1007/978-3-319-54558-5_7
Laamarti F, Eid M, El Saddik A (2014) An overview of serious games. Int J Comput Game Technol 2014. https://doi.org/10.1155/2014/358152
Layton D (2007) Revaluing the T in STS. Int J Sci Educ 10(4):367–78. https://doi.org/10.1080/0950069880100404
Lim SB, Yigitcanlar T (2022) Participatory governance of smart cities: insights from e-participation of Putrajaya and Petaling Jaya, Malaysia. Smart Cities 2022 5(1):71–89. https://doi.org/10.3390/SMARTCITIES5010005
Mangnus AC, Vervoort JM, Renger W-J, Nakic V, Rebel KT, Driessen PPJ, Hajer M (2022) Envisioning alternatives in pre-structured urban sustainability transformations : too late to change the future ? Cities 120:103466. https://doi.org/10.1016/j.cities.2021.103466
Norwegian Ministry of Environment (2008) Norwegian planning and building Act—Act of 27 June 2008 No. 71 relating to planning and the processing of building applications (the planning and building act) (the planning part)." https://www.regjeringen.no/en/dokumenter/planning-building-act/id570450/
Peffers K, Vegas L, Tuunanen T, Rothenberger MA (2007) a design science research methodology for information systems research. J Manag Inf Syst 24(3):45–77. https://doi.org/10.2753/MIS0742-1222240302
Plüschke-Altof B, Loewen B, Müüripeal A, Sooväli-Sepping H (2024) Balancing participatory planning and planning for resilience in nature-based solutions. A case of transformative agency? Geogr Ann, Ser B Hum Geogr 106(4):422–440. https://doi.org/10.1080/04353684.2023.2258931
Prabowo BN, Salaj AT, Lohne J (2024) Urban heritage facility management: a conceptual framework for the provision of urban-scale support services in Norwegian world heritage sites. Heritage 7:1372–1399. https://doi.org/10.3390/heritage7030066
Preston S, Mazhar MU, Bull R (2020) Citizen engagement for co-creating low carbon smart cities: practical lessons from Nottingham city council in the UK. Energies 13(24):6615. https://doi.org/10.3390/EN13246615
Pretty JN (1995) Participatory learning for sustainable agriculture. World Dev 23(8):1247–1263. https://doi.org/10.1016/0305-750X(95)00046-F
Reed MS (2008) Stakeholder participation for environmental management: a literature review. Biol Conserv 141(10):2417–2431. https://doi.org/10.1016/J.BIOCON.2008.07.014

Rowe G, Frewer LJ (2000) Public participation methods: a framework for evaluation. Sci Technol Hum Values 25(1):3–29. https://doi.org/10.1177/016224390002500101

Schleicher K, Schmidt C (2020) Citizen science in Germany as research and sustainability education: analysis of the main forms and foci and its relation to the sustainable development goals. Sustainability 12(15):6044. https://doi.org/10.3390/SU12156044

Scurati GW, Ferrise F, Bertoni M (2020) Sustainability awareness in organizations through gamification and serious games: a systematic mapping. In: Proceedings of the NordDesign 2020 Conference, NordDesign 2020. https://doi.org/10.35199/norddesign2020.1

Susi T, Johanenesson M, Backlund P (2007) Serious games effects: an overview. Procedia Soc Behav Sci 174:2240–44. https://doi.org/10.1016/j.sbspro.2015.01.881

Salaj T, Alenka SG, Senior C, Xue Y, Lindkvist C (2020) An interactive tool for citizens' involvement in the sustainable regeneration. Facilities 38(11–12):859–870. https://doi.org/10.1108/F-09-2019-0099

United Nations (2019) World urbanization prospects the 2018 revision. Department of Economic and Social Affairs, Population Division, New York

Vågsland V, Palm A, Arntzen K, Opinions (2022) UNG2023 Rapport Og Målgruppestudie Om Unge i Norge. Oslo

Wagner A, Gałuszka D (2020) Let's play the future: sociotechnical imaginaries, and energy transitions in serious digital games. Energy Res Soc Sc 70:101674. https://doi.org/10.1016/J.ERSS.2020.101674

West M, Yildirim O, Harte AE, Ramram A, Fleury NW, Carabias V (2019) Enhancing citizen participation through serious games. In: 24th International conference on urban development, vol 4, pp 881–88

Wolff A, Valdez A-M, Barker M, Potter S, Gooch D, Giles E, Miles J (2017) Engaging with the smart city through urban data games, pp 47–66. https://doi.org/10.1007/978-981-10-1962-3_3

Woolgar S (1990) Configuring the user: the case of usability trials. Sociol Rev 38(1):58–99. https://doi.org/10.1111/J.1467-954X.1990.TB03349.X

Chapter 21
Evolving Trends: BIM Application and Development in the Digitalized AECO Sector of Serbia

Mirjana Devetaković, Đorđe Đorđević, and Đorđe Mitrović

Abstract This chapter examines actual trends in the digitalisation of the AECO (Architecture, Engineering, Construction and Operations) sector in Serbia. The application and development of BIM (Building Information Modeling) technology is studied as the main indicator of the process of digitalisation. This research investigates a sample consisting of 12 video-recorded company presentations within the series of events called "BIM in Practice, Serbia", organised by the Serbian Chamber of Commerce and Industry and its Group for BIM Technologies. The sample has been created "by difference" and explores the most advanced examples of BIM usage in the Serbian market. The cases within the observed sample are analysed qualitatively. A summarisation given in the Discussion section indicates significant results in the BIM application on the Serbian market, highlighting the profiles of BIM-intensive companies and their role in the Serbian AECO sector, the range of BIM applications, activities in the national and international BIM knowledge community, etc.

Keywords AECO · Serbia · Building information modelling (BIM) · Architecture · Trends

21.1 Introduction

Nowadays, many fields of overall human activity are evolving quickly and dramatically. Among others, medicine, pharmaceutics, engineering, robotics, media, education, and many more could be mentioned. The AECO (Architecture, Engineering, Construction and Operations) sector is not an exception (Digital transformation set in motion 2022). The evolving forces in different fields are various and heterogeneous, but digitalisation could be considered as a common denominator for almost all sustainable development domains (Digital Acceleration Agenda and UNDP 2023).

M. Devetaković (✉) · Đ. Đorđević · Đ. Mitrović
Faculty of Architecture, University of Belgrade, Belgrade, Serbia
e-mail: mirjana.devetakovic@arh.bg.ac.rs

M. Vukmirović et al. (eds.), *Digital Dreams, Urban Realities*, The Urban Book Series,
https://doi.org/10.1007/978-3-031-76353-3_21

Focusing this study on the AECO sector, we chose digitalisation as one of the strongest and most powerful evolving forces, influencing design philosophy and methodology (Manzoor et al. 2021), construction processes, component fabrication, building operations, edifice twinning (https://www.teamcad.rs/index.php/en/bim/digital-twins), circularity, etc. Here, we observe the implementation of BIM (Building Information Modeling) technology, considering it the best and most comprehensive indicator of the process of digitalization in the field (Eastman et al. 2018).

21.1.1 BIM Technology Implementation as an Indicator of Digitalisation in the AECO Sector

The digitalisation of the AECO sector dates back to the late 60 s when computers started to be functional enough to support basic and sophisticated design processes (Eastman et al. 2018). After a period of restricted computer usage in digitally privileged engineering and academic institutions in the 70 s, the appearance and spreading of personal computers in the 80 s brought software seeds that are nowadays considered the origins of BIM technology (https://www.archdaily.com/302490/a-brief-history-of-bim). It is important to mention here that computers entered design offices during the 90 s, making this period a time of democratisation of the CAD (Computer Aided Design) software and making the majority of AECO participants digitally literate. The CAD software of that time was an excellent, efficient and widespread replacement of a traditional drawing board. However, using computers instead of drawing boards did not fundamentally change the AECO processes (https://images.autodesk.com/apac_grtrchina_main/files/aec_bim.pdf). Designers and engineers of that time worked on independent documents inherited from the traditional practice—plans, cross-sections, elevations, details, schedules, etc. Although the CAD way of working included sophisticated 3D models, their usage was predominantly for visualisation purposes, replacing expensive analogue 3D models.

The introduction of BIM technology in the late 90 s and early 2000s is firmly connected with the increasing power of personal computers, capable of processing larger and more complex amounts of geometric and alphanumeric information. The introduction of BIM technology is considered a fundamental change in the field of the AECO sector (https://images.autodesk.com/apac_grtrchina_main/files/aec_bim.pdf). The central structure in BIM technology is a 3D model consisting of intelligent elements created to represent building components from reality (Sampaio et al. 2023). A BIM model, therefore, consists of preprogrammed walls, columns, stairs, slabs, doors, windows, furniture, etc. Being intelligent, except for the graphical information, the BIM model components are characterised by a considerable range of additional information, mainly alphanumeric, suitable for various modes of future computing and representation. Adding and future deriving information from a BIM model makes it not only 3D (static three-dimensional information) but also the 4D (added time component in the process of design and construction), 5D (added financial component in the process of construction), 6D (added sustainability aspects), 7D (added facility management aspects in the operation stage), etc. BIM technology is considered to be present in the entire life cycle of a building, from its conception to its demolition (Eastman et al. 2018).

Implementation of the BIM technology presumes involvement and direct access of all participants in the conception, design, construction and operation process, making the central model the so-called "single source of truth" (Martino Di Giuda et al. 2020), facilitating complex communication and change tracking. In this aspect, BIM relies on cloud technology, supporting common data environments for the entire building body of information (https://www.autodesk.com/autodesk-university/article/ISO-19650-Common-Data-Environment-and-Autodesk-Construction-Cloud).

21.1.2 Research Methodology

The presented study is empirical, meaning that it relies on an examination of a real sample. In this case, the examined sample consists of 12 cases of BIM implementation, presented by Serbian companies, within the series of events titled "BIM in Practice Serbia". The events are organised by the Serbian Chamber of Commerce and Industry from April to December 2022. The format of the events was online, and all the materials were recorded and made available on the YouTube channel of the organising institution (YouTube Channel of the Chamber of Commerce and Industry of Serbia). The BIM implementation cases were selected for presentation, tending to promote the most advanced examples on the Serbian market and, therefore, educate the audience and stimulate other market players to dive into an exciting world of technological advancement. This means that the sample has been created "by difference" and is looking for various representative BIM usages. The audience for each event consisted of 30–50 professionals, either already familiar with the BIM technology or interested in commencing its implementation.

The results of this study indicate the current state of digitalisation in the Serbian AECO sector, which is represented by the most advanced examples of BIM technology usage. In a generalisation of the results, it is necessary to be aware that they don't apply to the entire Serbian market, and this is the reason why the study is meant to illustrate the evolving trends.

21.2 Characteristics and a Review of the Examined Sample

In this section the examined sample has been described, considering its occurrence, structure, contents and main characteristics. As previously said, the sample consists of 12 company presentations, structured into six sessions, realised from April to December 2022 (Industry 2023). An overview of the examined sample is given in

Table 21.1. In the table, the sessions are listed chronologically, with the links to the video recording on YouTube.

Shortly after its establishment, the BIM Technology Section within the Association for Construction Industry of the Chamber of Commerce and Industry of Serbia initiated a series of public presentations that will continue until the present times under the title BIM in Practice Serbia. The first in the series of events was organised in person at the International Construction Fair in Belgrade in April 2022. After the great success of the event, attracting a significant professional audience, it was evident that among Serbian professionals, there's a considerable demand for information on experiences in using BIM technologies. After this live event, the series continued in an online format, allowing professionals from all over the country to attend (https://www.gradnja.rs/webinar-bim-u-praksi-srbija-privredna-komora-srbije/).

In the following subsections, the units of the examined sample are analysed qualitatively. As said earlier in the text, the main unit in this study was a company presentation. Each of the presentations lasted approximately 30 min. Every presentation commenced with a short introduction about the presenting company. There, it needs to be stressed that all examined presentations included a general explanation of BIM technology, each from a different perspective. This made all the presentations equally useful for experienced BIM users as well as for BIM novices.

21.2.1 BIM in Practice—The Session at the Belgrade Construction Fair, April 2022

This is the first and only session that has not been recorded. In this session, three companies presented their cutting-edge results in the application and development of BIM:

- ATI Project
- BEXEL Consulting
- TeamCAD

The results presented in this session belong to the most advanced examples in the field of digitalisation in Serbia. The session attracted a significant audience, consisting of professionals already familiar with the BIM technology, as well as those interested in its initial application (Fig. 21.1).

ATI Project an international design company with one of its offices in Belgrade, presented the ongoing hospital projects in Tallinn (Estonia), Copenhagen (Denmark) and Odense (Denmark), highlighting the complexity and applied solutions in BIM modelling. The modelling has been completed based on the company's internal standards and specially developed scripts and plug-ins complementing the existing commercial software (Autodesk, Revit).

Table 21.1 A chronologically systematised examined sample

Event No. and date	Company	Web	Profile	Serbian/ International	BIM realization presented
01 April 2022	ATI project	https://atiproject.com/en/progetto/ati-headquarters-belgrade/	Architectural design	Int Ser	Hospital
	TeamCAD	https://teamcad.rs/	Software distribution, IT development	Ser Int	Housing, digital tween
	BEXEL consulting	https://bexelmanager.com/	Software development, BIM consulting	Ser Int	BEXEL manager software
	No recording available				
02 May 2022	DNEC	https://dnec.com/	Structural design/ consultancy	Ser Int	Bahrain Int. sports city; SKY Venture
	ArcoSite	https://arcosite.com/	Architectural design	Ser	A housing project in Belgrade
	https://www.youtube.com/watch?v=sRFaxKvxsZM				
03 June 2022	CMD cons engineering	https://cmdcons.com/	MEP design, engineering and construction	Ser Int	Chemical laboratory in Essen, Germany
	Integral group	https://www.introba.com/	MEP design, engineering and construction	Int Serbia	Elektrovojvodina head office, Viceroy hotel Kopaonik
	https://www.youtube.com/watch?v=LOXB5YBQW7M				
04 October 2022	Energoprojekt ENTEL	https://www.ep-entel.com/	Engineering, design and project management	Serbia	Power plant Kolubara
	KAPAPROJEKT	https://kapaprojekt.com/en/	Design, consulting, supervision	Serbia	Airport Nikola Tesla, Belgrade
	https://www.youtube.com/watch?v=2s13pvMDi1s				

(continued)

Table 21.1 (continued)

Event No. and date	Company	Web	Profile	Serbian/ International	BIM realization presented
05 November 2022	Arhi.Pro	https://www.arhipro.com/	Architectural design, BIM services, engineering, construction, furniture manufacturing	Serbia Int	Galerija, Belgrade; various buildings at Belgrade Waterfront; Boka place, Porto Montenegro; industrial hall Šimanovci
	https://www.youtube.com/watch?v=AABBfIszWQg				
06 December 2022	FINVEST	https://www.finvest.rs/en/home/	Investment, Construction and development	Serbia	"The Cerak Gardens" housing
	Delta real estate	https://deltarealestate.rs/en/	Investment, real estate management	Serbia Int	Delta house; sava center
	https://www.youtube.com/watch?v=iFyECbjGfLc				

Fig. 21.1 BIM in practice—Belgrade construction fair, April 2022

BEXEL Consulting is a software development and BIM consulting company with a head office in Belgrade and offices in Slovenia, the USA, Germany, the UK and India. BEXEL Manager, an award-winning software created by BEXEL Consulting, is now distributed globally and localised into 15 languages. It is estimated that at the moment, about 350 billion US$ of investments have been managed with the software. At the BIM in Practice event, major functionalities of the software were explained together with basic and advanced BIM principles.

As the third presenter, the TeamCAD Company presented its web-based digital tween (Shahzad et al. 2022) development (Fig. 21.2) based on the Autodesk Forge technology (https://aps.autodesk.com/blog/what-forge). The example is available online and represents a digital housing building tween that supports the real estate business (https://bimrealestate.herokuapp.com/).

21.2.2 DNEC and ArcoSite

DNEC is a company specialized in structural design, with the head office in Belgrade and offices in Abu Dhabi and Bangalore (India). The DNEC engineers presented two interesting BIM related projects, both realized on the international market: Bahrain International Sports City and SKY Venture Hall in Abu Dhabi.

The BIM usage of the Multipurpose Arena within the Bahrain International Sports City was a considerably complex object in terms of structural design (Fig. 21.3). The process of the design was divided between two groups, the design group and the BIM group. The collaboration process was supported by the BIM 360 CDE (common data environment) and completed in the Autodesk Revit software, with the significant application of custom-made plug-ins. In the presentation, the focus was on the working process organisation and dealing with the building complexity (Fig. 21.4).

The presentation of the BIM usage of the SKY Venture Hall in Abu Dhabi demonstrated the workflow performed in the Tekla software, which specialises in structural design. The software was applied in a very complex and demanding job, reaching the LOD400 level of detail, which presumes modelling practically all the detailed elements of the structure. In this example, DNEC BIM engineers also developed special plug-ins that help automate complex geometric repetitive jobs. The BIM model was used for CNC production of a range of different structural elements.

The second company presenting in this observed session was **ArcoSite**, a Belgrade-based architectural office oriented predominantly to the Serbian domestic market and specialising in architectural and interior design.

As a result of the Graphisoft ArchiCAD technology, the architect from ArcoSite decided to outline some basic BIM principles and present a typical workflow for a housing building project. Special attention was paid to the so-called "open BIM" principles, allowing the office to exchange BIM contents with offices using software other than ArchiCAD. The BIMx technology (https://en.wikipedia.org/wiki/BIMx) was also demonstrated, offering the possibility of reviewing the model for all participants in the design process, including the ones not familiar with the main authoring software (Fig. 21.5).

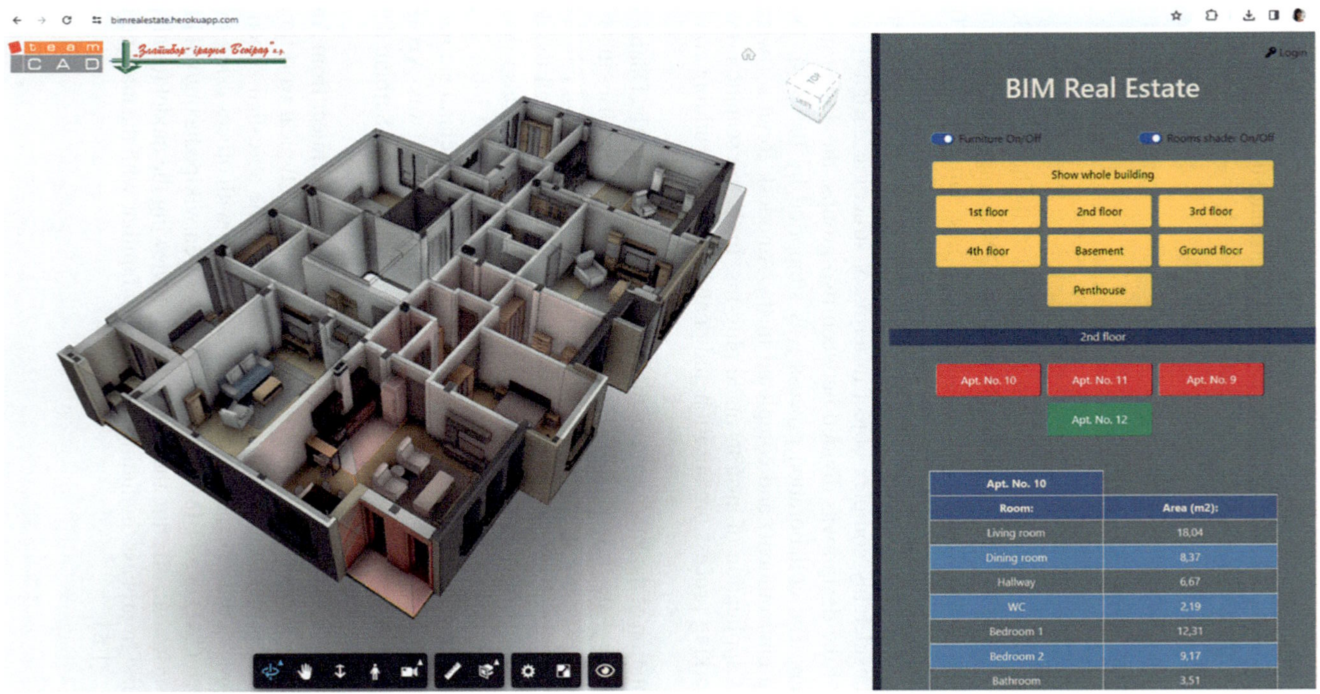

Fig. 21.2 BIM real estate—a digital tween supporting the real estate business (TeamCAD). *Source* https://bimrealestate.herokuapp.com

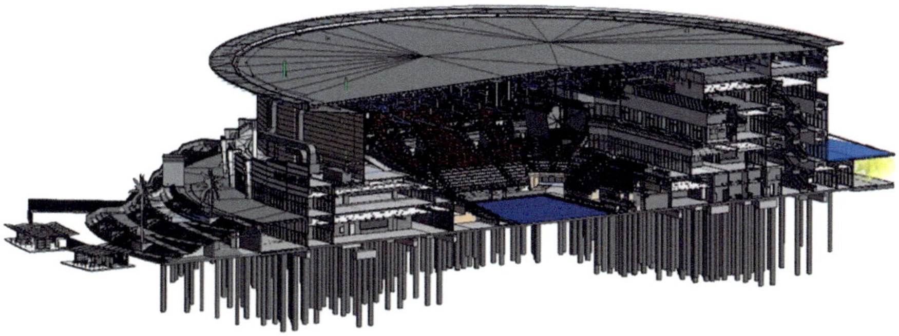

Fig. 21.3 DNEC, Bahrain international sports city, multipurpose arena, BIM model axonometric cross-section. *Source* https://www.youtube.com/watch?v=sRFaxKvxsZM

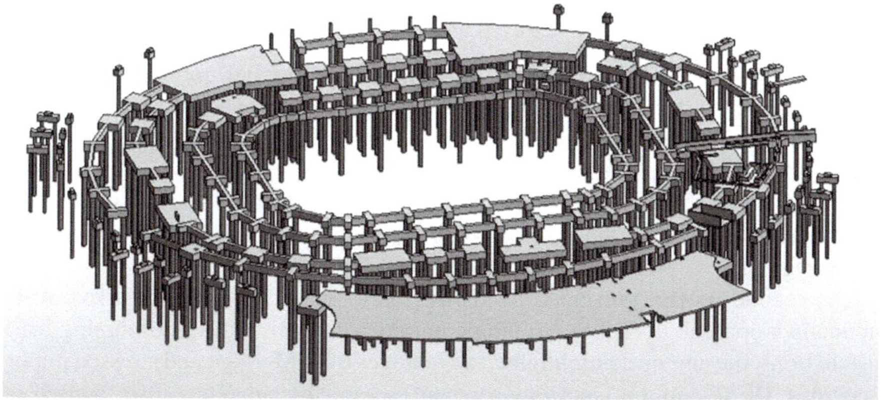

Fig. 21.4 DNEC, Bahrain international sports city, multipurpose arena, foundation part of the structural model. *Source* https://www.youtube.com/watch?v=sRFaxKvxsZM

21.2.3 CMD Cons Engineering and Integral Group

The third online event within the BIM in Practice Serbia was dedicated to MEP (Mechanical, Electrical, Plumbing) design.

CMD Cons Engineering is a Belgrade-based company that is acting on the global market, with jobs completed in more than 25 countries. The company specialises in MEP design, which is always designed in accordance with local regulations. At the beginning of the presentation, the engineers of CMD Cons stressed the advantages of using BIM technology in MEP design. They presented a design of electrical installations of the University Research Laboratory in Essen, giving the specific characteristics of the BIM usage and illustrating the process demonstrating a live project.

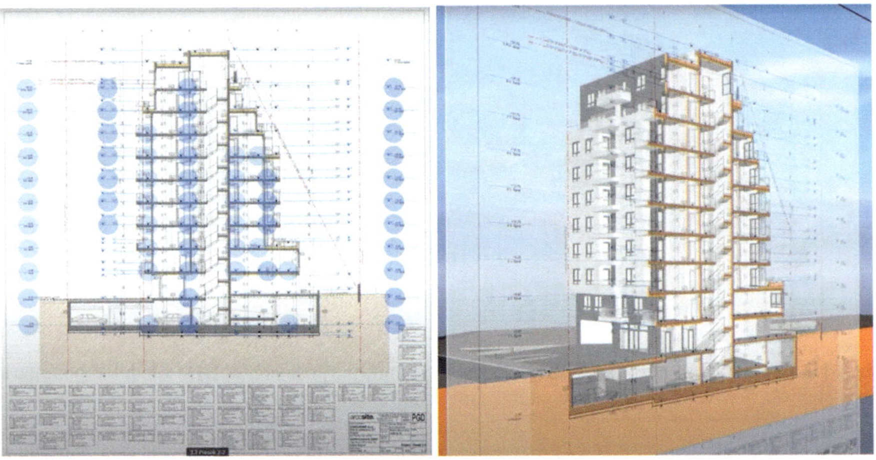

Fig. 21.5 ArcoSite—demonstration of the Graphisoft BIMx technology for BIM model review

The presentation of the **Integral Group**, a multinational corporation with an office in Belgrade, focused on two interesting projects. The very beginning of their BIM application is connected with the reconstruction project of the Elektrovojvodina Head Office, and the audience was informed about the main issues that occurred in the initial BIM usage. The lessons learned were highlighted, the main of which were related to the importance of in-house family creation and careful information modelling.

The second project presented by the Integral Group was Hotel Viceroy at the mountain Kopaonik in Serbia. Solving complex technical issues in designing MEP installations, the engineers highlighted advantages of BIM usage and stressed importance of CDE (Common Data Environment) establishment. They also focused on usefulness of Dynamo scripting in automation of repetitive tasks on the projects. Finally, the IG BIM academy was presented, as a platform for internal BIM education within the Group, initiated and realized by Belgrade office.

21.2.4 Energoprojekt ENTEL and KAPAPROJEKT

After a brief introduction to the advantages of BIM technology, the **Energoprojekt ENTEL** presentation was focused on the design of industrial complexes, predominantly power plants (Fig. 21.6). BIM modelling of such complex objects required careful strategic planning of the modelling process and combination of various BIM software, such as Revit, Plant 3D and Advance Steel. It is evident that all the software belongs to the Autodesk family, so the issues of interoperability were not so critical.

The **KAPAPROJEKT** is an architectural studio with a head office in Niš and an office in Belgrade. They presented the BIM technology application in the project of

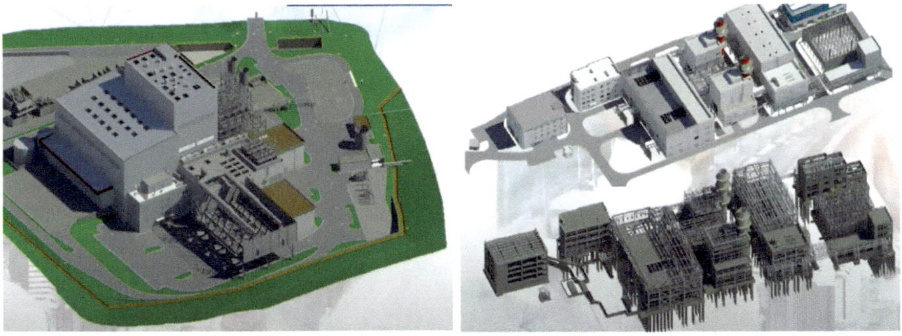

Fig. 21.6 Energoprojekt ENTEL—BIM models of industrial objects. *Source* https://www.youtube.com/watch?v=2s13pvMDi1s

reconstruction of the Airport Nikola Tesla in Belgrade (Fig. 21.7). Special attention in this presentation was paid to the phasing process, as well as to the level of detail of the modelling (LOD). The modelling was performed according to a detailed BEP (BIM Execution Plan) provided by the investor. The main modelling tool was the Revit software, and some tasks were realised in Navisworks. It is interesting that this model reached 3D and 4D, but also 5D and 6D levels of modelling, meaning that except for the geometry (3D), the chronology of future construction (4D) has been modelled, as well as schedules, quantities and costs (5D) and future facility management (6D).

21.2.5 Arhi.Pro

As one of the biggest Serbian architectural companies with two offices in Serbia and three offices in Eastern Europe, **Arhi.Pro** is a leader in BIM application in the region. In the presentation they primarily focused on internal BIM standards development that dates back in 2016, and based on UK BIM Standards (Fig. 21.8). It is interesting that various processes of automatization, supported by Dynamo visual scripting, are also regulated within the office BIM standards.

Regarding the software, Arhi.Pro relies on Revit and Navisworks, and recently on BEXEL Manager. Support for Dynamo scripting and some Python programming was also mentioned.

BIM experts from this company went through several large-scale projects, commenting on the particular issues of BIM usage: Belgrade Waterfront Gallery shopping centre, several housing developments also on Belgrade Waterfront, touristic complex Portonovi in Montenegro, etc. Particular attention was paid to a building in the Belgrade industrial zone, for which a BEXEL Manager was applied for the

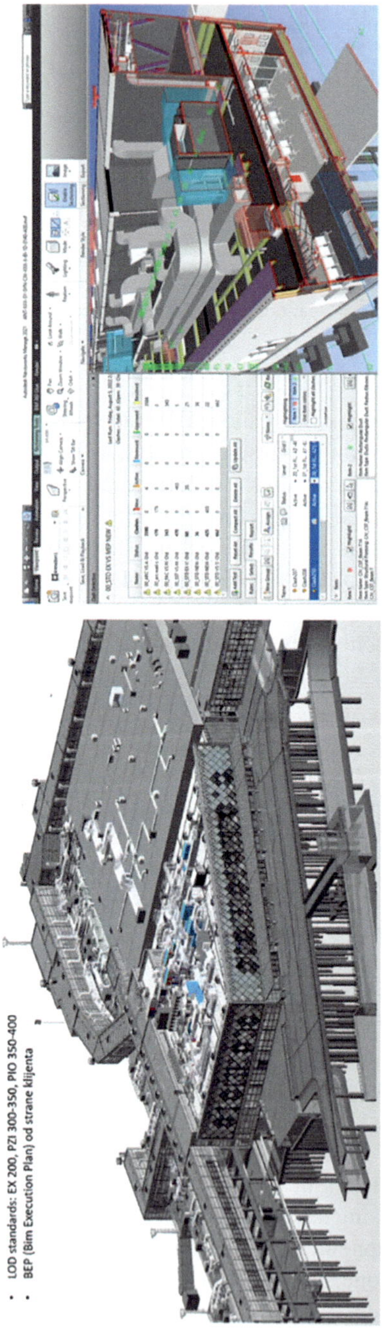

Fig. 21.7 KAPAPROJEKT—the BIM model of the Airport Nikola Tesla reconstruction, https://www.youtube.com/watch?v=2s13pvMDi1s

Fig. 21.8 Arhi.Pro—the book of BIM standards and an example of an integrated model. *Source* https://www.youtube.com/watch?v=AABBfIszWQg

control of construction. The presentation concludes with a Power BI visual representation (https://www.microsoft.com/en-us/power-platform/products/power-bi) of the construction data.

21.2.6 FINVEST and Delta Real Estate

FINVEST is a company that acts as an investor on the Serbian market, completing also design and construction activities. It specialises in mixed-use (housing and office) buildings.

After a brief introduction of the company, FINVEST engineers presented the housing project "Vrtovi Ceraka" (The Cerak Gardens, 77,000 m^2) and highlighted the main BIM implementation targets, a preliminary model on which was based quantity take-offs (QTO), then the creation of 4D and 5D representations and finally a system of construction monitoring. For the BIM activities, FINVEST engaged the Serbian company WOB (World of BIM), which specialised in BIM consultancy. In it is possible to see a BIM-based monitoring system created in the Forge technology (https://aps.autodesk.com/blog/what-forge) (Fig. 21.9).

Delta Real Estate as part of Delta Holding group, is a leading regional real estate company with a portfolio (offices, hotels, shopping malls, congress centres…) of more than 500,000 m^2. The BIM technology has been introduced by the Delta Real Estate innovation sector with the aim of modernising business processes. It is interesting that this company started their BIM journey with an education in using BIM management tools (Bexel Manager, https://bexelmanager.com/).

The pilot project in which the technology firstly has been applied is Delta House, is the head quarter of the Delta group, a building awarded the LEED certificate, as a highly sustainable. There's also an ongoing project, the reconstruction of the Congress Center Sava in Belgrade, in which the BIM technology has been fully applied. In both cases the focus of the BIM usage was on the investment control. The BIM model of the Congress Center Sava is certainly one of the most intriguing BIM models in the region.

21.3 Discussion of the Analysis Results

The review and analysis of the examined sample, given in the previous section, indicates that the Serbian companies demonstrate a significant level of BIM maturity. In this section a systematization of the analysis results is given.

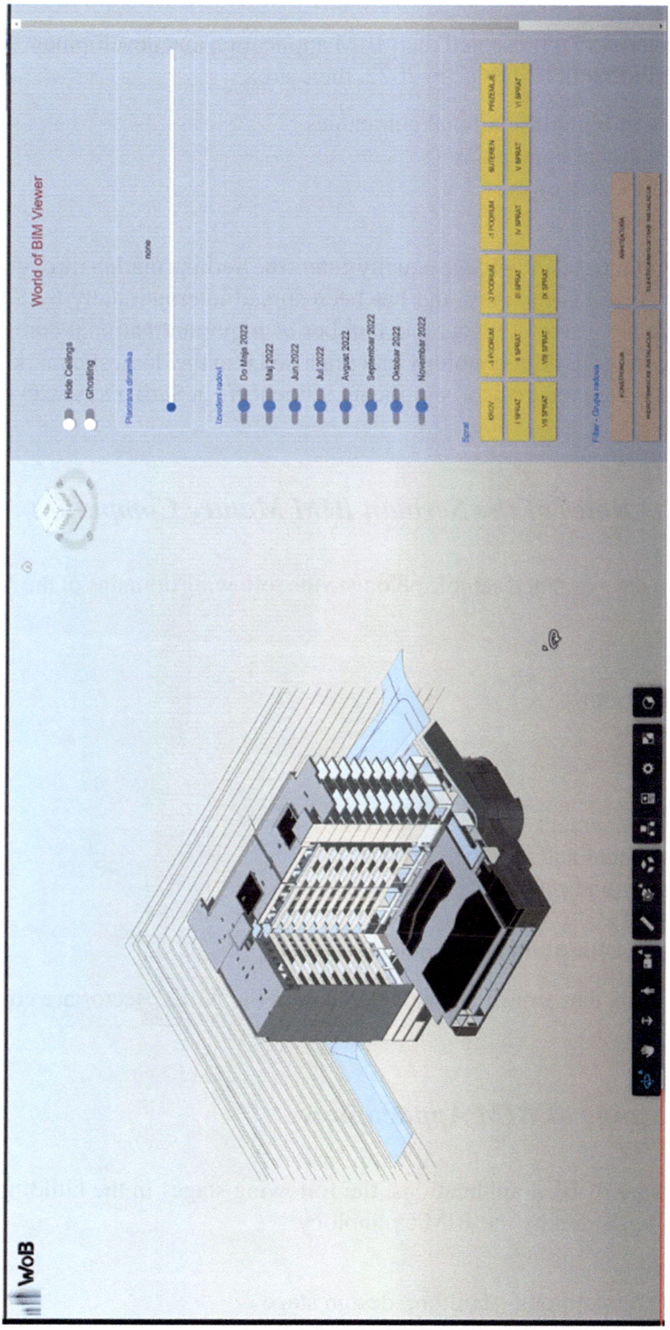

Fig. 21.9 FINVEST, the Cerak gardens—construction dynamics control in Forge technology by WOB. *Source* https://www.youtube.com/watch?v=iFyECbjGfLc

21.3.1 Profile of the Serbian BIM Mature Companies

Among the companies that presented their BIM application and development results within the "BIM in Practice Serbia" in 2022, there are:

Serbian offices of big multinational companies
Serbian multinational companies
Serbian national companies
Serbian start-ups

This indicates that BIM knowledge equally enters the Serbian market through local offices of multinational companies and has been spread internationally by Serbian companies. There are also a considerable number of important national companies that have fully acquired BIM technology and practised it in the domestic market. All the companies listed are regarded as significant influencers in Serbia's AECO sector.

21.3.2 AECO Roles of the Serbian BIM Mature Companies

The companies in the examined sample belong to the following domains of the AECO sector:

Architectural design
Infrastructure design
Structural design.
MEP design
Construction
Real Estate Management
Project management and software production
BIM consultancy and services
Equipment supply
BIM software reselling and development

This fact confirms that almost all the fields within the AECO sector are covered.

21.3.3 The Range of BIM Applications

Regarding the range of BIM applications, the following stages in the building life-cycle have been supported by the BIM technology:

Conceptual stage
Architectural competitions—building design stage
Building design and documentation stage
Building construction stage

Building operation stage
Building reconstruction

The examples examined represent all stages recognised as the life-cycle of a building, starting from conception through building design and documentation, fabrication, construction and operation. The observed sample does not contain the BIM application in the process of building demolition.

21.3.4 Adaptability to International Regulations

Since many of the companies of the examined sample operate in different international markets, they demonstrate considerable adaptability in acquiring international design regulations and operating accordingly.

21.3.5 In-House BIM Education

It is interesting that many companies in Serbia have established in-house educational systems related to BIM technology. Some of them are even leaders in BIM education within their wider international business contexts.

21.3.6 BIM Knowledge Community Activities

The companies that participated in the "BIM in Practice Serbia" represent important members of a large Serbian BIM knowledge community. Their knowledge of community activities could be systematised in the following way:

Conferences

(Association BIM Serbia Conference, TeamCAD Corporate BIM Conference (TeamCAD BIM How to Conference 2023), Baldini Studio Corporate BIM Conference, BIM in Infrastructure Conference (https://www.gradnja.rs/konferencija-o-digitalizaciji-gradjevinske-industrije-10-novembra-u-beogradu/) (https://www.gradnja.rs/konferencija-bim-implementacija-8-novembar-beograd-teamcad/).

Seminars

(BEXEL Consulting Corporate Days, University of Novi Sad BIM Day).

BIM Info Days.

(Kruševac, Kragujevac, Valjevo, Niš, Novi Sad…).

Apart from the BIM technology-related events, there are two more activities in which the companies from the observed sample actively participate:

BIM Standardization, led by the Institute for Standardization of Serbia.

The Working Group creating National BIM recommendations.

21.3.7 BIM Software Development

An important characteristic of the actual Serbian BIM scene is certainly the presence of the Bexel Manager software development. This outstanding software represents the BIM technology in its very essence, being the tool aimed at supporting the construction and operation processes. The software has been developed for more than a decade by BEXEL Consulting, the Serbian company specialising in BIM consulting. It imports BIM models created in standard authoring tools and supports BIM 3D, 4D, 5D and 6D processes. According to the software creators, BEXEL Manager is a BIM solution providing the most comprehensive set of features for advanced 3D visualisation, model data management, clash detection, automated quantity take-offs (QTO), cost management, smart scheduling and progress monitoring. As previously highlighted, the software has been localised in 15 languages and sold globally (Fig. 21.10).

It is important to stress that the BEXEL Consulting Company, with its global BIM expertise, has been the main driving force behind several activities in the Serbian AECO sector:

- Establishment and support of BIM Serbia Association
- Leading role in the BIM Technology Group of the Serbian Chamber of Commerce and Industry
- Organisation of the BIM Serbia annual regional conferences
- Specialised BEXEL manager-related seminars
- Support to University teaching and BIM-related curricula development
- Establishment of the building SMART Serbia Chapter
- BIM consulting to the Ministry of Construction and Infrastructure of Serbia

The presence of a successful software development company on a relatively small national market, has certainly been a huge inspiration for other market players to adopt the technology and to even contribute to its progress.

21.3.8 International Export of BIM Services

The analysed sample indicates that the majority of Serbian companies that are presented with BIM services export to the global market. Having in mind that it is a strategy of Serbian Government to increase export of ICT services, participating that way in the global knowledge-based economy, the BIM intensive companies give

Fig. 21.10 BEXEL manager—the Serbian software on the global market. *Source* https://bexelmanager.com/

its significant contribution. Working on the international market, Serbian companies acquire more advanced standards and engineering culture and improve their adaptability, which is highly desired in the global AECO sector.

21.3.9 Spreading the BIM Technology on the Serbian Market

The Serbian BIM-intensive companies, acting on the global market, have naturally acquired the technology, and they apply it to all projects, be they international or national. Conscious of the advantages of BIM technology in the entire construction process, such companies apply it regardless of investor requirements.

It needs to be said that more and more investors come with the BIM requirements, explicitly defined in the so called requirement for information, contractual documents that represents the base for developing BIM execution plans.

Finally, there's an indication from the Serbian Ministry of Construction, Transport and Infrastructure (https://www.mgsi.gov.rs/lat/aktuelnosti/vesic-od-2028-godine-svi-projekti-u-srbiji-u-bim-tehnologiji), that BIM technology is going to be mandatory from January 2028, for all buildings exceeding 10000 m^2.

21.4 Conclusions

Interested predominantly in the evolving trends in the digitalisation of the AECO sector, this cross-sectional study examined the situation in Serbia, represented by the usage of the BIM technology among the companies that participated in the series of events titled "BIM in Practice Serbia" in 2022. The time range and the companies are selected by excellence and leadership in BIM technology application, as well as by their readiness to share their achievements with a wider professional community.

The results obtained indicate that there are considerable accomplishments in BIM technology in Serbia, ranging from advanced usage cases to software development for the global market and from university curricula enrichment to national standardisation activities. The observed cases include a wide range of various AECO sector players: architectural and structural designers, constructors, MEP engineers, component producers, investors, software creators and distributors. Many of the examined companies have developed their internal BIM education systems. Some of them have developed internal BIM standards and actively participate in the national BIM standardisation processes. Finally, the BIM services delivered internationally tend to become interesting export goods, fitting into the Serbian Government's strategy of intensifying companies' participation in the global digital economy.

The BIM knowledge community in Serbia demonstrates significant vitality. There are numerous relevant events, like "BIM in Practice Serbia", BIM Info days, annual BIM Serbia conferences, software developer and resellers seminars, and many others. The companies whose presentations have been analysed in this study are frequent participants and organisers of such events. There's also a tight connection between the BIM-intensive companies and Serbian universities, resulting in frequent delivery of lectures, training and free student software licences.

Relying on the results of this study, one has to be deliberate in possible generalisation considering the overall Serbian AECO market, and one could not conclude that the entire context is highly digitised. A vast majority of design and construction companies in Serbia still do their business in a traditional way. It is possible, however, to conclude that there's an evident growing trend in BIM technology acquisition among Serbian companies. Although mainly applied to international projects, BIM technology is increasingly required within the domestic market. The statement from the Serbian Ministry of Construction, Transport and Infrastructure that BIM is going to be a mandatory requirement from 2028 (https://www.mgsi.gov.rs/lat/aktuelnosti/vesic-od-2028-godine-svi-projekti-u-srbiji-u-bim-tehnologiji) confirms the readiness of the professional context to fully support the digitalisation of the Serbian AECO sector.

The example of the Serbian AECO sector suggests that even relatively small markets of developing countries, could significantly advance in the process of digitalization, by actively participating in the global arena, and intensively spreading the international experiences in their national practice.

References

Allen J, Yee A. ISO 19650, the common data environment, and autodesk construction cloud. Autodesk. (Online). Available: https://www.autodesk.com/autodesk-university/article/ISO-19650-Common-Data-Environment-and-Autodesk-Construction-Cloud. Accessed 01 Apr 2024

BIM u praksi Srbija: Dokle je stigla digitalna transformacija građevinarstva. 29 Nov 2022. (Online). Available: https://www.gradnja.rs/webinar-bim-u-praksi-srbija-privredna-komora-srbije/. Accessed 28 Mar 2024

BIMx. 10 Dec 2023. (Online). Available: https://en.wikipedia.org/wiki/BIMx. Accessed 04 Apr 2024

Chand S, Busecan G. What is forge? Autodesk. 03 June 2022. (Online). Available: https://aps.autodesk.com/blog/what-forge. Accessed 02 Apr 2024

SDG Digital Acceleration Agenda, UNDP (2023) Available: https://www.sdg-digital.org/. Accessed 08 Apr 2025

Digital transformation set in motion (2022) FIRST (11):4–7. Available: https://www.arup.com.hk/marketing/FIRST_magazine-Foresight_Innovation_Research_Shaing_Training/Issue%2011/FIRST-11.pdf. Accessed 08 Apr 2025

Digital Twins. (Online). Available: https://www.teamcad.rs/index.php/en/bim/digital-twins. Accessed 28 Mar 2024

Eastman C, Sacks R, Lee G, Teicholz P (2018) BIM handbook, a guide to building information modeling for owners, designers, engineers, contractors, and facility managers, 3rd edn. Wiley

Industry FC (2023) Publication "BIM in practice Serbia" (in Serbian language). Chamber of Commerce and Industry of Serbia. https://api.pks.rs/storage/assets/BIM%20U%20PRAKSI%20SRBIJA2.pdf

Konferencija o BIM implementaciji 8. novembra u Beogradu. 24 Oct 2022. (Online). Available: https://www.gradnja.rs/konferencija-bim-implementacija-8-novembar-beograd-teamcad/. Accessed 28 Mar 2024
Konferencija o digitalizaciji građevinske industrije 10. novembra u Beogradu. 20 Oct 2022. (Online). Available: https://www.gradnja.rs/konferencija-o-digitalizaciji-gradjevinske-industrije-10-novembra-u-beogradu/. Accessed 28 Mar 2024
BEXEL Manager. BEXEL Consulting. (Online). Available: https://bexelmanager.com/. Accessed 19 May 2024
Manzoor B, Othman I, Pomares JC (2021) Digital technologies in the architecture, engineering and construction (AEC) industry—a bibliometric—qualitative literature review of research activities. Int J Environ Res Public Health 18(11):6135
Martino Di Giuda G, Pattini G, Seghezzi E, Schievano M, Paleari F (2020) The construction contract execution through the integration of blockchain technology. In: Digital transformation of the design, construction and management processes of the built environment. Springer Open, pp 27–36
Power BI. Microsoft. (Online). Available: https://www.microsoft.com/en-us/power-platform/products/power-bi. Accessed 03 May 2024
Building Information Modeling in Practice (2003) Autodesk. (Online). Available: https://images.autodesk.com/apac_grtrchina_main/files/aec_bim.pdf. Accessed 01 Apr 2024
Quirk V. A brief history of BIM. 07 Dec 2012. (Online). Available: https://www.archdaily.com/302490/a-brief-history-of-bim. Accessed 01 Apr 2024
BIM Real Estate. (Online). Available: https://bimrealestate.herokuapp.com/. Accessed 28 Mar 2024
Sampaio Z, Sequeira P, Gomes AM, Sanchez-Lite A (2023) BIM methodology in structural design: a practical case of collaboration, coordination, and integration. Buildings 13(31) (Special Issue: Smart and digital construction in AEC industry)
Shahzad M, Shafiq MT, Douglas D, Kassem M (2022) Digital twins in built environments: an investigation of the characteristics, applications, and challenges. Buildings 12(1):120 (Special Issue: Digital Twin in the AEC Industry—Advances and Challenges)
TeamCAD BIM How to? Conference 2023. Belgrade
VESIĆ: OD 2028. GODINE SVI PROJEKTI U SRBIJI U BIM TEHNOLOGIJI. 02 Nov 2023. (Online). Available: https://www.mgsi.gov.rs/lat/aktuelnosti/vesic-od-2028-godine-svi-projekti-u-srbiji-u-bim-tehnologiji. Accessed 12 May 2024
YouTube Channel of the Chamber of Commerce and Industry of Serbia. Available: https://www.youtube.com/@PKSMedia/playlists. Accessed 08 Apr 2025

Chapter 22
Sustainable Urban Aesthetics: Unraveling Complexities and Contradictions

Ana Kisjan, Dubravko Aleksić, and Tijana M. Vujičić

Abstract This study addresses the aesthetic implications of sustainable architecture through a critical analysis of a broad spectrum of philosophical traditions—from Romanticism to contemporary ecological discourses. The research points to a tendency towards excessive fragmentation in the field of sustainability, where ecological, technological, and aesthetic aspects are often separated into isolated domains instead of being viewed through an integrative prism. The study highlights the need to overcome entrenched dichotomies between technological progress and humanistic values in architecture. Instead, the work advocates for a holistic synthesis that harmoniously unites ecological responsibility, technological innovations, and deeper values related to the human condition and quality of life. By integrating various theoretical perspectives, the research offers an enriched conceptual framework that views technology and nature as inseparable elements in a complex interrelationship. The contribution of this study lies in a deeper understanding of the inherent complexity of sustainability, opening up space for further research and critical dialogue.

Keywords Sustainable architecture · Aesthetics · Holistic approach · Technological advancement · Humanism

22.1 Introduction: Between Utopia and Reality

In an era when we face ubiquitous ecological challenges and technological advancements, the aesthetics of sustainable design becomes not just a matter of artistic expression but also of ethical and ecological responsibility (Brady et al. 2009). Instead of settling for established, superficial understandings of sustainability, which often

A. Kisjan (✉)
Faculty of Architecture, University of Belgrade, Belgrade, Serbia
e-mail: 2021_41008@edu.arh.bg.ac.rs

D. Aleksić · T. M. Vujičić
Faculty of Architecture, Civil Engineering and Geodesy, University of Banja Luka, Banja Luka, Bosnia and Herzegovina

M. Vukmirović et al. (eds.), *Digital Dreams, Urban Realities*, The Urban Book Series,
https://doi.org/10.1007/978-3-031-76353-3_22

serve as justifications for existing, deeply ingrained practices, this work seeks to examine and characterise the gaps behind the simplified conceptions of "sustainable architecture" aesthetics. Employing critical literature analysis as a methodological tool, a broad range of philosophical debates—from Romanticism to contemporary ecological discourses—has been explored, aiming to understand how these movements have shaped our conception of sustainability and frequently overlooking its deeper layers of complexity.

In this context, the methodological approach of this research is based on a detailed analysis of literature centred around three key groups of sources. In the first group, "Philosophical and theoretical analyses," we illuminate discussions that provide a foundation for understanding the connections between the aesthetics of sustainable design, humanity, and nature, highlighting how these relationships shape our approach to architecture. "History and genealogy of sustainable aesthetics," as the second group, allows us to trace the evolution of sustainable aesthetics over time, understanding key moments and movements that have influenced its development and contemporary understanding. Finally, as the third group, we have included "Practical Examples" that practically demonstrate how theoretical principles and aesthetic values of sustainable practices are manifested in specific projects.

By studying the aesthetic principles of sustainability and their integration with societal, technological, and philosophical challenges, this work aims to shed light on the inherent complexity and contradictions often encountered in theoretical readings. The ambition is the re-conception of spaces that reflect not only ecological sustainability but also complexity, aesthetics, and the inherent interconnectedness of all life forms, thereby challenging reductionist narratives. The goal is to inspire in-depth philosophical reflections on the fundamental values that underpin our understanding and engagement, laying solid foundations for future research and development in this vital area of architecture.

22.2 Genealogy of Sustainable Aesthetics: Traces Through Time

In the late 1960s, "environmental aesthetics" began to evolve as a confluence of aesthetic and ecological philosophies, creating an interdisciplinary bridge between these fields. Although considerations of the aesthetic dimensions of nature have deep roots in the philosophical tradition, it was the ecological activism of the twentieth century that lent a new dimension and direction to this field, emphasising the aesthetic implications of ecological issues (Brady et al. 2009). This discipline deepens the philosophical analysis of aesthetics by focusing on interactions with natural environments, their components, and natural phenomena, thus distancing itself from the classical study of art. Although initial attention was directed towards unchanged natural landscapes, over time, the field expanded to include 'hybrid' spaces—those shaped or modified by human action, such as gardens, as well as entirely human-created environments (Brady et al. 2009).

The field of ecological aesthetics, rich in intellectual diversity, can be understood through three key dimensions: philosophical analyses of the aesthetic valuation of nature, theoretical and practical discourses on landscape design and aesthetics (including the impact of Romantic literature and poetry), and early concepts of nature conservation and ecological writing (Brady 2003). Although serious philosophical debates about the aesthetics of nature only became prominent in the eighteenth century, aesthetic appreciation of natural landscapes likely played an important role in various cultures throughout human history. This aesthetic sensibility is not merely an isolated phenomenon but is often deeply integrated into religious beliefs, cultural practices, and even moral frameworks (Budd 2002).

To thoroughly grasp what aesthetic experience and value mean within the context of ecology and the environment, it is imperative to combine a revisit to the historical perspectives that have sculpted the contemporary aesthetics of the environment we are familiar with today. This approach underscores the importance of understanding the evolution of these concepts to fully appreciate their significance in the current ecological discourse.

In the domain of philosophical aesthetics, the modern discipline traces its origins to the early eighteenth century, marked by significant contributions from individuals like Lord Shaftesbury, Joseph Addison, and Alexander Gottlieb Baumgarten. These philosophers broadened the scope of aesthetics to encompass nature as an independent object of aesthetic experience, moving away from earlier discourses that largely centred on art. Moreover, they introduced more expansive categories of aesthetic value, such as the *sublime* and the *picturesque*, thereby enriching and revitalising the discussions about nature at that time (Carlson 2000).

The narrative of aesthetic appreciation underwent notable transformations over the centuries, reflecting shifts in philosophical perspectives and societal values. The eighteenth century marked a significant shift towards subjectivity in aesthetic theory, moving the focus from beauty as an objective quality to the relational dynamics between the aesthetic object and the observer. This shift was especially evident in Kant's revolutionary concept of "disinterested judgment," which moved from moral philosophy into aesthetics, emphasising the non-instrumental nature of aesthetic experience. In this context, "disinterested judgment" indicates that a judgment is made without any hidden motive or interest. Kant argued that beauty is not merely what brings pleasure since pleasure can be influenced by personal interests. Instead, beauty is what brings disinterested pleasure, and this disinterestedness allows for the universality of judgment (Kant 1790). This idea has profoundly influenced modern conceptions of aesthetic value as a non-instrumental form of appreciation. This paradigm shift redirected focus from objective qualities, such as harmonious proportions, to subjective experiences, highlighting the emotional and sensory responses that aesthetic objects elicit (Guyer 1979). Kant's aesthetic theory is crucial for ecological aesthetics, positioning nature as the apex of aesthetic experience. His work influenced Romanticism and its subsequent impact on 19th-century transcendentalists in North America and the Hudson River School of Landscape Painters. Nature imitation was a dominant paradigm in art dating back to ancient times, but this association became bi-directional by the closing decades of the eighteenth century. Art principles were incorporated into garden design, and eventually, gardens evolved into places where people could enjoy both natural and artistic beauty. The concept of "*picturesque*", as formulated by writers like William Gilpin, Sir Uvedale Price, and Richard Payne Knight in the eighteenth century, was especially influential. This aesthetical category differentiated between beautiful and sublime things, with an appreciation for attributes such as roughness, irregularity, and cryptic aspects. This art-centred approach presented new questions about both aesthetics and ethics of ecology aesthetics, which were just emerging (Carlson Berleant 2004).

Transitioning into the nineteenth century, this evolving dialogue experienced a significant decline in the philosophical interest in the aesthetics of nature, primarily due to Hegel's emphasis on the supremacy of art over nature. Nevertheless, in North America, the discourse ventured down a different path, tilting more towards a "wild aesthetic" that impacted the conservation movement at the end of the nineteenth century due to widespread deforestation (Carlson Lintott 2008). Figures such as Henry David Thoreau (Thoreau 1963) and John Muir, along with artists from the Hudson River School, celebrated the untamed, sublime qualities of nature. This wild aesthetic was not just about remote enjoyment but involved a more intimate, participatory engagement with the environment (Muir 1894). It was an aesthetic deeply connected to ethical and spiritual values, emphasising nature as a source of moral and spiritual renewal. Aldo Leopold, a key figure in ecological ethics, expanded this line of thinking. His "land ethic" combined ecological understanding with aesthetic sensibility (Aldo 1949).

This ethical, aesthetic approach was further developed in contemporary ecological aesthetics, especially in the concept of "positive aesthetics," which asserts that all nature, untouched by human intervention, possesses inherent aesthetic value (Saito 1998).

22.3 Dialectics of Space: Aesthetics and Sustainability in Dialogue

The twentieth century witnessed a further shift away from nature as art moved towards abstraction and expressionism. Philosophers found themselves challenged to redefine art in light of avant-garde movements, which pushed the boundaries of traditional artistic forms. This shift posed challenges and opportunities for the field of ecological aesthetics, which seeks to recenter the aesthetic value of nature within philosophical discourse (Berleant 2002).

The development of ecological aesthetics is a complex, dynamic process which has evolved greatly, especially during the last two decades of the twentieth century. A more inclusive and comprehensive approach to aesthetics was initiated by Ronald Hepburn's "Contemporary Aesthetics and the Neglect of Natural Beauty", written in 1966, which laid the foundation for it. He challenged the traditional paradigms that limited aesthetics solely to art, highlighting that nature is not only worthy of aesthetic consideration but essential for a full understanding of the aesthetic experience. Among Hepburn's key contributions was his focus on the uniqueness and specificity of the aesthetic experience in the context of nature (Hepburn 1966). His work served as a catalyst for a new generation of philosophers and theorists, including Allen Carlson and Arnold Berleant, who further developed and critiqued Hepburn's ideas. Carlson, for example, advanced the discussion with his science-based approach, arguing that scientific knowledge can enrich our aesthetic experience of nature (Berleant and Carlson 2004).

This approach, while innovative, wasn't without its critics; theorists such as Thomas Heyd and Emily Brady challenged Carlson's reliance on scientific knowledge, pointing out that aesthetic experience can be much broader and more complex (Brady 2007).

Inspired by Hepburn's work, Arnold Berleant formed his theory of "engagement," which gave importance to immediate sensory experience and emotional reaction in our aesthetic relationship with nature. This was a major divergence from the classical approach of subject-object dualism, thereby allowing for the development of a multifaceted and all-inclusive approach to aesthetics. The strong subject-object dichotomy dissolves as the observer becomes a partaker in an aesthetic affair (Berleant 2002).

One of the most fascinating elements of ecological aesthetics is its increasingly deep relationship with ethical considerations, especially in sustainable design and architecture. This intersection has opened doors for considering new aesthetic principles that are not only material or technological but also ethical and ecological (Saito 1998). Beyond ethical principles, knowledge becomes central to how we experience and value the natural environment. Saito emphasises that while our aesthetic experience may begin with a superficial sensory impression—such as looking or touching—deeper understanding and appreciation often require additional context (Saito 1998). Scientific knowledge or cultural narratives, such as myths and legends, can enrich this context. In his work "Land Ethic," Aldo Leopold highlights this notion, suggesting that aesthetic values, such as beauty and harmony, can directly influence our efforts to protect nature (Aldo 1949). However, there is ongoing debate over how much knowledge, or what kind of knowledge, should influence our aesthetic appreciation of nature. Some, like supporters of "scientific cognitivism," argue that scientific knowledge can enrich and deepen our aesthetic experience. Meanwhile, others, like "noncognitivism," believe that personal experiences and emotional responses are equally, if not even more, important (Brady et al. 2009). Berleant, in his idea of "aesthetics of engagement," goes a step further, arguing that a true aesthetic experience of nature demands an active, engaged approach. Rather than relying solely on scientific knowledge, he proposes that aesthetic valuation should be based on direct, multisensory interaction with nature, where every sense plays a role in shaping our experience (Berleant 1997).

This debate leads to different theoretical approaches to the aesthetics of nature. Some, like Hepburn, advocate for a pluralistic approach, where different methods and perspectives are combined to achieve a richer and deeper understanding of the aesthetic values of nature (Hepburn 2001). These pluralistic views, a mix of ideas critically examining and combining various approaches, offer a new perspective and possibility for further exploration and comprehension of the aesthetics of nature.

22.4 Ecological Evolution: Mapping Sustainable Consciousness

Following World War II, the world was in transition. Industrial development and urbanisation were not just symptoms of post-war reconstruction—they were catalysts for changes with far-reaching ecological implications (McNeill 2000). In this context, the publication of Rachel Carson's "Silent Spring" in 1962 was not merely a literary event but a driving force for a change in general awareness. Carson managed to articulate what many could not express: that our ambition for progress had jeopardised the very foundation of our existence—nature (Carson 2002).

The first Earth Day in 1970 and subsequent global ecological movements were not just expressions of collective concern. They served as platforms for mass mobilisation and awareness-raising, as well as for articulating new ethical and aesthetic values that placed nature at the centre of our moral and aesthetic universe (Rome 2013). These movements did not just enhance ecological awareness; they also created space for a deeper understanding of nature as an entity with inherent value, independent of its utilitarian function for the human species.

The 1987 Brundtland Commission report, titled "Our Common Future," was not just another report. It was a manifesto for a new development paradigm that recognized the inseparable connection between ecological sustainability and social justice. This concept of sustainable development became a cornerstone of contemporary and future discourses on sustainability, as well as on quality of life (Brundtland 1987).

The United Nations Conference on Sustainable Development in 1992 in Rio de Janeiro and the World Summit on Sustainable Development ten years later were not merely diplomatic gatherings. They were milestones in formulating and establishing a global consensus on the path towards sustainable practices. These meetings highlighted the complex connection between poverty and sustainability, pointing to the imperative of addressing poverty as a prerequisite for achieving sustainability (United Nations 1993).

Sustainability, at its core, is not just about conserving resources; it is about enhancing the quality of life for all. This is the vision articulated by Dr. Brundtland, Valero, and Brasier in their works, emphasising not only the technical and economic aspects of sustainability but also its deeply rooted ethical dimension. They called on us to recognise our collective responsibility towards future generations and to work towards creating a world where sustainability is not just an abstract ideal but a concrete reality achievable through strategic resource management and innovative "green engineering" (Brady et al. 2009).

Understanding the historical context of ecological practices is crucial for a deeper comprehension of the current scene, as it helps us recognize how our conceptions and practices of sustainability have evolved and changed over time. These events have shaped our ecological awareness as well as our collective imagination of what is possible. A shift has occurred in the understanding of the human role and its engagement towards nature.

22.5 Challenges and Paradoxes in the Age of Globalization

Sustainability is a concept that has evolved over time, and its complexity has become increasingly evident as our needs and understanding of the world around us have changed. Starting from the German concept of "Nachhaltigkeit" (Ulrich 2012), which focused on maintaining a balance between the logging and replenishing of forests, sustainability has become much more than mere management of natural resources (Rolston 1998). Globalisation, which has enabled rapid technological development, has also brought a series of challenges and paradoxes that architecture faces today. These challenges are not merely technical in nature but extend to cultural and ethical dimensions that have become increasingly important in the context of sustainability (Sloterdijk 2011).

The first challenge is how sustainable architecture can align with the dominant model of the market economy, which often encourages excessive consumption and commodification. Herein lies a paradox: while sustainable architecture strives to minimise its impact on the natural environment, it is often subject to commodification and aesthetic consumption. This is particularly pronounced in the context of "green engineering," where sustainability is often reduced to technical performance and energy efficiency, neglecting broader ethical and aesthetic implications (Gill 2011). The second challenge is how sustainable architecture can contribute to a broader quality of life and social cohesion. Here, the role of aesthetics is crucial. The aesthetics of sustainability affects not only the visual and sensory experience of space but also the deeper meaning and value we attach to our environment. In this sense, aesthetics becomes a means to articulate and promote sustainable values, as well as to critically examine dominant norms and practices (Publishers 2011). The third challenge is how to overcome the gap between individual and collective interests in the context of sustainable architecture (Jencks 1997). According to Kenneth Frampton, this gap is best illustrated by the contrast between the avant-garde belief that architectural creativity is the product of subjective experience and the ecological stance that emphasises the need for deeper engagement in preserving ecological balance (Frampton 2011). The question arises: how can we reconcile these two seemingly opposed approaches?

In light of these challenges, it is worthwhile to consider a multidisciplinary approach that, in addition to technical and economic aspects, also turns towards ethical and aesthetic dimensions. This implies a new model of architectural practice that is focused not only on the "green" design as we know it today but also on promoting sustainable values that integrate technological, economic, ethical, and aesthetic practices. In this sense, sustainable architecture ceases to be merely a technical challenge where aesthetics is the outcome of solutions but rather takes on the role of an emancipator of a sustainable way of living.

22.6 Complexities of Sustainable Logics

In considering the aesthetic implications and sustainability, there is a significant gap in various approaches, both in theory and in practice. Susan Heghan described this gap by saying, "The intellectual spectacle present in one is absent in the other. The intellectual consistency present in the second case is lacking in the first" (Hagan 2001).

Works like "Six Contradictory Logics of Sustainable Architecture" by Simon Guy and Graham Farmer represent milestones of their time, offering a pluralistic approach that transcended the then-dominant ecological and technical debates. They analyse the multifaceted nature of green building, identifying six different logics: eco-technical, eco-centric, eco-aesthetic, eco-cultural, eco-medical, and eco-social. Eco-technical logic is oriented toward the future and uses the latest technologies to achieve sustainability, while eco-centric logic strives for symbiosis with nature, often using recycled or natural materials practice. On the other hand, eco-aesthetic logic emphasises iconic buildings that are visually impressive but not necessarily sustainable. Eco-cultural logic values local traditions and culture, whereas eco-medical logic focuses on health and quality of life. Lastly, eco-social logic is community-oriented and often developed through a collaborative design process (Guy and Farmer 2001).

In Guy's and Farmer's categorisation of sustainable architecture, there is a subtle but deeply ingrained contradiction that cannot be easily overlooked. This categorisation, while useful in mapping different approaches and philosophies, can inadvertently create false dualisms that are fundamentally antithetical to a holistic understanding of sustainability. For example, separating "eco-technical" from "eco-aesthetic" architecture might suggest that technological innovation and aesthetic expression are mutually exclusive, which is far from reality. Such categorisation can, in fact, discourage synthesis between technology and aesthetics, which is key to creating architectural solutions that are both functional and beautiful, reflecting the complex relationships between humans and their environment. Moreover, there is a risk that this division encourages fragmentation in the field of sustainable architecture, where each category develops in isolation instead of striving for integrative models that encompass a broad spectrum of ecological, social, and aesthetic issues. While Guy's and Farmer's categorisation provides a valuable framework for discussion, it also poses challenges that call for deeper consideration and, perhaps, revision to get closer to the true essence of sustainable architecture.

Sustainable architectures within these classifications typically span an extensive array of methodologies and ecological configurations, ranging from high-tech, energy-efficient structures like Norman Foster's Commerzbank headquarters to budget-conscious, low-tech solutions such as the Cotton Tree Housing by Australia's Clare Design studio. The "low-tech" methodology particularly highlights a communal socio-cultural lifestyle, manifesting an approach acutely sensitive to ecological considerations. Conversely, the high-tech perspective, employing technology as the foremost tool alongside the so-called digital methodology towards nature, often overlooks the significance of quantitative and empirical components in the formulation of architectural discourse (Lee 2011).

To gain a more profound comprehension of the six logics of sustainable architecture introduced by Guy and Farmer, we can conceptualise them through two fundamental categories: "Techno-Centrism" and "Humanism." This division is not intended as a mere classification but as a framework for grasping the essential philosophical and practical tensions that sculpt the field of sustainable architecture. Within the "Techno-Centrism" category, eco-technical and eco-aesthetic logic represent approaches focused on technological innovations and design solutions. These logics are driven by the belief that technological advancement, whether in terms of energy efficiency or aesthetic expression, will be crucial for achieving sustainability. Conversely, the "Humanism" category encompasses eco-centric, eco-cultural, eco-medical, and eco-social logic. These approaches prioritise the human dimension and interaction with the environment, whether it concerns health, culture, communal living, or harmonious coexistence with nature.

This division into "Techno-Centrism" and "Humanism" aids in illuminating the fundamental tensions within different approaches to sustainable architecture. On one hand, we have technological advancement as a potential solution to ecological challenges, while on the other, there is a humanistic approach that aims for a holistic understanding of sustainability, where technology and innovation must be balanced with social and ecological needs.

22.7 Discussion and Concluding Considerations

This study relies on a critical analysis of literature, spanning a wide range of philosophical discussions—from the influence of the Romantic movement to contemporary ecological discourses—exploring how these movements have shaped our understanding of sustainability in architecture. In concluding this study, we find ourselves at the crossroads of historical considerations and contemporary challenges, where the aesthetics of sustainable architecture is revealed as a field rife with contradictions and paradoxes. The excessive fragmentation of sustainable architecture has occurred due to a growing tendency to dissect complex issues into isolated perspectives, thereby losing the comprehensiveness that encompasses ecological, technological, and aesthetic aspects. While technological advancement offers solutions for energy efficiency, we often observe neglect of broader ethical and aesthetic implications. Conversely, the humanistic approach underscores the essential connection between humans and their environment but faces obstacles in practical application. The highlighted tension between technological and humanistic approaches underscores the need for developing new integrative perspectives. These insights into the current state of practice serve as arguments for overcoming reductionist paradigms and a call for a holistic approach that unifies different perspectives and value systems. A holistic approach implies that technological progress and humanistic ideals are not opposing forces but complementary drivers in a synthesis that integrates ecological responsibility, technological innovation, and deeper, humane values. In the end, this study doesn't offer conclusive answers; it invites us to re-examine our previous assumptions and engage in ongoing contemplation about how to live in harmony with nature, not merely from it.

References

Aldo L (1949) A sand county almanac, and sketches here and there. Oxford University Press, Oxford

Berleant A (1997) Living in the landscape: toward an aesthetics of environment. University Press of Kansas, Lawrence

Berleant A (2002) Environment and the arts: perspectives on environmental aesthetics, 1st edn. Ashgate, Aldershot, UK, Burlington, VT

Berleant A, Carlson A (eds) (2004) The aesthetics of natural environments. Broadview Press, Peterborough, Canada

Brady E (2003) Aesthetics of the natural environment. Edinburgh University Press, Edinburgh, UK

Brady E (2009) Environmental aesthetics. In: Callicott J, Frodeman R (eds) Encyclopedia of environmental ethics and philosophy. Macmillan Reference, USA, Detroit, pp 313–321

Brady E (ed) (2007) Introduction to 'environmental and land art': a special issue of ethics, place and environment. Ethics Place Environ 10(3):257–261

Brundtland G (1987) Report of the world commission on environment and development: our common future. United Nations General Assembly

Budd M (2002) The aesthetic appreciation of nature. Oxford University Press, Oxford

Carlson A (2000) Aesthetics and the environment: the appreciation of nature. Art and Architecture. Routledge, London

Carlson A, Berleant A (eds) (2004) The aesthetics of natural environments. Broadview Press

Carlson A, Lintott S (eds) (2008) Nature, aesthetics, and environmentalism: from beauty to duty. Columbia University Press

Carson R (2002) Silent spring. Houghton Mifflin, Boston

Frampton K (2011) Urbanization, and its discontents: megaform and sustainability. In: Lee S (ed) Aesthetics of sustainable architecture. 010 Publishers, Rotterdam

Gill G (2011) The aesthetics of architectural consumption. In: Aesthetics of sustainable architecture, pp 26–41. 010 Publishers, Rotterdam

Guy S, Farmer G (2001) Reinterpreting sustainable architecture: the place of technology. J Architectural Educ 54(3):140–148

Guyer P (1979) Kant and the claims of taste. Cambridge University Press, Cambridge

Hagan S (2001) Taking shape: a new contract between architecture and nature, 1st edn. Routledge

Hepburn R (1966) Contemporary aesthetics and the neglect of natural beauty. In: Williams B, Montefiore A (eds) British analytical philosophy. Routledge and Kegan Paul, London

Hepburn R (2001) The reach of the aesthetic: essays on art and nature. Ashgate, Burlington, VT

Jencks C (1997) The architecture of the jumping universe: a polemic: how complexity science is changing architecture and culture. Academy Press

Kant I (1790) Critique of the power of judgment. Cambridge University Press, Cambridge, UK

McNeill JR (2000) Something new under the sun: an environmental history of the twentieth-century world. W.W. Norton & Company

Muir J (1894) A near view of the high sierra. In: Muir J (ed) The mountains of California. The Century Company, New York

Lee S (ed) (2011) Aesthetics of sustainable architecture. 010 Publishers, Rotterdam

Rolston H III (1998) Aesthetic experience in forests. J Aesthetics Art Criticism 56(2):157–166 (Special ed. Environmental Aesthetics)

Rome A (2013) The genius of earth day: how a 1970 teach-in unexpectedly made the first green generation. Hill and Wang

Saito Y (1998) Appreciating nature on its own terms. Environ Ethics 20(3):135–149

Saito Y (1998) The aesthetics of unscenic nature. J Aesthetics Art Criticism 56(2):101–111 (Special edition: Environmental Aesthetics)

Sloterdijk P (2011) Bubbles: spheres volume I: microspherology. Semiotext (E), Los Angeles

Thoreau HD (1963) Walden and civil disobedience. New American Library, New York

Ulrich G (2012) Sustainability: a cultural history. Green Books

United Nations (1993) Report of the United Nations conference on environment and development in Rio de Janeiro 3–14 June 1992

Part IX
Education: New European Bauhaus Design Curricula

Education: Nurturing the Next Generation of Urban Innovators

As the fields of urban planning, design, and architecture continue to evolve, the role of education in preparing future professionals has never been more critical. Contemporary education in these fields is at a crossroads, where traditional approaches are being reimagined to meet the complex and dynamic challenges of the 21st century. With cities facing unprecedented pressures from climate change, social inequality, and rapid urbanization, there is a growing need for educational models that are not only innovative, but also deeply aligned with the principles of sustainability, inclusivity, and beauty. The New European Bauhaus initiative offers a visionary framework that integrates these principles into the built environment. This section explores how contemporary education is evolving to align with these ideals and nurture the skills, values, and creative thinking necessary to craft cities in the future.

One of the defining features of contemporary education in these fields is its emphasis on interdisciplinary and collaborative learning. Urban challenges are inherently complex and require solutions from a wide range of disciplines, including environmental science, sociology, economics, and technology. Educational programs are increasingly being designed to break down silos between these disciplines, encouraging students to work together on projects that integrate diverse perspectives. This collaborative approach not only mirrors the realities of professional practice, but also fosters innovation by exposing students to a broad spectrum of ideas and methods (Boyer and Mitgang 1996).

Another cornerstone of modern education in urban planning, design, and architecture is the integration of project and experiential learning. Students are now more frequently engaged in real-world projects where they can apply theoretical knowledge in practical settings. This hands-on experience is crucial for developing the problem-solving and critical thinking skills necessary for successful urban planning and design (Kolb 1984). Such initiatives also align with the growing emphasis on participatory design and the co-creation of spaces that reflect the diverse needs and aspirations of communities (Salama and Wilkinson 2007).

Contemporary education now centrally focuses on sustainability and resilience. As cities grapple with the effects of climate change and resource depletion, educational programs are increasingly incorporating sustainability into every aspect of their curriculum. From the selection of materials and construction techniques to the planning of green infrastructure and sustainable mobility systems, students are being trained to design and plan urban environments that are both resilient and ecologically sound (Farr 2007; Kibert 2016). The integration of frameworks such as Leadership in Energy and Environmental Design (LEED) into academic programs is a testament to the growing importance of sustainability in urban education.

Furthermore, integrating digital technologies such as Building Information Modeling (BIM) and Geographic Information Systems (GIS) transforms the educational landscape. These tools are now a staple in urban planning and design curricula, equipping students with the technical skills needed to innovate in the digital age (Becerik-Gerber and Kensek 2009). The rise of virtual and augmented reality further enhances the learning experience, offering new ways to visualize and interact with urban spaces (Batty 2024).

In a globalized world, contemporary education balances global perspectives with a deep understanding of the local context. Students are encouraged to learn from best practices around the world while developing sensitivity to the unique cultural, social, and environmental conditions of the places where they will work (Healey 2010). International exchanges, study tours, and collaborations with global institutions are becoming integral parts of the educational experience, providing students with a broad and nuanced understanding of urban challenges and solutions (Sandercock 1998).

Finally, the ethical dimensions of urban planning, design, and architecture have been increasingly emphasized in contemporary education. Students are taught to consider the social implications of their work, striving to create inclusive and equitable spaces (Agyeman 2005), (Fainstein 2010). This ethical focus is closely aligned with the New European Bauhaus's vision of cities that are not only sustainable, but also socially just and culturally vibrant.

As we explore the education section of this book, we delve into how contemporary educational practices shape the next generation of urban innovators. These chapters highlight innovative curricula, teaching methodologies, and experiential learning opportunities that prepare students to become leaders in urban planning, design, and architecture. By integrating sustainability, inclusivity, and beauty into their education, these future professionals are equipped to create urban environments that meet the challenges of the 21st century while inspiring and enriching the lives of those who inhabit them.

This section not only examines how current educational approaches are evolving but also provides a roadmap for how education can continue to adapt to the needs of a rapidly changing world. By aligning with the principles of the New European Bauhaus, contemporary education is laying the foundation for a new era of urban design, visionary, inclusive, and deeply committed to creating beautiful, sustainable cities for all.

References

Agyeman J (2005) Sustainable communities and the challenge of environmental justice. NYU Press
Batty M (2024) Inventing future cities The MIT Press, USA
Becerik-Gerber B, Kensek K (2009) Building information modeling in architecture, engineering, and construction: emerging research directions and trends, J Prof Issues Eng Educ Pract 136(3):139–147. 10.1061/(ASCE)EI.1943-5541.0000023
Boyer EL, Mitgang LD (1996) Building community: a new future for architecture education and practice, Princeton, NJ
Fainstein SS (2010) The just city. Cornell University Press
Farr D (2007) Sustainable urbanism: urban design with nature. Wiley, USA
Healey P (2010) Making better places: the planning project in the twenty-first century. Red Globe Press
Kibert CJ (2016) Sustainable construction: green building design and delivery. Wiley, USA
Kolb DA (1984) Experiential learning: experience as the source of learning and development, no. 4. Prentice Hall, Englewood Cliffs, NJ
Salama AM, Wilkinson N (2007) Design studio pedagogy: horizons for the future. Urban International Press, Gateshead
Sandercock L (1998) Towards Cosmopolis: Planning for multicultural cities. Wiley, USA

Chapter 23
Creating a Better Future: Contextual Learning on Structural Design from Architecture Students' Perspective

Aleksandra Nenadović and Jelena Milošević

Abstract The education of architects should enable acquiring knowledge and skills essential to function effectively in an interdisciplinary team in practice dedicated to sustainability. Integrating theoretical and practical knowledge within the course curriculum is one way to accomplish the above. Lecture-based courses in structural design within the Undergraduate Studies of Architecture at the University of Belgrade combine theoretical and practical knowledge within the traditional learning model by applying theoretical knowledge to individual non-contextual problem-solving. In the final year of the study, structural design learning is based on contextual problem-solving in teams in the studio. This paper aims to present the applied approach of a combination of two learning models and to assess, using the survey, the influence of this approach on competence development. It will also point out the key shortcomings and suggestions for improvements from students' perspectives. A total of 230 final-year undergraduate architecture students participated in the survey. The analysis of the survey showed that the applied approach of early team contextual learning in the field of structural design contributes to competence development but indicated the need for constant improvement of the architectural education methodologies and processes to respond to growing complex requirements in architectural practice.

Keywords Integrated design · Architect education · Structural design · Contextual learning · Competency development

23.1 Introduction

In order to respond to complex requirements of sustainability-oriented architectural practice, architects must have the comprehensive abilities to rethink, criticise, and sense (Bereiter and Scardamalia 1993). In addition, they must have technical expertise, social awareness, and a bias toward innovation (Crawley et al. 2007). This

A. Nenadović (✉) · J. Milošević
Faculty of Architecture, University of Belgrade, Belgrade, Serbia
e-mail: aleksandra@arh.bg.ac.rs

M. Vukmirović et al. (eds.), *Digital Dreams, Urban Realities*, The Urban Book Series,
https://doi.org/10.1007/978-3-031-76353-3_23

combined set of abilities, knowledge, skills, and attitudes should enable productivity, entrepreneurship, and excellence (Crawley et al. 2007). Transversal skills are particularly emphasised among the above because they barely match the level needed to do the job (Cedefop). In this context, practical knowledge connected with the unique context of an authentic working situation (Lave 1996) is gaining importance. Therefore, scholars should seek the best means to integrate theoretical and practical knowledge within the course curriculum (Yong 2012) in ways that foster a more synthetic, concurrent, and dynamic integration of design issues, with a constant awareness that design is a creative activity, involving imagination, intuition, and deliberate choice (Arup and Partners 1986).

Integration of theoretical (know-what) and practical (know-how) knowledge, as the basis of competency-based architectural education (Spiridonidis 2008) in most subjects at the University of Belgrade—Faculty of Architecture (UB-AF), is achieved through work in the studio, in which students learn to develop design proposals with the help of their tutors (Wilkinson and Salama 2007). The studio as a pedagogical construct involves three basic functions: information-seeking, new knowledge-generating, and decision-making. In addition, it simulates design practice based on users' needs and environmental concerns (Nenadović 2014), that is, real-world projects, in order to integrate education, research, and practice into a single process (Incedayi 2005). Conversely, theoretical subjects in structural design are lecture-based, where students apply theoretical knowledge to individual non-contextual problem-solving. This model emphasises teachers' guidance and related coded knowledge transfer to students (Eraut 2000). This approach is preferred for resource efficiency but criticised for its rigidity and pedagogical ineffectiveness, especially for design-related learning (Aparicio and Ruiz-Teran 2007).

Bearing in mind the stated goals of architect education, one of the approaches that is gaining more importance when it comes to the engineering aspects of the discipline, including the field of structural design, is contextual learning. This type of learning should enable students to gain experience through solving real-world problems, which can give them real knowledge and provide conditions for knowledge valorisation. The key strategies of contextual learning in the technical–technological field are learning in context (Crawford 2001; Detri and Dadan 2019; Berns and Erickson 2001) and teamwork (Detri and Dadan 2019). The mentioned strategies should contribute to reaching a higher level of knowledge (Detri and Dadan 2019), understanding the meaning of certain learning content (Johnson 2002) and its critical evaluation, developing curiosity (Detri and Dadan 2019), understanding concepts (Crawford 2001), understanding the design logic (Easterday et al. 2017), motivation for self-regulated learning and self-efficacy (Lam et al. 2012), authentic assessment of achievement (Johnson 2002; Detri and Dadan 2019), and formation of adequate attitudes towards the professional technical area.

Based on the previously stated starting points, a new structural design learning model is applied to the Undergraduate Studies of Architecture. This model is a combination of two learning models—lecture-based and studio-based. Students first acquire competencies in lecture-based courses through individual non-contextual learning, and then in the final year, they apply and improve their competencies by solving real-world problems (Whitehead 2015a, b) within studio design courses (Chance et al. 2016). This project-based learning (Mills and Treagust 2003; Graham 2010) aims to bridge the gap in knowledge transfer between theory and practice, that is, to better prepare students to solve emerging problems under different contexts. This paper aims to present the applied approach of combining two models of structural design learning, individual non-contextual and team contextual, i.e., the means and methods of these specific learning formats. Also, a study aims to assess, using a survey, the influence of applied approach on the development of cognitive abilities, knowledge, skills, and motivation of architecture students, representing the key elements of competency-based architectural education (Spiridonidis 2008) and the key indicators of teaching success. In addition to the above, a study aims to get students' free comments on shortcomings and suggestions for improvement, especially considering that limited effort has been directed toward seeking input from students in order to derive key topics related to developing strategies for teaching and learning (Cook-Sather et al. 2009).

23.2 Structural Design Learning Model

The means and methods of the specific learning model, which represents a combination of two learning models, individual non-contextual and team contextual, applied within the Undergraduate Studies of Architecture at the University of Belgrade, are presented.

23.2.1 Lecture-Based Structural Design Learning Model

Lecture-based structural design courses begin with the basic issues of statics and strength of materials and continue with the behaviour and analysis of structures, after which the design of timber, steel and reinforced concrete structures is introduced. In all courses, students acquire theoretical and practical knowledge in the traditional way. During lectures, students must understand, memorise, and reproduce the theoretical data. Then, students apply theoretical knowledge to individual non-contextual problem-solving during exercises that follow the lectures. They solve generic numerical examples related to dimensioning of structural elements. This process is guided by teachers who transfer coded knowledge to students (Eraut 2000).

23.2.2 Studio-Based Structural Design Learning Model

In the final year of Undergraduate Studies, within the studio-based course named "Studio Design—Project Development," students apply and improve the acquired theoretical knowledge in structural design through contextual learning by working on specific real-world projects and by sharing informal learning experiences through teamwork. This contextual, team learning-by-doing approach combines theoretical and practical knowledge, refocusing on more applied forms of learning and teaching (Graham 2010). The studio combines three methods of education. First is a teacher-centred method in the function of theoretical knowledge transfer. Second is a student-centred method in the function of search, collection, and aggregation of knowledge, especially in the phase of concept creation, whereby the design is seen as a competition (Anthony 1991), and finally, a teamwork-cantered method that implies the team's creation, construction, and development of knowledge, in order to encourage innovation. Teamwork should prepare students for practice, given that effective architectural practices are associated with a culture of teamwork and collaboration (Nicol and Pilling 2000).

Students in the studio first work on building concepts that comply with sustainability requirements. Work starts with critically rethinking examples from architectural practice while simultaneously comprehending the given design program and the environment, after which concepts are created. After tutors select the best design proposals, students collaborate in teams to develop the project, including the design of the building structure. Then, through consultations with tutors and independent teamwork, by applying knowledge from previous years of study, students work on positioning and calculating structural elements and structural details in a process that simulates design practice. Since two fields of research are integrated into the studio, architecture and structural design, the participation of tutors from both disciplines is foreseen. The process occurs through a design charrette, i.e., through focused and collaborative brainstorming sessions that encourage the exchange of ideas and information and enable the creation of truly integrated design solutions (Nenadović et al. 2017). Previous is a framework in which students integrate architectural and structural aspects of design into their design proposals (Herr et al. 2012; Herr and Tu 2011) and develop a structural sensitivity, not typically taught in lecture-based modules (Herr and Tu 2011; Herr 2013).

23.3 Structural Design Learning Model from Students' Perspective

In order to evaluate students' structural design learning experience, that is, to find out how students rate the importance of certain activities in lecture-based and studio-based courses and to find out to what extent students felt their learning fostered competence development, the survey was conducted. A quantitative research strategy is applied to collect information from a representative sample (Ponto 2015). A total of 230 final-year undergraduate architecture students anonymously responded to questions (Check and Schutt 2012). The questionnaire is formed with a modified Likert scale. The students could answer with none (0) or give a rating from 1–5, that

Table 23.1 Students' evaluation of specific learning activities

Learning activities in			
Lecture-based courses		Studio-based course	
How would you rate the importance of:			
Lectures		Work with tutors	
Exercises		Independent work	

is, to express their minimal (1) to maximal (5) strength of feeling (Katajavuori et al. 2006). In addition to answering the questions, students could give free comments about courses in terms of shortcomings and suggestions for improvement.

23.3.1 Students' Evaluation of Specific Learning Activities

Results of students' evaluation of specific learning activities in structural design lecture-based and studio-based courses are presented in Table 23.1.

23.3.2 Students' Evaluation of Competence Development

The following indicators to assess the influence of the learning model on competence development in the technical–technological field are analysed:

- Element 1—Abilities—Indicators: cognitive reasoning (Ul-Haq 2015) and intuition and creativity (Ul-Haq 2015; Sternberg 1998)
- Element 2—Knowledge—Indicators: understanding of knowledge (Detri and Dadan 2019); comprehending the design logic (Easterday et al. 2017), understanding the meaning of the subject matter (Johnson 2002)
- Element 3—Skills—Indicators: research and analysis, problem-solving, organisation, communication, independent learning, and teamwork skills (Cedefop), self-assessment and self-criticism (Johnson 2002; Detri and Dadan 2019)
- Element 4—Motivation—Indicators: motivation for the acquisition of new knowledge (Detri and Dadan 2019) and self-engagement (Lam et al. 2012; Britt et al. 2007).

Results of students' evaluation of the development of abilities, knowledge, skills, and motivation in structural design lecture-based and studio-based courses are summarised in Table 23.2.

23.3.3 Students' Comments on Lecture-Based Courses and Studio-Based Courses

Students' free comments on lecture-based courses, in terms of shortcomings and suggestions for improvement, are summarised in Table 23.3.

Students' free comments on studio-based courses, in terms of shortcomings and suggestions for improvement, are summarised in Table 23.4.

23.4 Discussion

The content analysis of the survey showed that, from architecture students' perspective, the traditional engineering-based teaching method in lecture-based courses is quite ineffective in providing the knowledge needed for architects to work in practice, as indicated by researchers (Ihsan and Ceng 2006; Dhanorkar and Tarar 2017; Fahmi et al. 2012). Students are aware that theoretical courses are necessary, but they do not provide a sufficient basis for understanding the subject matter and its practical application (Guthrie 2012). They think that theory is at a very preliminary level and that lack of a strong foundation about basic laws limits their ability to solve concrete problems. They think that the theoretical basis should be better explained in order to solve problems (Dhanorkar and Tarar 2017). They think that too much material is presented per semester without sufficient explanation of the logic and the way in which things should be practically done. "Things should be logically explained, not just to convey information." They think that theoretical topics are mainly analysed without any interaction with structural design in real-world situations (Fahmi et al. 2012; Ihsan and Ceng 2006). They emphasise that literature has no connection to practical application and that material can be learned superficially. There is a lack of applied examples and laboratory work that involves active student participation (Ihsan and Ceng 2006), which causes them to show less interest in lecture courses (Ünay and Özmen 2006). Superficial and rote learning gives them low motivation. They do just enough to pass the exams (Ihsan and Ceng, 2006). They attach more importance to exercises than to lectures in lecture-based courses, but they consider their content to be inadequate. There is a lack of good explanation. The exercises are mechanically rewritten. There is a lack of gradual transition from abstract theory to concrete problems. They think they need to know about structural application rather than structural calculations and that structural design should be learned by observing and doing instead of listening (Fahmi et al. 2012; Öztaş

Table 23.2 Students' evaluation of competency development

	Structural design learning in	
	Lecture-based courses	Studio-based course
Element 1 *Abilities*	*To what extent did you improve the following abilities:*	
Reasoning	7 17,5 38,6 28,1	15,8 28,9 47,4
Intuition	12,3 31,6 28,1 21,1	6,6 21,1 34,2 32,9
Creativity	24,6 26,3 10,5 22,8 8,8 7	9,2 17,1 25 39,5
Element 2 *Knowledge*	*To what extent did you improve the understanding of:*	
Subject matter	14 35,1 22,8 21,1	11,8 22,4 32,9 7
Design logic	15,8 28,1 21,1 28,1	7,9 17,1 35,5 34,2
Learning goals	24,6 26,3 10,5 22,8 8,8 7	6,6 9,2 22,4 22,4 7
Element 3 *Skills*	*To what extent did you improve the following skills:*	
Research and analysis	12,3 14 22,8 22,8 14 14	9,2 25 27,6 27,6
Problem-solving	7 10,5 19,3 28,1 21,1 14	13,2 21,15 30,3 7

(continued)

Table 23.2 (continued)

	Structural design learning in	
	Lecture-based courses	Studio-based course
Organization	10,5 12,3 19,3 19,3 28,1 7	6,6 9,2 13,2 19,7 25 26,3
communication	21,1 21,1 21,1 19,3 14	9,2 11,8 11,8 23,7 22,4 7
Independent learning	19,3 8,8 8,8 26,3 15,8 21,1	6,6 10,5 7,9 13,2 28,9 32,9
Teamwork		6,6 9,2 6,6 26,3 17,1 34,2
Element 4 Motivation	*To what extent did you improve the motivation for:*	
Self-engagement	15,8 15,8 8,8 28,1 14 7	7,9 7,9 17,1 26,3 35,5
Knowledge acquisition	14 8,8 22,8 14 19,3 21,1	7,9 6,6 17,1 30,3 34,2
	■ 0 ■ 1 ■ 2 ■ 3 ■ 4 ■ 5	

Table 23.3 Students' comments on lecture-based courses

	Structural design learning in Lecture-based courses	
	Shortcomings	Suggestions for Improvement
Lectures		
	Too much material is presented within one semester	Only essential material should be included
	It is difficult to follow the lectures because they are incomprehensible	We need to learn more theory/Teachers should better explain theoretical basis
	The subjects are designed so that the material can be learned superficially	The matter should be logically explained and not just to convey information
	Without a good and strong foundation, it is difficult to build on knowledge and really understand problems, which is the opposite of superficial and rote learning	I think that it is necessary first to master the basic laws governing this field, which I think was not the case in these courses, so that students could solve problems
	The lack of adequate literature is a problem due to which we make mistakes regarding the basic principles even in the third year of study	Adequate and more extensive literature on theory should be provided/There is a lack of literature with gradual transition from theory to concrete problems
Exercises		
	The exercises are based on mechanical rewriting from collection of exercises/The exercises are done only to get points	The exercises should not be reduced to just one specific task/It is necessary to introduce practical work on concrete problems
	On exercises we only work on tasks from collection of exercises, without detailed explanation/On exercises there is lack of explanation how things actually work	Teachers should give better and clearer, logical explanation of how certain things really work, rather than just transmit information/We need better explanation in the collection of exercises, which implies gradual transition from abstract to concrete problems
	The course cannot produce enough practical knowledge/The exercises as such are not enough to understand all needed for studio project, and later for the practice	While explaining the exercise, teachers should draw parallels with other variations that are possible/It is necessary to draw a parallel with practice

and İpekçi 2015). They think practical work on concrete problems should be introduced, drawing a parallel with practice. Intuition and creativity are underdeveloped in lecture-based courses while understanding the design logic and learning objectives is insufficient. Lecture-based courses do not contribute enough to the motivation for self-engagement, the acquisition of new knowledge, and the development of research analytical and communication skills.

The studio-based course is intended to function as a place where all the knowledge acquired in lecture-based courses students should apply in a design process. In that context, students give equal importance to the work with tutors and to independent work in the studio. However, from the student's perspective, there are factors that prevent the system from working as intended, as indicated by other researchers (Ihsan

Table 23.4 Students' comments on the studio-based course

	Structural design learning in Studio-based couse	
	Shortcomings	Suggestions for Improvement
Work in studio		
	In order to work on structural design during project development, the knowledge from theoretical subjects is required, which students have not sufficiently acquired/When it comes to theoretical subjects, students concentrate only on passing the exam, but when all of that should actually be applied, there is not enough understanding on the logic and the ways in which things should be practically done	It is necessary to increase the fund of teaching hours of theoretical subjects as a basis for work in the studio/It is necessary to change the concept of theoretical subjects by reducing the math and increasing the logic, in order to make work in the studio meaningful/The building structure design in the conceptual phase should be prioritized
	The problem is the lack of appropriate literature which also includes the real-life examples/Everything that was done in the studio was done with self-initiative, supported by modest knowledge acquired in theoretical subjects	Better availability of literature/sources that provide a link between theory and real-life problems is needed
	Too many commitments, too little time/The classes should be organized more efficiently	Stop forcing unnecessary things when the semester is short/Harmonize learning objectives with the students' capabilities
	Most of us don't deal with creativity, but instead move towards simpler projects to secure points, which are difficult to achieve due to large amount of work	It is necessary to set aside more time for project development than for the development of concept
	The subject program cannot produce knowledge that can be directly applied in practice. Again, this is certainly a better way than theoretical subjects/You will learn something only when you get a job, not at the faculty	Both students and teaching staff should be concerned and work on improving learning, which would provide better foundations for practice

and Ceng 2006). Students think that knowledge from theoretical subjects is not sufficiently acquired. Also, they are unable to fully relate lecture courses to design in the studio (Öztaş and İpekçi 2015; Wetzel 2012; Ünay and Özmen 2006; Dhanorkar and Tarar 2017; Ihsan and Ceng 2006), that is, unable to utilise the acquired knowledge in their design work (Fahmi et al. 2012). They think that there is a missing link between theory and real-life problems (Whitehead 2015a, b). They think that the concept of theoretical subjects should be changed by reducing the math and increasing the logic. They think that there is a lack of appropriate literature which includes real-life examples. In lecture-based courses, only simple structures are introduced, after which they should design complex ones within the studio (Dhanorkar and Tarar 2017; Fahmi et al. 2012; Ihsan and Ceng 2006). They think they can't be creative enough due to a lack of understanding of structural application (Fahmi et al. 2012). In the studio, they are asked to design large buildings containing spaces with large spans, which is beyond the scope of their present skills (Ihsan and Ceng 2006). For this reason, they see the material as too extensive, with insufficient time to master it. They think teachers should harmonise learning objectives with student's capabilities. The above is one of the reasons why they go for simpler solutions, which somewhat limits their creativity (Didem 2021). They think more time should be given to the development of the project than to the design concept. Regardless of the opinion that work in a studio cannot produce knowledge that can be directly applied in practice, they think that it contributes to understanding design logic and learning goals and to the development of reasoning, intuition, and especially creativity, research and analysis, organisation, communication, independence in learning, and teamwork skills. However, they think teachers should further improve the structural design learning model in order to improve students' understanding of the subject matter and students' problem-solving skills.

23.5 Conclusion

The applied approach of combining two models of structural design learning, individual non-contextual in lecture-based courses and team contextual in studio-based courses, where theoretical knowledge, research, and practical work integrate through real-world problem-solving, improves architecture students' ability to transfer knowledge between theory and practice. The analysis of the survey showed that the applied approach of early team contextual learning contributes to the development of critical and creative thinking, understanding of the meaning of the subject matter, development of research and analysis, organisational, communication, independent learning, and teamwork skills, and motivation to the acquisition of new knowledge and self-engagement, which are all important elements of competency-based architectural education. On the other hand, the applied approach contributes to a lesser extent than expected to the understanding of structural design subject matter, indicating the need to further explore the potentials and limitations of the learning model in order to find ways to bring this complex matter closer to architecture students. The above also indicates the need for constant improvement of the architects' education methodologies and processes, including project-based, problem-oriented, situation-oriented, and work-based learning, so that architects can respond to growing complex requirements in integrated practice.

Acknowledgements This research was funded by the Ministry of Education, Science and Technological Development of the Republic of Serbia, grant number 451-03-68/2020-14/200090. The research was done under the research lab of the University of Belgrade, Faculty of Architecture - Laboratory for Innovative Structures in Architecture (LISA).

References

Anthony KH (1991) Juries on trial: analysis and critique of design juries and studios. Van Nostrand Reinhold, New York
Aparicio AC, Ruiz-Teran AM (2007) Tradition and innovation in teaching structural design in civil engineering. J Prof Issues Eng Educ Pract 133(4):340–349
Ove Arup and Partners (1986) Ove Arup and Partners: 1946–1986. St. Martin's Press, London
Bereiter C, Scardamalia M (1993) Surpassing ourselves: an inquiry into the nature and implications of expertise. Open Court, Chicago
Berns RG, Erickson PM (2001) Contextual teaching and learning: preparing students for the new economy. National Dissemination Center for Career and Technical Education, Columbus, OH
Britt TW, Dickinson JM, Greene-Shortridge TM, McKibben ES (2007) Self-Engagement at work. In: Nelson D, Cooper CL (eds) Positive organizational behavior. SAGE Publications, London, pp 143–158
Chance S, Marshal J, Duffy G (2016) Using architecture design studio pedagogies to enhance engineering education. Int J Eng Educ 32(1):364–383
Check J, Schutt RK (2012) Survey research. In: Check J, Schutt RK (eds) Research methods in education. Sage Publications, Thousand Oaks, CA, pp 159–185
Cook-Sather A, Clarke B, Candon D, Cushman K, Demetriou H, Easton L (2009) Learning from the student's Perspective: a sourcebook for effective teaching. Routledge, New York
Crawford ML (2001) Teaching contextually: research, rationale, and techniques for improving student motivation and achievement in mathematics and science. CCI Publishing Inc., Waco, TX
Crawley EF, Malmqvist J, Östlund S, Brodeur DR (2007) Rethinking engineering education: the CDIO approach. Springer Science + Business Media, New York
Detri KT, Dadan R (2019) Contextual teaching and learning,g to develop critical thinking and practical skills. J Phys Conf Ser 1233(1)
Dhanorkar SS, Tarar A (2017) A comprehensive and integrated pedagogical approach for teaching structures in architecture. Int J Eng Res Technol 10(1):72–86
Didem GY (2021) Teaching the principles of structural design: rethinking the methods for architecture students. In: 4th International conference of contemporary affairs in architecture and urbanism, Alanya, Turkey
Easterday MW, Lewis DR, Gerber E (2017) The logic of design research. Learn Res Pract 4(2):1–30
Eraut M (2000) Non-formal learning and tacit knowledge in professional work. Br J Educ Psychol 70:113–136
Fahmi MM, Aziz AA, Elhadi MAS (2012) The integration of structural knowledge in studio design projects: an assessment curriculum in architecture course. J Sci Technol 13:59–71
Graham R (2010) UK approaches to engineering project-based learning. Gordon-MIT Engineering Leadership Program. https://www.rhgraham.org/resources/MIT-White-Paper---UK-PjBL-April-2010.pdf. Last accessed 4 Jan 2023
Guthrie JB (2012) Structural engineering for architecture and construction management students: teaching methods and changing needs. In: 2012 ASEE annual conference & exposition. American Society for Engineering Education, Washington DC
Herr CM (2013) Qualitative structural design education in large cohorts of undergraduate architecture students. Glob J Eng Educ 15(2):96–102
Herr CM, Fischer T, Brown A, Millard SG (2012) Form and formalism: on the future role of structural design in architectural education in China. Adv Mater Res 450–451:257–262
Herr CM, Tu YH (2011) Teaching structural design. ArS Arch Sci 4:59–74
Ihsan U, Ceng I (2006) Building structure design as an integral part of architecture: a teaching model for students of architecture. Int J Technol des Educ 16:253–271
Incedayi D (2005) Architect as a facilitator, the changing education (of architecture). In: Harder E (ed) Writings in architectural education. EAAE, Copenhagen, pp 112–129

Johnson EB (2002) Contextual teaching and learning: what it is and why it's here to stay. Corwin Press, INC, Thousand Oaks, CA

Katajavuori N, Lindblom-Ylänne S, Hirvonen J (2006) The significance of practical training in linking theoretical studies with practice. High Educ 51:439–464

Lam SF, Wong BPH, Yang H, Liu Y (2012) Understanding student engagement with a contextual model. In: Christenson SL, Reschly AL (eds) Handbook of research on student engagement. Springer Science + Business Media, New York, pp 403–419

Lave J (1996) Teaching, as learning, in practice. Mind Cult Act 3(3):149–165

Mills JE, Treagust DF (2003) Engineering education—is problem-based or project-based learning the answer? Australasian J Eng Educ 3:2–16

Nenadović A, Đukanović Lj, Radivojević A (2017) Integrated design in the process of architectural education. In: Bijedić Dž, Krstić-Furundžić A, Zečević M (eds) Proceedings of the 4th international academic conference on places and technologies 2017: keeping up with technologies in the context of urban and rural synergy. University of Sarajevo—Faculty of Architecture, Sarajevo, pp 408–416

Nenadović A (2014) Integrated design of structural systems based on the application of ferrocement. Doctoral dissertation, University of Belgarde, Belgrade

Nicol D, Pilling S (2000) Architectural education and the profession—Preparing for the future. In: Nicol D, Pilling S (eds) Changing architectural education: towards a new professionalism. Taylor & Francis, London, pp 1–21

Öztaş SK, İpekçi CA (2015) Innovations for structural system education in architecture. In: Proceedings of the CBU international conference on innovation, technology transfer and education 3. CBU Research Institute, pp 263–270

Ponto J (2015) Understanding and evaluating survey research. J Adv Pract Oncol 6(2):168–171

Spiridonidis C (2008) Towards a competences based architectural education in Europe. In: Spiridonidis C, Voyatzaki M (eds) New directive—new directions... What is the academic direction of our schools in this new context? Charis, Thessaloniki, pp 55–78

Sternberg RJ (1998) Cognitive mechanisms in human creativity: is variation blind or sighted? J Creative Behav 32(3):159–176

Ul-Haq S (2015) Intuition and creativity. https://www.researchgate.net/publication/282332743_Intuition_and_creativity?channel=doi&linkId=560ce45d08ae73e7a6a30d46&showFulltext=true. Last accessed 23 Feb 2023

Ünay Aİ, Özmen C (2006) Building structure design as an integral part of architecture: a teaching model for students of architecture. Int J Technol des Educ 16:253–271

Wetzel C (2012) Integrating structures and design in the first-year studio. J Architect Educ 66(1):107–114

Whitehead R (2015a) Comprehensive and creative conclusions: enhancing structural design educational opportunities in labs for architecture students. In: Proceedings of the AEI 2015 conference. ASCE, Reston

Whitehead R (2015b) Rebuilding a framework for learning: rethinking structural design instruction in an architectural curriculum. In: Ingraffea N, Libby M (eds) Proceedings of the structures congress 2015. ASCE, Reston

Wilkinson N, Salama AMA (2007) Design studio pedagogy: horizons for the future. The Urban International Press, Gateshead, UK

Yong T (2012) The mode of theoretical knowledge and practical knowledge combination: the significance of internship. World J Educ 2(4):55–63

Chapter 24
Tracing the Path for the New European Bauhaus: From University Curricula to the Innovative Ideas on Ecological Transition

Aleksandra Stupar, Ivan Simić, Vladimir Mihajlov, and Marija Negovan

Abstract The introduction of the New European Bauhaus project (NEB) in September 2020 underlined the importance of the goals defined by the 2030 Agenda for Sustainable Development (SDGs), especially Goal 11, which focuses on sustainable cities and communities. Within this framework, the imperatives of social inclusion, fairness and equality are promoted as the cornerstones for future sustainable development and climate resilience. Simultaneously, the NEB supports ecological, digital and social transition within the scope of architecture and urbanism. Bringing the European Green Deal into the lives of all citizens, this initiative should be implemented through design, delivery and dissemination with the support of three main principles—multi-level and global transformation, involvement and inclusivity, as well as transdisciplinarity. Considering all these elements which define the NEB as one of the integrative European cultural policies directly linked to architecture and urban planning, it is necessary to include them in university curricula. The University of Belgrade—Faculty of Architecture has already structured its programs around the SDGs emphasised in the NEB, especially in the Master program in Architecture_Model Urbanism, entitled Sustainable City. In order to present the specificities of the educational process conducted within this framework, the chapter will provide an insight into the project-oriented approach applied during one of the obligatory courses (Sustainable City 1—Transformations) and the Master diploma project, which includes several courses/phases of the final master/graduation assignment—from thematic research to thesis and project. In line with the NEB agenda, this approach enabled the updating of a general study content toward ecological transition while supporting the students' work through the stimulation of critical thinking and inquiry, creativity and innovation, research and investigation.

Keywords New European Bauhaus · Ecological transition · Sustainability · University curricula · Cities

A. Stupar (✉) · I. Simić · V. Mihajlov · M. Negovan
University of Belgrade—Faculty of Architecture, Belgrade, Serbia
e-mail: stupar@arh.bg.ac.rs

M. Vukmirović et al. (eds.), *Digital Dreams, Urban Realities*, The Urban Book Series,
https://doi.org/10.1007/978-3-031-76353-3_24

24.1 Introduction

Since the beginning of the twenty-first century, contemporary society has been dramatically exposed to a number of turbulences caused by climate extremes, political and economic conflicts, wars, natural catastrophes, migrations, inequality, poverty, gender misbalance and global pandemics. Considering the growing share of the urban population (currently more than 50% of the total population), cities around the world have strengthened their role as the major nodes of interaction, exchange and innovation but also confirmed their enormous capacity for generating a wide range of problems which affect us on all spatial scales. Pervading our everyday lives and influencing the future of human habitat(s), these problems trigger a number of fears which have to be addressed through innovative responses. While shaping further urban development and spatial transformations, new and updated tools will simultaneously broaden our knowledge and experiences, which have to be transmitted and/or adjusted globally.

Having in mind that urban space represents the basic and most stimulating setting for the creation, verification and implementation of novel social, technological, environmental and spatial remedies, the position of architecture, urban planning and design has been reconsidered, especially after the challenges met during COVID-19. As a result, the New European Bauhaus (NEB) was launched in 2020 as a project based on both the development goals set by the 2030 Agenda for Sustainable Development (UN 2015) and the different policies targeting sustainability, climate resilience and culture (European Commission 2021). During the same year, the EU adopted the Green Deal (EU 2020a), a document defining a vision of Europe as the first climate-neutral continent, and the New Leipzig Charter (European Commission 2020), representing the broadest European policy for sustainable urban development which highlights the role of integrated approach in strategical thinking and implementation, on all spatial levels. The application of these documents was accompanied by the approval of the financial program NextGenerationEU as a temporary instrument providing 800 billion euros for the economic recovery from the coronavirus pandemic, as well as for the further actions leading to a greener, more digital and resilient future.

In line with these activities within the EU framework, the NEB initiative underlines climate shift as the most impactful threat to sustainable development and focuses on urban nodes as the main polygons for the demonstration of both the problems and solutions. The elements of sustainability included in Goal 11 are used as a starting point ensuring the achievement of more resilient societies by providing opportunities for business and employment, safer and more affordable housing, efficient and accessible public transport and green public spaces (UN 2015). Therefore, the importance of urban planning, urban design, and architecture in this process has been acknowledged, and it is a necessity for the inclusiveness and participatory character of governance.

Although universities should act as the important drives of change guided by the SDGs and the NEB postulates, the education process applied in architectural and planning schools has its own pace. Often, the transition toward the new models of understanding, learning and application of emerging social processes and environmental concepts lags behind official declarations, which is reflected in professional practice. Consequently, it has become imperative to update the university curricula according to the contemporary paradigms of general sustainability and resilience while creating an educational setting able to stimulate creative insights, critical thinking, research and innovation (Salama 2016).

The University of Belgrade—Faculty of Architecture and its programs, accredited and verified by both the National Entity for Accreditation and Quality Assurance in Higher Education (NEAQA) and the Royal Institute of British Architects (RIBA), has accepted the same approach, which has been positively validated by students, educators and professionals in the field of architecture, urbanism and allied disciplines. The latest accreditation cycle, conducted by both the NEAQA and the RIBA, was successfully finalised in 2021/22, strengthening the dedication to the 2030 SDGs and the NEB agenda while providing a learning setting for a necessary ecological transition.

Considering the above-mentioned trends, this chapter will focus on the evolving synergy between the global environmental imperatives, the EU development strategies and the updated pedagogical process applied in architectural and planning education. After the introduction part, the elements connecting the NEB and the ongoing changes in the university curricula will be elaborated, as well as the dominant themes connecting theory and practice. The third part highlights the case study of the University of Belgrade—Faculty of Architecture through one of its master programs and two selected courses. The conclusion discusses the effects of the applied changes and provides recommendations for future innovations in the educational process, which should serve as catalysts for the general change of environmental awareness and the faster implementation of ecological transition.

24.2 New European Bauhaus: Educating for the Future

The NEB represents one of the culture-oriented EU policies that tackle the issues of sustainable development, aesthetics, and inclusiveness through architecture and urbanism, which includes planning and designing urban spaces. Fostering an integrative approach to the anticipated transition on several levels—social, ecological and technological, the NEB agenda should facilitate activities which target the effects of numerous challenges caused by the extreme manifestations of climate shift and the deterioration of the environment, further amplified by the global pandemic. Its origin could be traced back to the 2018 Davos Declaration (Davos 2024), which highlighted the importance of the environmental aspect of building/construction. Emphasised as one of eight basic principles (besides governance, functionality, economy, diversity, context, the sense of place and beauty), the protection of the environment, directly and indirectly, acknowledged the important role of architects and planners in the sustainability of our future. In line with the updated perception of these professions, the continuous adjustment of university education to the latest environmental imperatives has become a requirement (Salama 2009). It has been concluded that the inflexible and outdated pedagogical approaches have to be redefined through new university curricula relying on the paradigms of general sustainability and resilience. However, the mode of change has remained questionable, causing discussions among educators and academics. The main dilemma was related to the actual outcome of the updated design pedagogy—whether it should respond to the needs of the contemporary profession, providing the best results and/or what would be its effects on contemporary society and its transformations.

Initiatives targeting the process of architectural and planning education occurred between the mid-1960s and late 1990s when the New Design Pedagogy was introduced in order to improve the ongoing situation and prepare future professionals to address accumulating ecological problems. In this phase, the importance of more sustainable urban communities was underlined, generating new approaches to design techniques, models and tools. The result was the application of new methodologies in the design process—case study model, community-based design learning model and participative curriculum model. Furthermore, a comprehensive report published by the MIT School of Architecture and Planning (Beinart 1981) discussed the effects of contemporary education on society and its transformations, while many other contributions shaped a new discourse on design pedagogy, stimulating the debates on the current architectural education and practice. Some of them (e.g. Dutton 1991) emphasised the importance of intensified interaction between architectural educators and students regarding the political, social and cultural aspects of urban/architectural space while positioning the architectural pedagogy within a critical analysis of society.

Teymur (1992) concentrated on the role and importance of architectural education for international practice and global exchange, discussing the impact of various theories in design teaching. The local (UK) context represented the focus of the first general history of architectural education, provided by Crinson and Lubbock (1994). The authors also considered the relationship between the built environment and architects' education through the lens of urban history. Meanwhile, new trends were analysed by Salama (1995), who highlighted the role of the design studio and the innovative concepts that were applied to this form of architectural/planning teaching.

Between the mid-1990s and mid-2010s, some new ideas and considerations appeared, leading to Critical inquiry and Process-oriented design pedagogy (Salama 2016). One of them is related to the importance of civil initiatives, which is a trend based on the ideas of a more liberal, flexible, and integrated curriculum based on actual problems (Boyer and Mitgang 1996). The reality of professional challenges and their relation to architectural education was also highlighted (Nicol and Pilling 2000), while Harriss and Widder (2014) underlined the importance of successful, evidence-based live projects in curricula.

The 2020s shifted a focus to the evolving problems of urban communities, opening a stage for the implementation of Interchangeable design pedagogies, Community-based design pedagogy and live project studios recently promoted by the NEB.

In line with the latest educational trends, the schools of architecture and planning also embraced new learning philosophies and digital tools, which can facilitate the assessment of performative criteria—including daylight, shading, noise, air and water qualities, biodiversity health, comfort, user appreciation, energy, water and waste, as well as the related services (e.g. shared mobility). Although their application can significantly increase the capacity and scale of passive design techniques, they have to be used in a rational and systematic manner in order to incorporate the Green Deal goals and the NEB strategy.

24.3 Between Theory and Practice: Themes and Paradigms

The education of future architects and planners has to provide a balance between the theoretical frameworks that are occurring, practical knowledge, and technological skills. The shifting focus of global imperatives, innovative responses and professional trends has always influenced the content, structure and prevailing methodologies in university curricula reflecting the dominant paradigms. Environmental issues and their socio-economic repercussions have raised significant (and justified) attention from architectural/planning educators and professionals for more than three decades. The NEB initiative also emphasises the causal relation between all these aspects, supporting an integrative approach to the upcoming phase of urban restructuring. The ecological transition has a threefold role in this process—representing an aim, instrument and pre-condition for a wider multidisciplinary action targeting both the society and the built environment.

The concepts of sustainability and resilience have been gradually embraced as a general foundation for the anticipated change of environmental awareness, becoming a significant feature of the reformative processes conducted in architectural and planning schools all over the world. The meaning of the term sustainable development has undergone a significant change under the influence of the resilience theory. Defined as "the system's capacity to absorb disturbance and reorganise itself" during the occurring changes, resilience also indicates the capability to simultaneously maintain the same function, structure and identity (Folke et al. 2010: p. 20). From the perspective of resilience, sustainability should focus on the system's capacity to create and test possibilities for maintaining adaptive potential while avoiding reducing variability in system dynamics and/or optimising system performance in order to keep its equilibrium (Holling 2001). Considering that cities represent complex, resilient, and open systems constantly exposed to unpredictable external influences, their equilibrium is always challenged and cannot be perceived as static. As a result, new definitions and international documents dealing with sustainable development have recognised resilience as an essential feature of sustainability, which supports development but with fundamentally different implementation mechanisms.

Another concept gaining attention in university curricula is circularity, representing a model of resource production and consumption based on sharing, leasing, reusing, repairing, refurbishing and recycling existing materials and products for as long as possible (OECD 2020). Although the transition from a linear to a circular economy is a huge challenge, the benefits would be significant. Therefore, the application of different policies targeting local public services (e.g. transport, waste, water and energy systems) would contribute to the circulation of products and materials, the reduction/elimination of pollution, and, consequently, the regeneration of the natural environment. Cities, as the complex and most important nodes of global interaction and innovation, represent key actors in the process implementation, creating a setting for the improvement of general well-being, economic growth and environmental quality. OECD (2020) underlines several steps which would facilitate circularity - Rethink (stimulating local communities and raising their awareness), Regenerate (by increasing green infrastructure), Reduce (strategies focused on the resource footprint reduction), Reuse (establishing sharing platforms) and Recover (resources and landfills).

The importance of green infrastructure is also highlighted in various documents, and for the last two decades, it has represented the main conceptual framework for green spaces in the EU. Defined as a strategically planned network of natural and semi-natural areas with other environmental features, green infrastructure is designed and managed to deliver a wide range of ecosystem services while also enhancing biodiversity (EU Commission 2013). The network has a multi-scalar character functioning continuously from the largest to the smallest spatial units, which are interrelated (i.e. from regions to plots). The sufficient connectivity to ecological corridors provides environmental services in urban areas, which is a prerequisite for ensuring biodiversity, social and territorial cohesion, sustainable development and overall human well-being (Simic et al. 2017). These principles of green infrastructure have a profound influence on the design process in architecture and urbanism, especially during the initial and conceptual phases, which have to consider the spatial and functional possibilities for the integration of green infrastructure, as well as the characteristics of the built environment (e.g. urban typologies).

The NEB agenda includes the above-mentioned concepts and themes, underlining their importance for the future of our habitats. Supported by low and high-technology solutions in the participatory process of co-creation, the activities promoted by the NEB initiative are structured around design, delivery and dissemination. While aiming toward sustainability and resilience and respecting the principles of inclusion and cohesion, the transformation should be considered on all spatial levels in an interdisciplinary manner. The role of architects and planners becomes especially important in creating and/or protecting the sense of belonging, increasing physical and digital accessibility and designing a wide range of sustainable spatial solutions which integrate the principles of circularity, affordability and zero pollution (European Commission 2021).

24.4 Local Insight: University of Belgrade—Faculty of Architecture

The education of architects and planners at the University of Belgrade—Faculty of Architecture has a tradition of over 170 years. Due to the specific position of Belgrade as an important node in Southeast Europe, the students are adjusted to a very complex contextual setting, which is beneficial for both their future professional practice or academic advancement on local, regional and global levels. The study programs that were conducted provide an integrated approach to the fields of architecture and urbanism, supported by all four departments (Architecture, Urbanism, Architectural Technologies, and History and Theory of Architecture and Art). Based on the synergy of technical and technological sciences, arts and humanities, and social sciences, the programs create a solid foundation for the future practice and research of young professionals and academicians capable of addressing the emerging challenges of the contemporary environment and society.

Master Studies in Architecture (also available in the English language for international students) offers four modules from the first semester (Architecture, Urbanism, Architectural Technology, Structural Engineering), resulting in the title of Master of Architecture—M. Arch, with reference to the selected module. The studies have also been validated for RIBA Part 2 Status (since 2015), further enhancing the possibilities for an international career. They are structured around several study units, which include Studio (design project, seminar), Module compulsory course, Elective courses and Theoretical discourse (provided by all departments), Common compulsory course (sociology/philosophy/economy) and Master diploma project—consisting of Thematic research, Master thesis, Master design and Master final project. The hybrid model of teaching is also enabled, based on the Teams and Moodle platform.

24.4.1 Toward Sustainable City: Methods

Due to the fact that the design-based curriculums represent the main part of the study program, the ideas of the NEB framework are integrated into the educational process through both theoretical/research-oriented courses and design-studio work. However, the module Master in Architecture_Urbanism, entitled Sustainable City, emphasises the theoretical and practical dimensions of sustainable imperatives even more (Stupar 2017). Therefore, the Department of Urbanism's courses focus not only on the relationship between people and the built environment and the contemporary problems of settlements and society but also on a wide range of themes targeting sustainable and resilient development and sustainable communities. Consequently, this module provides knowledge and skills based on a comprehensive understanding of natural and socio-spatial contexts, policy frameworks, spatial structures and ongoing processes, using an integrated, problem-based approach in all courses.

Considering these specificities, two compulsory courses were selected for the analysis—Sustainable City 1—Transformations, situated in the first semester, and the Master diploma project, positioned in the fourth semester. Their structure and applied methodology will be elaborated, highlighting the dominant paradigms and themes, the impact and the demonstration of the NEB framework, and the public perception and validation of results.

Within the sustainable and environmentally conscious educational agenda of the program, the first course is theoretical and research/project-oriented, while the other one uses the same methods along the design-driven studio approach (depending on the phase). The project-oriented approach implemented in both cases supports active student participation in the actual problems of local communities. In accordance with the experimental and innovative game-based curriculum design applied in studies of urban surroundings (Chen and Hoffman 2017) and the method of reflexive thinking and journaling (Kelly 2006), the curricula of both courses use predictive learning and future design in order to study the potential impact of changes on a selected environment.

The education process can be explained by Kolb's learning cycle (Kolb 2024), which connects concrete experience, reflective observation, abstract conceptualisation and active experimentation. During the first step, students study and discuss the main ecological, social and technological issues affecting the contemporary living environment. This phase directly influences their environmental awareness, leading to reflexive thinking and observations regarding the occurring problems, challenges, time horizons, available methods and tools, as well as the possible responses/solutions which might improve the current state of highlighted problems in an innovative way. Abstract conceptualisation directly triggers the process of urban transformation by studying the cases selected by students, considering different aspects of environmental sustainability and resilience, addressing the identified problems and evaluating the provided solutions. The step of active experimentation refers to the creative phase, i.e. students' proposals (programs, projects) and recommendations for the improvement of a selected spatial polygon.

In line with the concept of Thomas S. Kuhn, on a general paradigm shift, the students are instructed to take into account the possibilities for sustainable growth and transformation that are attractive to society (Polak 1973). This framework influences further decisions and ensures the successful materialisation of plans and projects in accordance with the NEB postulates. Therefore, both courses recognise the importance of green infrastructure in the process of transforming cities into climate-resilient living environments. This is achieved by introducing nature-based solutions (NBS), which represent the latest conceptual iteration of green infrastructure. The projects selected as the NEB Awards finalists in the category "Reconnection With Nature" (NEB 2024) are included in exemplary case studies providing insights into a successful integration of nature, society, ecosystems and built features. The biophilic approach in urban and architectural design is stimulated, as is the concept of circularity (urban recycling, the adaptive reuse of buildings).

The courses include the use of project databases (exemplary case studies, students' contributions/results of different phases, catalogues), extending the range of implementation possibilities for final outcomes/projects. They contain solutions for a number of specific urban problems related to the different aspects of anticipated environmental sustainability. Therefore, a project database, as a method used in many educational institutions, represents a valuable source of knowledge and experiences used during both courses. The benefits of this approach have already been verified, and the research of McNamara (2010), conducted at 86 universities in the United States which implemented databases on sustainability project initiatives, indicated that further steps/strategies in upgrading this method should be focused on the reinforcement of student involvement and the increased number of participants.

Due to an increased digital capacity, the use of online teaching tools has also become an important support for the educational process which follows the NEB imperatives. The Teams platform provides the open access to the teaching material within course/studio, facilitates a continuous monitoring of the outcomes produced and uploaded by course participants, and enables an interactive and efficient exchange between educators and students.

24.4.2 Sustainable City 1—Transformations: Experiences

The course entitled Sustainable City 1—Transformations (Associate Professor Dr Vladimir Mihajlov/Assistant Professor Dr Ivan Simic) is focused on the process of urban transformations as the main mechanism of urban development. It addresses the adaptation to a sustainable urban future by considering different phenomena that trigger the change. Students are trained to critically understand origins and effects, mastering and applying the obtained knowledge related to the theories of urban design, the planning of communities and the influence of urban design and development on the contemporary built environment. The course also provides insight into the current planning policies and legislation, as well as their relevance to design.

The SDGs and the basic principles of the NEB framework are incorporated into the course curricula, highlighting the themes of green urban transformation—from the various models of sustainable cities and lifestyles, greening strategies and climate adaptation to resilient urban design and the use of sustainable materials in public space. Since the integral approach promoted by the NEB implies expertise from many disciplinary fields (e.g. landscape ecology, urbanism, biology, hydrology, geology, transportation, social sciences, etc.), a multidisciplinary approach is necessary, especially in the early phases of conceptualisation. Therefore, the course curriculum also includes specialised guest lectures.

Teaching consists of ex-cathedra lectures on the theoretical aspects of urban development and transformations, independent students' research (group work) implemented through debates and a pilot project proposal. Each group studies a particular phenomenon and creates a critical-argumentative stance, which is publicly presented and discussed. The final phase of the course is conceived in the form of a Civil

Initiative Project Proposal (Fig. 24.1), applicable for different funding sources (e.g. civil society organisations, local initiatives, individuals and other interested parties). Including the recommendations for the improvement of selected urban environments, the pilot project tackles different problems in various forms of public innovations, tactical urbanism, urban revitalisation, climate change adaptation or business-driven sustainable solutions. The exam also tests and verifies students' ability for the process of grant application targeting affordable urban transformations with ecological purposes (e.g. tree planting, recycling, waste discount and reuse, urban green infrastructure preservation, defensive mangrove forest, experiencing vegetarianism, business start-up programs, etc.).

The study process ends with the Civil Initiative Project Proposal Catalogue, which is established on the Teams platform. Providing an insight into different, up-to-date competitions and publicly announced calls from different organisations (e.g. A Guide to Potential Funding Sources in Serbia, Belgrade Open School Competition, 2024), this database provides an active involvement in the current practice and the problems of urban communities. Additionally, it could be used for the survey of international and domestic funding sources, with basic information on the types and fields of support, potential users or the geographical area of operation.

During the school 2021/22, the project New Page—Make a Change, proposed by students Emilija Drndarski and Sara Brkic (supervisor: Associate Professor Dr. Vladimir Mihajlov), won the 1st award at the 2021 CirEkon and EIT Food Competition for Engaging Citizens in Circular Economy.

24.4.3 Master Diploma Project: Horizons

The Master in Architecture_Urbanism program concludes with the Master diploma project, which combines the processes of research, formulation, conceptualisation, design, and materialisation of a complex urban and architectural program. During four steps—Thematic research, Master thesis, Master design and Master final project—students work independently, under the supervision of a mentor and the support of a mentoring committee, combining theoretical and practical knowledge and skills. The first three phases include systematisation and presentation of results, which are internally verified. The thesis is based on the previously conducted thematic research, and its outcome represents the formulation of a project theme and program structure elaborated by the project. During the project phase, the research is done by design (including analytical and generic studies), modelling, shaping and structuring, leading to the final concept and project presented through graphical drawings, models and a master portfolio. The Master Final Project defence is public, and all projects are included in a Yearbook. The best-selected thesis and projects from all studios are evaluated by an external jury and nominated for the selected national and international awards.

The principles listed in the NEB initiative have been frequently tackled by diploma projects directly or indirectly. However, the overall theme of the studio entitled NS

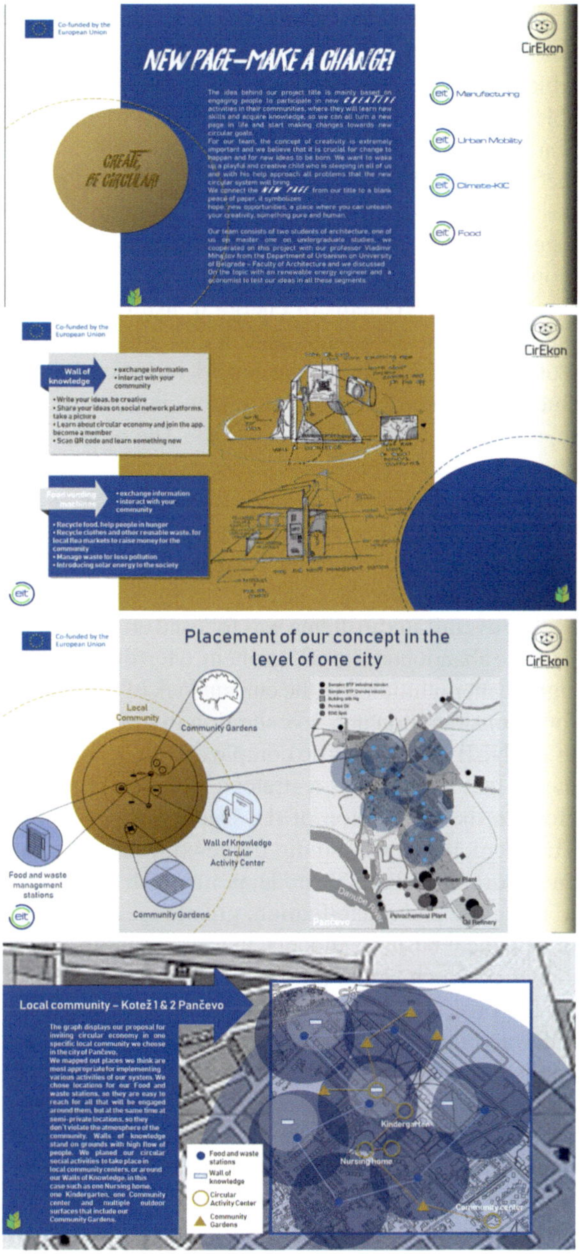

Fig. 24.1 The project proposal for improving the sustainability of the urban environment in Pancevo, tested by developing a small pilot project (Students: Emilija Drndarski and Sara Brkic. Supervisor: Associate Professor Dr. Vladimir Mihajlov)—1st award at the 2021 CirEkon and EIT Food Competition for Engaging Citizens in Circular Economy

2030: ECO-HORIZONS (supervised by Professor Dr Aleksandra Stupar), which was conducted during the school year 2021/22, enabled the testing of several concepts which integrate the environmental, social and technological dimensions of sustainable and resilient transformations, driven by nature-based solutions and circularity. Starting from a thematic research, a number of themes were studied through their theoretical background, selected cases, critical evaluation and contextual specificities. The topics included the issues of urban regeneration, industrial archaeology, regenerative architecture, climate adaptation, and flexible space. The concepts of biophilia, biomimicry, biomorphism, and circular economy were also highlighted, as well as the problems of energy transition, the remediation of degraded sites and the integration of animals and insects into living and working environments. The work on a thesis included the consideration of the general thematic framework of the studio and the selected site, research on contextual limitations, and the elaboration of selected paradigms and concepts. The next step was a study of identified urban phenomena and representative cases. According to the obtained knowledge, both a thesis and a project focus were formulated, as well as a program structure for the following phase. The research by design was applied during a project elaboration, while the final contribution integrated a concept, the regulation and design of an urban area and specific content.

The overall studio theme was focused on the part of the industrial zone of Novi Sad, with the abandoned site of Agrohem, a fertiliser manufacturer. The site was exposed to significant pollution, but the current lack of activities represented a unique possibility for reconsideration and regeneration in line with the SDGs and the paradigms listed in the NEB initiative. The complex, with a visually striking paraboloid structure of the ex-factory, also has a strategic position at the entrance of the city of Novi Sad, close to the connection of the Danube River and the canal Danube-Tisa-Danube. Therefore, the overall studio framework suggested a transformation of this area into an important sustainable node, in line with the Gaia hypothesis and the ideas related to recycling and rehabilitation, green regeneration through biophilia, the activation of new local business models, the carbon footprint reduction and the introduction of activities which would promote sustainable ideas.

The project by Milica Brajovic, entitled The park of ecological and psychological renewal_biophilic design, uses the misbalance of the surroundings as a trigger for the application of biophilic design to everyday spaces (Fig. 24.2). Adjusted to the local climate specificities, the elements of biophilia are incorporated into all spatial levels, while the symbiosis of context, natural cycles and existential principles represents the main design theme. The integration with nature, regeneration and the preservation of resources are also used as steps toward the strengthening of environmental awareness. The anticipated activities in the area stimulate community cohesion while directly and indirectly contributing to the preservation, renewal and sustainability of a newly created mini-ecosystem. The planned educational programs encourage cooperation between educators and students (all age groups), increasing the interest in the natural environment and the sustainability of everyday life.

The paraboloid structure of the ex-Agrohem factory was transformed according to the principles of biophilic design, while some surrounding buildings were renovated

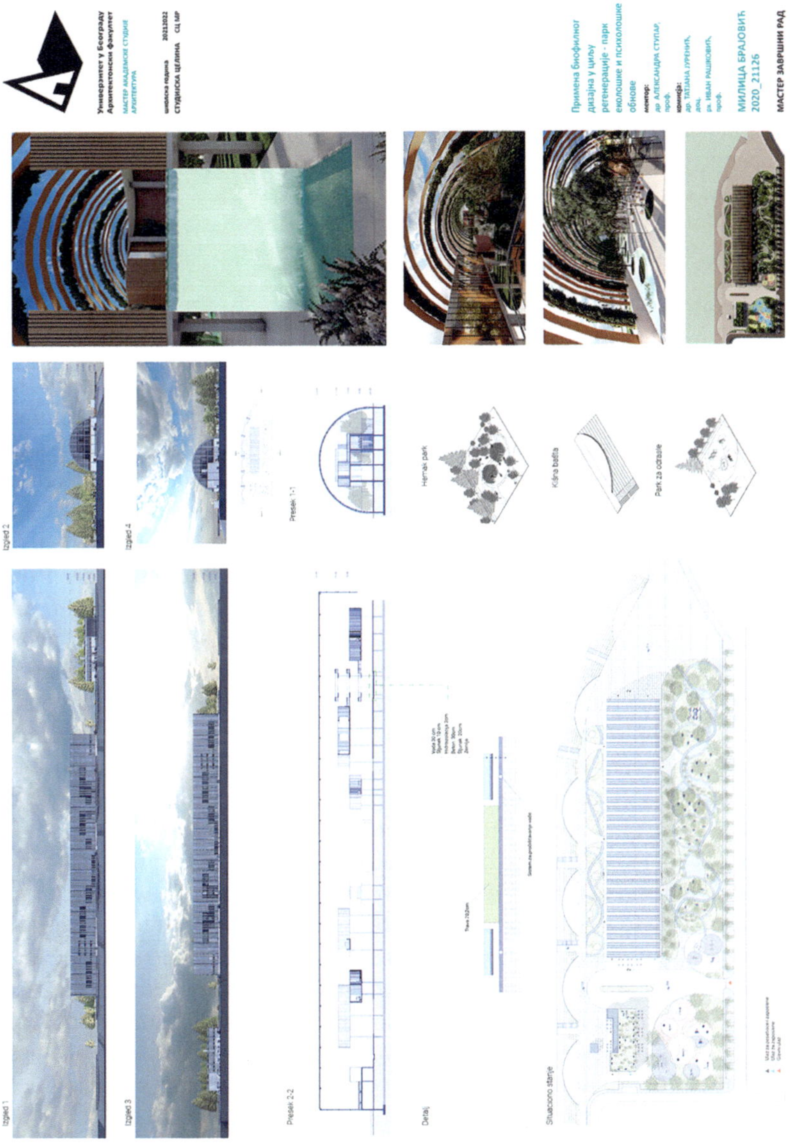

Fig. 24.2 NS 2030: ECO-HORIZONS. The park of ecological and psychological renewal_biophilic design—Student Milica Brajovic. Mentor: Professor Dr. Aleksandra Stupar. (Committee: Professor Ivan Raskovic, Associate Professor Dr. Tatjana Jurenic. External advisor: Dr. Manuel Collado Arpia, University of Alcalá, Madrid. Assistant: Dr. Marija Negovan)

or removed, allowing space for green areas. The influx of natural light and ventilation contributes to different recreational activities and relaxation practices situated in the main unit, while users benefit from the proximity of natural systems. The network of water surfaces ensures higher quality and space comfort but also reflects circularity since it provides the maintenance of indoor and outdoor vegetation.

The implementation process is anticipated to take several phases in accordance with the postulates of economic and environmental sustainability.

After the public presentation, the project received the prize of the 2022 International Urban Planner Exhibition of Serbia (student category) and the annual award 'Milorad Macura', granted by the Institute for Architecture and Urbanism of Serbia for the best master diploma project in architecture (2023).

24.5 Conclusion

The education of future architects and urban planners represents a challenging task in a world saturated by numerous contradictions, extreme situations and fast-changing imperatives. However, the values embedded in the globally adopted frameworks of sustainability and resilience provide a preferred direction supported by numerous visions, strategies, programs, plans and policies operating on all spatial levels. The signposts defined by the 2030 SDGs and the recently introduced NEB initiative interlink the importance of modern society, urban spaces and technology in an effort to raise environmental awareness and general well-being.

The modifications of the university curricula conducted in architectural and planning schools represent a step forward since they introduce themes and paradigms important for sustainable futures but also address general attitudes and the responsibilities of students, which could have an important impact both on our environment and the process of ecological transition. The provided pedagogical stimuli also contribute to a deeper understanding of emerging problems influencing the perception and evaluation of roles, obligations and anticipated actions.

The experiences gained during the implementation of the master's program at the University of Belgrade—Faculty of Architecture indicate that the new curriculum acts as a catalyst for expanding the research beyond classrooms into real urban and natural surroundings. Furthermore, students are encouraged to connect reflexive thinking and motivation through cultural, ecological and technological aspects of a selected context. They are stimulated to act in a socially responsible manner, actively participate in a learning process, question their empirical views and the nature of cities and, consequently, provide future-oriented proposals for sustainable communities and a resilient environment.

The application of digital tools facilitates the transmission of knowledge and experiences, while the use of social media, although not sufficiently tested in the educational process, could target various aspects of digital citizenship. Consequently, students could play an important role in civic engagement on social media, expressing their opinions and influencing the pace and range of sustainability. However, the path toward ecological transition is not an easy one, but the education process, based on the principles comprised in the NEB initiative, could provide a necessary interchange of theory and practice, as well as innovative and creative insights and proposals, which could be tested and verified in the years to come.

References

Beinart J (1981) Analysis of the content of design. 1st edn, Architecture education study. CECSA-Consortium of East Coast Schools of Architecture, Boston, MA

Belgrade Open School Competition. Accessed Feb 8, 2024. https://transformator.bos.rs/kalendar/

Boyer EL, Mitgang LD (1996) Building Community: a new guide for architectural education and practice, 1st edn. Carnegie Foundation for the Advancement of Teaching, Princeton, NJ

Chen KH, Hoffman J (2017) Serious play: transforming futures thinking through game-based curriculum design. J Futures Stud 22(2):41–60

Crinson M, Lubbock J (1994) Architecture: art or profession? Three hundred years of architectural education in Britain. Manchester University Press, Manchester, UK

Davos declaration. Accessed Feb 8, 2024. https://whc.unesco.org/en/news/1773

Dutton TA (1991) The hidden curriculum and the design studio: toward a critical studio pedagogy. In: Dutton TA (ed) Voices in architectural education. Bergin and Garvey, New York, pp 165–194

EU Commission (2013) Green infrastructure. Accessed Feb 8, 2024. https://environment.ec.europa.eu/topics/nature-and-biodiversity/green-infrastructure_en

European Commission (2020) New leipzig charter-the transformative power of cities for the common good. Accessed Mar 6, 2024. https://ec.europa.eu/regional_policy/whats-new/newsroom/12-08-2020-new-leipzig-charter-the-transformative-power-of-cities-for-the-common-good_en

European Commission (2021) New European bauhaus-beautiful, sustainable, together. Accessed Feb 8, 2024. https://new-european-bauhaus.europa.eu/system/files/2021-09/COM(2021)_573_EN_ACT.pdf

Folke K, Carpenther S, Walker B, Scheffer M, Chapin T, Rockstrom J (2010) Resilience thinking: integrating resilience, adaptability and transformability. Ecol Soc 15(4). http://www.ecologyandsociety.org/vol15/iss4/art20/

Harriss H, Widder L (eds) (2014) Architecture live projects: pedagogy into Practice (1st ed). Routledge. https://doi.org/10.4324/9781315780764

Holling CS (2001) Understanding the complexity of economic, ecological, and social systems. Ecosystems 4(5):390–405

Kelly P (2006) What is teacher learning? A socio-cultural perspective. Oxf Rev Educ 32(4):505–519. https://doi.org/10.1080/03054980600884227

Kolb's learning styles and experiential learning cycle. (2024). https://www.simplypsychology.org/learning-kolb.html. Accessed Mar 6, 2024

McNamara KH (2010) Fostering sustainability in higher education: a mixed-methods study of transformative leadership and change strategies. Environ Pract 12(1):48–58

NEB Reconnecting with nature. Accessed Feb 8, 2024. https://prizes.new-european-bauhaus.europa.eu/finalists/applications/champions/reconnecting-with-nature

Nicol D, Pilling S (2000) Changing architectural education: toward a new professional role. Spon Press, London, UK

OECD—The circular economy in cities and regions: Synthesis Report 2020. Accessed Feb 6, 2024. https://www.oecd.org/regional/the-circular-economy-in-cities-and-regions-10ac6ae4-en.htm

Polak F (1973) The image of the future (Trans and Abr.: Boulding E) Jossey-Bass, San Francisco

Salama AM (1995) New trends in architectural education: designing the design studio. Tailored Text and Unlimited Potentials

Salama AM (2009) Transformative pedagogy in architecture and urbanism, 1st edn. Umbau-Verlag, Solingen, Germany

Salama AM (2016) Spatial design education: new directions for pedagogy in architecture and beyond. University of Strathclyde, Glasgow, United Kingdom, Department of Architecture

Salama AM (2015) spatial design education: new directions for pedagogy in architecture and beyond, 1st edn. Ashgate Publishing Ltd, London, United Kingdom. ISBN 978-3937954-042

Simić I, Stupar A, Djokić V (2017) Building the green infrastructure of Belgrade: the importance of community greening. Sustainability 9(1183). https://doi.org/10.3390/su9071183

Stupar A, Mihajlov V, Simic I (2017) Towards the conceptual changes in architectural education: adjusting to climate change. Sustainability 9(8):1355. https://doi.org/10.3390/su9081355

Teymur N (1992) Architectural education: issues in educational policies and practice, 1st edn. Question Press, London, UK

UN (2015) Transforming our world: the 2030 agenda for sustainable development. Resolution adopted by the general assembly on 25 September 2015, 42809, 1–13. Accessed Feb 6, 2024. https://new-european-bauhaus.europa.eu/system/files/2021-09/COM(2021)_573_EN_ACT.pdf

Chapter 25
Methodologies for Using LEED in Academic Architectural Education

Nataša Ćuković Ignjatović and Dušan Ignjatović

Abstract LEED (Leadership in Energy and Environmental Design) rating system was introduced into the academic curricula of the Faculty of Architecture in Belgrade in the early 2010s as a single course designed to introduce the basics of green building and preparatory lectures for LEED GA (LEED Green Associate) professional accreditation exam. Over time, a range of methodologies have evolved and developed for specific learning objectives and outcomes for various study programs and course structures. Like all major green building rating systems, LEED constantly evolves, which makes its incorporation into academic curricula a challenging task. The paper presents methodologies that use LEED-based tools and criteria as an educational approach that goes beyond the introduction to basic principles of green building, empowers the students to assess the environmental impact of their design decisions from the very initial design stages, introduces a wide range of design strategies that can improve the environmental features of their design and encourages them to communicate green features of their design in a variety of ways. In 2022, the student's class work won first prize at The New Green Voice international design competition, which encompasses the use of LEED, verifying the relevance and quality potentials of stated methodologies.

Keywords LEED · Green architecture · Building certification · Sustainable architecture · Academic education

25.1 Introduction

LEED (Leadership in Energy and Environmental Design) is a green building certification system developed by the U.S. Green Building Council (USGBC; founded in 1993) in the late 1990s. By early 2000, it had become rather visible internationally and was considered the most widely used green building rating system in the world. LEED uses an elaborate set of tools and criteria for evaluation, aiming to provide a

N. Ćuković Ignjatović (✉) · D. Ignjatović
University of Belgrade—Faculty of Architecture, Belgrade, Serbia
e-mail: natasa@arh.bg.ac.rs

M. Vukmirović et al. (eds.), *Digital Dreams, Urban Realities*, The Urban Book Series,
https://doi.org/10.1007/978-3-031-76353-3_25

"framework for healthy, efficient, and cost-saving green buildings" (USGBC—LEED rating system homepage 2024).

Green architecture is present in the curricula of Belgrade's Faculty of Architecture ever since 1990 (the specialization "Bioclimatic Design" in the 4th and 5th year of studies). When the collaboration with USAID started in 2010, the general topics of green and sustainable architecture were already well-known to a number of lecturers and senior students (Cukovic Ignjatovic et al. 2011).

The project of introducing LEED to the Serbian professional and academic environment was carried out in three steps, executed by distinguished LEED educators and practitioners, and members of USGBC were directly involved in the creation and modifications of the LEED rating system. The first step was a course introducing the basics of green design and LEED certification, as an initial course providing basic knowledge relevant to obtaining LEED GA (LEED Green Associate) professional accreditation. A number of senior students, teaching assistants and professors from the University of Belgrade—Faculty of Architecture attended the course, and three staff members took the formal exam and obtained LEED GA professional accreditation. This step was followed by the "training of trainees" component that enabled the selected teaching staff to deliver verified LEED courses, making the Faculty of Architecture in Belgrade the first academic institution outside the United States qualified for embedding LEED in academic curriculum and delivering LEED GA preparatory course within teaching green architecture design principles. The final step was an extensive course aimed at training a group of professionals and academics to actively participate in the actual process of green building certification, leading to obtaining LEED BD + C (Building Design and Construction) or LEED EBOM (Existing Buildings Operation and Maintenance) Accredited Professional title. Aside from extensive theoretical classes, the training involved performing the actual certification-related tasks, analyses and calculations relevant to a selected number of projects that were in the process of certification at the time. The same group of teaching staff from the Faculty of Architecture took part in this, and as a result, one of the teachers managed to prepare and pass the LEED BD + C Accredited Professional exam, successfully renewed the credentials and maintained the professional accreditation.

What started as a single course designed to introduce the basics of green building and preparatory lectures for the LEED Green Associate professional accreditation exam has evolved in time and developed a range of methodologies for specific learning objectives and outcomes for various study programs and course structures. The paper presents the unique methodologies developed within Master's and Specialized study programs, as well as key learning outcomes and internationally recognised results.

The methodology and findings are presented in two main chapters: the first chapter explains how the LEED system was embedded into course methodology, while the second chapter presents how the LEED system was used simultaneously as a learning material and as a design-oriented tool, encouraging students to explore the effectiveness of various design strategies and simulate integrative design approach. The conclusion sums up the results and provides guidelines for future improvements in the educational process.

25.2 LEED System and Course Methodology

25.2.1 LEED System Goals and Assessment Categories

LEED rating system goals (Fig. 25.1) are set as an attempt to provide a holistic answer to the question, "What should a green and sustainable project accomplish?" (Owens et al. 2013). They focus on issues where impacts of the built environment can be addressed and where changes in design practice, operation and maintenance of buildings can result in various environmental benefits. The formulation of LEED system goals helps communicate clearly to the students the general idea of a wide scope of interconnections between the results of their professional actions and our environment.

Building's environmental impact, however, is not the same regarding all system goals. LEED impact categories derive from system goals and their "weight" is set to illustrate the scope and the reach of building's environmental impact. As the relevant knowledge evolves, so do the impact categories, and current versions of LEED put in focus operational and embodied carbon, whilst the "weighting" of impact categories is presented in Fig. 25.2.

Green building certification systems introduced methodologies to quantify benefits that had previously been considered too difficult to quantify, mainly by providing a consistent system that "translates" general goals to a set of measurable parameters and evaluation criteria. In the case of LEED, these parameters and criteria (defined in "credits") are grouped into "categories" (Fig. 25.2) and reflect the impact categories and their weightings.

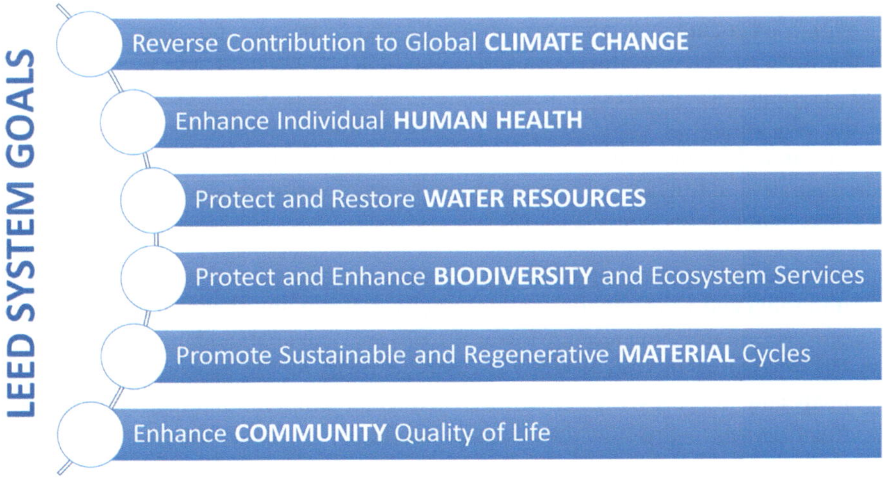

Fig. 25.1 LEED system goals (per LEED v4.1 (USGBC—LEED rating system homepage 2024))

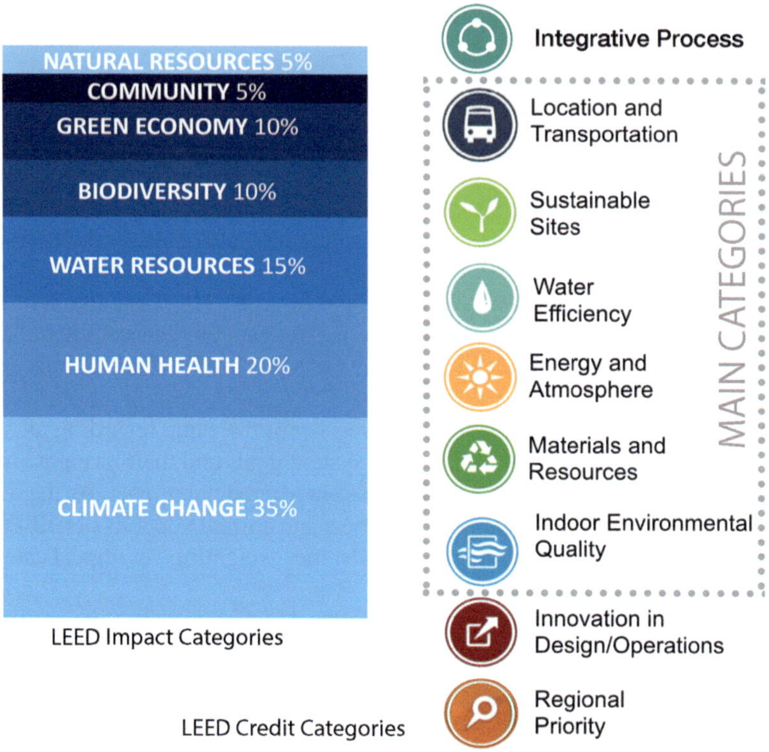

Fig. 25.2 LEED impact categories (left) and credit categories (right), as per LEED v4.1

Within each category, a variety of criteria and respective indicators are defined in the form of "credits" that earn the points that are counted into the final score and "prerequisites" that need to be met in order to pursue any certification level. In the educational process, students are guided towards understanding processes, design strategies and synergies that lead toward sustainable building rather than administrative and technical issues related to certification procedures. Each credit has three main components (Fig. 25.3):

1. *Intent*—The intent of a LEED credit provides a clear understanding of why it is important and what benefits it aims to deliver. It explains the overarching goal or purpose, establishing relation towards LEED system goals and impact categories. Understanding the intent helps students to focus on the broader sustainability goals and outcomes associated with each credit and relevant design approach.
2. *Requirements*—Specific set of design strategies and verification methods and tools that help achieve the general goal stated in "Intent". Requirements define metrics—evaluation and criteria that are used to assess a variety of relevant

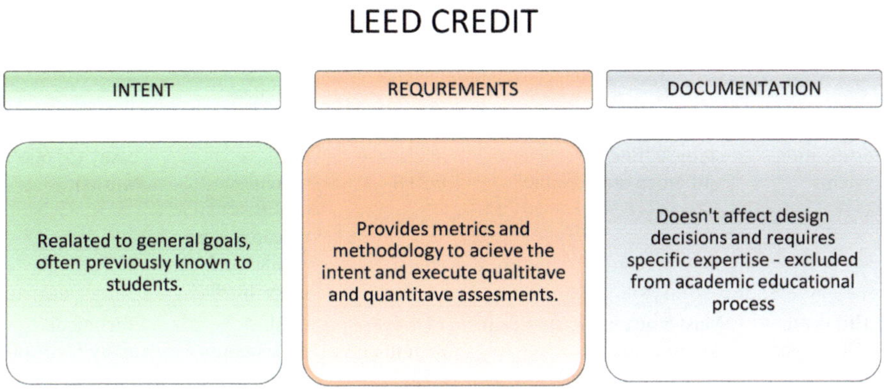

Fig. 25.3 LEED credit components and their relevance to academic architectural education

building characteristics. Students are strongly encouraged to explore requirements and use the proposed methodology to quantify the expected outcomes of their design proposals.

3. *Documentation*—In the actual process of certification, compliance with credit requirements needs to be presented in the form of technical documentation, in the form required by the rating system. Since LEED certification is entirely done online, documenting often demands highly trained and experienced experts. Students are given general information, but this step doesn't reflect the design process and, as such, is not relevant to their expected learning outcomes.

25.2.2 LEED in Relation to Various Study Programs

The initial step of introducing LEED in academic curricula at the University of Belgrade—Faculty of Architecture was adjusting the curriculum of the typical LEED GA (Green Associate) course to the relevant study program's requirements and dynamics. LEED GA is a basic professional accreditation that demonstrates general knowledge of green building practices and how to support others working on LEED projects. The content of the LEED GA course was embedded in several academic courses at the Faculty of Architecture (Table 25.1), each time adjusting the methodology to the course's position within the curricula, relevant previous knowledge of green design principles and technologies and the possibility to refer to the current or recent design studio. Students are encouraged to check (where possible) if their project meets LEED requirements for certain credits and explore design strategies that would help their project meet sustainable goals.

Table 25.1 Courses with LEED-related content

Course title	Study program	Course volume	Student activities	Final work
Green building certification systems	Specialized academic studies "Energy efficient and green architecture"	7 weeks/28 h in class/4 ECTS	Homework and in-class activities: discussions, research, case studies and assessments	Technical essay (5–6000 words)
LEED and WELL systems in design process	Master academic studies in Architecture	14 weeks/ 28 h in class/ 3 ECTS	Oral presentations, simple case studies and basic project analysis, infographic, discussions	Technical essay (3–5000 words)
LEED and WELL systems and architectural design	Master academic studies in Architecture—Architectural Technologies	14 weeks/ 28 h in class/ 2 ECTS	Oral presentations, handouts	Poster presentation

25.2.3 Specialised Studies Course

Within Specialised academic studies "Energy efficient and green architecture", LEED is integrated into the course "Green building certification" (mandatory course, 1st year, 2nd module). Within this course, LEED is thought to have complete LEED GA-related content, extended with certain elements relevant to LEED AP (LEED Associate Professional) professional accreditation. At the same time, the course was in the same study module as Design Studio (2012–2018), and LEED concepts, design strategies, and assessments were tested in parallel with the design process. Currently, students use a project (preferably from their professional work) and assess it as accurately as possible. Initial tasks done during classes refer to familiarising oneself with the LEED rating system and relevant resources like LEED User, LEED Credit library, LEED Project directory, etc. The intent and requirements of each credit are presented and explained in detail, followed by a discussion of the relationship between local practice and challenges that might occur in real-life situations. For credits with rather simple calculations, the students are expected to follow the requirements to the detail and provide valid quantitative and qualitative assessments, while for the credits that require specific expertise and/or complex calculations and simulations, students provide a critical overview. Since the complete study program is focused on green building and energy efficiency, students are already acquainted with general concepts and certain tools. The course emphasises the logic behind the green building certification systems, the challenges of implementation, "greenwashing", and "LEED brain" (Schendler 2010; Owen 2010). Final work is done in the

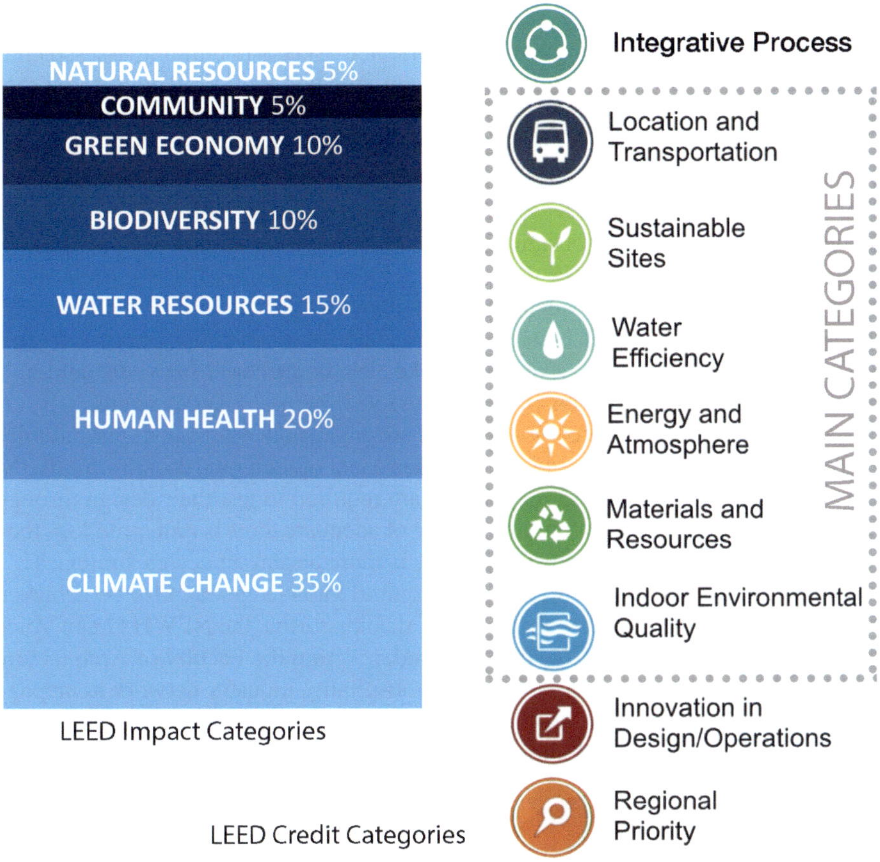

Fig. 25.4 Sample pages from final work: Specialised studies course "Green building certification", student Bojana Čanković, prof. Nataša Ćuković Ignjatović

form of an extensive essay supported by schemes, infographics, calculations and assessments. Figure 25.4 going through all LEED categories and relevant credits.

25.2.4 Master Studies Course

LEED-related content was initially introduced in 2012 as an elective course, "LEED and environmental aspects of architectural design", offered to the students in 3rd semester of Master's studies in Architecture and 9th semester of Integrated single-cycle-5-year studies in architecture. From the very beginning, the students were very interested in the course, which resulted in rather selective admission and ambitious participants. In the new accreditation cycle in 2021, the course continued with the updated title "LEED and WELL systems in the design process" since WELL Building Standard[1] has emerged, adding more user-oriented aspects to building assessment.

The main goal of the course is to acquaint students with systems for environmental certification of buildings and trends in contemporary architectural practice that are related to them. During the course, students, through theoretical and practical work, become familiar with the basic principles and strategies of LEED and WELL certification systems, as well as with the possibilities for implementing the acquired knowledge in the design process. The emphasis is, however, on LEED, where the complete LEED GA content is included, while WELL is introduced in general, highlighting the equivalences and analogies with some LEED credits, as well as selected elements of WELL standard ("preconditions" and "optimisations") that refer directly to architectural design and make clear connections between building's design, materials and operation and health and well-being of its occupants.

Aside from ex-cathedra lectures, students are engaged in discussions, case studies of good practice, bad practice and critical assessment of their own designs. Similarly, like in Specialised studies course, students are required to test their design through some calculations, but since their knowledge of adequate tools is rather modest, they perform some basic calculations while there is more accent on design strategies.

Direct application and verification of acquired knowledge can also be achieved through work at the international architectural competition, the NEWH Green Voice Design Competition (Fig. 25.5). In 2022, students from the Faculty of Architecture in Belgrade won the first prize (NEWH the hospitality industry network homepage 2024) with their class project on this course, which verified the relevance and quality potentials of stated methodologies.

25.2.5 Master Studies Seminar

A seminar titled "LEED system and architectural design" was introduced in 2012, complementing the design studio in Architectural Technologies (3rd semester of Master studies in Architecture), aiming to establish an interactive connection between the principles and design strategies of sustainable architecture and the design process. The metrics and the basis of green building certification systems are embedded in the design process and reflect on the overall outcome of the project within the design studio. In 2018, the WELL standard was added to the curriculum, providing the students with the latest trends in the field of green architecture, human health, comfort and well-being and the course title was modified to "LEED and WELL systems and architectural design".

[1] WELL Building Standard™ was launched in 2014, introducing a "set of strategies—backed by the latest scientific research—that aim to advance human health through design interventions and operational protocols and policies and foster a culture of health and well-being" (WELLv2™ homepage 2024).

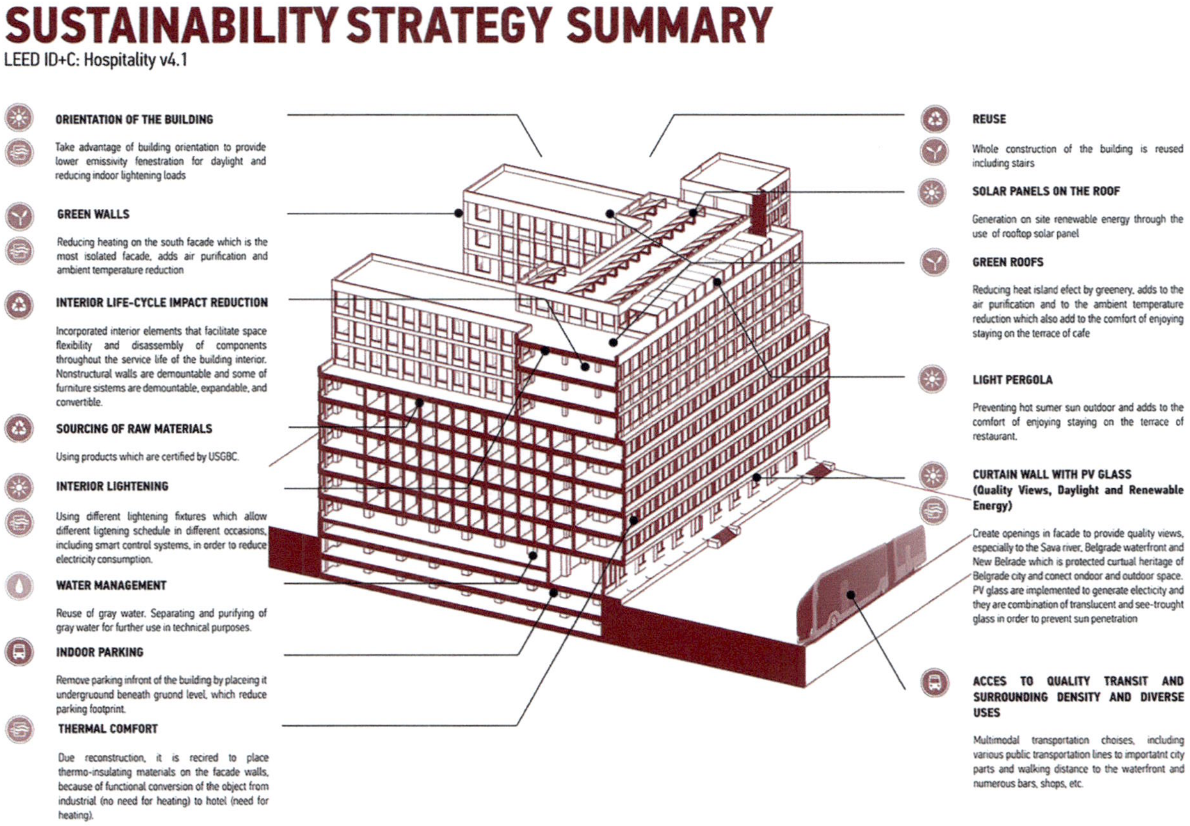

Fig. 25.5 Sample page from NEWH Green Voice Design Competition work (Master studies course "LEED and WELL systems in the design process", students Zorana Jović and Anja Pantić, prof. Nataša Ćuković Ignjatović

During the seminar, through theoretical and practical work, students become familiar with the principles of the LEED certification system and WELL standards. From the complex structure of certification systems, elements that can be successfully checked during work on a studio project have been singled out. The emphasis of the work is on the adoption and practical application of design strategies and technical solutions that achieve high-value ecological qualities of the building. The first part of the course is dedicated to LEED, mainly categories LT (Location and Transportation) and SS (Sustainable Sites), while the second part introduces the elements of the WELL standard, the overlaps and differences in relation to LEED as well as the features that are overlooked by LEED. The final work is presented as a board, explaining the key green/sustainable features of the studio project (Fig. 25.6), but the most valuable outcomes were, in fact, reflected in added value to the projects done within the design studio and students' capacity to verify and argument the environmental benefits of their design proposal.

25.3 The Role of Design Project in the Learning Process

Design studios play a key role in contemporary architectural academic education, as they bring into play analytical and theoretical thinking deployed in focused research, facing practical problems in the process of shaping our built environment. As Salama states (Salama 2021), "the question of the relationship between the design education provided, the skills required for successful practice and for addressing the matter of knowledge is, obviously, of paramount importance". Therefore, the metrics and assessment methodologies from the LEED certification system are used as specific didactic tools, making the general issues of green and sustainable architecture tangible and quantifiable while connecting them to design strategies and effective design solutions.

The common denominator of all three courses is establishing an active relation to an actual design project in the process of learning. A similar approach is used in LEED Lab, where student's understanding of sustainable design is developed by evaluating existing buildings through green assessment (Andrasik 2022). While LEED Lab uses the LEED O + M (Operations and Maintenance) rating system, the courses developed in Belgrade follow the LEED for Building Design and Construction (BD + C) rating system. The main difference is that LEED O + M assesses existing buildings while LEED BD + C assesses new construction and major renovation during the design and construction phases. This is crucial for establishing a high level of relevance as all the above mentioned study programs are focused on architectural design rather than on facility management.

When possible, assessments are made for an ongoing project (design studio or professional commission) to simulate the non-linear process of design for sustainability. The type of design project is chosen in relation to general goals and level of study program (Table 25.2) while keeping in mind the actual previous knowledge of students and accessibility of relevant tools required in the learning process. The

Fig. 25.6 Master studies seminar "LEED and WELL systems and architectural design", presentation boards. Students (from left to right): Ilija Nikolić, Katarina Pejić, Bojan Stojiljković; Prof. Nataša Ćuković Ignjatović, design studio M03AT, prof. Dušan Ignjatović

aim of conducting assessments is dual: acquiring the knowledge related to the green building certification process and learning about the environmental effects of various design-related decisions, even in situations where there is no imposed environmental imperative nor requirements for building certification. Current national legislation in Serbia addresses only the energy efficiency of buildings, taking into calculation only energy needed for heating (Službeni glasnik Republike Srbije 2011), while other environmental issues are not directly addressed at the building level. This makes the tools and methods provided by green building certification systems even more interesting, as there are no practical or formal references and benchmarks that would provide some information on a building's environmental performance.

The integration of learning LEED and adopting green building design principles is made in a way that requires an understanding of all relevant topics, prioritising the ones that are related to architectural design rather than procurement processes or

Table 25.2 Projects used for LEED-related analysis and assessments

Course title	Study program	Design project
Green building certification systems	Specialized academic studies "Energy efficient and green architecture"	Ongoing current practice or specialized design studio
LEED and WELL systems in design process	Master academic studies in Architecture	Design studio project, various topics
LEED and WELL systems and architectural design	Master academic studies in Architecture—Architectural Technologies	Design studio—Architectural Technologies

facility management policies. For each major LEED category (see Fig. 1b), the basic process requires several steps:

1. Identification of potentials and limitations of the site and architectural program relevant to the building's performance in a certain category.
2. Identify project design features that have beneficial effects; comment on how they can be quantified and/or verified in accordance with the LEED system or national regulations.
3. Identification of design features and strategies that are not applied in the project but would have beneficial effects; comment on how they can be successfully integrated into the final architectural design.
4. Identification of design strategies that are not applicable to a given project (with proper argumentation).
5. Identification of strategies that are outside the student's professional domain; who are the stakeholders and/or specific professionals that should be involved?

Throughout the process, students use the latest version of the LEED certification system, which is available through the online tool "LEED Credits Library" (LEED credit library homepage (LEED BD+C, v4.1) 2024) and official certification guides relevant to LEED BD + C (USGBC 2024). The role of the design project in the learning process is tailored in accordance with the specifics of each study program and the course's position within the study program structure. The specialised studies course is the most demanding in terms of in-depth analysis, accuracy, and technical presentation. The master studies course is more flexible, allowing students to combine their own design with relevant examples of good practice, while the seminar includes mainly elements that can be related to their specific task in the studio.

At the specialised studies level, students mainly come from professional practice, aiming to broaden their expertise in the field of energy efficiency and green architecture. Other courses within the curriculum equip them with the tools and knowledge primarily related to energy efficiency, EPC procedures, performance simulations and measurements. This means that they are capable of making rather accurate assessments themselves. At the same time, they are in contact with detailed design documentation, and they can gather and analyse plenty of relevant data. The more they are involved in professional practice, the more they are likely to gain knowledge relevant to LEED AP professional accreditation. The relevance of the Integrative Process (IP) is easily communicated to the students, and the accent is on the means of effective moderation, active engagement of stakeholders, organising design charrettes, etc. The innovation (ID) category is explored as technical (i.e., "exemplary performance"), a connection to some WELL features, and the source for expanding a palette of green design features. Regional Priority (RP) is explored in relation to national codes. NEWH Green Voice Design Competition would be a good alternative, but the current schedule is rather unfortunate—classes start in the very final weeks of the competition, and there is not enough time to complete a rather demanding task.

The course in 3rd semester of Master studies is open for students from various architecture study programs, meaning that their studio projects are usually focused on different research areas, often not dealing with green/sustainable approaches. Some designs are very conceptual, but the students are always strongly encouraged to critically address their designs and to explore the paths of including green design principles in the process. The nature of design imposes variation in approach, but throughout the years, it is evident that all five steps can be successfully completed regardless of the studio methodology. The NEWH Green Voice Design Competition ends almost at the same time as the fall semester, so the students can include the competition work entirely in their curricular activities and submittals. The competition task is very demanding and extends beyond the course's curriculum, so therefore, it is voluntary, with an alternative way available to complete formal course obligations in case they do not finish the full competition submittals.

The seminar on 3rd semester of Master studies accompanies a design studio, M03AT, which is focused on green design principles and technologies. Lectures, assessments, and discussions are placed as complementary material to the studio, and the learning process is strongly related to the design process. The focus is on effective communication of green design features, using the LEED system as a tool for assessing and presenting the qualitative aspects of the project. The five-step assessment process is applied, but calculations and quantitative assessments are limited to selected credits from LT (Location and Transportation) and SS (Sustainable Sites) categories. The credits that have equivalence or high compatibility with WELL optimisations are also highlighted during the lectures, but students are free to choose their own path in improving and documenting the qualities of their studio project. Taking part in the NEWH Green Voice Design Competition is also an option, but it turned out to be too demanding for the seminar: the students who entered the competition did not complete the whole submittal.

25.4 Conclusions

At the Faculty of Architecture, for more than a decade, LEED has been embedded in various courses, both as content that is being taught and as a tool for learning about green architecture, validating and communicating diverse design features. The methodology of relating to the student's design project throughout the course has produced high-quality results that were verified through an international LEED-specific design competition. The 5-steps process has proven to work on diverse types of projects, from conceptual design studios to complex professional design tasks. Apart from obtaining the knowledge relevant to LEED GA professional accreditation, these courses develop specific skills that are beneficial for the effective implementation of green design principles and communicating their values and expected benefits regardless of whether the building is going to be certified or not.

References

Andrasik P (2022) LEED lab: a model for sustainable design education. Routledge, New York

Cukovic Ignjatovic N, Ignjatovic D, Stankovic B (2011) Tools and methods for energy efficiency evaluation in the process of architectural design. In: Cosic G (ed) Instalacije & Arhitektura, zbornik radova. Arhitektonski fakultet Univerziteta u Beogradu, Belgrade, pp 221–226

LEED credit library homepage (LEED BD+C, v4.1). https://www.usgbc.org/credits. Last accessed 02 Mar 2024

NEWH the hospitality industry network homepage was last accessed on 21 Feb 2024

Owen D (2010) Green Metropolis: why living smaller, living closer, and driving less are the keys to sustainability. Riverhead Books, New York

Owens B, Macken C, Rohloff A, Rosenberg H (2013) LEED v4 Impact category and point allocation process overview. US Green Building Council, Washington

Salama A (2021) Transformative pedagogy in architecture and urbanism. Routledge, Oxfordshire

Schendler A (2010) Getting green done: hard truths from the front lines of the sustainability revolution. Public Affairs, New York

Službeni glasnik Republike Srbije (2011) Rulebook on energy efficiency of buildings. Official Gazette of RS, No. 61/2011. Official Gazette of RS, Belgrade (in Serbian)

USGBC (2024) LEED v4.1 building design and construction: getting started guide for beta participants. US Green Building Council, Washington

USGBC—LEED rating system homepage. https://www.usgbc.org/leed. Last accessed 21 Feb 2024

WELL v2™ homepage. https://v2.wellcertified.com/en/wellv2/overview/. Last accessed 15 Mar 2024

Part X
Education: Living Labs and Collaborative Innovation Ecosystems

Chapter 26
Building Community Through Active Urban Design: DANUrB+ Student Workcamp in Golubac, Serbia

Milorad Obradović, Aleksandra Đukić, Branislav Antonić, Jelena Marić, and Nikola Mitrović

Abstract Urbanism is among the professions in which public and community awareness is of great importance. All its segments, including urban design, therefore strongly rely on the participation of the local community. This stance implies that participation and community-led design are the core of higher education in urban design; hence, the curriculum in urban studies usually has several courses with this topic on their agenda. Nevertheless, they usually consider the participation process in the first phase of urban design, where students collect data from citizens, local experts or stakeholders through surveys, interviews and questionnaires. Making an on-site urban design workcamp with the local community in an open public space is rarely implemented, as it is simply more demanding for organisation and resources. This research presents this specific form of student workcamp in the local community. It is done as the activity of the DANUrB+ INTERREG Project in Golubac, Serbia, in June 2022. The idea behind the workcamp was to raise awareness among both students and the local community in Golubac about the importance of open public space and its urban design through a small-scale 'design and build' urban project. However, the final objective of conducting the workcamp was even more significant—to train both groups in the elements of urban participation through the design process, such as mutual understanding, democratic awareness, increased independence, self-reliance

M. Obradović (✉) · A. Đukić · B. Antonić · J. Marić · N. Mitrović
University of Belgrade—Faculty of Architecture, Belgrade, Serbia
e-mail: info@urbanfactory.rs

A. Đukić
e-mail: adjukic@afrodita.rcub.bg.ac.rs

B. Antonić
e-mail: antonicb@arh.bg.ac.rs

J. Marić
e-mail: jelena.maric@arh.bg.ac.rs

N. Mitrović
e-mail: nikola.mitrovic@arh.bg.ac.rs

M. Vukmirović et al. (eds.), *Digital Dreams, Urban Realities*, The Urban Book Series,
https://doi.org/10.1007/978-3-031-76353-3_26

and compromise negotiation. The aim of this research is to show how the whole process was planned and implemented.

Keywords Workcamp · Urban design · Community building · Urban participation · Open public space · Student camp

26.1 Introduction: Urban Participation Through Workcamps

Urban design is a creative process in an environment with a focal point to physically shape human settlements and their components—streets, urban blocks, public squares, urban quarters, waterfronts, etc. This implies the cooperative and multidisciplinary character of urban design, dealing with all aspects of developing and maintaining human settlements—social, economic, ecological, cultural, technological, communicative, spatial, etc. negatively (Han et al. 2021; Madanipour 1997). Designing for the human environment means designing for the local community—practically, urban design makes places for people (EP 2000). This human-based nature of urban design is essential, as one of the prime visions of urbanism is to make our settlements more lively, safe, sustainable and healthy (Gehl 2013). Hence, urban design is among creative professions where the involvement of the community and general public matters more (Djukić 2021).

In line with the previous premise of the human dimension of urban design, community participation in urban design is of great importance. Its final goal is to make a compromise between all included stakeholders in the urban design process so it produces a design solution that is acceptable and just for all involved sides (Foroughi et al. 2023; Tato 2021). This is primarily driven by a humane primordial tendency to improve their (urban) surroundings (Moughtin 2003). The forms of community participation in urban design vary. Such participants can act from rather passive to very active roles, formally or informally, as well as in the different stages of the whole process. However, each mode of urban participation underlines that it is a significant segment in the professional ethos of urban designers (Münster et al. 2017).

Significant and continual community participation in urban design presents an imperative for the teaching and learning process in urban design. Indeed, the modern high-education curriculum in urban design and wider urban studies highlights the inclusion of public and community participation, which is usually implemented through several faculty courses where this topic is directly or indirectly taught and learned. In this consideration, cooperation between the community and students during the urban design process is an especially important experience (Oh and Minho 2022). Combrinck and Venter (2020) discuss that students in urban design thereby develop "the capacity to engage with the complexity and temporal fluidity of urban citizenship" (p. 1). The other issue is how to implement such goals and intentions into the learning-teaching process, which is always challenging (Borhani and Kalantar 2017).

The practical side of the faculty courses with urban participation as a topic usually includes public audits, surveys, questionnaires, and interviews, which students conduct as on-site or online training with citizens, experts and/or stakeholders. The forms that entail more the active involvement of students in urban participation are much rarer. One of them is a WORKCAMP. A workcamp represents active volunteering where volunteers work and live together with the local community on a short-term basis and for a not-for-profit cause (Sherrard et al. 2006). The focus group of workcamp participants is younger adults. For example, if a workcamp is organised within a higher education programme, volunteers are actually students, while teaching staff can be senior volunteers, mentors or supervisors.

Workcamps, in their modern sense, are not a very old social institution. Their history began in the 1920s in Western Europe, and they became more common after World War II (SCI 2018). They have become more frequent and global in the last decades (Lough and Tiessen 2018). Nevertheless, the aims behind a workcamp are still the same. First, there are always multiple participants, particularly related to participants and volunteers, and their deeper involvement or more active participation, such as mutual understanding, the rise of democratic awareness, increased independence, and self-reliance (Burghardt 2015).

Although international workcamps are a relatively widespread type of volunteering and can be found across the world, not many of them are directly incorporated within formal higher education. They are also rare in urban studies despite workcamps often have tasks related to space, such as an archaeology, the restoration of monuments, environmental protection (EYP, no data). This is simply because a workcamp is more demanding (organisationally, financially, time-consuming) to be included into formal higher education than other, aforementioned forms of participation.

This paper presents a case of a building camp for students organised within the DANUrB+ INTERREG Project (2020–2022) in Golubac Town in Eastern Serbia in June 2022. The aim of the camp was to materialise a small-scale 'design and build' urban project that connects with the revealed underused local heritage or cultural potential as a tangible implementation of an intangible potential. The paper intends to show how this process involved both students and the local community in jointly developing, designing, and building a small-scale public square on the Danube Riverfront in the centre of Golubac Town.

26.2 Methodology

This paper is organised as a single-case study, where the case is the DANUrB+ student building camp in Golubac, held on June 21–24, 2022. Relevant methodology involves a more precise research organisation, where theoretical fundamentals are first settled. In this paper, the essence of workcamps is already explained in the introductory section. Then, the case of the building camp is presented through its contextual framework, preparations and on-site realisation, with a spotlight on undertaken on-site student activities.

The method of single-case study emphasises a clear theoretical framework to analyse the case (Gustafsson 2017). In this research, the theoretical fundamentals about workcamps as a form of urban participation state several important criteria to be explained and discussed through the analysis of the single case:

(1) The special topic, organisation and aim of a workcamp;
(2) The goal of a workcamp related to a not-for-profit cause;
(3) In situ cooperation between local community and workcamp volunteers;
(4) The confirmation of ultimate goals for volunteers in a workcamp as the mode of urban participation: the understanding of the purpose of the workcamp, the raise of (democratic) awareness—cooperation with community, and increased independence and self-reliance.

26.3 Workshop Context

Contextual presentation refers to the organisational, territorial and temporal elements of the implemented workcamp—a building camp for students realised in Golubac.

26.3.1 Preconditions from DANUrB+ Project

DANUrB+ Project was implemented within the EU INTERREG Danube Transnational Programme. The full title of the project was "DANube Urban Brand+ Building Regional and Local Resilience through the Valorization of Danube's Cultural Heritage". The main aim of the project was "to reactivate underused cultural heritage and resources in shrinking settlements of Danube River's peripheral and border regions, to create new possibilities to make its towns and regions attractive again" (INTERREG Danube Programme 2020).

The duration of the programme was 2.5 years, from July 2020 to December 2022, whereas the building camps for students were organised in Danubian cities and towns in the last project—2022—year due to their character of a final and more complex activity to be implemented.

The building camps for students were organised within the working package WP.T4 "Actions—putting the heritage of the Danube into action: local, regional and international projects". The activities related to the preparation and organisation of the building camps are labelled as the thematic deliverable 4.4.1., under the title "Building camps for students to materialise valorisation processes together with local communities". This deliverable was specially formed for six university partners from Bratislava, Budapest, Belgrade, Timisoara, Sofia and Bucharest to carry a part of their education activities in situ in a small city or town which has the status of a strategic associate partner in the project. According to the project application, "a building camp is a final stage of a participatory process with the help of professionals (universities and NGOs) among local stakeholders. The aim of each camp is to realise and materialise a small-scale 'design and build' project which connects with the revealed underused heritage or cultural potential as a tangible implementation of an intangible potential. During these activities, partners shall include local craftsmen and materials as well. Therefore, the building camps will also function as easily adaptable small-scale interventions, which strengthens the local community and contributes to further improvement of such local activities. A total of 6 building camps will be organised with the involvement of international and local (regional) students between the age of 16 to 25 by the leading of the universities, 2 in each section" (INTERREG Danube Programme 2020).

26.3.2 Territorial Context: Golubac, Serbia

Golubac Town in Eastern Serbia was intentionally chosen by the project partner—the University of Belgrade—Faculty of Architecture—as a polygon for the urban-design building camp for student, as the town is a good reflection of contrasts in urban and regional development in Serbia.

The location of Golubac on the western entrance to the longest gorge of the Danube and Europe—the Iron Gates—was both a pro and con during history. This can be easily shown by the last three decades of the post-socialist transition. This borderland town particularly suffered during the international isolation of Serbia and the collapse of the socialist economy during the 1990s and 2000s. As a consequence, Golubac's economy was devastated, especially its core sectors—industry and tourism. The town also lost more than 15% of its population during the period 1991–2011, especially younger and more educated people. Finally, Golubac Municipality was enlisted among the so-called' devastated areas' in Serbia, i.e., the municipalities whose development level was under 50% of the national average (Ministry of Public Administration and Local Self-Government 2009–2015).

Since the 2010s, Golubac has witnessed significant economic recovery, mainly driven by the rise of cultural and religious tourism. The main driver of international tourism was the Danube River as a cultural and cruising route (Fig. 26.1), with a new important tourist destination in Golubac—the medieval Golubac Fortress (Fig. 26.2), which was completely restored in 2019. For local tourists, the rise of nearby Tumane Monastery as a pilgrimage site has been equally important. These two local attractions further fuelled the development of other smaller sites, as well as the general refurbishment of Golubac Town and its surroundings. Local experts in tourism highlight that the numbers of tourists as well as tourist accommodation ('tourist beds') are more than tenfold bigger than just one decade ago.

The problem with this development is that it has not been adequately reflected into limited socio-economical benefits for local population due to local capacities (Antonić et al. 2021). Local municipal institutions are pretty weak due to the aforementioned long-term 'brain drain' to exploit this new development in a proper way.

Fig. 26.1 Two major tourist attractions in Golubac seen during student fieldwork: The first one, the Danube, which is the widest in its entire flow (6–7 km wide) in the front of the town (Author: N. Mitrović)

Fig. 26.2 The second one, restored Golubac Fortress (Author: B. Antonić)

Such development generates problems in local infrastructure and low-wage jobs for local people. Therefore, Golubac is still demographically shrinking despite this ‘tourism boom’.

26.4 Case Study: DANUrB+ Building Camp for Students in Golubac, Serbia

As a rather complex activity in the DANUrB+ Project, the building camp for students in Golubac was organised as the final event of a much longer process in higher education, done between the University of Belgrade—Faculty of Architecture and the Municipality of Golubac. Therefore, even though the duration of the Golubac camp was just four days, the preparations for it lasted almost one year. It is worth mentioning that the whole activity was not just organisationally demanding; the camp also required significant finances.

26.4.1 Workshop Preparations

The preparations started with several meetings between local representatives and the project researchers from the University of Belgrade, which were organised during the late 2021-early 2022. During them, the basic in situ research of several possible sites in the town was conducted, to offer students a variety of them in prospective selection process.

The work with students to prepare all conditions for the building camp lasted one semester, from late February to Early June 2022. It was realised within the faculty course STUDIO 06 URBANISM—SEMINAR. First, students visited, explored and documented the site for the camp in Golubac with teaching staff (researchers) on March 9, 2022 (Fig. 26.3). In addition, they spoke with local experts and citizens relating to their opinions and experiences with this site and its surroundings. During this workshop, students, researchers, staff and local experts jointly decided on the most appropriate site in Golubac for the camp. It was a small concrete plateau—an old cinematic concrete stage for performances—surrounded with urban greenery and next to the “Golubac” Hotel and close (30–40 m) to both the Danube and the town’s main square. Despite this excellent location, the site was in a poor state before the building camp. Technical measurements and photo documentation of the site were also done there.

During the next three months, students developed their projects. Initially, most of them proposed quite radical and resource-demanding designs. Then, the teaching staff worked together to customise these initial visions to the level that would be implementable. Students preliminarily targeted several objectives: suitable urban furniture, optimal micro-location, communication paths, and opportunities for urban

Fig. 26.3 Students visited the site for the camp in Golubac on March 9, 2022 (Author: N. Mitrović)

renewal. Then, the project development encompassed many elements, including urban furniture design, space analysis, material sourcing, and construction, as well as the customisation of limitations (budget, timeframe, space-related issues, available/recycled materials). Their preliminary project 2D drawings were submitted as a colloquium in April 2022. The next step for students was to develop these drawings into small-scale 3D models (Fig. 26.4), which influenced all of them to further simplify.

The overall design. In late May, teaching staff and students chose the most adequate projects for the camp. During the early June, all material and tools were bought for the camp, as a final preparation step.

26.4.2 Workshop Realisation

The Danubian Pavilion project exemplifies the power of architectural innovation and community engagement in revitalising the neglected public square in Golubac. Spearheaded by architecture students from the Faculty of Architecture in Belgrade, the initiative aimed to infuse the area with sustainability, historical sensitivity, and community spirit. Respecting the square's history while embracing sustainable urban development, the planning and execution phases of the project were marked by a commitment to ecological responsibility. The initiative involved various stages, such as urban furniture design, space analysis, material sourcing, and construction. Despite financial constraints, the students creatively utilised recycled materials and repurposed existing infrastructure to rejuvenate the square, demonstrating resourcefulness and a sustainable approach to design.

Fig. 26.4 Small-scale 3D models were a critical element for students to simplify the initial projects so to be implementable (Author: J. Marić)

Community engagement was central to the project, with Golubac residents actively participating in the development process, ensuring the pavilion reflected the community's identity and aspirations. This collaboration aimed to transform the abandoned square into a vibrant, inclusive space, underscoring the significance of community involvement in urban renewal projects.

26.4.3 *Space Analysis*

During the initial phase of the project, the students were dedicated to extensive research and analysis of basic local needs. They sought to identify suitable urban furniture solutions and conducted comprehensive space analyses, considering factors such as the optimal location, communication paths, and opportunities for urban renewal. This section delves into the students' efforts to develop innovative design concepts while remaining mindful of the town's historical identity and available resources. The most important action is to organise the common activity and work in larger groups with tasks of the management of duties and assignments, and eventually, the establishment of their roles. All of them spoke about organising tasks, distributing them among team members, and ensuring that everyone has a specific role to contribute effectively to the project or task at hand. The plan was to use 42 years of the abandoned public square at the bank of the Danube, where the river spilt over time and made layers of dirt where there are only grass and wild bushes

Fig. 26.5 Old cinematic stage (Author: M. Obradović)

Fig. 26.6 Clearing the ground of an old public square (Author: M. Obradović)

now. It was hard even for older members of Golubac to remember that there was a pavement and some sort of public square. The old cinematic concrete stage is located on the side, and it could be recycled for new usage, even if it was damaged a little bit after some time without the purpose (Figs. 26.5, 26.6, 26.7 and 26.8).

26.4.4 Material Sourcing and Budget Constraints

The project-encountered financial constraints necessitate the creative use of recycled materials. This action could be very important considering that local citizens are very much capable of joining action together, but when it comes to financial costs, all initiatives tend to break down. Therefore, it had to be presumed that locals are not capable of doing the most they can, so recycling of all attributes sometimes has

Fig. 26.7 Collecting the recycled materials (Author: M. Obradović)

Fig. 26.8 Working on the urban furniture (Author: M. Obradović)

more power in action in future development of attachment of users of space, with each other and with space itself. Very often, it is a case with more expensive projects but without links to the local community and their needs. The students sorted and repurposed various items, including wooden pallets, wooden cylinders and wires, aiming to make a new and vibrant urban mobility at the abandoned concrete stage, which had fallen into disrepair over the course of several decades.

Plan had the intention of joining those two squares and making the new one recognisable with a new identity. Considering the limited budget and certain donations in materials, the only possible way to make a difference was to use as much of the existing infrastructure as possible. This section delves into the challenges faced in acquiring materials and the strategies employed to maximise the use of recycled infrastructure.

Fig. 26.9 Building a central sculptural figure—The Portal (Author: M. Obradović)

26.4.5 Construction and Student Involvement

The students actively participated in all stages of the construction process, gaining valuable hands-on experience on a regular construction site. The main idea was to involve every student in the construction stages, which included dirt and send excavation, armature reinforcement, concrete work, assembly of mobiliary and finalising with paintwork. This section highlights the educational benefits of involving students in real-world construction scenarios, enabling them to acquire practical skills and insights into fundamental construction techniques. There, students had a chance to get familiar with rebars and making concrete with their hands—just cement, sand, and water. After making the urban furniture, the concrete pedestal was the most important for the main wooden installation in the very centre of a new space—The Portal (Fig. 26.9).

Almost all of the budget was spent on laser cutting the main wooden installation in the shape of a big portal. It has the purpose of having a reference point, a central figure, and an artistic form. Everything else was made of cheaper recycled materials, but the main figure was built to last much longer, considering the possibility of temporary architecture being destroyed by time and weather without regular maintenance. Other forms of public furniture were wooden chairs, wooden staircases to connect with the former stage, a higher elevation, and a small public shelf for books, which represented an open public library.

26.4.6 Community Engagement and Participation

Central to the success of the open space project was the active involvement and engagement of the local community. The students organised a lot of small tasks that ordinary citizens could involve themselves in, allowing citizens to contribute to the project's transformation. Furthermore, they implemented participatory practice by inviting residents to suggest names for the new public space, cultivating a sense of pride and ownership. This section explores the positive impact of community engagement as it fosters a greater sense of belonging and responsibility among the local population. In the end, furniture was polished and painted in different colours in order to create more identity and slightly more propelling space for younger generations, considering that the Municipality of Golubac has a dramatically decreasing number of young people. In order to involve other citizens in the project as a good participation practice, students chose four possible names and made a post on Facebook for everybody in Golubac to participate and give their suggestions for the name of the new public space. After some time, every citizen could feel that he belongs to the project. After some time, the consensus was given on the new name of the public square—The Danubian Pavilion. That small action gave everybody a sense of belonging to the project, which eventually made locals spend time there with their children to watch and protect their new place. The most important goal was for an activity to address another activity by participating from locals.

26.4.7 Sustainability Measures and Safety Enhancements

To ensure the usability and safety of the newly revitalised public space, the students incorporated sustainability measures such as the installation of solar lamps. These lamps not only enhanced safety but also engaged the younger population through interactive light displays. This section highlights the role of sustainable design in creating an inclusive and environmentally conscious public space. To implement safety measures as every public space should have, more than ten solar lamps were placed in the square at night. That way, we brought people to use the space even at night, which minimised vandal behaviour. The youngest kids find it very entertaining for them to have an opportunity to communicate with the square itself because whenever they step close to the solar lamps, the light turns on. The best light show was for a person entering through the portal with a brighter light (Fig. 26.10).

Fig. 26.10 Night lights as a model of interactivity (Author: M. Obradović)

26.5 Discussion

The Danubian Pavilion initiative stands as a testament to the transformative power of innovation in urban design and community involvement, particularly in the revitalisation of a once-forgotten public square in Golubac through a student workcamp. This ambitious camp, spearheaded by over 30 architecture students from the Faculty of Architecture in Belgrade, not only aimed to rejuvenate the space but also to weave the fabric of sustainability, historical awareness, and communal connectivity into the very core of its design ethos (Fig. 26.11).

At the heart of the conception of the Danubian Pavilion was a profound respect for the square's historical narrative, coupled with a forward-thinking approach to sustainable urban development. The students and faculty members embarked on a journey that was as much about honouring the past as it was about writing a new chapter for the future of public spaces in Golubac. The project and camp objectives were meticulously crafted to ensure that the transformation of this town square would serve as a beacon of community-centric design, where every element echoed the voices and aspirations of the local populace. The chronology of the project reveals a meticulous planning and execution phase, underscored by an unwavering commitment to sustainable design principles. From the initial site analyses to the final touches of the pavilion, each step was imbued with an ethos of ecological responsibility and adaptability. The use of environmentally friendly materials and energy-efficient construction methods not only minimised the ecological footprint of the pavilion but also set a precedent for future small urban co-design projects (Fig. 26.12).

The active participation of Golubac residents in the planning and development phases fostered a sense of ownership and pride among the local community. Workshops, public consultations, and interactive design sessions became the avenues through which the voices of the citizens shaped the project's trajectory. By transforming a neglected area into a bustling hub of activity and social interaction, the

Fig. 26.11 Urban furniture with a small open library (Author: M. Obradović)

Fig. 26.12 Completed composition with a new public space (Author: M. Obradović)

Danubian Pavilion has redefined the parameters of community space and urban liveability. The square now stands as a vibrant testament to the power of collaborative innovation, serving not only as a leisure-recreational area but as a symbol of local resilience and communal spirit. To sum up, the Danubian Pavilion Project embodies the essence of sustainable, community-focused urban renewal. It showcases how the synergy between architectural creativity, historical consciousness, and community engagement can breathe new life into public spaces, fostering a sense of belonging and identity among its inhabitants.

26.6 Conclusions

The discussion underlines that the case study of the student workcamp in Golubac, Serbia, is an inspiring blueprint for possible endeavours aimed at revitalising urban environments through the lens of sustainability and urban co-design. Its main visible outcome, the Danubian Pavilion, has become a new urban hotspot at the local level; it attracts the locals and visitors of Golubac and provides a safe and engaging environment for various activities. However, the student work has additional outcomes, which are less visible but equally significant. They are related to the process of urban participation and co-creation in urban design. This workcamp involved and connected the local community with local institutions, such as local tourist organisations and municipal libraries, as well as academics, such as students, teaching staff, and researchers, by providing adequate results with a limited budget and time. Hence, it could be considered a tool for urban participation and co-design, teaching methods (trans-disciplinarity, exchange ideas, real-life simulation), placemaking, cooperation between different stakeholders, and increasing awareness of public space or socialising.

To conclude, both the student workcamp in Golubac and the Danubian Pavilion, as its main outcome, present pilots for local and regional levels, which can be replicated in similar places in Serbia or along the Danube and other riversides. The main recommendations from the explained Golubac case are:

The success of the whole project cannot be attributed only to the student innovative design solutions and the issues related to used materials and on-site implementation, but also their efforts in community engagement through on-site urban participation. Thus, the involvement of local community is essential for the success of projects of this scale. In addition, the local community should be involved as early as possible.

The format of a workcamp is a good tool to implement other novel approaches at the local level, such as the use of recycled materials or fast-track creativity in local urban co-design. Locals in Golubac have become aware that this a solution for some urban places which are less than the main town square or park but which also need attractive and comfortable urban design, such as a local beach or the meeting points along the Danubian promenade in Golubac.

The utilisation of urban identity, heritage, tradition and culture is a very important element in linking any urban design intervention to the local cultural and social fabric. The Danubian Pavilion serves as an example of how local identities, spaces, and symbols can be repurposed to foster co-design innovation and sustainable design—a gate between the town and its river as one of its main symbols, a form of a small urban forum for socialisation typical for Serbian heritage towns, etc. Therefore, the use of local identity is crucial for mutual understanding and the future maintenance of such interventions.

The project in Golubac also emphasises the importance of a flexibility. By embracing creative problem-solving, resourcefulness, the reuse of existing, albeit neglected infrastructure and inclusive design practices, the project showed that approach can speed up the overall realisation of the project, as well as to overcome certain problems, such as limited budget or different views on the final design solution.

The final conclusion is based on how the form of a (student) workcamp can be a sustainable tool to address old identities, spaces, and symbols to form a novel place for local people through urban participation and co-design. Probably the most important aspect of this project is the creation of the idea of how to involve everybody, citizens, public sector or external visitors (workcamp attendees), to participate, where everybody has a chance to share knowledge and dedication with the common goal of connection and having a place where to connect.

Acknowledgements This paper is done to present the results of the INTERREG Danube Project "Cultural Heritage of the Danube as a Tool to Develop Local Opportunities and Common Brand of Underprivileged Communities—DANUrB+" (2020–2022). The authors of this paper also want to thank (1) the representatives and experts of the Municipality of Golubac, the Tourist organisation of Golubac (director: Jelena Petrović) and "Veljko Dugošević" National Library in Golubac (director: Marija Živanović Ilić); (2) sponsors: "Arhi.pro" Architectural Bureau, Belgrade, and "Novo Kulturno Naselje" non-governmental organisation, Novi Sad; and (3) students-participants in the building camp: Kristina Anđelić, Katarina Vratonjić, Jana Vukosavović, Milica Vučić, Lazar Dimitrijević, Jovana Dragutinović, Sanja Jevremović, Matija Jevtić, Ana Jovanović, Emilija

Jović, Milica Kapa, Lenka Kostić, Marina Maljković, Milica Mašulović, Teodora Miletić, Marija Milić, Matea Milinković, Veljko Milivojević, Milica Milutinović, Filip Nikolić, Ana Nićiforović, Anastasija Peković, Lazar Petrović, Sofija Pribičević, Marija Radulović, Neda Ristanović, Emilija Stanisavljević, Sanja Stevanović, Iva Timotijević, Danilo Fatić, Nina Čolić and Milica Crnčević.

References

Han D, Song Y, Wang E, Liu H, Fang R (2021) Multiple dimensions of urban design development from a practice perspective: a case study of an institute in Nanjing. Front Architect Res 10(1):79–91. https://doi.org/10.1016/j.foar.2020.11.005

Madanipour A (1997) Ambiguities of urban design. Town Plann Rev 68(3):363–383. https://doi.org/10.3828/tpr.68.3.2365658h658v0157

English Partnerships—EP (2000) Urban design compendium. Housing Corporation, London

Gehl J (2013) Cities for people. Island Press, Washington, DC

Djukić A (2021) Обликовање отворених јавних простора (Design of open public space). Faculty of Architecture, Belgrade

Foroughi M, de Andrade B, Pereira Roders A, Wang T (2023) Public participation and consensus-building in urban planning from the lens of heritage planning: a systematic literature review. Cities 135:104235. https://doi.org/10.1016/j.cities.2023.104235

Tato B (2021) Making participation relevant to design. Harvard: Cambridge, MA. https://www.gsd.harvard.edu/course/making-participation-relevant-to-design-spring-2021/

Moughtin C (2003) Urban design: street and square. Architectural Press, Oxford

Münster S, Georgi C, Heijne K, Klamert K, Noennig JR, Pump M, van der Meer H (2017) How to involve inhabitants in urban design planning by using digital tools? In Zanni-Merk C, Frydman C, Toro C, Hicks Y, Howlett R, Jain L (eds) Procedia computer science, vol 112. Elsevier, Amsterdam, pp 2391–2405. https://doi.org/10.1016/j.procs.2017.08.102

Oh J, Minho S (2022) Evaluation of citizen—student cooperative urban planning and design experience in higher education. Sustainability 14(4):2072. https://doi.org/10.3390/su14042072

Combrinck C, Venter R (2020) Embedded participation in the architectural curriculum towards engendering urban citizenship in graduates. Image Text 34:1–29. https://doi.org/10.17159/2617-3255/2020/n34a2

Borhani A, Kalantar N (2017) Advancing community design education. In: Proceedings of international conference on engineering and product design education. https://www.designsociety.org/group/18/Search+DS+publications

Sherrard Sherraden M, Stringham J, Costanzo Sow S, Moore McBride A (2006) The forms and structure of international voluntary service. Center for Social Development, Washington, DC

Service civil international—SCI (2018) Words about deeds—Publication about 100-year SCI. SCI, Antwerpen

Lough B, Tiessen R (2018) How do international volunteering characteristics influence outcomes? Perspectives from partner organizations. Voluntas 29:104–118. https://doi.org/10.1007/s11266-017-9902-9

Burghardt A (2015) Social media and pre-project inter-cultural dialogue in international work-camps: potential and reality. Interculture J 14(24):37–52. https://www.ssoar.info/ssoar/handle/document/45246

European youth portal—EYP (no data) What is a work camp? Choose the best for you. Europa. Eu. Accessed Feb 6, 2024. https://youth.europa.eu/es/article/45/26619_en/span/span

Gustafsson J (2017) Single case studies vs. multiple case studies: a comparative study. Accessed Feb 8, 2024. http://urn.kb.se/resolve?urn=urn:nbn:se:hh:diva-33017

INTERREG Danube Programme (2020) DANURB+ project. Accessed Feb 8, 2024. https://www.interreg-danube.eu/approved-projects/danurb_plus

Ministry of public administration and local self-government (2009–2015) Закон о регионалном развоју (Law on regional development). National Gazette No 2009/51, 2010/30, 2015/89, Belgrade

Antonić B, Djukić A, Marić J (2021) Planning cultural tourism in Serbia: Golubac town in the Danube iron gates region. In Frenţ C, Niculescu AC, Colceru I (eds) 50 years of research in the field of tourism in Romania: past, present and future. INCDT, Bucharest, pp 56–58

Chapter 27
Creative Education for Inspiring Students: New Approaches in Teaching Methodology

Eva Vaništa Lazarević, Jelena Marić, and Milena Vukmirović

Abstract The modern times of new technology rise, social media and artificial intelligence, accompanied by a hectic lifestyle full of high stress, are reflected in almost every aspect of life, including Academia, Universities and teaching methods. Also, the COVID-19 pandemic presented a severe issue in the education process. Within this situation, it is essential to determine how to enhance students' enthusiasm at the University level, especially regarding the education in architecture. The methodology relies on the focused theoretical background research and the overview of the student course outcomes. University professors have to adapt and transform the teaching methodology. The elective course, "Architecture in Context," is presented as a case study at the master level of studies at the University of Belgrade, Faculty of Architecture. Creative techniques and ever-changing curricula of the course showed how the usage of innovative technology, involvement of students in scientific research and conferences, and collaboration of academia and practice could produce higher motivation levels among students and more quality results. This paper represents the possible benefits of creative education, which is one of the crucial preconditions for the future methodology of architecture teaching for creating beautiful cities.

Keywords Creative education · Teaching methodology · Architecture in context · Academia and practice · University of Belgrade · Faculty of Architecture

---

E. V. Lazarević · J. Marić (✉)
University of Belgrade—Faculty of Architecture, Belgrade, Serbia
e-mail: jelena.maric1989@yahoo.com

M. Vukmirović
Department of Landscape Architecture and Horticulture, University of Belgrade—Faculty of Forestry, Belgrade, Serbia

M. Vukmirović et al. (eds.), *Digital Dreams, Urban Realities*, The Urban Book Series,
https://doi.org/10.1007/978-3-031-76353-3_27

27.1 Introduction: Education Challenges and Creative Approaches in Teaching

Being part of the education process at the University level comes with complex tasks, and teachers have to overcome many issues regarding the teaching process apart from the education level and knowledge of a particular subject. Transferring knowledge and skills has always been challenging due to the ever-changing student structure and interests. Nowadays, these issues have become even more challenged by the modern lifestyle. The rise of technology, sedentary lifestyle with high stress levels, constant usage of different kinds of social media and gaming, and use of artificial intelligence tools today have influenced the level of attention among students at all levels, especially at the University level.

Technology has led to significant efficiency in the education and learning processes and has enabled more productivity among students, lowering students' attention span and creativity. Also, many students misuse the technology. Further, the situation with architecture students' questionable creativity and enthusiasm worsens with the challenges during COVID-19. These challenges have influenced the mental health of students, and the majority of education processes have been drastically changed. The pandemic state has introduced online teaching as an obligatory means of teaching while respecting the social distancing restrictions (Babbar and Gupta 2021). This seriously struggled the education system (Shaheen 2010).

This paper examines whether different innovative approaches to architecture teaching, transformative curricula, and course tasks and methods can improve overall student engagement with the subject, teamwork, or the quality of the results. Using the University of Belgrade- Faculty of Architecture elective course, we give an extended overview of the creative teaching methodology and outcomes.

27.1.1 Creative Teaching Methodologies_an Overview

The most significant creative discoveries, which represent a critical contribution to modern art, science, culture, and education, begin with the pursuit of not potentials but problems in a specific time and environment. Once a situation is defined, it consists of questions such as how the problem can be solved, when it could be resolved, and how individuals can participate in its resolution. It is known that estimates of the number of creative individuals in the population hover around a few per cent, while the prevailing opinion is that only 1% of adults are creative or manage to realise and manifest their creativity. Creativity and education have always been inseparable factors. Education as a process has to be creative, starting from the very beginning of education in elementary school and going up to the university level. This includes a variety of science, educational approaches, methodologies, and different topics or interests (Lin 2011). In this paper, we are focusing on education at the university, on the master's level of studies in architecture and urbanism.

At different educational stages and topics, there are different challenges to overcome (Skaalvik and Skaalvik 2016). Some sciences require mostly traditional lectures, while others need laboratory analysis or specific artistic exercises. However, the creative approach is crucial in each situation (Hadzigeorgiou et al. 2012). As a concept, creative thinking presents a basis and an initial idea for creative education that has been adopted into many subject curricula at the university level (Craft 2001; Tapinos 2016). Furthermore, the constant close connection between academia and practice benefits students and future experts in overcoming the distance between learning processes and working in practice. Contemporary practice of urban design and architecture education highlights the importance of community participation. In this consideration, cooperation between the community and students during the urban design process is a significant experience (Oh and Minho 2022). Constant participation from different stakeholders and students in design implementation is a modern practice of urban design teaching. Combrinck and Venter (2020) state how students develop "the capacity to engage with the complexity and temporal fluidity of urban citizenship" (p. 1). The other issue is implementing such goals and intentions into the learning-teaching process, which is always challenging (Borhani and Kalantar 2017).

There are many creative approaches regarding teaching methodology in education at the university level that can help engage students and foster creativity (Bell 2010). One of the most used includes (1) Project-Based Learning: This approach involves project design and developing concepts that require students to work collaboratively to solve a real-life problem or create a tangible product-usable design proposal. Project-based learning allows students to apply their knowledge and skills in real-life circumstances way, which can be very engaging for students; (2) Problem-solving and Flipped Classroom: In a flipped classroom, students attend lectures or read handout materials before the course starts, and active learning time is spent engaging in activities and discussions related to the material. Students are encouraged to use their knowledge to solve real-life problems; (3) Active Learning: Active learning involves getting students involved in the learning process through various activities such as group discussions, debates, and similar exercises. One of the most recent methodologies and the most creative approaches is (4) Gamification: Gamification involves incorporating game-like elements into the learning process, such as challenges, imagination, points, typology and mapping. This approach can make learning more engaging and fun, encouraging students to take risks and be more creative in their thinking processes.

Different creativity exercises can also produce creative thinking and brainstorming effects among students. Finally, incorporating creativity exercises into the curriculum can be a great way to encourage students to think creatively. These simple exercises can be as simple as brainstorming sessions or as complex as design thinking workshops or focus groups. Allowing students to explore and experiment with their ideas makes them more likely to come up with truly innovative solutions and think "outside the box". Creative visualisation is one of the most effective exercises for young architects. Architecture and urbanism as a research field have a unique learning and research methodology based on the sensitive nature of the artistic approach that requires unique techniques used in teaching processes and courses. Architecture faculties often use creative methods to involve students in creating quality results regarding thinking processes and design.

Considering Architecture and urban design, out of the mentioned learning techniques, the most used one is Project-based learning (PBL). In PBL, students pre-learn the class material. During the active stages of the course, while attending the classes at the faculty, students are encouraged to use the knowledge they acquired and solve the predefined problems, in this case, connected to architecture and urban design. Mutual collaboration and creative thinking, including several learning techniques, are needed for the PBL method to be successful among students. PBL allows students to develop critical thinking, problem-solving, collaboration, and communication skills. It also allows students to apply their knowledge creatively in practical conditions.

27.2 Methodology

This paper overviews the one course selected as a case study. It presents an example of creative and ever-changing teaching methodology and unique approaches to learning.

The relevant methodology involves a focused and precise research organisation, where theoretical fundamentals in literature and good practice analysis of creative education as a concept and what it includes are first settled.

In the second phase of the research, the main idea and methodology of the course selected as a case study—"Architecture in Context"—as an elective course for higher education are presented.

In the third phase, selected examples of students' work as course outcomes are presented and discussed with a particular focus on the benefits of each presented course characteristic on the quality of the results and the skills of students to work as a group, collaborate with relevant stakeholders, or meet the requirements of a professional exhibition or scientific conference.

Finally, the last section of the paper summarises the results and the overall goal of the research.

The next chapter presents the results of the selected students' school year from 2015/16 until 2023/24.

27.3 The Elective Course "Architecture in Context"

"Architecture in Context" is an elective course prepared and accredited for students in the 1st year of Master Academic studies or the 4th year of Integrated Academic studies at the University of Belgrade, Faculty of Architecture. It has been developed by Professor Eva Vaništa Lazarević, and co-authors Assistant Professor Jelena Marić and Associate Professor Milena Vukmirović have worked as mentors on the course. The course has been developed and ongoing for more than 15 years at the Faculty, and each year, students are more than happy to join the course. Around 30 students each year choose a course that lasts for 15 weeks of active learning, and it also requires passive learning while students work with mentors in the form of undirected consultations. During that period, the class is divided into three main phases. The first phase consists of interactive lectures and explanations of the course structure, timetable, and ideas since the main tasks and curricula change almost yearly. In the second phase, students develop their concepts, visions, projects, and designs in various forms, from the essay type to presentations or projects, advertising campaigns or models. Students work closely with their mentors in this stage and are encouraged to use modern technology and tools.

The course explores architecture in different spatial, social, cultural, and economic contexts. Also, one of the positive aspects from the creativity point of view is that each school year, the focused topic and methods used in this course change, adapting to current global issues, like pandemics, technical advances, social media, and student interests.

One of the main topics ranges from analysing spatial characteristics of historical city centre areas using multi-criteria analysis to tackling global issues like climate change or urbanisation. Urban, historical, and architectural heritage, such as the Belgrade fortress, were also researched. Students expressed interest in different topics regarding urban issues and city transformation and growth using different media, such as podcasts. In order to deal with the topic of Architecture on different levels, as opposed to the global issues and understanding of the city and space, the students also explored Architecture in detail. In collaboration with international companies, students have developed project proposals for the creative usage of doors and windows in Architecture in the form of a student competition. During the last year, students developed severe marketing and branding campaigns using the "Open AI" artificial intelligence tool, which was entirely new.

The course explores architecture within diverse contexts. Specific objectives include examining architecture from various perspectives, encompassing both spatial, social, and economic dimensions and architectural dialogue, media, and publications. One notable aspect is its adaptability; the course content and methods evolve annually to address current global issues, such as pandemics or technological advancements, as well as student interests. This adaptability ensures relevance to contemporary events and trends at both global and individual levels.

27.4 Results—students' Work and Outcomes

Significant results were achieved over the last eight years of the course Architecture in Context. UB Faculty of Architecture students are familiar with creatively presenting their work. An integral part of almost every subject at the UB Faculty is the final exhibitions, where students present their design projects. At the end of each semester, several studio design subjects have large exhibitions in the classes and web exhibitions (Fig. 27.1). This is different for elective courses since they require less

Fig. 27.1 Student exhibition at the University of Belgrade Faculty of Architecture in a studio design class of the school year 2023/24 (Author: Jelena Marić)

time, and students are less involved in the design than in the study and the research. However, "Architecture in Context" is a rather unique course since the students have each year a unique presentation of their work. In this section, the selection of the course outcomes is presented. Since the overall topic and methodology changes every school year.

27.4.1 The School year 2015/16_Scientific Conference "Real Corp."

During the spring semester of 2015, the course "Architecture in Context" focused on contemporary architectural issues within the urban setting of historical significance. One is the Municipality of Belgrade, the Vračar area, known for its historical architecture and the density of a central municipality in a Capital city. Over two decades, this area gained popularity among investors, leading to the adoption of new detailed urban plans. Consequently, new buildings emerged within the familiar and "old" ambience, sparking discussions and critiques regarding their quality, appearance, materials, proportions, height, and overall compatibility with the existing physical context. In light of these frequent discussions among students, the project provided a framework for researching the extent to which the new structures either aligned with or deviated from the context, employing an evaluative approach based on a set of criteria.

The analysis encompassed the following dual criteria: (1) order and unity, (2) ratio and proportion, (3) scale and hierarchy, (4) symmetry and rhythm, (5) detail and texture, and (6) harmony and beauty. These criteria served as the foundation for qualitative and quantitative research, as each analysis assigned a score ranging from 1 to 5 based on how well the buildings fit into the context according to the observed criteria. In the final phase of the research, each student was required to identify a suitable and commendable example of good practice and provide a final assessment of the quality and character of the proposed and implemented interventions. The overall research culminated in a catalogue of new buildings, accompanied by critical reviews, which contributed professionally to the current built stock in Belgrade. Consequently, the course fostered critical thinking skills among students regarding built objects and trained them to consider the contextual aspects when formulating their intervention proposals (Fig. 27.2).

27.4.2 The School year 2017/18_BAB Student Workshop "Belgrade Calling 4 SPACE"

The school year 2017/18-course students were excited to join the workshop and international conference "Belgrade calling 4 SPACE" as a central part of the Balkan Architectural Biennale event organised every two years since 2013. Moreover, it has been an essential architectural event in Belgrade. Students had an opportunity to meet some of the most established architectural professors worldwide and work closely with them. Every student developed a strategic plan for the future of Belgrade development, with a focus on crucial topics such as climate change, social and ecological sustainability, and rapid urbanisation; students especially presented the social, economic and ecological sustainability approaches they took in their works, and the knowledge of expert observation of changes in a city. They successfully identified issues and potentials of the capital city and developed a comprehensive study of good and bad practice examples. Also, some students focused on the possibilities of urban renewal and revitalisation. In contrast, others dealt with the critical issues of climate change adaptation and technology revolution in tomorrow's megacities. Their presentations (Fig. 27.3) were impressive for all the participants of the Balkan Architectural Biennale in 2017.

27.4.3 The School year 2020/21_Publication "Belgrade Fortress—Heritage as a Potential"

During the school year of 2020/21, students, alongside mentors, developed a Publication in participation with PE "Belgrade Fortress" and the director Marija Kuprešanin.

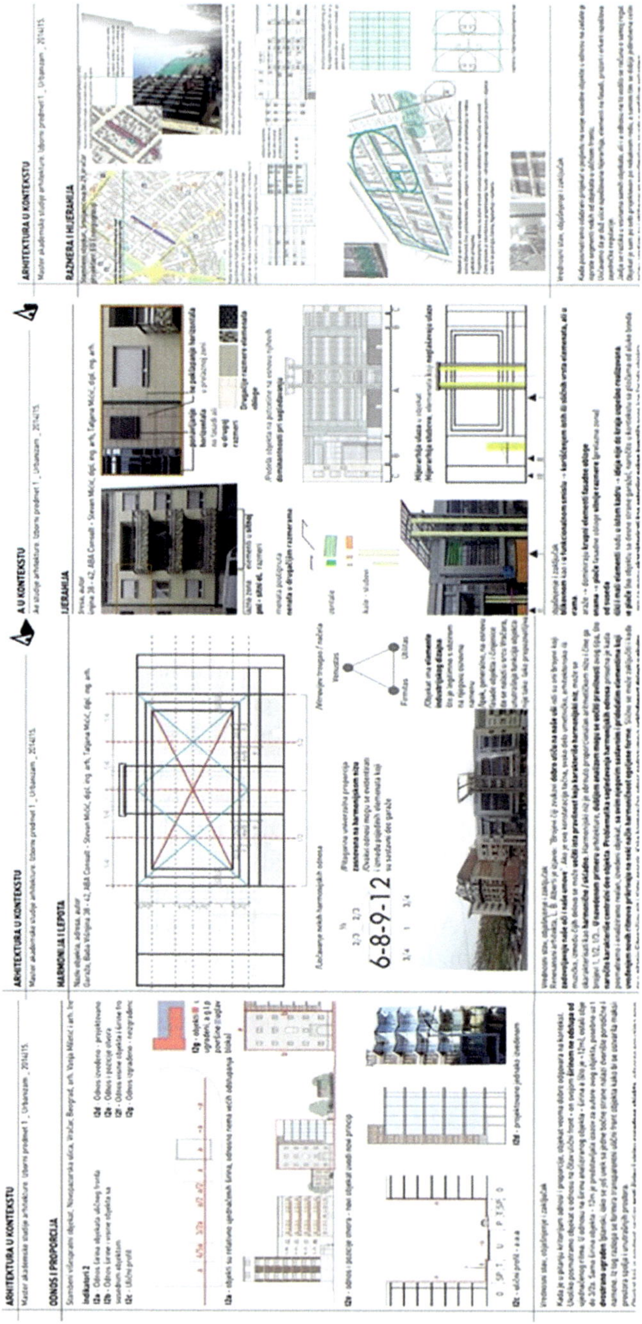

Fig. 27.2 Part of the student research study that was published in the conference paper at the Real Corp conference 2016 (authors: Eva Vanista Lazarevic, Milena Vukmirović)

Fig. 27.3 Student workshop and presentation as a part of the BAB Balkan Architectural Bienalle, 2017, in collaboration with dr Tatjana Mrdjenovic (Author: Jelena Marić)

Fig. 27.4 Cover of the publication "Belgrade Fortress/Heritage as a potential" (Author: Jelena Marić)

In this class, students were pleased to collaborate closely with the public institutions, and they had an opportunity to have expert lectures and guided field research on the Belgrade fortress. Their work consisted of four main steps: (01) identifying the existing stages of the Fortress, the spatial and functional characteristics, and its historical values and timeline. (02) The second stage included good practice examples where students developed a unique database with similar examples of fortresses around Europe and, following the predefined set of criteria, they evaluated the selected examples, thus gaining knowledge and specific insight into the fortress's preservation and heritage importance from the social and cultural aspects. In the third (03) stage, and based on the previous research, students proposed measurements for improvement of the current state of Belgrade fortress concerning historical heritage and the importance of the area not only for Belgrade but also for the whole region. Finally, at the end of the semester, students, alongside mentors, developed a 246-page (Fig. 27.4) long publication with visual elements that are important and usable for students and teaching courses and the PE Belgrade Fortress and citizens. The publication contains text algorithms, mapping, diagrams, photos, and creative schemes.

27.4.4 Workshop School year 2021/22_Podcast—"Evina Klasa"

During the school year of 2021/22, students and teachers were overwhelmed with the concept of online education. The COVID-19 pandemic has transformed our everyday lives, and we have had to deal with structural changes in the work department. Due to the online classes that could not present the meaning of architecture in context, in order for students not to lose their attention to the subject, the mentors suggested using an informal method for teaching a class—the podcast.

Considering the limited number of students in one class, each week, a different group of students prepared a whole podcast episode for the podcast called "Evina Klasa" (Figs. 27.5 and 27.6), and each week, another episode was put on the YouTube channel. This approach was somewhat unusual, and students were more enthusiastic about being a part of it.

Each podcast episode dealt with serious social issues, such as the future of Architecture as a profession in Serbia, and spatial topics, such as traffic and transportation issues, sustainability challenges, social housing, and the development of high-rise buildings. Other topics included informal settlements, large music festivals, and urban areas for youngsters in capital cities.

Also, each week, students chose a special guest appropriate for the selected topic. It was a unique experience for the students, professors, and guests, who represented experts in urbanism and architecture (Fig. 27.7).

27.4.5 The Student Design Competition, the School year 2022/23_"Poetics of Portals"

During the school year of 2022/23, students took part in the design competition organised by the Faculty of Architecture and the "Hormann" company. The main topic of the competition was to develop creative and unique ideas for implementing various construction elements like doors and windows into housing or public function buildings, even in open public spaces. Although it has long been clear that cooperation between academia and practice is helpful and necessary in education, it rarely happens. In this case, the initiative was made by the German company "Hormann", which specialised in the highest quality industrial doors. This collaboration resulted in the student competition. The students chose the jury: Prof. Zoran Djukanović, UB Faculty and Arch. Maja Lalić, since the topics were urban recycling of industrial heritage, new housing and building structure and façade forms, and social activism. Representatives of Hormann, director Bojan Simović and Arch. Nina Mijatović was also included in the jury from the aspect of expertise in practice. Students were divided into ten groups representing unique design concepts on the abovementioned topics (Fig. 27.8). The first-prized work of Luka Stevanović offered his software that could enable users to design their buildings. The great effort of the entire group of students on the course and the extraordinary enthusiasm created by the competitive spirit were noticed. The event of the competition followed the student exhibition at the Faculty of Architecture (Figs. 27.9 and 27.10), and it was covered and promoted by different local media and social media.

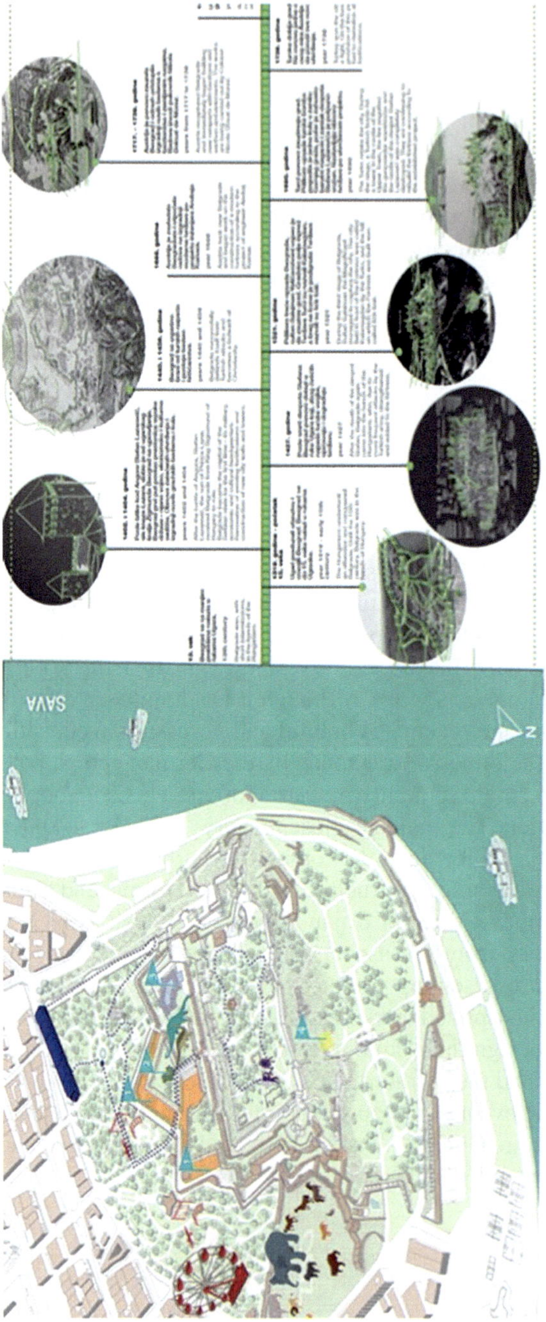

Fig. 27.5 Student research, historical timeline and mapping as part of the publication "Belgrade Fortress/Heritage as a Potential" in collaboration with Marija Kupresanin (Author: Jelena Marić)

Fig. 27.6 The series of podcasts, students: Lazar Dimitrijevic, Lazar Petrovic, Ana Jovanovic; with Dušan Letić (EXIT festival, Novi Sad, Serbia) (Author: Eva Vanista Lazarevic, Jelena Marić)

Fig. 27.7 Podcast episodes posted on YouTube; students: Lazar Dimitrijevic, Lazar Petrovic, Ana Jovanovic; (Author: Eva Vanista Lazarevic, Jelena Marić)

27.4.6 The Student Design Competition, the School year 2023/24_"AI and Architecture Branding"

In the school year of 2023/24, students took part in another organised student design competition also for the abovementioned German company "Hormann" for the second year in a row, based on the success of the student competition for architectural design and concept in 2023. In 2024, although the concept of the competition stayed the same, the topic changed. The mentors wanted to emphasise the importance of branding in architecture and the importance of project presentation for the students. Hence, the task for the students was to develop a unique branding and marketing campaign with creative ideas and concepts for the construction material and mainly the quality door of the Hormann company. Students developed many successful ideas

Fig. 27.8 Student with mentors in a visit to the Hormann company in Belgrade 2023, in collaboration with Bojan Simović and Nina Mihajlović (Author: Eva Vaništa Lazarevic, Jelena Marić)

for the company's promotion based on the specific user groups, the specific quality of the brand, or the possible implementation into architectural buildings. In addition to the design proposals, students had another requirement—to use the "Open AI" tools of artificial intelligence that were available for students. All of these requirements were challenging for students but also quite interesting, and they managed to develop a set of exciting and usable marketing campaigns based on the identity of the brand and modern technology and social media. Students prepared an exhibition alongside the presentation (Fig. 27.11). The jury consisted of Associate Professor Ivan Lalić, Assistant Professor Ivana Rakonjac, and the Representatives of Hormann, given three prizes and two honourable mentions (Fig. 27.12).

27.5 Discussion and Conclusion

This paper presents the potential benefits of creative education, which is one of the crucial prerequisites for future teaching methodology, especially in architecture, where creative projects and creative thinking are expected from its students. In order to generate these ideas and concepts, we must adapt to current global issues and transform the way we teach to be competitive in addressing the collective attention deficit disorder caused by today's modern and hectic way of life. Regarding the presented example of outcomes from an elective course at the University of Belgrade—Faculty of Architecture, we can conclude that bringing a new and edited curriculum each year can result in gathering new ideas and engaging students more qualitatively. Students were highly motivated and demonstrated better overall activity and exceptional teamwork throughout the process of active teaching each school year. However, this approach includes more engagement for students and mentors in the class organisation and preparing collaboration with different stakeholders from

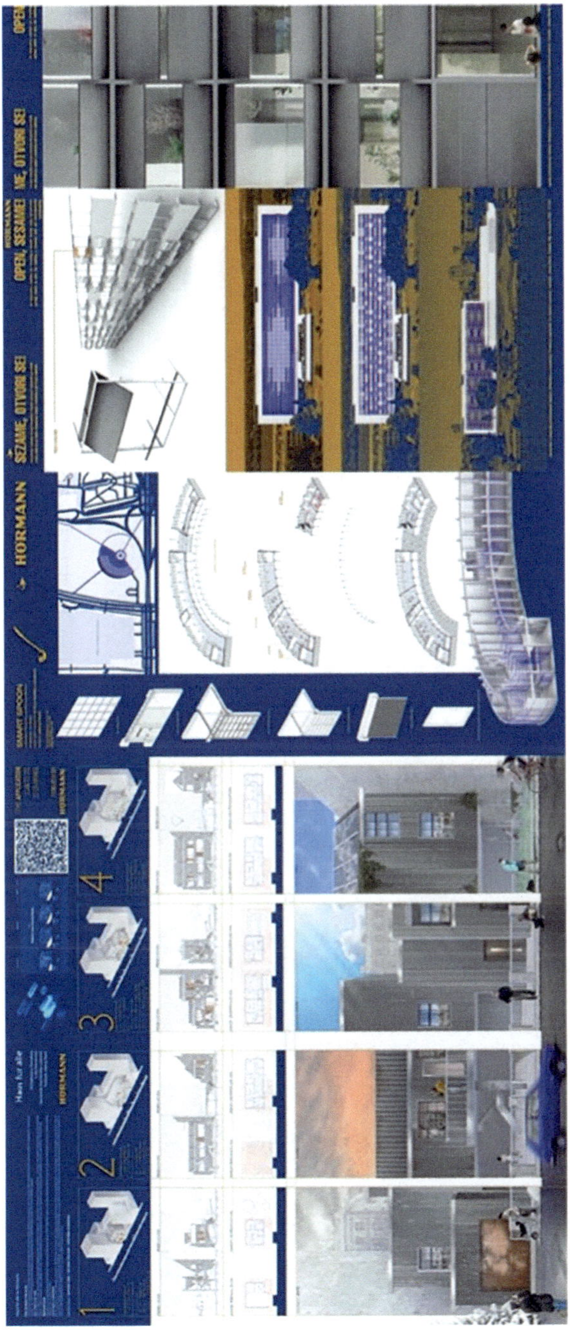

Fig. 27.9 Part 1 Winning concepts of student work for the design competition, 2023, 1st prize: Luka Stevanovic, second prize: Jasmin Rastic, Simona Radojević, Marija Itov, third prize: Illija Jovanović (Author: Eva Vaništa Lazarevic, Jelena Marić)

Fig. 27.10 Part 2Winning concepts of student work for the design competition, 2023, 1st prize: Luka Stevanovic, second prize: Jasmin Rastic, Simona Radojević, Marija Itov, third prize: IIlija Jovanović (Author: Eva Vaništa Lazarevic, Jelena Marić)

Fig. 27.11 Posting the exhibition of student work for the design competition "AI and architecture branding", 2024, 1st prize winners Aleksa Vučković and Sonja Vuković giving the interview for national media; (Author: Eva Vanista Lazarevic, Jelena Marić)

architectural practice and public institutions. This paper could contribute to future research in transforming educational methods and approaches in the architectural context, making education more based on creative thinking and approaches.

The paper highlights the contemporary challenges facing academia, particularly in architecture education, amidst technological advancements, societal changes, and the disruptions caused by the COVID-19 pandemic. It underscores the importance of enhancing student engagement and enthusiasm in university-level education, especially within the architecture discipline.

The methodology employed in this study involves thoroughly exploring theoretical frameworks and analysing outcomes from a specific course, "Architecture in Context," offered at the University of Belgrade, Faculty of Architecture. This elective course is a case study illustrating how innovative teaching approaches can positively impact student motivation and educational outcomes.

In conclusion, this paper advocates integrating creative educational approaches as essential for future architecture teaching methodologies. The case study of the "Architecture in Context" course demonstrates the tangible benefits of embracing innovation and collaboration in higher education.

Moving forward, universities and educators must continue to adapt and evolve their teaching methodologies to meet the needs of students in an ever-changing academic landscape. Educators can inspire future architects to envision and create beautiful, sustainable cities by prioritising creativity, technology integration, and interdisciplinary collaboration. This emphasis on creative education not only enhances student motivation but also equips them with the skills and mindset needed to tackle complex challenges in architecture.

Fig. 27.12 Winning concepts of student work for the design competition, 2024, 1st prize: Aleksa Vučković and Sonja Vuković, second prize: Marko Stajkovac and Nikolina Đuričković, third prize: Mina Matković, Danijela Pantić, Ognjen Bojić and Mila Jeremić (Author: Eva Vanista Lazarevic, Jelena Marić)

References

Aljughaiman A, Mowrer-Reynolds E (2005) Teachers' conceptions of creativity and creative students. J Creat Behav 39:17–34

Bell S (2010) Project-based learning for the 21st century: Skills for the future. Clearing House J Educ Strat Issues Ideas 83:39–43

Babbar M, Gupta T (2021) Response of educational institutions to COVID-19 pandemic: an inter-country comparison. Policy Futures Educ. https://doi.org/10.1177/14782103211021937

Borhani A, Kalantar N (2017) Advancing community design education. In: Proceedings of international conference on engineering and product design education. https://www.designsociety.org/group/18/Search+DS+publications

Collins MA, Amabile TM (1999) Motivation and creativity. In: Sternberg RJ (ed) Handbook of creativity. Cambridge University Press, Cambridge

Combrinck C, Venter R (2020) Embedded participation in the architectural curriculum towards engendering urban citizenship in graduates. Image Text 34:1–29. https://doi.org/10.17159/2617-3255/2020/n34a2

Craft A (2001b) An analysis of research and literature on creativity in education. Qualification Curriculum Authority. http://www.ncaction.org.uk/creativity/creativity_report.pdf

Djukić A (2021) Обликовање отворених јавних простора (Design of open public space). Faculty of Architecture, Belgrade

Hadzigeorgiou Y, Fokialis P, Kabouropoulou M (2012) Thinking about creativity in science education. Creat Educ 3:603–611. https://doi.org/10.4236/ce.2012.35089

Oh J, Minho S (2022) Evaluation of citizen-student cooperative urban planning and design experience in higher education. Sustainability 14(4):2072. https://doi.org/10.3390/su14042072

Le T (2018) Project-based learning in 21st century: a review of dimensions for implementation in university-level teaching and learning. In: 4th ICEAC international conference on english across cultures. Bali, Indonesia

Lin Y (2011) Fostering creativity through education—a conceptual framework of creative pedagogy. Creat Educ 2:149–155. https://doi.org/10.4236/ce.2011.23021

Skaalvik E, Skaalvik S (2016) Teacher stress and teacher self-efficacy as predictors of engagement, emotional exhaustion, and motivation to leave the teaching profession. Creat Educ 7:1785–1799. https://doi.org/10.4236/ce.2016.713182

Shaheen R (2010) Creativity and education. Creat Educ 1:166–169. https://doi.org/10.4236/ce.2010.13026

Tato B (2021) Making participation relevant to design. Cambridge, MA, Harvard. https://www.gsd.harvard.edu/course/making-participation-relevant-to-design-spring-2021/

Tapinos E (2016) The limitations impacting teachers' understanding of creative thinking. Creat Educ 7:1404–1409. https://doi.org/10.4236/ce.2016.710145

Chapter 28
The DANUrB Experience: Combining Digital Tools and Analogue Actions to Create a Network of Communities Connected by the Cultural Heritage of the Danube

Bálint Kádár

Abstract Stakeholder and community involvement in tourism planning and heritage management are relatively new practices. Still, new ICT tools for cooperation and inclusion open up ample opportunities and, therefore, many published cases in these fields. When heritage and tourism are the basis of creating a sizeable cultural ecosystem, such digital tools are indispensable, as ICT effectively connects parties in very different geographic locations. However, how effective are the online community involvement tools for distant stakeholders' cooperation among abstract cultural notions? Can online platforms and social media replace live interactions? The DANUrB Interreg DTP project aimed to create a transnational brand connecting the settlements along the Danube by demonstrating their unique yet similar cultures, united by the cultural flows along the river in history. The project worked in 7 countries, reaching 10,000 people on site, while only on Facebook more than 5 million page visits have been registered. The various online and analogue tools used in the project permitted the analysis of the usability of the different methods. While digital tools were essential to organise data and work for project partners and to provide news and generate interest for a broad audience, local stakeholders could only be engaged by analogue participatory techniques, always requiring personal contact with project partners to be actively involved in the co-creation process of a new brand, in tourism destination building and heritage management.

Keywords Community engagement · Heritage management · Tourism destination · Digital tools · Participatory tools

---

B. Kádár (✉)
Budapest University of Technology and Economics, Budapest, Hungary
e-mail: kadarb@urb.bme.hu

M. Vukmirović et al. (eds.), *Digital Dreams, Urban Realities*, The Urban Book Series,
https://doi.org/10.1007/978-3-031-76353-3_28

28.1 Introduction

28.1.1 The Challenge of Using Large Cultural Ecosystems to Enhance the Potential of Localities

Cooperation is beneficial for the common good of nearly any community. Many studies try to understand which social rules and actions can help a group act for the common good (Katz et al. 2021). Such studies focus on different social groups or small communities, but what about larger territorial or cultural systems where the study group does not even have a chance to meet in reality? Cooperation between institutions, governmental bodies or interest groups can be understood even in large cultural ecosystems like tourism (Baggio 2011; Studzieniecki and Soares 2017; Weidenfeld et al. 2011). In tourism or cultural cooperation, even the interregional or cross-border scale is well understood by major stakeholders, like regional tourism organisations and major regional service providers (Bjeljac and Curcic 2006; Rădoi 2020; Radu et al. 2010; Vujko and Gajic 2014). However, local community members are not directly involved in regional or interregional cooperation. Interregional or large-scale cultural cooperation is complex for these individuals and local stakeholders. How could members and stakeholders of a local community be part of a large-scale cultural ecosystem if the basis of such lies in abstract cultural ideas? How can institutions and activists promoting an interregional ecosystem and its cultural values reach out with their agenda to local communities and convince them that cooperation at such a meta-level benefits their localities? How do old-fashioned analogue communication techniques perform concerning digital tools created to reach distant audiences? The DANUrB project provided evidence of the techniques used to involve local communities in the significant cultural ecosystem defined by the historical connections along the river Danube.

28.1.2 The Potential of the Danube as a Transnational Cultural Promenade for Communities

Universities along the river Danube started cooperating and gaining EU funding in the first and second calls of the Interreg Danube Transnational Programme under the name DANUrB. In 2017, universities partnered with municipalities, NGOs, tourism organisations, and other professional institutions from seven countries to create the foundations for a DANube Urban Brand. The main idea behind this cooperation was the creation of a Danube Cultural Promenade, where the settlements along the river could cooperate by finding similarities in their rich cultural heritage, tangible and intangible.

The DANUrB cultural network aimed to strengthen the Danube regional cultural identity and to create a common brand by fostering transnational cultural ties between the settlements along the Danube and by exploring the unused or hidden cultural and social capital resources for a better economic and cultural return (Benkő et al. 2019). The main goal of the project is to create a comprehensive spatial-cultural network, a "European cultural promenade" connecting all communities along the Danube, unifying these into one tourism destination brand, offering thematic routes and development possibilities that can increase the number of visitors and can prolong their stay in the region. DANUrB aimed to reveal the underused cultural heritage and resources (industrial heritage, the heritage of the communist past, fortifications, traces of commerce, living cultural traditions) in towns, organised spatially by the Danube, connecting these to form viable cultural collaborations and thematised tourism products according to a spatio-cultural strategy, that re-brands previously neglected local resources to a unified cultural destination, bringing better access to culture for locals, and better economic outputs based on the increased interest of visitors, finally reaching the unexplored small towns from the large cities of culture to along the river. This project mapped more than 500 hidden heritage assets, working in 12 towns or twin towns from both sides of the river, contacting 300 stakeholders, creating an online guided tour application for 12 tours, a strategy, a platform, and many events gathered along the notion of the international Danube Days. The partnership led by the Department of Urban Planning and Design of the Budapest University of Technology and Economics found a balance to work scientifically and planning-based both at an interregional scale and the local urban scales and to work on-site with local stakeholders, gathering their needs and potentials.

Out of the findings of this project, the second application was written and won under the name DANUrB+. Between 2020 and 2022, the partnership focused on the underprivileged communities by the river, excluding the prosperous cities and including only shrinking cities and regions. The new project aimed to reactivate underused cultural heritage and resources in shrinking settlements of the Danube River's peripheral and border regions to create new possibilities to make its towns and regions attractive again. Initial DANUrB tools were found to be most useful in peripheral and border regions, where the decaying socio-economic situation can only rely for development on the special resources of the Danube. The mapping of the shrinking situation of these regions was done by the project partners at the Belgrade University Faculty of Architecture (Djukić et al. 2022). DANUrB+ created a dense network of stakeholders and projects along the Danube, implementing EUSDR actions in the peripheral and border regions along the river, which could count only on the potential of the Danube to stop socio-economic shrinkage. The main objective was the capacity building for local stakeholders to enable them to cooperate locally and interregionally to valorise their Danube-related heritage with local actions under a unified brand strong enough to increase local prosperity and international tourist attractiveness. The novelty of DANUrB+ was to bring down-to-earth strategical goals to real stakeholders with action plans and actions usable in peripheric situations parallelly in all sections of the Danube and to brand these initiatives in inclusive and effective ways.[1]

[1] https//www.danurb.eu

The main achievement of these projects came from their main challenges: how to effectively work with communities that are very distant from each other and are fueled by the same concept, aiming to connect them into a large-scale cultural ecosystem. Eighteen partners worked parallelly in 12, later in 24 localities, using a unified methodology and the same tools from the Delta until the Slovakian-Hungarian borderline. DANUrB managed to actively involve many stakeholders and move entire communities into action for the cultural values of the Danube, disseminating ideas of the benefits of transnationality in Danubian culture.

The partnership and the researchers working on the projects could try out many methods of community involvement and transnational networking while trying to build an interregional cultural promenade. Many digital tools have been used and developed specifically, as well as on-site personal methods of stakeholder involvement, community building, and place-making. Which tools are the most suitable for building such a large cultural ecosystem? The answer lies in the comprehensive usage of analogue and digital tools.

28.2 Literature Review

28.2.1 Interregional Cultural Networks and Tourism Systems

Large cultural systems are often organised based on their cultural offers to visitors; therefore, they are tourism systems. Large-scale tourism destinations have fewer studies and literature than urban or smaller-scale rural destinations, as these are harder to define if the components of such tourism systems are settlements with their own population and cultural heritage, tied together by some cultural construct; nature-based tourism is more often analysed as a destination system in large scales (Orsi and Geneletti 2013). Most large-scale cultural destinations are defined by natural elements, such as Yellowstone Park in the US (Mings and Mchugh 1992). However, in large-scale natural destinations, like the Alps (Tissot 2011) or the Pyrenees mountain region (Blasco et al. 2014) in Europe, tourism combines natural and cultural heritage. Such large-scale destination systems are cross-border regions where the coherence of the tourism system is compromised by the boundary shielding effect (Peng et al. 2016).

Such destination networks were first analysed as multi-destination pleasure trips (Lue et al. 1993), and many studies defined territorial systems where tourist mobility defined the structure of the destination system (Asero et al. 2016; Kádár and Gede 2021). Other network analyses focused on functional aspects where some settlements of attractions show more centrality than others based on their amenities and services (Lee et al. 2013). However, few similar studies analyse the network of stakeholders working together in a large region for a well-functioning unified tourism destination (Scott et al. 2008). A large-scale tourism destination can become a cultural network or, even better, a cultural ecosystem only if a network of stakeholders or entire communities can define themselves as cooperating parties for an interregional cultural goal.

28.2.2 Analogue Tools for Stakeholder and Community Involvement

Not all community engagement practices are equally meaningful to the community. The tools for stakeholder or community involvement can be categorised as follows according to the Edinburgh Beltane—Beacon for Public Engagement (Faulkner 2011), from the one-way communication towards more inclusive participatory tools in 5 levels:

Informing

One-way communication, like press releases.

Consulting

Collecting the opinion of stakeholders, but without granting engagement in the process, like in a public hearing or with data collecting surveys.

Involving

Two-way interaction, working actively with stakeholders, like in participatory workshops, festivals, and urban walks, organised centrally, but to involve the community.

Collaborating

Cooperation with stakeholders in the development of a project, involving their skills and goods to co-create a project. Joint working groups and committees, a network-building process for cultural ecosystems, the creation of new products, including the services of different stakeholders, or festivals co-organized by the owners/inhabitants of a site are examples.

Empowering

Delegating the final decision-making to stakeholders, such as creating a local cultural or tourism ecosystem with an ownership and management structure that fully consists of members of the local community.

Higher levels of engagement are more difficult to achieve in areas where professional guidance and higher levels of knowledge are needed and where the community does not often meet physically. Fields where stakeholder involvement was a challenge were heritage management and tourism (Aas et al. 2005). Public participation in tourism planning started to become a practice in the early 1980s, and gamification techniques were already used for community involvement (Loukissas 1983). Community-based and planned tourism destinations became a positive way to develop the visitor economy (Haywood 1988; Jamal and Getz 1995; Murphy 1988). Planning tourism destinations and ecosystems requires a negotiation between involved stakeholders, each with a professional or motivational connection to tourism.

Community involvement in heritage management became a topic much later, only at the end of the 1990s. Before, the strong professional competencies needed to work with heritage (archaeology, museology, building and conservation skills) excluded local businesses, communities, or even municipalities from this field. As the set of heritage assets recognised as valuable became wider, the role of the community who can relate to it locally became important, as stakeholders involved in such processes could help the conservation and management of such sites where the state or large organisations did not have enough capacities (Hodges and Watson 2000). This became especially true and beneficial in the case of small heritage sites (Grimwade and Carter 2000). The role of communities in heritage management became widely acknowledged only in the past decades (Deacon and Smeets 2013).

Heritage must be managed properly to bring benefits, these can be not only economic but also social, environmental and cultural, but these benefits will be sustainable only with a participatory approach in the management procedures themselves (Hribar et al. 2015). Recent studies highlight the challenges facing heritage tourism operators and recommend that these organisations focus on inter-stakeholder group collaboration, participative decision-making, responsibility and benefits sharing, and building an institutional power structure to involve hosting communities for sustainable operations (Li and Hunter 2015).

With appropriate methods, not only participatory management but also co-creation of heritage valorisation is possible (Grcheva and Oktay Vehbi 2021). Cooperation between archaeologists and aboriginal people brought the valorisation of culturally significant landscapes to locals even when there were few tangible archaeological artefacts (Prangnell et al. 2010).

Cultural festivals are the most widely used tools for community involvement in heritage sites, and such festivals can lead to the co-creation of heritage-related branding that is useful also for a tourism destination (Kádár and Klaniczay 2022).

28.2.3 Digital Tools for Stakeholder and Community Involvement

With the diffusion of ICT and the internet, new possibilities emerged in community engagement, promising tools which can easily involve and even engage anyone in a process online. Even though the optimism for the use of new online participatory tools was evident even among scholars (Evans-Cowley and Hollander 2010), early diffusion of digital tools for community involvement was impeded by the digital divide (Rogers 2001) bought by the indifferences in accessibility to digital tools and by the widespread digital illiteracy. Even though the accessibility to the internet and digital tools has become widespread, there is still a digital divide in all societies, and it is not decreasing due to many reasons, not only educational and infrastructural but also motivational ones (Helsper and Reisdorf 2017).

The promise for improved involvement was strong in tourism and heritage management, where new tools and travel motivations were born (Bohlin and Brandt 2014; Jacobsen and Munar 2012). Social media usage opened new research topics and possibilities for tourism (Kádár 2014; Kádár and Gede 2013; Shoval and Isaacson 2007), but also new consumption models (Kádár and Gede 2022; Weaver 2008). Social media opened new possibilities in heritage management, like community involvement in the digital mapping of heritage assets, but also posed new challenges to cultural institutions and museums (Giaccardi 2012). By simply making digital content, people will not engage more. In fact, digital tools in culture, like the digitalisation of museums and cultural heritage, can further increase the inequalities in cultural consumption, excluding social groups like the elderly (Gallistl et al. 2020; Mihelj et al. 2019).

While communication and sometimes consultation are effective through ICT, higher levels of involvement remain a challenge. Using information technology and online games to involve communities in design questions resulted in some successful examples of participatory design when used in joyful and interesting ways (Hanzl 2007; Poplin 2012; Westerberg and Fanny 2015). In fact, literature shows how gamification helps achieve active involvement even online (Hassan and Hamari 2020; Thiel et al. 2016; Thiel and Fröhlich 2017).

28.3 Methods and Tools

28.3.1 Digital Tools Used in DANUrB

The DANUrB project realised from the start the need for ICT to undertake a mission in such a large geographical area, but also because digital tools in tourism destination management were considered essential for a successful valorisation process. The COVID-19 pandemic hit the second project period hard, but the partnership was already in full control of online communication tools. Therefore, transnational work could continue. To create a Danube Cultural Promenade, the DANUrB projects used the following online tools:

28.3.1.1 Interior Communication Tools—Collaborating with and Empowering Partners

The online communication tools inside the partnership proved essential, especially in the second project period when the partnership had to start DANUrB+ during the COVID-19 pandemic. Even before, the project had a well-organized Google

Workspace area, where all information was accessible in a structured and up-to-date way, and project partners communicated through online meetings on platforms like Skype, but only for one-to-one meetings. The kick-off meeting of DANUrB+ had to be held online through the Microsoft Teams platform; only this way, all partners and stakeholders could meet and start the work. Online meetings have proven effective to some extent; however, the main topic has always been speculation about the first possible date for a real-life partner meeting. After the lockdowns of COVID-19, the partnership kept the online Teams meetings as communication tools between personal partner meetings. Therefore, the number of large-scale meetings in DANUrB+ exceeded those in the first DANUrB project.

28.3.1.2 Project Communication Tools—Informing

The regular communication channels consisted of the Interreg project webpage, the DANUrB webpage, and regular newsletters. These provided general information, but the newsletters became an important channel for reaching target groups and stakeholders, especially to invite them to in-person events.

28.3.1.3 Social Media Tools—Informing with Some Involvement

The project used mainly Facebook pages, Instagram, and Twitter.

The project's Facebook page was given a special priority, as this was the most effective communication tool to connect stakeholders and communities interested in the project by posting regularly about cultural programs and heritage-related information from all over the Danube. Many from outside the project followed DANUrB on Facebook, as it became the Facebook page with the most multicultural content from the river Danube, covering news from Germany until the Delta, from 9 countries. The largest achievement of the DANUrB Facebook page lies not only in its 5400 followers and over 1.3 million page visits but also in the fact that it continues to be active with new regular content even after the project closes (Fig. 28.1).

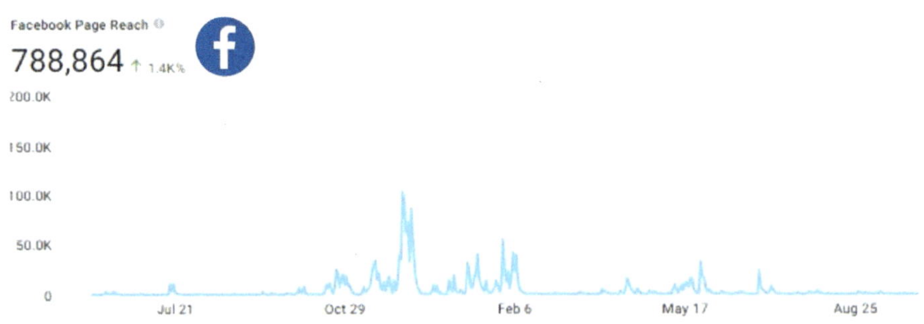

Fig. 28.1 DANUrB Facebook page reach between 2020 and 2021

28.3.1.4 Stakeholder Involvement Tools—Consulting

The most interactive online tools have been utilised to personally involve stakeholders in co-creating the large-scale ecosystem and local heritage valorisation projects. Stakeholder surveys, mainly online questionnaires, have been used to get data from target groups, but the main stakeholder involvement tool also became a communication and research tool; this was the DANUrB Platform.

28.3.1.5 Project Platform—Involving (and Informing)

An innovative side-project for cultural ecosystem management was the creation and constant development of the DANUrB Platform. This is a map-based online database and workspace with a user management system that allows different access levels to different user groups. Stakeholders from 7 countries have been registered to this platform, curated by the University partners of DANUrB. The web-based interface of the platform allowed ease of use; therefore, any stakeholder could upload new heritage items and create new projects related to such heritage items by simply placing these, connecting them on a map, and filling out forms. An openly accessible version of the DANUrB Platform was made available on www.danurb.eu, making it the largest available dataset on the linear spatial system of the Danube, showing more than 1000 heritage and stakeholder data points all related to the cultural ecosystem defined by the river (Fig. 28.2).

28.3.1.6 Tourism Marketing Tools—Informing

DANUrB, as a project, didn't develop from scratch apps or info pages targeting tourists, as many projects doing so failed to reach relevant numbers of tourists this way. The project instead included PocketGuide as a project partner and an established service provider for tourists. The PocketGuide app is a map-based audio tour offering guided tours along predefined routes in hundreds of cities on all continents. The partnership developed tours for 15 destinations along the Danube, offering these tours for free to anyone. Small towns or twin towns connected by bridges have never been included in such widely used online tourism services (Fig. 28.3).

28.3.1.7 Research Tools—Consulting and Informing

DANUrB academic partners created online databases for the settlements of the Danube, gathering data on their heritage, socio-economic and morphological conditions and history, and the behaviour of travelers visiting different destinations along the river. Research results have been published in the DANUrB Atlas, DANUrB Book, and scientific papers (Fig. 28.4).

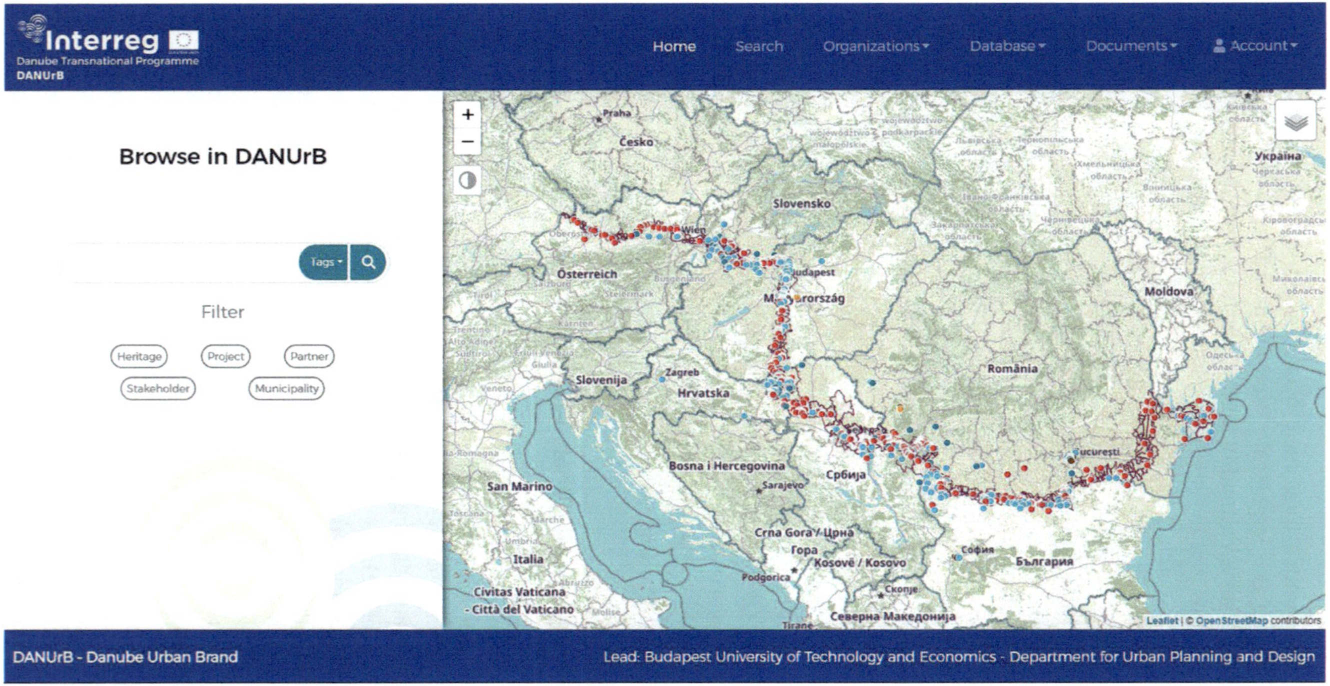

Fig. 28.2 The DANUrB Platform, an interactive tool gathering all heritage and stakeholder data from along the Danube,allowing stakeholders to interact with local heritage resources and each other

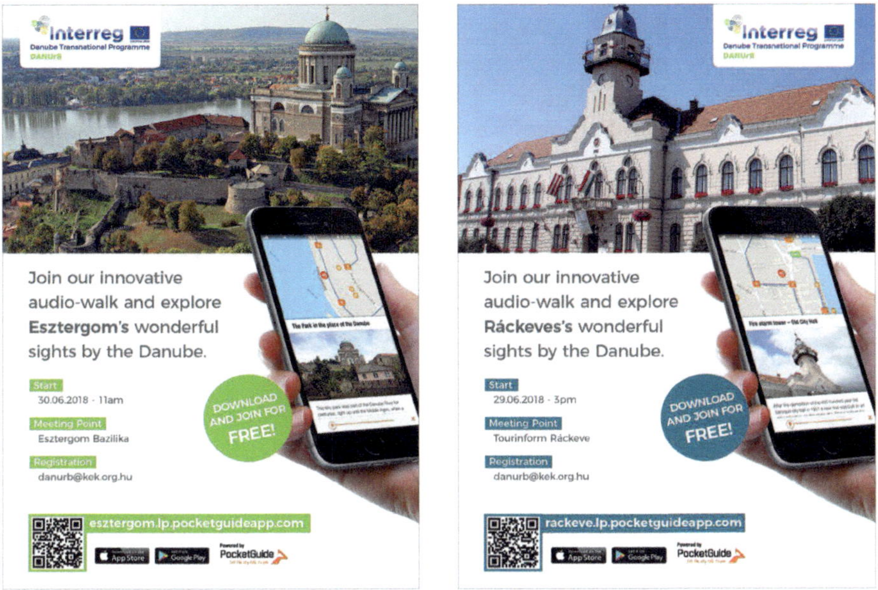

Fig. 28.3 The promotional material of two of the PocketGuide tours,available for free on the global tourist platform

28.3.2 Analogue Tools Used in DANUrB

The project built heavily on online tools, partly because only these guaranteed the transnational communication channels needed in the 7 countries involved and partly because of the effectiveness in reaching many people simultaneously. However, the whole work methodology was based on an involvement strategy defining how to create heritage valorisation projects locally in 12 action areas simultaneously along the same values coherent to the Danube Urban Brand (Fig. 28.5). Such methodology could only by carried out by personal interaction between project partners and stakeholders.

Therefore more than 70 events were held during the first DANUrB and 150 events during the second DANUrB+ project period (Fig. 28.6). 10,000 people have been reached personally all along the Danube, which is only a fraction of those reached online, but the live interactions guaranteed the full understanding of the DANUrB's goals, and many of these events served as workshops and activities with real stakeholder involvement and co-creation of project results.

The main analogue techniques for community and stakeholder involvement were the following:

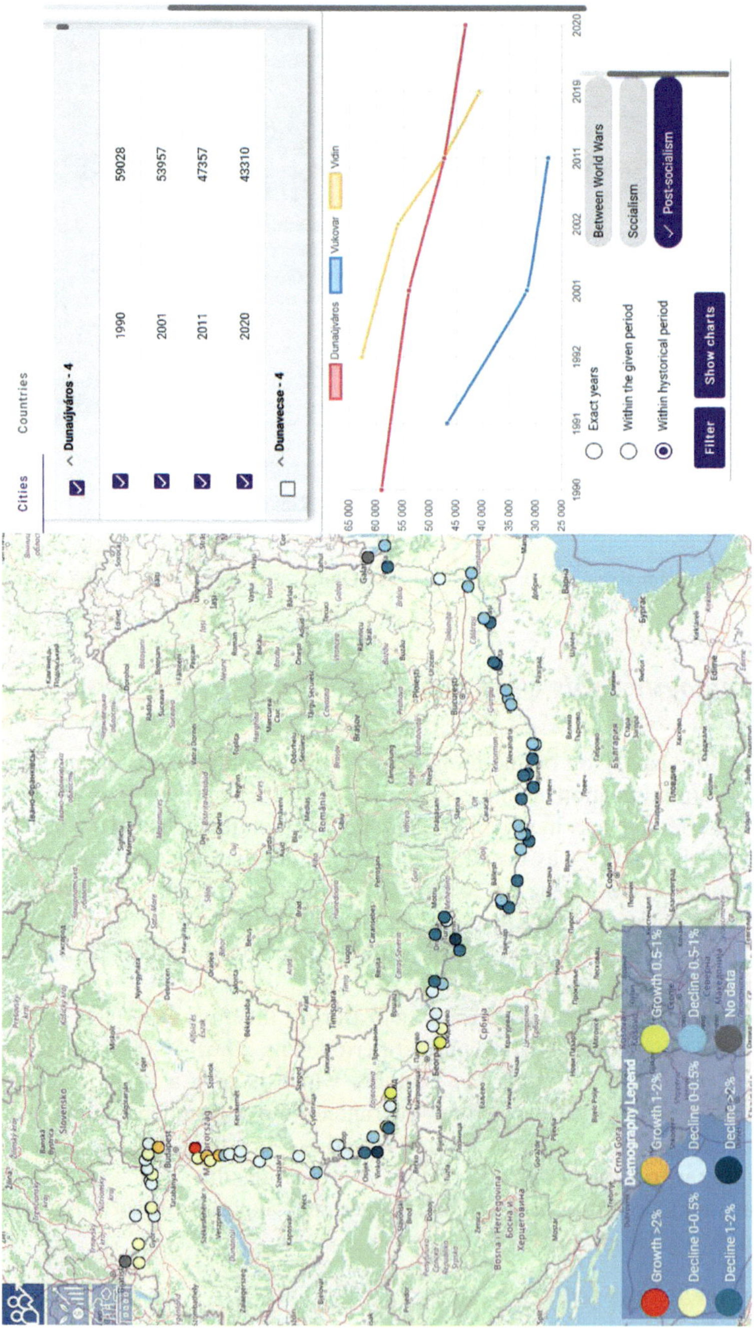

Fig. 28.4 Shrinking cities in the DANUrB Digital Atlas (compiled by UBGD-FA)

Fig. 28.5 Involvement strategy presented by project partner KÉK

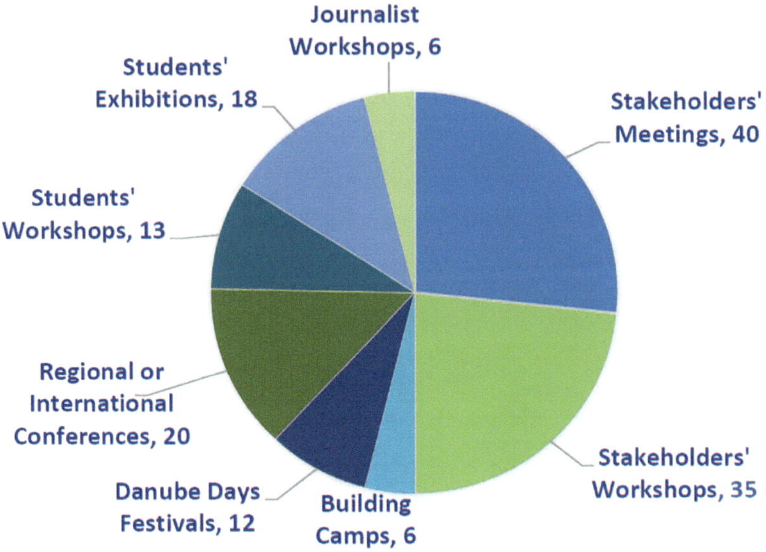

Fig. 28.6 DANUrB+ live events between 2020–2022 (compiled by NTC BG Guide)

28.3.2.1 Festivals and Other Cultural Events—Involving and Collaborating

Festivals and cultural events are good community tools to involve people in abstract cultural topics, and the common celebration of the Danube-related cultures was such a topic. The Danube Days Festivals became a brand during DANUrB, held in various towns in many countries, at the same time between June and July each year—the most successful celebrations became local traditions, celebrated even without the project's contribution since initiation. Cultural festivities have been successful tools in settlements for informing (showing the public that the Danube is a valuable part of their lives with rich cultural offers), for consulting (workshops and surveys have been conducted during the festivals), and for involving (the programs have been put together to connect the local population with their Danube related heritage and to make them participate actively in such cultural events, and also they involved many local initiatives to give them opportunities to show themselves to the public). But DANUrB applied a methodology to make festivals tools for collaboration, as the Danube Days were never totally new constructs but an organisational boost for smaller local cultural initiatives under a new umbrella brand to upscale their cultural activities and festivals. An example is the first Danube Days festival connecting Esztergom (HU) and Sturovo (SK), celebrating the common culture of the two towns on the opposite banks of the Danube by uniting many local initiatives and small events into a two-day long event (Fig. 28.7).

Fig. 28.7 Danube days festival from 2022 in Vidin, BG

28.3.2.2 Exhibitions and Conferences—Informing and Consulting

Exhibitions of projects and conferences on Danube-related cultural topics had the major task of informing a more refined audience on specific or professional topics. However, these were also events aiming to be platforms for active consultation with stakeholders from different fields on the project topics. The DANUrB Strategy and Action plans, and also the local heritage valorisation plans, have been consulted with local and professional stakeholders during exhibitions and especially conferences (many times tied together), inviting local or expert presenters to make an exchange of ideas and therefore to develop DANUrB results accordingly (Fig. 28.8).

28.3.2.3 Educational and Training Programs—Empowering

The DANUrB educational programs had a special place among the project activities and outputs, as these were meant to empower local young people and local stakeholders to use the potential given by Danube's cultural ecosystem and heritage to act locally for their own and their community's prosperity. Two target groups have been defined. First, the young students who learn in primary and secondary schools in small and medium towns by the Danube, and second, the local entrepreneurs and stakeholders who provide a service in such towns. Young people can become local entrepreneurs and cultural stakeholders in their original communities only if they learn about their regions' potential, giving them a value proposition that can compete with prospects from larger cities or other regions. Local entrepreneurs and cultural stakeholders can find prosperity options in their activities for themselves and their communities only if they learn about advanced and specialised methods

Fig. 28.8 The 2017 DANUrB traveling exhibition in Komarno, SK

to provide better services based on the valorisation of their local heritage and the internationally competitive values it has in a trans-Danubian context. Twelve schools from 6 countries were involved in the DANUrB school programs, and 12 training programs have been conducted on tourism branding, business development and local identity for local stakeholders, including entrepreneurs (Fig. 28.9).

Fig. 28.9 DANUrB educational program in Kolarovo, Slovakia (NDC and STU)

28.3.2.4 Student Workshops and Building Camps—Collaborating and Empowering

Student workshops targeted primarily architecture and planning students of the Danube region, empowering them to act responsibly in small and medium cities in the cultural context of the Danube. This way, students—usually trained in fictive or urban contexts—learnt about the geographical, cultural and socio-economical context of smaller settlements along the Danube and learnt to plan and design in such contexts. Students have also been thought to collaborate with local stakeholders in these tasks, as these workshops always had an on-site workshop program with participatory design and planning elements. Students have finally been collaborating with other students from other countries and working on similar topics, as many international workshops have been organised, forming mixed groups of students from the countries of the Danube region. These workshops secondarily targeted these regions' communities, as students practised the participatory methods to involve them, especially relevant in the building camps where students had to design and implement urban furniture and public space improvements for communities collaborating with them.

28.3.2.5 Stakeholder Meetings—Consulting and Involving

To assess the local needs and resources and target the involvement of stakeholders, DANUrB organised 3–4 stakeholder meetings in each action area. These meetings selected the final circle of stakeholders capable of undertaking cultural valorisation projects (Fig. 28.10).

Fig. 28.10 Ráckeve 2022: Building camp, a participatory planning workshop with stakeholders, and the Danube days festival combined to ensure maximum community involvement in the valorisation project of a former clubhouse by the Danube

28.3.2.6 Workshops with Stakeholders—Involving, Collaborating and Empowering

The project aimed to co-create culture-based projects with local stakeholders, and the workshops involving them were crucial parts of the methodology. These served as a collaborative platform for the creation of new project ideas and project action plans, empowering stakeholders to organise themselves even after the project implementation.

28.4 Results and Discussion

28.4.1 The Benefits and Limits of Digitalisation

Communication was always the main bonding glue for cultural ecosystems; the exchange of knowledge and the sharing of commonly used solutions were only possible between communities that were connected. Large cultural ecosystems were made possible by large infrastructures in history; the fleets of triremes and quadriremes on the Mediterranean Sea and the network of built roads in the Roman Empire were the most effective communication channels in early history. The Danube provided a similarly effective waterway that connected goods and cultures across different empires, building cultural connectivity still visible in the era when highways and aeroplanes redefined the geography of effective ecosystems. This system has been torn apart by the geopolitical processes of the twentieth century, leaving communities by the Danube isolated from each other by borders and their cultural identities threatened by global influences. However, in the twenty-first century, the authentic cultures and heritage values once connected by the river have the potential for an outstanding tourism system that could reconnect the region and revive the cultural ecosystem.

As shipping became the slowest connection method, the limitless connection speed of ICT tools became the main promise to re-connect this valuable historic cultural ecosystem along the Danube. Digitalising and sharing online the common values of cultures defined by this river was the main tool for the cultural ecosystem creation of the DANUrB Interreg projects. Online databases shared by the website and the DANUrB Platform, globally accessible tourist guidance all along the river on PocketGuide, and an active social media channel on Facebook connecting communities along the river provided ICT tools that proved to be effective in showing the culture of the Danube as an integrated system in the global space. These tools proved to local communities that they are integrated parts of a large transnational cultural ecosystem, and the number of online visits exceeding a million proved that such an ecosystem generates global interest. ICT tools have proven to be effective communication channels for bonding cultural ecosystems. Therefore, these can be considered as the contemporary versions of trade routes and naval infrastructures, the first great infrastructures uniting cultural uses on a large scale.

The analogy between physical and digital infrastructures of trade and culture has a limit, proven to be real by the DANUrB project. Physical connectivity allows a personal presence that could induce real actions locally to connect different communities into a network, which is essential to involve a critical mass of people from a large geographic space in a cultural ecosystem. Personal action was often military intervention imposed by weapons, but at other times, culture was helped by trade or the personal diffusion of cultural innovation, like in the early diffusion of Christianity. In the digital age, the commodity of ICT tools promises the minimalisation of personal action, offering interactive and cooperative methods online. DANUrB tested many such methods, partly because these promised cost-effective ways to work with local communities and partly because of necessity during the COVID-19 pandemic. Digital surveys resulted in tools for consulting only if someone could convince stakeholders locally to fill these out. Online conferences—like the kick-off conference of DANUrB+ in 2020—have been attended only by project partners directly involved. The DANUrB Platform, an interactive tool for involvement and co-creation perfected in 5 years to be easily usable for project creation by stakeholders without centralised action, proved to be effective only if stakeholders have been personally educated to use it and if projects have been initiated at a personal level during live workshops: only after these data have been filed out, not one project have been initiated by online interaction on this platform connecting hundreds of stakeholders. From the five levels of involvement, ICT tools proved to be great for Informing, and only after the necessary diffusion of online meetings during COVID-19 were these effective tools for Consulting and sometimes for Collaborating, especially between project partners who knew each other personally and used tools such as Microsoft Teams to communicate exactly in the same way as they would do in a personal interaction. Digital tools have never worked for Involving or Empowering stakeholders in DANUrB's cultural ecosystem; however, the project successfully empowered local communities with personal on-site interaction to use the digital database and tools created uniquely for the Danube for better tourism management and communication.

28.4.2 The Responsibility of Professionals to Act Locally

The DANUrB partnership could be effective in giving content and strong local connections to the large cultural ecosystem of the Danube Cultural Promenade because it fostered collaboration and co-creation between the most local and most global actors in Danubian culture. Professionals from municipalities, NGOs and local cultural institutions knew that personal interaction is the key to cultural creation, while academic partners and cultural entrepreneurs working in an international space knew how to position elements of local culture globally for effective heritage management and a united tourism system. While local stakeholders and partners have been empowered to upscale their local actions for larger systems, benefiting from the transfer of knowledge and from the transnational network, the academic and global partners had the biggest lessons learnt. A large-scale cultural and tourism ecosystem can be well funded theoretically, can be derived from centuries of historical connections, and can have the promise to be socio-economically beneficial to all communities taking part if the elements—stakeholders and communities—of the system are not involved in it and are not empowered to keep the cultural ecosystem alive at long term. Such levels of involvement and empowerment are only possible to achieve through local action, and consultation and cooperation are only possible in the online digital space if involvement has been achieved through personal actions. The on-site work of local partners, members of NGOs and local municipalities has been essential to creating a Danube Cultural Promenade, but to make the Involvement and Empowering of these partners possible and to show to local communities the global values and transnational relevance of their culture the global actors—in this case, academics from universities and professionals from the capital cities of the region—had to go on-site and work personally with locals. The creators of PocketGuide tours and the DANUrB strategy had to make on-site consultations, and the tourism experts and university teachers had to conduct in-person workshops and training for locals to empower them. Universities could finally validate real-life theories and teach their students—future professionals—to work in real locations instead of virtual global spaces. The DANUrB project did empower local cultural actors to think globally, but it also empowered professionals with globally renowned knowledge to act locally.

28.4.3 The Complexity of Cultural Ecosystems

The conclusions of the five-year-long activities for a large-scale cultural ecosystem testing digital and analogue tools of community involvement can be summarised as follows: complex cultural ecosystems need complex mixed methods of digital and analogue tools for community involvement and heritage and tourism management. ICT communication tools are essential for project organisation, data management, and the dissemination of information to communities connected by the cultural ecosystem. Heritage management and tourism ecosystem creation can be supported by ICT tools, but these are not enough to create and manage such processes. Higher levels of involvement are impossible to achieve effectively only using ICT; on-site presence is needed even for the simple task of Consulting about the local needs or project development. On-line surveys provide only general data, while on-site consultations may deepen the understanding of local needs and opinions. Involving stakeholders and larger portions of the local community in common cultural processes and local heritage management always needs local presence and participatory methods, first of all, because the idea of belonging to an abstract community defined by cultural uses is too complex to convey by non-interactive communication, and interaction online is not effective in the Involvement phase, when the reality of sharing common values and goals must be set up. The same is true for collaboration and empowerment; these involvement levels need high levels of trust and very intense communication and co-creation processes. Such an ecosystem's network must be based on connections at a deeply personal level.

However, the role of ICT cannot be reduced to Informing when building a cultural ecosystem on a transnational scale. If the community and stakeholders participate in a community sharing common values, the on-site presence of partners who work on such involvement is granted. Collaborating with distant members of the same abstract cultural community is possible through online tools. DANUrB project proved that a meaningful once-a-year personal involvement is enough to maintain good online collaboration between partners and stakeholders. Online tools are essential to give up-to-date information in a geographically balanced way to all members of a large-scale cultural community, and after COVID-19, most stakeholders can collaborate through meeting apps and online workspaces that are refined enough to facilitate all forms of collaboration. A good ICT ecosystem effectively empowers a large-scale cultural ecosystem to act jointly and be present in real-world cultural geographies as an active entity. If a cultural ecosystem has the vital power to maintain personal contact between key stakeholders regularly, and if these stakeholders (partners) are motivated and active enough to involve personally other members of such cultural network regularly, then ICT tools can ensure the reality of a large-scale community even through transnational space.

Acknowledgements The DANUrB and DANUrB+ projects have been co-funded by the European Union ERDF, ENI and IPA funds.

References

Aas C, Ladkin A, Fletcher J (2005) Stakeholder collaboration and heritage management. Ann Tour Res 32(1):28–48. https://doi.org/10.1016/j.annals.2004.04.005

Asero V, Gozzo S, Tomaselli V (2016) Building tourism networks through tourist mobility. J Travel Res 55(6):751–763. https://doi.org/10.1177/0047287515569777

Baggio R (2011) Collaboration and cooperation in a tourism destination: a network science approach. Curr Issue Tour 14(2):183–189. https://doi.org/10.1080/13683500.2010.531118

Benkő M, Gregor P, Vitkova L (eds) (2019) Book on unexplored cultural heritage in communities by the Danube: DANUrB 2017–2019, Gasset

Bjeljac Z, Curcic N (2006) Tourism in the Serbian, Romanian and Hungarian borderline area as part of cross-border cooperation. Geographica Pannonica 10:73–77. https://doi.org/10.5937/geopan0610073b

Blasco D, Guia J, Prats L (2014) Tourism destination zoning in mountain regions: a consumer-based approach. Tour Geogr 16(3):512–528. https://doi.org/10.1080/14616688.2013.851267

Bohlin M, Brandt D (2014) Creating tourist experiences by interpreting places using digital guides. J Herit Tour 9(1):1–17. https://doi.org/10.1080/1743873X.2013.799173

Deacon H, Smeets R (2013) Authenticity, value and community involvement in heritage management under the world heritage and intangible heritage conventions. Heritage Soc 6(2):129–143. https://doi.org/10.1179/2159032x13z.0000000009

Djukić A, Kádár B, Stan A, Antonić B (eds) (2022). D+ Atlas: atlas of hidden urban values along the Danube. Univerzitet u Beogradu—Arhitektonski fakultet

Evans-Cowley J, Hollander J (2010) The new generation of public participation: internet-based participation tools. Plan Pract Res 25(3):397–408. https://doi.org/10.1080/02697459.2010.503432
Faulkner W (2011) Dialogue in public engagement: a handbook, Edinburgh Beltane
Gallistl V, Rohner R, Seifert A, Wanka A (2020) Configuring the older non-user: between research, policy and practice of digital exclusion. Soc Incl 8(2):233–243. https://doi.org/10.17645/si.v8i2.2607
Giaccardi E (ed) (2012) Heritage and social media. Routledge, UK
Grcheva O, Oktay Vehbi B (2021) From public participation to co-creation in the cultural heritage management decision-making process. Sustainability 13(16):9321. https://doi.org/10.3390/su13169321
Grimwade G, Carter B (2000) Managing small heritage sites with interpretation and community involvement. Int J Herit Stud 6(1):33–48. https://doi.org/10.1080/135272500363724
Hanzl M (2007) Information technology as a tool for public participation in urban planning: a review of experiments and potentials. Des Stud 28(3):289–307. https://doi.org/10.1016/j.destud.2007.02.003
Hassan L, Hamari J (2020) Gameful civic engagement: a review of the literature on gamification of e-participation. Gov Inf Q 37(3). https://doi.org/10.1016/j.giq.2020.101461
Haywood KM (1988) Responsible and responsive tourism planning in the community. Tour Manage 9(2):105–118
Helsper EJ, Reisdorf BC (2017) The emergence of a "digital underclass" in Great Britain and Sweden: changing reasons for digital exclusion. New Media Soc 19(8):1253–1270. https://doi.org/10.1177/1461444816634676
Hodges A, Watson S (2000) Community-based heritage management: a case study and agenda for research. Int J Herit Stud 6(3):231–243. https://doi.org/10.1080/13527250050148214
Hribar MŠ, Bole D, Pipan P (2015) Sustainable heritage management: social, economic and other potentials of culture in local development. Procedia Soc Behav Sci 188:103–110. https://doi.org/10.1016/j.sbspro.2015.03.344
Jacobsen JKS, Munar AM (2012) Tourist information search and destination choice in a digital age. Tourism Manage Perspect 1(1):39–47. https://doi.org/10.1016/j.tmp.2011.12.005
Jamal TB, Getz D (1995) Collaboration theory and community tourism planning. AnnMs of Tourism Res 22(1):186–204
Kádár B (2014) Measuring tourist activities in cities using geotagged photography. Tour Geogr 16(1):88–104. https://doi.org/10.1080/14616688.2013.868029
Kádár B, Gede M (2013) Where do tourists go? Visualizing and analysing the spatial distribution of geotagged photography. Cartographica Int J Geogr Inf Geovisualization 48(2):78–88. https://doi.org/10.3138/carto.48.2.1839
Kádár B, Gede M (2021) Tourism flows in large-scale destination systems. Ann Tour Res 87:103113. https://doi.org/10.1016/j.annals.2020.103113
Kádár B, Gede M (2022) The measurable predominance of weekend trips in established tourism regions—the case of visitors from Budapest at waterside destinations. Sustainability (Switzerland) 14(6). https://doi.org/10.3390/su14063293
Kádár B, Klaniczay J (2022) Branding built heritage through cultural urban festivals: an Instagram analysis related to sustainable co-creation, in Budapest. Sustainability 14. https://doi.org/10.3390/xxxxx
Katz L, Finestone L, Paskevich DM (2021) Competition when cooperation is the means to success: understanding context and recognizing mutually beneficial situations. Cogent Psychol 8(1). https://doi.org/10.1080/23311908.2021.1878984
Lee S-H, Choi J-Y, Yoo S-H, Oh Y-G (2013) Evaluating spatial centrality for integrated tourism management in rural areas using GIS and network analysis. Tour Manage 34:14–24. https://doi.org/10.1016/j.tourman.2012.03.005

Li Y, Hunter C (2015) Community involvement for sustainable heritage tourism: a conceptual model. J Cult Heritage Manag Sustain Dev 5(3):248–262. https://doi.org/10.1108/JCHMSD-08-2014-0027

Loukissas PJ (1983) Public participation in community tourism planning: a gaming simulation approach. J Travel Res 22(1):18–23

Lue C, Crompton JL, Fesenmaier DR (1993) Conceptualization of multi-destination pleasure trips. Ann Tourism Res 20(2):289–301. https://doi.org/10.1016/0160-7383(93)90056-9

Mihelj S, Leguina A, Downey J (2019) Culture is digital: cultural participation, diversity and the digital divide. New Media Soc 21(7):1465–1485. https://doi.org/10.1177/1461444818822816

Mings RC, Mchugh KE (1992) The spatial configuration of travel to Yellowstone National Park. J Travel Res 30(4):38–46. https://doi.org/10.1177/004728759203000406

Murphy PE (1988) Community-driven tourism planning. Tour Manage 9(2):96–104

Orsi F, Geneletti D (2013) Using geotagged photographs and GIS analysis to estimate visitor flows in natural areas. J Nat Conserv 21(5):359–368. https://doi.org/10.1016/j.jnc.2013.03.001

Peng H, Zhang J, Liu Z, Lu L, Yang L (2016) Network analysis of tourist flows: a cross-provincial boundary perspective. Tour Geogr 18(5):561–586. https://doi.org/10.1080/14616688.2016.1221443

Poplin A (2012) Playful public participation in urban planning: a case study for online serious games. Comput Environ Urban Syst 36(3):195–206. https://doi.org/10.1016/j.compenvurbsys.2011.10.003

Prangnell J, Ross A, Coghill B (2010) Power relations and community involvement in landscape-based cultural heritage management practice: an Australian case study. Int J Heritage Stud 148–163. https://doi.org/10.4324/9781315875064-11

Rădoi I (2020) European capital of culture, urban tourism and cross-border cooperation between Romania and Serbia. J Balkan near East Stud 22(4):547–559. https://doi.org/10.1080/19448953.2020.1778881

Radu SA, Bianca DA, Nicoleta DA (2010) Cross-border cooperation in the Danube-lined Romanian/Bulgarian border-space. geographical considerations. Geogr Pannonica 14(2):67–75

Rogers EM (2001) The digital divide. Convergence 7(4):96–111

Scott N, Cooper C, Baggio R (2008) Destination networks. Four Australian cases. Ann Tourism Res 35(1):169–188. https://doi.org/10.1016/j.annals.2007.07.004

Shoval N, Isaacson M (2007) Tracking tourists in the digital age. Ann Tour Res 34(1):141–159. https://doi.org/10.1016/j.annals.2006.07.007

Studzieniecki T, Soares JRR (2017) Interregional tourism cooperation: a Europe case study. Holos 4:135. https://doi.org/10.15628/holos.2017.5263

Thiel S-K, Fröhlich P (2017) Gamification as motivation to engage in location-based public participation?. Springer, Cham, pp 399–421. https://doi.org/10.1007/978-3-319-47289-8_20

Thiel S-K, Reisinger M, Röderer K, Fröhlich P (2016) Playing (with) democracy: a review of gamified participation approaches. JeDEM EJournal EDemocracy Open Gov 8(3):32–60. https://doi.org/10.29379/jedem.v8i3.440

Tissot L (2011) From Alpine tourism to the "Alpinization" of tourism. In: Zuelow EGE (ed) Touring beyond the nation: a transnational approach to European tourism history. Ashgate, pp 59–78. https://doi.org/10.1080/01402390.2011.569130

Vujko A, Gajic T (2014) Opportunities for tourism development and cooperation in the region by improving the quality of tourism services—the 'Danube Cycle Route' case study. Econ Res Ekonomska Istrazivanja 27(1):847–860. https://doi.org/10.1080/1331677X.2014.975517

Weaver A (2008) When tourists become data: consumption, surveillance and commerce. Curr Issue Tour 11(1):1–23. https://doi.org/10.2167/cit338.0

Weidenfeld A, Butler R, Williams AW (2011) The role of clustering, cooperation and complementarities in the visitor attraction sector. Curr Issue Tour 14(7):595–629. https://doi.org/10.1080/13683500.2010.517312

Westerberg P, Fanny von H (2015) Using Minecraft for youth participation in urban design and governance. United Nations Human Settlements Programme

Author Index

M. Vukmirović et al. (eds.), *Digital Dreams, Urban Realities*, The Urban Book Series,
https://doi.org/10.1007/978-3-031-76353-3

GPSR Compliance

The European Union's (EU) General Product Safety Regulation (GPSR) is a set of rules that requires consumer products to be safe and our obligations to ensure this.

If you have any concerns about our products, you can contact us on ProductSafety@springernature.com

In case Publisher is established outside the EU, the EU authorized representative is:

Springer Nature Customer Service Center GmbH
Europaplatz 3
69115 Heidelberg, Germany

Batch number: 08831926

Printed by Printforce, the Netherlands